T0392733

Nanobiotechnology for Food Processing and Packaging

Nanobiotechnology for Food Processing and Packaging

Edited by

Jay Singh

Assistant Professor, Department of Chemistry, Institute of Science, Banaras Hindu University, Varanasi, Uttar Pradesh, India

Ravindra Pratap Singh[†]

Assistant Professor, Department of Biotechnology, Faculty of Science, Indira Gandhi National Tribal University, Amarkantak, Madhya Pradesh, India

Ajeet Kumar Kaushik

Assistant Professor of Chemistry, NanoBioTech Laboratory, Department of Environmental Engineering, Florida Polytechnic University, Lakeland, FL, United States

Charles Oluwaseun Adetunji

Applied Microbiology, Biotechnology and Nanotechnology Laboratory, Department of Microbiology, Edo State University Uzairue, Iyamho, Edo State, Nigeria

Kshitij RB Singh

Department of Chemistry, Institute of Science, Banaras Hindu University, Varanasi, Uttar Pradesh, India

ACADEMIC PRESS

An imprint of Elsevier

elsevier.com/books-and-journals

[†] *Ravindra Pratap Singh (Deceased) has served as the editor throughout the development of this book.*

Academic Press is an imprint of Elsevier

125 London Wall, London EC2Y 5AS, United Kingdom
525 B Street, Suite 1650, San Diego, CA 92101, United States
50 Hampshire Street, 5th Floor, Cambridge, MA 02139, United States

Notices
Knowledge and best practice in this field are constantly changing. As new research and experience broaden our understanding, changes in research methods, professional practices, or medical treatment may become necessary.

Practitioners and researchers must always rely on their own experience and knowledge in evaluating and using any information, methods, compounds, or experiments described herein. In using such information or methods they should be mindful of their own safety and the safety of others, including parties for whom they have a professional responsibility.

To the fullest extent of the law, neither the Publisher nor the authors, contributors, or editors, assume any liability for any injury and/or damage to persons or property as a matter of products liability, negligence or otherwise, or from any use or operation of any methods, products, instructions, or ideas contained in the material herein.

ISBN: 978-0-323-91749-0

For Information on all Academic Press publications visit our website at https://www.elsevier.com/books-and-journals

Publisher: Nikki P Levy
Acquisitions Editor: Nina Bandeira
Editorial Project Manager: Akanksha Marwah
Production Project Manager: Surya Narayanan Jayachandran
Cover Designer: Greg Harris

Working together
to grow libraries in
developing countries

www.elsevier.com • www.bookaid.org

Typeset by Aptara, New Delhi, India

*In loving memory of the **Late Professor Ravindra Pratap Singh**, whose insightful guidance and unwavering support shaped this book into its finest form. Your dedication to excellence and commitment to nurturing the written word will forever inspire us. Your legacy lives on through these pages. Sorrow fills our hearts, for we miss you dearly.*

Contents

[†] deceased.

PART 2 Food Processing

CHAPTER 5 **Nanobiotechnological utility for the removal of food contaminants: Physicobiochemical** **97**

Xiaoyi Liu, K. M. Faridul Hasan and Shaofeng Wei

CHAPTER 12 Common techniques in food processing technologies **223**

Abel Inobeme, John Tsado Mathew, Alexander Ajai
(Ikechuku), Charles Oluwaseun Adetunji,
Jonathan Inobeme, Munirat Maliki,
Mathew Adefusika Adekoya, Elija Shaba (Yanda),
Olori Eric, Sadiq Akhor (Oshoke) and Chinenye Eziukwu

PART 3 Food packaging

CHAPTER 19 Perspectives for carbon-based nanomaterial and its antimicrobial films in food applications 367

*Eli José Miranda Ribeiro Júnior, Marcos Túlio da Silva,
Alexandre Gonçalves Pinheiro and Stephen
Rathinaraj Benjamin*

Contributors

Mariam M. Abady
Organic Metrology Group, Division of Chemical and Medical Metrology, Korea Research Institute of Standard and Science, Daejeon, Yuseong-gu, Republic of Korea; Department of Bio-Analytical Science, University of Science and Technology, Daejeon, Yuseong-gu, Republic of Korea; Department of Nutrition and Food Science, National Research Centre, Dokki, Cairo, Egypt

Anna Abdolshahi
Food Safety Research Center (Salt), Semnan University of Medical Sciences, Semnan, Iran

Mathew Adefusika Adekoya
Department of Physics, Edo State University Uzairue, Edo State, Nigeria

Charles Oluwaseun Adetunji
Applied Microbiology, Biotechnology and Nanotechnology Laboratory, Department of Microbiology, Edo State University Uzairue, Iyamho, Edo State, Nigeria

Alexander Ajai (Ikechuku)
Department of Chemistry, Federal University of Technology, Minna, Nigeria

Sadiq Akhor (Oshoke)
Department of Chemistry, Edo State University Uzairue, Edo State, Nigeria

Wasim Akram
School of Studies in Pharmaceutical Sciences, Jiwaji University, Gwalior, Madhya Pradesh, India

Prabha Arya
Department of Biochemistry, Deshbandhu College, University of Delhi, New Delhi, India

Mamoni Banerjee
Bio-Research Laboratory, Rajendra Mishra School of Engineering Entrepreneurship, Indian Institute of Technology Kharagpur, Kharagpur, West Bengal, India

Stephen Rathinaraj Benjamin
Drug Research and Development Center (NPDM), Department of Physiology and Pharmacology, Federal University of Ceará-UFC, Fortaleza, Ceará, Brazil

Nilay Bereli
Department of Chemistry, Biochemistry Division, Hacettepe University, Beytepe, Ankara, Turkey

Monika Bhattu
University Centre for Research and Development, Chandigarh University, Mohali, Punjab, India

Merve Çalışır
Department of Chemistry, Biochemistry Division, Hacettepe University, Beytepe, Ankara, Turkey

Duygu Çimen
Department of Chemistry, Biochemistry Division, Hacettepe University, Beytepe, Ankara, Turkey

Anirban Dandapat
University School of Automation and Robotics, Guru Gobind Singh Indraprastha University, East Delhi Campus, Delhi, India

Adil Denizli
Department of Chemistry, Biochemistry Division, Hacettepe University, Beytepe, Ankara, Turkey

Santanu Dhara
Biomaterial and Tissue Engineering Laboratory (BMTE), School of Medical Sciences & Technology (SMST), Indian Institute of Technology Kharagpur, Kharagpur, West Bengal, India

P K Dutta
Department of Chemistry, MNNIT Allahabad, Prayagraj, Uttar Pradesh, India

Arezoo Ebrahimi
Food Safety Research Center (Salt), Semnan University of Medical Sciences, Semnan, Iran

Olori Eric
Department of Chemistry, Edo State University Uzairue, Edo State, Nigeria

Muhammed Erkek
Department of Chemistry, Biochemistry Division, Hacettepe University, Beytepe, Ankara, Turkey

Chinenye Eziukwu
Department of Chemistry, Edo State University Uzairue, Edo State, Nigeria

Geetanjali
Department of Chemistry, Kirori Mal College, University of Delhi, New Delhi, India

K. M. Faridul Hasan
Simonyi Károly Faculty of Engineering, University of Sopron, Sopron, Hungary

Sahar Imtiaz
Department of Microbiology, University of Health Sciences, Lahore, Pakistan

Sila Imtiaz
Department of Microbiology, University of Health Sciences, Lahore, Pakistan

Abel Inobeme
Department of Chemistry, Edo State University Uzairue, Edo State, Nigeria

Jonathan Inobeme
Department of Geography, Ahmadu Bello University, Zaria, Nigeria

Ramakant Joshi
Department of Pharmaceutics, ShriRam College of Pharmacy, Morena, Madhya Pradesh, India; School of Studies in Pharmaceutical Sciences, Jiwaji University, Gwalior, Madhya Pradesh, India

Vineeta Kashyap
Department of Biochemistry, Deshbandhu College, University of Delhi, New Delhi, India

Noorkamal Kaur
Department of Food Processing Technology, Sri Guru Granth Sahib World University, Fatehgarh Sahib, Punjab, India

Monika Kaurav
KIET School of Pharmacy, KIET Group of Institutions, Ghaziabad, Uttar Pradesh, India

Ajeet Kumar Kaushik
NanoBioTech Laboratory, Department of Environmental Engineering, Florida Polytechnic University, Lakeland, FL, United States

Kumai Kiran
Department of Biotechnology, Sir J.C. Bose Technical Campus Bhimtal, Kumaun University, Nainital, Uttarakhand, India

Addanki P. Kumar
The University of Texas Health Science Center at San Antonio, TX, United States

Ganesh Kumar
Sita Ram Kashyap College of Pharmacy, Rahaud, Chhattisgarh, India

Krishan Kumar
Department of Chemistry, D.C.R. University of Science and Technology, Sonipat, Haryana, India

Santosh Kumar
Department of Chemistry, Harcourt Butler Technical University, Kanpur, Uttar Pradesh, India

Ram Sunil Kumar L
Department of Chemistry, Kirori Mal College, University of Delhi, New Delhi, India

Xiaoyi Liu
School of Public Health, The Key Laboratory of Environmental Pollution Monitoring and Disease Control, Ministry of Education, Guizhou Medical University, Guiyang, China

Umesh R. Mahajan
Department of Polymer and Surface Engineering, Institute of Chemical Technology, Mumbai, Maharashtra, India

Munirat Maliki
Department of Chemistry, Edo State University Uzairue, Edo State, Nigeria

Padmavati Manchikanti
Plant Metabolic Pathway Laboratory, Rajiv Gandhi School of Intellectual Property Law (RGSOIPL), Indian Institute of Technology Kharagpur, Kharagpur, West Bengal, India

John Tsado Mathew
Department of Chemistry, Ibrahim Badamasi Babangida University Lapai, Niger State, Nigeria

Shashank T. Mhaske
Department of Polymer and Surface Engineering, Institute of Chemical Technology, Mumbai, Maharashtra, India

Sunita Minz
Department of Pharmacy, Indira Gandhi National Tribal University, Amarkantak, Madhya Pradesh, India

Biswajit Mishra
Department of Medicine, Warren Alpert Medical School of Brown University, Providence, RI, United States

Elaine Gabutin Mission
PressTech, Instituto de Bioeconomia de la Universidad de Valladolid, Valladolid University, Valladolid, Spain

Dina Mostafa Mohammed
Department of Nutrition and Food Science, National Research Centre, Dokki, Cairo, Egypt

Jyoti Darsan Mohanty
Department of Polymer and Surface Engineering, Institute of Chemical Technology, Mumbai, Maharashtra, India

Pooja Mongia
Department of Pharmaceutics, Delhi Institute of Pharmaceutical Sciences and Research University, New Delhi, India

Gunjan Nagpure
Department of Biotechnology, Faculty of Science, Indira Gandhi National Tribal University, Amarkantak, Madhya Pradesh, India

Arunadevi Natrajan
Department of Chemistry, PSGR Krishnammal College for Women, Coimbatore, Tamil Nadu, India

Ankur Ojha
Department of Food Science and Technology, National Institute of Food Science Technology Entrepreneurship and Management, Sonipat, Haryana, India

Olaniyan Olugbemi
Laboratory for Reproductive Biology and Developmental Programming, Department of Physiology, Rhema University Aba, Abia State, Nigeria

Erdoğan Özgür
Department of Chemistry, Biochemistry Division, Hacettepe University, Beytepe, Ankara, Turkey

Veena Pande
Department of Biotechnology, Sir J.C. Bose Technical Campus Bhimtal, Kumaun University, Nainital, Uttarakhand, India

Megha Pant
Department of Biotechnology, Sir J.C. Bose Technical Campus Bhimtal, Kumaun University, Nainital, Uttarakhand, India

Jhansi Lakshmi Parimi
Plant Metabolic Pathway Laboratory, Rajiv Gandhi School of Intellectual Property Law (RGSOIPL), Indian Institute of Technology Kharagpur, Kharagpur, West Bengal, India; Biomaterial and Tissue Engineering Laboratory (BMTE), School of Medical Sciences & Technology (SMST), Indian Institute of Technology Kharagpur, Kharagpur, West Bengal, India

Alexandre Gonçalves Pinheiro
Department of Physics, Faculty of Education, State University of Ceará-UECE, Planalto Universitário, Quixadá, Ceará, Brazil; Department of Physics, The University of Texas at Dallas, Texas, United States

Ekta Poonia
Department of Chemistry, D.C.R. University of Science and Technology, Sonipat, Haryana, India

Pawan Prabhakar
Bio-Research Laboratory, Rajendra Mishra School of Engineering Entrepreneurship, Indian Institute of Technology Kharagpur, Kharagpur, West Bengal, India

Madhulika Pradhan
Gracious College of Pharmacy, Abhanpur, Chhattisgarh, India

Rakesh Raj
DSEU Meerabai Maharani Bagh Campus, Delhi Skill and Entrepreneurship University, New Delhi, India

Narender Ranga
Department of Physics, D.C.R. University of Science and Technology, Sonipat, Haryana, India

Rohit Ranga
Department of Chemistry, D.C.R. University of Science and Technology, Sonipat, Haryana, India

Roopa Rani
Department of Sciences, School of Science, Manav Rachna University, Faridabad, Haryana, India

Shweta Rathee
Department of Food Science and Technology, National Institute of Food Science Technology Entrepreneurship and Management, Sonipat, Haryana, India

Eli José Miranda Ribeiro Júnior
Department of Pharmacy, Faculty of CGESP (Centro Goiano de Ensino Superior), Goiânia, Goiás, Brazil

Kantrol Kumar Sahu
Institute of Pharmaceutical Research, GLA University, Mathura, Uttar Pradesh, India

Arpit Sand
Department of Sciences, School of Science, Manav Rachna University, Faridabad, Haryana, India

Elija Shaba (Yanda)
Department of Chemistry, Federal University of Technology, Minna, Nigeria

Srishti Sharma
School of Studies in Chemistry, Pt. Ravishankar Shukla University, Raipur, Chhattisgarh, India; Department of Chemistry, Dr. Ghanshyam Singh P.G. College, Varanasi, Uttar Pradesh, India

Marcos Túlio da Silva
Instituto de Saúde e Biotecnologia, Universidade Federal do Amazonas, Espírito Santo, Coari – Amazonas, AM, Brazil

Jay Singh
Department of Chemistry, Institute of Science, Banaras Hindu University, Varanasi, Uttar Pradesh, India

Kamana Singh
Department of Biochemistry, Deshbandhu College, University of Delhi, New Delhi, India

Kshitij RB Singh
Department of Chemistry, Institute of Science, Banaras Hindu University, Varanasi, Uttar Pradesh, India

Namrata Singh
Department of Engineering Sciences, Ramrao Adik Institute of Technology, DY Patil Deemed-to-be University, Navi Mumbai, Maharashtra, India; Department of Chemistry, Faculty of Science, University of Hradec Kralove, Hradec Kralove, Czech Republic

Ram Singh
Department of Applied Chemistry, Delhi Technological University, New Delhi, India

Ravindra Pratap Singh[†]
Department of Biotechnology, Faculty of Science, Indira Gandhi National Tribal University, Amarkantak, Madhya Pradesh, India

Shalinee Singh
Department of Chemistry, Harcourt Butler Technical University, Kanpur, Uttar Pradesh, India

Shikha Kapil Soni
University Institute of Biotechnology, Chandigarh University, Mohali, Punjab, India

Andrew Lambert M. Tampoc
School of Chemical, Biological and Materials Engineering and Sciences, Mapua University, Manila, Philippines

Sushma Thapa
Department of Chemistry, Banaras Hindu University, Varanasi, Uttar Pradesh, India

Anurag Tiwari
Department of Applied Mechanics, MNNIT Allahabad, Prayagraj, Uttar Pradesh, India

Ashish Tiwari
Department of Chemistry, Government College Dumariya Jarhi, Surajpur, Chhattisgarh, India

Aykut Arif Topçu
Medical Laboratory Program, Vocational School of Health Service, Aksaray University, Aksaray, Turkey

Jaya Tuteja
School of Basic Sciences, Galgotias University, Greater Noida, Gautam Buddh Nagar, Uttar Pradesh, India

Meenakshi Verma
University Centre for Research and Development, Chandigarh University, Mohali, Punjab, India

---

[†] deceased.

Shaofeng Wei
School of Public Health, The Key Laboratory of Environmental Pollution Monitoring and Disease Control, Ministry of Education, Guizhou Medical University, Guiyang, China

Krishna Yadav
Raipur Institute of Pharmaceutical Education and Research, Raipur, Chhattisgarh, India

Editor biographies

Dr. Jay Singh

Dr. Jay Singh is currently working as an Assistant Professor at the Department of Chemistry, Institute of Science, Banaras Hindu University, Varanasi, Uttar Pradesh, India, since 2017. He received his PhD in Polymer Science from Motilal Nehru National Institute of Technology, Prayagraj, Uttar Pradesh, India, in 2010 and did MSc and BSc from Allahabad University, Uttar Pradesh, India. He was a postdoctoral fellow at National Physical Laboratory, New Delhi, India, Chonbuk National University, Jeonju, South Korea and Delhi Technological University, Delhi, India. Dr. Jay has received many prestigious fellowships like CSIR (RA), DST-Young Scientist fellowship, DST-INSPIRE faculty award, etc. He is actively engaged in the development of nanomaterials (CeO_2, NiO, rare-earth metal oxide, Ni, $Nife_2O_4$, Cu_2O, Graphene, RGO, etc.) based nanobiocomposite, conducting polymer and self-assembled monolayers based clinically important biosensors for estimation of bioanalaytes such as cholesterol, xanthine, glucose, pathogens and pesticides/toxins using DNA and antibodies. Dr. Jay has published more than 135 international research papers with total citations of more than 4800 and an h-index being 41. He has completed/run various research projects in different funding agencies. He has more than 10 edited books and has authored more than 50 book chapters of internationally reputed press for publications, namely Elsevier, Springer Nature, IOP, Wiley, and CRC. He is actively engaged in fabricating metal oxide-based biosensors for clinical diagnosis, food packaging applications, drug delivery, and tissue engineering applications. His research has contributed significantly toward the fundamental understanding of interfacial charge transfer processes and sensing aspects of metal nanoparticles.

Dr. Ravindra Pratap Singh

Dr. Ravindra Pratap Singh did his BSc from Allahabad University, Uttar Pradesh, India, and his MSc and PhD in Biochemistry from Lucknow University, Uttar Pradesh, India. He is currently working as an Assistant Professor in the Department of Biotechnology, Faculty of Science, Indira Gandhi National Tribal University, Amarkantak, Madhya Pradesh, India. He has previously worked as a scientist at various esteemed laboratories globally, namely Sogang University, Seoul, South Korea, Institute Gustave-Roussy (IGR), Paris, France, etc. His work and research interests include biochemistry, biosensors, nanobiotechnology, electrochemistry, material sciences, and biosensors applications in biomedical, environmental, agricultural, and forensics. He has to his credit several reputed national and international honors/awards. Dr. Singh has authored over 70 articles in international peer-reviewed journals, more than 70 book chapters of international repute, and has edited 14 books. He serves as a reviewer of many reputed international journals and is also a member of many international societies. He is currently also involved in editing various books, which will be published in internationally reputed publication houses, namely IOP Publishing, CRC Press, Elsevier, and Springer Nature. Moreover, he is a book series editor of "Emerging advances in bionanotechnology," CRC Press and Taylor and Francis Group. He is also actively involved in guest-editing special issues (SI) for reputed international journals, and one of the latest edited SI by Dr. Singh is "Smart and intelligent nanobiosensors: Multidimensional applications" for Materials Letters, Elsevier and "Smart and intelligent optical materials for sensing applications" for Luminescence, Wiley.

Professor Charles Oluwaseun Adetunji

Professor Charles Oluwaseun Adetunji is presently a faculty member at the Microbiology Department, Faculty of Sciences, Edo State University Uzairue (EDSU), Edo State, Nigeria, where he utilized the application of biological techniques and microbial bioprocesses for the actualization of sustainable development goals and agrarian revolution, through quality teaching, research, and community development. He is currently the Ag Dean for the Faculty of Science and the Head of the Department of Microbiology at EDSU. He is a visiting professor and  the executive director for the Center of Biotechnology, Precious Cornerstone University, Ibadan, Nigeria. He has won several scientific awards and grants from renowned academic bodies. He has published many scientific journal articles and conference proceedings in refereed national and international journals with over 370 manuscripts. He was ranked among the top 500 prolific authors in Nigeria between 2019 till date by SciVal/SCOPUS. His research interests include microbiology, biotechnology, postharvest management, and nanotechnology. He was recently appointed as the President and Chairman Governing Council of the Nigerian Bioinformatics and Genomics Network Society. He holds the position of the General/Executive Secretary of the Nigerian Young Academy. He is currently a series editor with Taylor and Francis, United States editing several textbooks on agricultural biotechnology, nanotechnology, pharmafoods, and environmental sciences. He is an editorial board member of many international journals and serves as a reviewer for many double-blind peer-reviewed journals such as Elsevier, Springer, Francis and Taylor, Wiley, PLOS One, Nature, American Chemistry Society, and Bentham Science Publishers.

Dr. Ajeet Kumar Kaushik

Dr. Ajeet Kumar Kaushik is an assistant professor of Chemistry at the Department of Environmental Engineering, Florida Polytechnic University, FL, United States. He has over 15 years of experience in exploring analytical techniques for the characterization of nanostructures, bio/chemical sensors fabrication confirmation, and nanomedicine optimization. Before joining Florida Polytechnic University in 2019, he worked as a faculty member at the Medical College of Florida International University, FL, United States.

Mr. Kshitij RB Singh

Mr. Kshitij RB Singh is a postgraduate in biotechnology from Indira Gandhi National Tribal University, Amarkantak, Madhya Pradesh, India. He is currently working in the laboratory of Dr. Jay Singh, Department of Chemistry, Institute of Science, Banaras Hindu University, Varanasi, Uttar Pradesh, India. He has more than 65 peer-reviewed publications to his credit, has edited 12 books, and has authored more than 60 book chapters published in internationally reputed press, namely Elsevier, IOP Publishing, Springer Nature, Wiley, and CRC Press. He is currently also involved in editing books with international publishing houses, including CRC Press, IOP Publishing, Elsevier, Wiley, and Springer Nature. His research interests include biotechnology, biochemistry, epidemiology, nanotechnology, nanobiotechnology, biosensors, and materials science.

Preface

Food processing, packaging, and safety mean that all foodstuffs must be protected from chemical, biological, physical, and radiation contamination through processing, management, purification, and transportation. To control the great economic losses from the spoilage of foods each year, people worldwide have been focused on processing and packaging food as protection from various microorganisms. In this direction, nanobiotechnology science is an emerging and conjugate field between interdisciplinary material science, nanotechnology, and biology. Nanobiotechnology's speedy development has transformed many food science domains, particularly those that engage in the processing, packaging, preservation, transportation, functionality, and other safety aspects of food. Food technology is regarded as a crucial industrial sector where nanobiotechnology will play an important role in the future. It is normally distinguished between two forms of nanofood applications: food additives (inside nanotechnology) and food packaging (outside nanotechnology). Nano-food additives are an example used to affect the shelf-life of food, consistency, taste, and nutrient composition, or even detect various food pathogens such as mycotoxin and provide functions such as food quality assurance. Another is food nanostructured ingredients encompass a wide area from food processing and packaging, and nanotechnologies are mainly considered to be of use to increase product storage stability, carriers for smart delivery of nutrients, anticaking agents, indicate spoilt ingredients or generally increase product quality, e.g., by preventing gas flow across product packaging. The food industry has also embraced the "nanobiotechnology" era in response to a rising need by consumers for healthier and nutritionally appealing, cost-effective products.

Nanobiotechnology can transform packaging material's permeability, increasing gas barrier properties, improving mechanical and heat-resistance properties, developing bioactive interfaces, and creating nanobiodegradable packaging materials. It also offers absolute food solutions from industrialized food processing to packaging. Nanostructured materials bring about a great difference not only in food superiority and safety but also in the health benefits that food delivers. Many food industries, researchers, scientists, and national and international food organizations develop novel techniques, methods, and products that directly apply nanobiotechnology in food science and food industries. Several reports and research papers confirm that nanostructured materials can successfully progress food safety by enhancing the usefulness of food packaging, shelf-life, and nutritional value as additives without altering the flavor and physicochemical characteristics of food items. Even though nanotechnology is still facing several challenges in using cost-effective processing operations to generate edible and low or nontoxic nano-delivery systems and enlarge effective formulations and managements safe for human beings and the environment. Therefore, due to the increased demands of these food industries, there have been increasing concerns regarding the development of biodegradable, biocompatible, safe, and nontoxic nanomaterials from food-grade additives using easy, eco-friendly, cost-effective, easy handling, and effective management strategies to be urgently nodded.

Therefore, the application of nanostructured materials has been discovered as an effective technique that could be sustainable, efficient, effective, and innovative to combat all the aforementioned challenges.

This book aims to focus on nanobiotechnology in food processing and food packaging, considering their reflection on food quality, safety, and management aspects. The book highlights various preparative methods and antimicrobial/antifungal activity, including the mechanism of the antimicrobial action of various nanobiocomposites and food toxin detection by utilizing nanobiosensor. The optimization of the different bioactive properties of these so-called nanobiocomposites films, coating, adhesive, and biocatalysts' role in improving the quality and shelf life of foods is also discussed. Moreover, detailed information is also provided on the possible food toxin detection, and food packaging and their mechanistic approach through materials such as nanomaterials, nanocomposites, carbon-based nanomaterials, polymer-based nanocomposites, and various binary and tertiary nanocomposites.

Jay Singh
Ravindra Pratap Singh
Ajeet Kumar Kaushik
Charles Oluwaseun Adetunji
Kshitij RB Singh

Acknowledgment

It gives us immense pleasure to acknowledge Professor Shri Prakash Mani Tripathi, Honourable Vice-Chancellor of Indira Gandhi National Tribal University, Amarkantak, Madhya Pradesh, India, Professor Sudhir K Jain, Honourable Vice-Chancellor, Banaras Hindu University, Varanasi, Uttar Pradesh, India, Institutes of Eminence (IoE), Ministry of Education, India, Professor Emmanuel O. Aluyor, Vice-Chancellor, Edo State University, Uzairue, Nigeria, and Professor Julius Kola Oloke, Precious Cornerstone University, Ibadan, Nigeria for providing constant assistance in all the possible ways. It is also our great pleasure to acknowledge and express our enormous debt to all the contributors who have provided their quality material to craft this book. We are grateful to our beloved family members (Babita Singh, Gioconda Kaushik, Juliana Bunmi Adetunji, Ranjana Verma, Aparajita RB Singh, Avani Jiya Kaushik, Emmanuel Ayomiposi Adetunji, Taiwo Jesugbemi Adetunji, Kehinde Jesugbemi Adetunji, and Shanvi Singh), who joyfully supported and stood by our side during the countless hours of our absence, ensuring the completion of this book. We extend our gratitude to Nina Bandeira, Aera Gariguez, and the entire publishing team for their patience and extra care in publishing this book.

Jay Singh
Ravindra Pratap Singh
Ajeet Kumar Kaushik
Charles Oluwaseun Adetunji
Kshitij RB Singh

Fundamentals

Introduction: Nanobiotechnology for food processing and packaging

1

Arunadevi Natrajan[a], Kshitij RB Singh[b], Sushma Thapa[c], Ajeet Kumar Kaushik[d], Jay Singh[b] and Ravindra Pratap Singh[e,†]

[a]*Department of Chemistry, PSGR Krishnammal College for Women, Coimbatore, Tamil Nadu, India,* [b]*Department of Chemistry, Institute of Science, Banaras Hindu University, Varanasi, Uttar Pradesh, India,* [c]*Department of Chemistry, Banaras Hindu University, Varanasi, Uttar Pradesh, India,* [d]*NanoBioTech Laboratory, Department of Environmental Engineering, Florida Polytechnic University, Lakeland, FL, United States,* [e]*Department of Biotechnology, Faculty of Science, Indira Gandhi National Tribal University, Amarkantak, Madhya Pradesh, India*

1.1 Introduction

The human population will increase globally and reach 10 billion in another 20 years. Agricultural production should be increased vertically to meet the basic needs of people. The food processing industry is one of the economy-based industries and produces $7 trillion annually. The market value of food processing was raised from $100 billion (2020) to $150 billion (2030). Due to globalization, people are aware of different types of foods available in all parts of the world and are attracted to tasting more processed food items. Many people started eating safe and hygienic processed food during the pandemic and postpandemic periods. Industries focusing on food processing have begun investing in advanced technological equipment to enhance productivity. This is a main influential factor for the increasing global need for the food processing industry.

Starting from the historical period, people have been struggling to preserve food. They adopted various methods to save fresh food in rock shelters or buried inside the soil in closed containers. Drying, dehydration, fermenting, roasting, mixing with vinegar and salt, and smoking were the few other techniques employed during the 17th century (Chellaram et al., 2014). The above-said process is anticipated to destroy or stop the growth of diseases causing microorganisms like bacteria, fungi, Norovirus, and Campylobacter without the knowledge of the scientific background behind it. Different countries adopted different methods e.g., Egyptians employed sun-dry, Romans familiarized with pickling, and Greeks introduced a coating of honey/sugar to preserve food materials (Pradhan et al., 2015).

---

† deceased.

Nanobiotechnology for Food Processing and Packaging. DOI: https://doi.org/10.1016/B978-0-323-91749-0.00019-8

Nanotechnology, a new and innovative technical method, can change the compound's physical, chemical, and thermal stability at the nano range. It is an interdisciplinary field; it covers areas like physics, chemistry, and biology which help researchers to understand the relationship between structure and properties. Nanobiotechnology is an integrated technique that couples with cellular processes to master the biological properties in various fields, from therapeutic to cultivation (Avella et al., 2005). It has a remarkable and rapid development in the food and allied industry. With the concern of a healthy and disease-free environment, budding researchers are striving hard to bring out cutting-edge technologies to facilitate better food quality and shelf life. Nanotechnology applications in the agricultural field will comprise the transmission of DNA or genetic cells to produce insect-resistant crops, encapsulated food material, upgraded shelf-life crops, and healthy fruits and vegetables.

Approaches to nanotechnology in the food industry are entirely different from other traditional methods. Industries based on food processing and packaging depend on various factors like raw material, working temperature, humidity, safety measures, and technical procedures. A few emerging areas in which nanobiotechnology plays a significant role are as follows:

- Nanoencapsulation for flavor enhancement.
- Improving the bioavailability of nutraceuticals.
- Production of new functional foods.
- To improve the food color, taste, and stability.
- Carriers for distribution of nutrients.

Nanotechnology helps to improve food quality without affecting/degrading the nutritional value. The nanoparticle-based food materials are enclosed with a few metal ions which are essential for the human body and are nontoxic. The physical, chemical, and thermal properties were found to be high and the shelf-life period was also extended due to nano-sized compounds. Nanotechnology plays an important role in all stages of food production, packaging, and delivery. Recently researchers found many novel processes and techniques for the implementation of nanotechnology in food sectors (Dasgupta et al., 2015). Taste, sensory attributes, and texture were enhanced with nano-structured food materials and decreased the microbial action on food. The usage of nanocarriers in the form of food additives has increased in the recent era without affecting the morphology of the food material (Gorantla et al., 2021).

Nanoencapsulation was regarded as one of the most constructive methods to annihilate excess bioactive compounds. Nanoencapsulation methods include emulsification, supercritical fluid, nanoprecipitation, and coacervation (Ezhilarasi et al., 2013). Active packaging with nanocarriers and natural antimicrobial substances helps to protect the food from degradation. Nanoencapsulation methods deliver numerous applications like fastening the process of solubility and absorption, fortification in contradiction of environmental causes, and deactivating all the blockades in food packaging (Bahrami et al., 2020).

Further, nanoemulsions are added to food products like fruit salad, beverages, and processed oils to improve the taste of the food. It delivers various tastes during

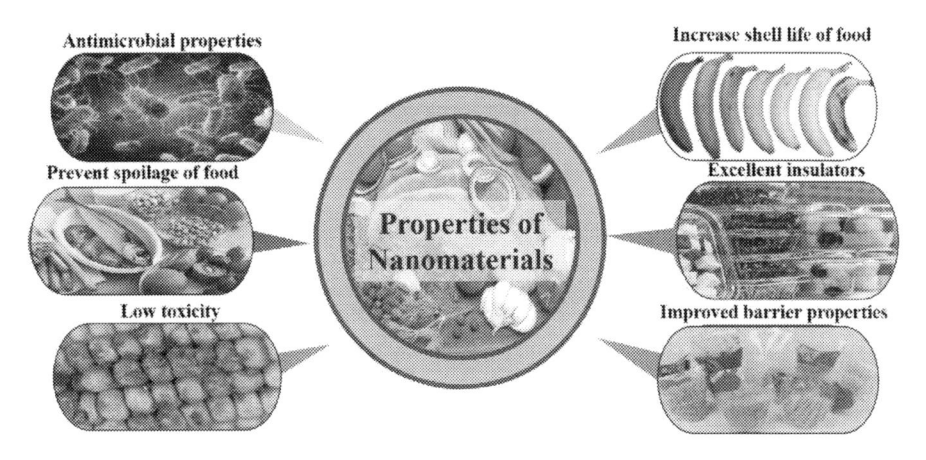

FIGURE 1.1

Properties of nanomaterials.

different stages like heat, changes in pH, and temperature. It hinders the enzymatic oxidation of the food and prevents moisture loss (Aswathanarayan and Vittal, 2019). Nanoemulsions possess a significant role in the food and allied industries by transporting nutraceuticals to the body and performing antimicrobial activity. Biodegradable coating and films can be prepared from nanoemulsion with active packaging and increasing the shelf life of foods (Aswathanarayan and Vittal, 2019).

1.2 Properties of nano biomaterials and nanocomposites

To maintain sustainability, packaging industries also find new solutions for recycling packaging products. The ultimate target is to diminish the utility of nondegradable plastics by substituting them with mechanically stable products with high efficacy for recyclability. In this aspect, nano biomaterials and nanocomposites occupy the dominant role with a multifunctional activity. The physical and thermal properties of the packaging materials can also be increased by adding nanofillers (Sarfraz et al., 2020). The enriched properties (Fig. 1.1) of nano biomaterials and nanocomposites have encouraged and provoked the progress of technologies that will increase the shelf-life period and develop ecologically benign and easily decomposable food packaging. Nanocompounds lead to the production of better-quality food materials, decreasing the environmental consequences owing to their nano size and morphology (Kuswandi et al., 2011). Further, the food processing industries upgraded the quality of packaging material by incorporating nanocomposites with high thermal stability, biodegradable, flexible, and user-friendly. With the help of nanomaterials, the functional activity of the packaging compounds was increased as follows:

• Smart packaging—employing nanosensors for the detection of gases, food safety, and product identification.

- Active packaging—incorporating nanobiocomposites encompassed with good antibacterial activity, e.g., silver.
- Improved packaging—improve gas barrier properties, polymer composites are impregnated with packaging material, e.g., clay.
- Bio-based packaging—mixing with nanobiomaterials to enhance the process of biodegradability.
- Edible packaging—bio-active edible packaging in the form of coating or films to sustain food nutrition.

The implementation of nanotechnology in the food and allied sectors focused mainly on two major areas: food processing and food packaging. In food processing, nanosized composites are used as (1) anticaking agents (zinc and silica)—to increase the consistency and hinder the lump and binding formation, (2) food additives (calcium and magnesium)—to enhance the nutritional availability of food and may be in the form of metal, polymer or carbon-based composites, and (3) gelling agents, stabilizes—act as a natural barrier and thickeners.

Nano-sized compounds comprise multifaceted physical, chemical, and thermal properties compared with large molecules. The large surface area of nanoparticles is the reason for the high energy and the melting point directly depends on the size of the particle. A decrease in the size of the particle decreases the melting point (Singh et al., 2018). Nanometal oxides in the form of titanium, silver, zinc, calcium, and silica are mostly used in food industries worldwide owing to their antibacterial activity (Rezić et al., 2017). Based on the nature of the metal, its properties, morphology, and reactivity will vary and thus, modify its application in food packaging. Silver (Carbone et al., 2016) and bio-polymeric materials usually have a larger surface area than others. The major aspect of adding nanomaterial/composites to the food packaging industry is to protect the food from spoilage due to attack by insects and microorganisms. They can be easily distributed, which will enhance the barrier properties and shelf life of the food.

Nanosilver has varied applications in the food industry (Cao et al., 2018). It was added as the main compound in the food packaging process due to its excellent antibacterial properties and its ability to affect the metabolic activity of bacteria. The silver-coated foil will release metallic ions due to exposure to oxidation and inhibit the growth of bacteria. Nano titanium dioxide is a nontoxic, permeable, odorless, and tasteless material that is particularly utilized in fruit packaging to decompose ethylene molecules produced during oxidation. It makes the fruit remain fresh and stable for an extended period. Nano zinc oxide, with high stability and catalytic activity, is also used for food packaging, particularly for packing breakfast and cereals due to its regulatory action in antibacterial function. Nano clay possesses good water absorption and tensile strength will increase the shelf-life of food particles. With nanoparticles, the continuous phases were disturbed and the physicochemical properties of the packaging were upgraded. One of the key purposes of mixing nano biocomposites is to progress the mechanical stability and to avoid the decomposition of the food during transport.

The development of food packaging using nanobiocomposites having smart/intelligent functions will help to regulate physical, chemical, mechanical, and in a few cases microbial growth within the food product (Kuswandi, 2017; Madhusudan et al., 2018). Based on optical properties and reactivity towards gases, photonic nanocrystals are used widely as nanosensors. The response of physical and chemical changes taking place inside the food, the presence of pathogens, and the gases responsible for food decomposition will be monitored regularly with the help of nanosensors.

Biopolymers as packaging materials have an impact on the chemical structure, morphology, molecular weight, and properties of the food along with the methodology adopted. The poor mechanical strength is the major problem of biopolymers which will be rectified by adding nano-sized molecules with good mechanical and thermal characteristics. The small-sized molecule has a greater influence on chemical reactions with biopolymers and results in the formation of nanobiocomposites with noteworthy features.

1.3 Factors influencing food processing

1.3.1 Particle size

The progress of biocompatibility depends on the size of the particle and nano-sized particles are more efficient than macro particles. Delivery of nutrients is high in nanoparticles owing to their large surface area and found to be more stable (Miar et al., 2020). The novel properties of the nanoparticles like small size, upgraded solubility, tailored made surface, and their functions open new pathways in food applications. The antibacterial activity also relies on size, and it was established that small-sized particles show better activity. The possibility of transfer of nanoparticles across the cell membrane is made possible due to the large surface area (Wang et al., 2017). Particle size plays a significant role in food packaging, for maximum packing the suspension viscosity should be less. To minimize the viscosity, the particle should be less and the size ratio should be high (Servais et al., 2002). Nano-sized particles discharge the encapsulated substances more readily than larger molecules, as a result, it enlightens the bioavailability of nutrients.

1.3.2 Roughness and zeta potential

An increase in roughness increases the surface area to mass ratio and thereby, initiates the strong adsorption of protein (Servais et al., 2002). Zeta potential has a remarkable influence on the nanoparticles. Further, more positively charged nanoparticles like metal ions will be attracted more strongly to the surface than negatively charged particles, and also generate more reactive oxygen species (ROS). In the case of negatively charged particles, more repulsion is there but due to molecular crowding, it shows minimal antibacterial activity (Arakha et al., 2015).

1.3.3 **Doping**

Doping is another factor that modifies and controls the interaction of nanoparticles. Doping was particularly done to hinder the aggregation process and increase the dispersion in the water medium. ROS increases as a result of doping and it was proven by doping halogen atoms in zinc oxide nanoparticles, and the experiential that an increase in antibacterial activity due to ROS than undoped particles (Podporska-Carroll et al., 2017). With doping atoms, the modification of the band gap is possible with increases in the photocatalytic activity of nanoparticles, thereby inducing the movement of ROS.

1.3.4 **Surface charge and pH**

The physicochemical properties of the nanosized molecules are highly affected by local ions and molecules even at low concentrations. The stability and reactivity of nanoparticles depend on surface charges, and it paves the way for nonionic collaboration with biological moieties and leads to unwanted modifications. Further, buffering effect by surrounding ligands, quenching effect, and fluorophore also influence the reactivity of nanoparticles. Due to surface charge and hydrophobicity effects, protein concentration will be high in the surface concentration of nanoparticles which will deplete bulk solutions. By preventing agglomeration, stabilization of nanoparticles is possible. In a few cases, electrostatic repulsion by like charges also shows favorable stabilization (Guarnieri et al., 2011). The pH is another factor that affects the activity of nanoparticles and the oxidation process will take place on the surface by modifying pH. Usually at a low pH of 2–4, mostly positive ions will get adsorbed on the surface, and at a high pH, negatively charged particles will adhere strongly.

1.4 **Application of nanobiotechnology in the food industry**

Food wastage is a major loss in the food industry, and it was reported by food and agriculture organizations that more than 1.3 billion metric tons of food material is wasted every year. The reason may be due to inadequate knowledge of postharvest procedures, storage, packing, and transport. Despite the increase in food production, suitable solutions should be introduced to avoid food wastage and future food crises. Further, food wastage occurs due to poor food quality, food spoilage through microorganisms and pathogens, and a shorter shelf-life period. The progression of nanotechnology in all major areas of the food industry from processing to packaging will decrease food wastage. The application of nanotechnology in food processing and packaging industries is depicted in Fig. 1.2.

1.4.1 **Food processing**

The food prepared by utilizing nanotechnology in various processes like harvesting, production, processing, and packaging is known as nanofood. Bioavailability, nutritional value, taste, texture, and shelf-life period were improved by using this

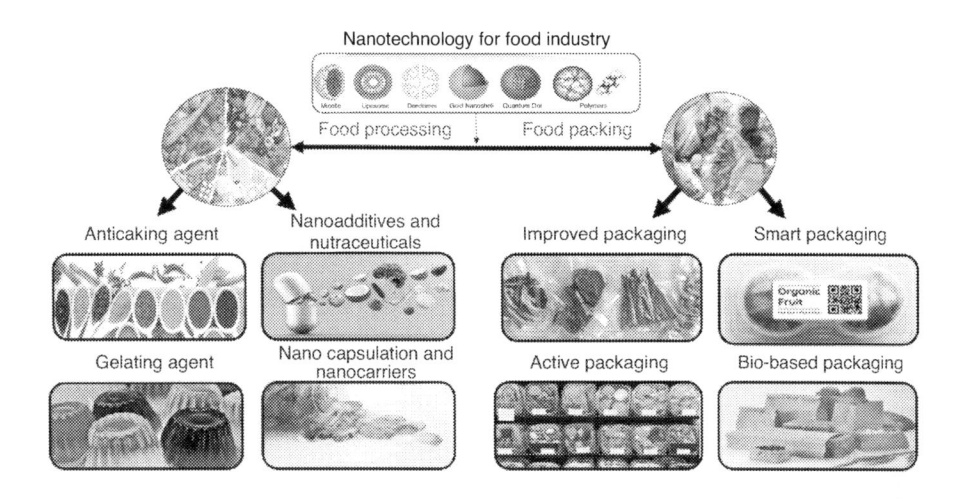

FIGURE 1.2

Application of nanotechnology in the food industry.

cutting-edge technology. The role of nano-ranged particles in the postharvesting process was remarkable. Appropriate food management encompasses numerous phases like processing, manufacturing, packaging, and preservation. This process converts raw food into semiprocessed or processed food. To prevent the absorption of moisture and harmful gases, thin nanocoating can be applied on the surface of fruits and vegetables without affecting the original taste of the food. Nanocoating will deliver additional taste, color, nutrition, and antioxidant to the products. Nanofiltration, a low-pressure driven membrane and approved technology was widely used in food manufacturing sectors like beverage, milk products, and oil processing units for filtration, concentration, refinement, deacidification/alkalization, and microbial reduction (Yadav et al., 2022; Vatai, 2000). Nano filters are used to remove the excess color in beetroot without affecting taste and help to remove the bacteria from dairy/milk products without boiling (Press, 2010). Nanoparticles like TiO_2 and SiO_2 are used as food additives in the form of coated sugars. Anticaking agents with nanoparticles will perform various actions like preventing moisture absorption, producing moisture-protective films on the external surface, and delivering smooth areas to reduce particle friction (Lipasek et al., 2012). Silicon oxide is used as a thickening agent in the processing of dairy products. Nanoparticles like cellulose and zinc oxide can be utilized as gelling agents in the packaging of chicken and cheese (Primozic et al., 2021). Nanoemulsions, a colloidal particulate induces the development of a large surface area and increases the bioavailability of active substances (Gasa-Falcon et al., 2020). Another prominent technique is nanoencapsulation where all nanoparticles are packed in nanostructures for the controlled release of the target molecule. It is mainly used as a nutritional supplement to prevent unpleasant odors and creates a forum for the dispersion of nutrients/supplements. This technique is used to increase the shelf life of milkshakes, tomatoes, fruit juices, and supplementary drinks.

1.4.2 Food packaging

Food preservation is the method of managing food to decrease decomposition and minimize the depletion of food quality by microorganisms. Freshly harvested fruits and vegetables are prone to early decomposition due to moisture content, ethylene, and photooxidation (Gaikwad et al., 2020). Proper food packaging is the solution to solve this issue. The purpose of packaging is to terminate the decay and deterioration process and to protect food material from external factors. Nanoparticles were found to be suitable candidates and demonstrated proven results in food packaging (Pradhan et al., 2015). Further, attention should be given to the nature of the material, antimicrobial agents, and contamination of nanoparticles with food. Freezing, fermentation, and drying are traditional methods for food preservation (Pradhan et al., 2015). Smart packaging and active packaging are two major types of food packaging. Active packaging improves the shelf-life period of processed food and it was designed in such a way that it will integrate the substance which engages oxygen or liberate antimicrobial agents from food (Majid et al., 2018). The assimilation of nanoparticles with a polymeric packaging system will increase the quality of packed food. Metals and metal oxide nanoparticles like silicon oxide, zinc oxide, calcium, etc. are used for active packaging based on the need. The nanoparticles will react directly or indirectly with the organic compound present inside the food. The incorporation of nanosilver ion in packaging will hinder bacterial growth and kills 80% of bacteria (Rhim et al., 2014). Different active packaging is available with silver ions in edible coating for the preservation of fruits and vegetables. For carrots, pears (Mohammed Fayaz et al., 2009), and asparagus spears (An et al., 2008), the silver packaging extends the shelf life to almost 25 days. Moreover, the presence of carbon nanotubes with organic ligands in packaging material protects cooked meat from oxidation. A decrease in microbial contamination and color change was the positive influence of nanoparticles (Dias et al., 2013).

Smart packaging comprises nanoparticles for the detection and identification of food contamination. Nanosensors with metal ions have a promising impact in monitoring the physical and chemical changes that occur during processing and preservation. It helps to detect the harmful toxins, bacteria, and pathogens present inside the food and helps to track the quality of food materials throughout the transport (Majid et al., 2018). The consumers will check the quality of food material through the attached nanosensors. Nanoparticles-based barcodes are also available which record all the responses related to changes taking place inside the food product. Nanosensors consist of two components, electronic and sensing components which detect all the physical changes and convert them into electronic format. The participation of nanoparticles in various forms like rods, wires, and crystals will increase the optical and electrical properties. Owing to their excellent sensitivity and selectivity, nanosensors are more preferred by food manufacturers and consumers than traditional sensors.

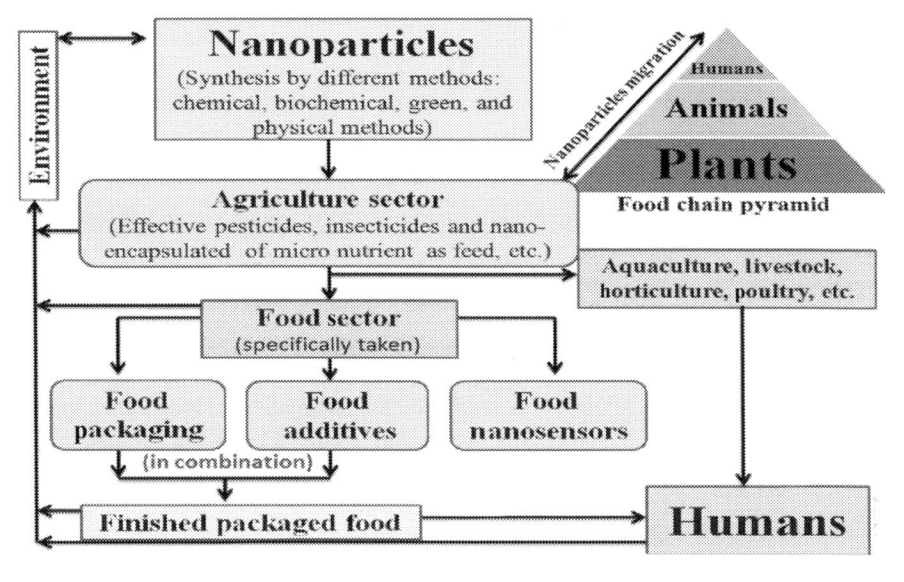

FIGURE 1.3

Possible migration pathway of nanoparticles.

(Reproduced with permission from Kumar et al., 2020).

1.5 **Challenges of nanobiotechnology in the food industry**

The development of nanotechnology in food-related sectors raises concerns about the side effects of nanoparticles on human health and also in ecology (Eleftheriadou et al., 2017). Nanoparticles have distinct physicochemical properties, size, and aggregation than large-sized molecules. Hence, the impact of these particles will be entirely different from similar large-sized molecules.

In addition, the utilization of nanotechnology in the food and agricultural field is of public concern. Due to its small size and large surface area, it will attract very reactive and unwanted substances in the food particles. The diffusion of nanoparticles to the food relies on the nature of the food and packing material. Fig. 1.3 shows how the nanoparticles are migrated to the human body through various sources. Public acceptance of nanotechnology-based food products relies upon safety issues. People consume food and beverages containing nanomaterials without the knowledge of health hazards. When it is settled in the gastrointestinal system, it will start aggregating in different body organs, leading to life-threatening diseases. Depending on the chemical nature of the nanoparticle, its application, and the environment, different issues will rise and the solution is not common to all. Even though nanotechnology has innumerable advantages for the food industries, the safety issues of small-sized particles cannot be ignored. Many reports have an in-depth demonstration of the toxic effects of nanoparticles and emphasis on the prospect of a transfer of nanoparticles from packaging to food and their influence on health issues (Bradley et al., 2011). More research should be carried out to investigate the toxic/ill effects of nanoparticles since the properties are entirely varied from macromolecules. The probabilities for bioaccumulation of nano-structured particles within body organs are higher. Anticaking agents based on silica particles are carcinogenic and affect the lung cells (Athinarayanan et al., 2014). The toxicity range varies depending on concentration, size, pH, temperature, pressure, and surface energy. Even though a large surface area is beneficial for the adsorption of essential ions, in some cases, it is a challenging process due to the integration of unwanted substances and chemical reactions.

The proclaimed advantages of nanotechnology will be apprehended by food industries only if they address the issues and problems concerned with safety. The guidelines mentioning the limit of usage of nanoparticles and their level of migration to the food particle were framed separately by the United States Food Drug Administration (He and Hwang, 2016) and the European Commission. The level should not exceed 10 mg/dm^2 and all the food industries should follow the rules and regulations framed by the concerned authorities (Hannon et al., 2015).

The existing regulation for the application of nanotechnology in food products is pigeonholed by several inconsistencies concerning side/ill effects. All the regulations framed worldwide failed to deliver standard management control of nanoparticles and associated risks in usage and disposal to the common public. The challenge also includes the lack of worldwide accepted rules that may ultimately fail to provide appropriate guidance in response to general public and occupational health risks associated with the manufacture, use, and disposal of nanomaterials.

Two primary concerns for the applicability of nanoparticles in the food industry are allergy and the discharge of toxic metals. Nanotechnology is used for food-allergen management (Pilolli et al., 2013) but few nanoparticles will induce pulmonary allergy (Ilves and Alenius, 2016). The use of nanometallic compounds along with food polymers will improve the properties and prevent oxidation but we should avoid the discharge of the toxic heavy metals. Accumulation of metallic compounds for a longer duration will severely affect human health. Mostly nanomaterials, like zinc (Fukui et al., 2012), silver (McShan et al., 2014), and copper (Karlsson et al., 2013) were more toxic materials and will damage DNA. The possibility of an oxidation reaction accompanied by side products will lead to the discharge of metal ions and affect the environment. The nanoparticles used for food additives will have direct contact with the human digestion system depending upon the concentration of nanomaterials in the food. The nano titanium oxide-coated chewing gum is largely available in the market for its sweetness but sometimes the titanium will be swallowed by the person and will get accumulated in the body (Chen et al., 2013). So proper awareness should be given to consumers, producers, and manufacturers and accurate clarity should be given about the nature of the nanoparticles.

The toxicology studies used for large molecules can be used for nano-sized particles but more emphasis should be given to size parameters (Bahadar et al., 2016).

Hence, the modified technology with proper fabrication and assimilation for the determination of toxicology will provide more persuasive results. The relation between toxicity and physiochemical characteristics of the nanometal ions should be studied in detail. Further, exhaustive characterization and *in vitro* and *in vivo* assessment should be carried out to determine the toxicity level.

1.6 Conclusion and prospects

Advancement and progression of nanotechnology have renovated vast areas of research and engineering including the food industry. The need for nanoparticles in fields like processing, packaging, safety, and functional food development was increasing every day. Recently the use of nano-sized particles in food industries has grown worldwide in various forms like additives, nanosensors, enhancers, active packaging, anticaking agents, and food processing. It upsurges both the overall quality of the food material and shelf-life. Furthermore, exciting results were obtained in the food preservation process with the help of nanomaterials owing to their size and sensitivity. It protects the spoilage of the food from moisture, gases, and photooxidation and carries bioactive compounds to various parts of the tissues. Nanoparticles show excellent antibacterial activity and are consumed as antimicrobial agents in the food packaging process. The distribution of nutritional supplements to the targeted area by employing the nanoencapsulation process will be an upcoming research area in the pharmaceutical sector. Further, nano-sized food particles deliver more taste, flavor, and texture, and inhibit microbial growth, and food wastage. Nanocarriers are used as a vehicle to carry food additives to food particles without affecting the basic properties and morphology. The encapsulation activity of nanoparticles was found to be more pronounced than in traditional methods. It controls the reaction of active particles with food, protects from decomposition, and proper release of nutrients. Integrating nanoparticles with food products will enlarge the oxygen barrier characteristics and diminish the transmission flow of traditional food packaging. Hence, the process of incorporating the small-sized particle with packaging films/coatings will help in the effectual packing and transport of food products.

The progress of nanotechnology in food processing and packaging was laying a new avenue every day but still, few challenges and consequences are addressed. The clarity regarding safety and ecological influence should be given more importance when mastering the advancement of technology in food-related processes and proper testing should be done before delivering to the commercial market for usage. A main consequence of the usage of nanotechnology is its toxicity and insufficient research towards assessment. Furthermore, research should be concentrated on delivering a suitable mechanism for the activity of nanoparticles with food. Several nanocompounds have different reactivity and can create adverse effects once migrated into foodstuff. Henceforth, it is significant to enhance the study of reactivity, migration, toxic levels, and permissible limits of nanoparticles.

References

An, J., Zhang, M., Wang, S., Tang, J., 2008. Studies on preservation of two cultivars of grapes at controlled temperature. LWT 41, 1100–1107.

Arakha, M., Pal, S., Samantarrai, D., Panigrahi, T.K., Mallick, B.C., Pramanik, K., Mallick, B., Jha, S., 2015. Sci. Rep. 5, 1–12.

Aswathanarayan, J.B., Vittal, R.R., 2019. Front. Sustain. Food Syst. 3, 1–21.

Athinarayanan, J., Periasamy, V.S., Alsaif, M.A., Al-Warthan, A.A., Alshatwi, A.A., 2014. Cell Biol. Toxicol. 30, 89–100.

Avella, M., De Vlieger, J.J., Errico, M.E., Fischer, S., Vacca, P., Volpe, M.G., 2005. Food Chem. 93, 467–474.

Bahadar, H., Maqbool, F., Niaz, K., Abdollahi, M., 2016. Iran. Biomed. J. 20, 1–11.

Bahrami, A., Delshadi, R., Assadpour, E., Jafari, S.M., Williams, L., 2020. Adv. Colloid Interface Sci. 278, 102140.

Bradley, E.L., Castle, L., Chaudhry, Q., 2011. Trends Food Sci. Technol. 22, 604–610.

Cao, G., Lin, H., Kannan, P., Wang, C., Zhong, Y., Huang, Y., Guo, Z., 2018. Langmuir 34, 14537–14545.

Carbone, M., Donia, D.T., Sabbatella, G., Antiochia, R., 2016. J. King Saud Univ. Sci. 28, 273–279.

Chellaram, C., Murugaboopathi, G., John, A.A., Sivakumar, R., Ganesan, S., Krithika, S., Priya, G., 2014. APCBEE Procedia 8, 109–113.

Chen, X.X., Cheng, B., Yang, Y.X., Cao, A., Liu, J.H., Du, L.J., Liu, Y., Zhao, Y., Wang, H., 2013. Small 9, 1765–1774.

Dasgupta, N., Ranjan, S., Mundekkad, D., Ramalingam, C., Shanker, R., Kumar, A., 2015. Food Res. Int. 69, 381–400.

Dias, M.V., De Fátima F. Soares, N., Borges, S.V., De Sousa, M.M., Nunes, C.A., De Oliveira, I.R.N., Medeiros, E.A.A., 2013. Food Chem 141, 3160–3166.

Eleftheriadou, M., Pyrgiotakis, G., Demokritou, P., 2017. Curr. Opin. Biotechnol. 44, 87–93.

Ezhilarasi, P.N., Karthik, P., Chhanwal, N., Anandharamakrishnan, C., 2013. Food Bioprocess Technol 6, 628–647.

Fukui, H., Horie, M., Endoh, S., Kato, H., Fujita, K., Nishio, K., Komaba, L.K., Maru, J., Miyauhi, A., Nakamura, A., Kinugasa, S., Yoshida, Y., Hagihara, Y., Iwahashi, H., 2012. Chem. Biol. Interact. 198, 29–37.

Gaikwad, K.K., Singh, S., Negi, Y.S., 2020. Environ. Chem. Lett. 18, 269–284.

Gasa-Falcon, A., Acevedo-Fani, A., Oms-Oliu, G., Odriozola-Serrano, I., Martín-Belloso, O., 2020. J. Funct. Foods 64, 103615.

Gorantla, S., Wadhwa, G., Jain, S., Sankar, S., Nuwal, K., Mahmood, A., Dubey, S.K., Taliyan, R., Kesharwani, P., Singhvi, G., 2021. Drug Deliv. Transl. Res. 12, 2359–2384.

Guarnieri, D., Guaccio, A., Fusco, S., Netti, P.A., 2011. J. Nanoparticle Res. 13, 4295–4309.

Hannon, J.C., Kerry, J., Cruz-Romero, M., Morris, M., Cummins, E., 2015. Trends Food Sci. Technol. 43, 43–62.

He, X., Hwang, H.M., 2016. J. Food Drug Anal. 24, 671–681.

Ilves, M., Alenius, H., 2016. Biomed. Appl. Toxicol. Carbon Nanomater. 16, 397–428.

Karlsson, H.L., Cronholm, P., Hedberg, Y., Tornberg, M., De Battice, L., Svedhem, S., Wallinder, I.O., 2013. Toxicology 313, 59–69.

Kumar, P., Mahajan, P., Kaur, R., Gautam, S., 2020. Mater. Today Chem. 17, 100332.

Kuswandi, B., 2017. Environ. Chem. Lett. 15, 205–221.

Kuswandi, B., Wicaksono, Y., Jayus, A.Abdullah, Heng, L.Y., Ahmad, M., 2011. Sens. Instrum. Food Qual. Saf. 5, 137–146.

Lipasek, R.A., Ortiz, J.C., Taylor, L.S., Mauer, L.J., 2012. Food Res. Int. 45, 369–380.

Madhusudan, P., Chellukuri, N., Shivakumar, N., 2018. Mater. Today Proc 5, 21018–21022.

Majid, I., Nayik, G.Ahmad, Mohammad Dar, S., Nanda, V., 2018. J. Saudi Soc. Agric. Sci. 17, 454–462.

McShan, D., Ray, P.C., Yu, H., 2014. J. Food Drug Anal. 22, 116–127.

Miar, M., Shiroudi, A., Pourshamsian, K., Oliaey, A.R., Hatamjafari, F., 2020. Exp. Mol. Pathol. 45, 147–158.

Mohammed Fayaz, A., Balaji, K., Girilal, M., Kalaichelvan, P.T., Venkatesan, R., 2009. J. Agric. Food Chem. 57, 6246–6252.

Pilolli, R., Monaci, L., Visconti, A., 2013. Trends Anal. Chem. 47, 12–26.

Podporska-Carroll, J., Myles, A., Quilty, B., McCormack, D.E., Fagan, R., Hinder, S.J., Dionysiou, D.D., Pillai, S.C., 2017. J. Hazard. Mater. 324, 39–47.

Pradhan, N., Singh, S., Ojha, N., Shrivastava, A., Barla, A., Rai, V., Bose, S., 2015. Biomed Res. Int. 12, 17.

Press, D., 2010. Nanotechnol. Sci. Appl. 18, 1–15.

Primozic, M., Knez, Z., Leitgeb, M., 2021. Nanomater 11, 292.

Rezić, I., Haramina, T., Rezić, T., 2017. Food Packag 22, 497–532.

Rhim, J.W., Wang, L.F., Lee, Y., Hong, S.I., 2014. Carbohydr. Polym. 103, 456–465.

Sarfraz, J., Gulin-Sarfraz, T., Nilsen-Nygaard, J., Pettersen, M.K., 2020. Nanomaterials 11, 10.

Servais, C., Jones, R., Roberts, I., 2002. J. Food Eng. 51, 201–208.

Singh, M., Lara, S., Tlali, S., 2018. J. Taibah Univ. Sci. 11, 922–929.

Vatai, G., 2000. Integr. Membr. Process. Bioconversions 24, 155–163.

Wang, L., Hu, C., Shao, L., 2017. Int. J. Nanomedicine 12, 1227–1249.

Yadav, D., Karki, S., Ingole, P.G., 2022. Food Eng. Rev. 14, 579–595.

Properties of nanomaterials for utilization in the food industry

2

Gunjan Nagpure [a], Shweta Rathee [b], Kshitij RB Singh [c], Ankur Ojha [b], Jay Singh [c] and Ravindra Pratap Singh [a,†]

[a] *Department of Biotechnology, Faculty of Science, Indira Gandhi National Tribal University, Amarkantak, Madhya Pradesh, India,* [b] *Department of Food Science and Technology, National Institute of Food Science Technology Entrepreneurship and Management, Sonipat, Haryana, India,* [c] *Department of Chemistry, Institute of Science, Banaras Hindu University, Varanasi, Uttar Pradesh, India*

2.1 Introduction

Nanotechnology is the production of nanostructured materials and their implementation in various fields. It involves a new science where diverse areas such as physics, chemistry, biology, material science, and engineering cover the nanoscale. Nanomaterial consists of particles with one or more external dimensions ranging between 1 nm and 100 nm for more than 1% of their number size distribution (Morris, 2011). Nanomaterials captured a market size of USD 8 billion in 2020 and are expected to grow at 14.1% from 2020 to 2028. Excellent properties drive its use in electronics, healthcare, aerospace, food, agriculture, and textile industries. They can be present in different morphologies (particles, tubes, wires, films, and flakes), compositions, dimensions, physicochemical and biological properties, resulting in improving the chemical reactivity, optical properties, superconducting potential, mechanical strength, antimicrobial potential, digestibility, bioavailability, and distribution (McClements et al., 2016).

Food grade nanomaterials application in different facets such as color (or flavor) additives, preservatives (antioxidant, antimicrobials), delivery of functional compounds, novel products formulations, nanofiltration (NF), screening of foods, fabrication of nanobiosensors (mycotoxins, microbial contaminants, antibiotics, pesticides, food additives, and dyes), and development of (active, smart/intelligent) packaging films (Rathee et al., 2021). They strengthen the functional component's bioavailability by modifying their bioaccessibility and barrier against environmental variations. They improve products' sensory properties like color, flavor, and texture by encapsulating food additives. They also help develop novel food packaging materials with enhanced

---

† deceased.

Nanobiotechnology for Food Processing and Packaging. DOI: https://doi.org/10.1016/B978-0-323-91749-0.00010-1

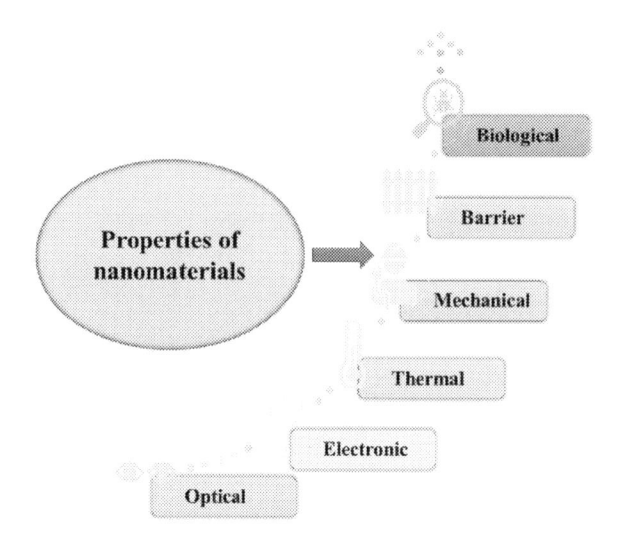

FIGURE 2.1

Represents the properties of nanomaterials.

thermomechanical, barrier, microstructural, and antimicrobial characteristics (Neme et al., 2021). Different nanobiosensors with improved detection limits offer a more straightforward and faster means of monitoring the undesirable contaminants present in food. However, this chapter provides a detailed overview of different food-grade nanomaterials, emphasizing their properties and applications in various areas of the food industry, and concludes with prospects.

2.2 Properties of nanomaterials

All nanomaterials show general properties such as size, surface area, shape, and reactivity. On the other hand, specific properties are owned by nanomaterials discussed in detail in this section (Sajid, 2022). The pictorial representation showing the properties of nanomaterials is highlighted in Fig. 2.1.

2.2.1 Electronic and optical properties

Silver nanomaterials possess unique optical properties due to the surface plasmon resonance phenomenon. They are biocompatible, easily functionalized, and tunable optical properties with prominent applications in treatment, labels in assay strips, and colorimetric sensors. For example, Harke et al. (2022) used vitamin B_{12} functionalized green silver nanoparticles to determine Fe^{+3} ions in food samples with a 2 mg/L limit of detection. It is a low-cost, portable, and sustainable method for rapid detection. Many other nanomaterials, such as metal oxides and sulfides, possess optical properties and are used in desired applications.

2.2.2 **Magnetic properties**

Magnetic nanomaterials are metals like iron, cobalt, and nickel that show superparamagnetic, high-field irreversibility, and manipulation with the applied external magnetic field. Their primary applications are in disease treatment, catalysis, and separation. For instance, a recent study on immunomagnetic nanoparticle-based lateral flow assay for detecting the peanut allergen Ara h_1 with a visual limit of 1 µg/g in chocolates agrees with an AOAC-approved ELISA kit (Yin et al., 2022). Another research group studied utilizing clay-modified green magnetic nanoparticles to treat agricultural wastewater generated by the guava paste production process (Mateus et al., 2021).

2.2.3 **Thermomechanical and barrier properties**

Thermomechanical properties of nanomaterials such as elastic modulus, hardness, movement law, interfacial adhesion, friction, and heating effect are of prime importance in different applications in the food industry. ZnO NPs incorporated in chitosan/carboxymethyl cellulose blend and polypropylene films improved thermomechanical stability and barrier properties (Youssef et al., 2016). Similarly, nanobiocomposite films made using konjac glucomannan/chitosan and mulberry anthocyanin extract improved thermomechanical and ultraviolet (UV)-vis light barrier properties (Sun et al., 2020).

2.2.4 **Biological properties**

Many metals and their oxides synthesized by green chemistry show excellent biological properties used in the food industry. Extracts and solutions obtained from plants, bacteria, fungi, and viruses show improved biological properties. Green synthesized silver and gold nanoparticles possess enhanced antibacterial, antifungal, and antiviral properties. For example, novel silver nanoparticles synthesized from *Forsythia suspensa* fruit water extract showed antibacterial activities against foodborne pathogens (Du et al., 2019). Another significant study utilizing safflower waste extract showed enhanced antibacterial activity against *Staphylococcus aureus* and *Pseudomonas fluorescens* (Rodríguez-Félix et al., 2021).

2.3 **Applications in the food industry**

Food nanotechnology shows numerous applications in eliminating chemical toxicants, delivering functional components, NF, and packaging, thus, enhancing food functionality. Fig. 2.2 displays the importance of nanomaterials in each sector of the food industry.

2.3.1 **Engineered nanomaterials in food**

Nanomaterials can potentially increase the effect of different food additives, including flavorings, colorants, antioxidants, and antimicrobial agents. These nanomaterials can

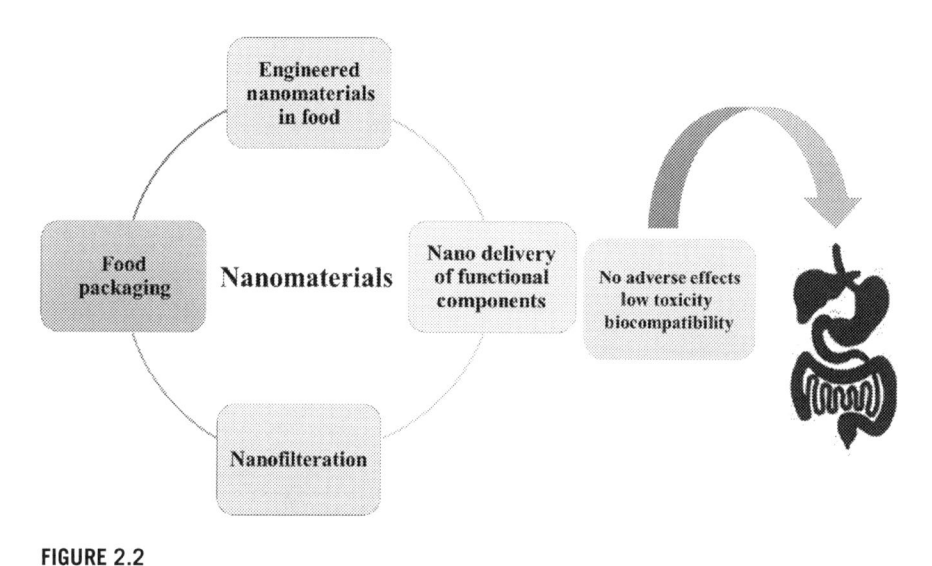

FIGURE 2.2

Represents the importance of nanomaterials in the food industry.

be directly added to foods as ingredients or applied as treatments to extend their functionality. Composition, size, shape, and charge affect nanomaterial functionality. Food nanoadditives include titanium dioxide (TiO_2, E171), silver (Ag, E174), gold (Au, E175), silicon dioxide (SiO_2, E551), iron oxide (Fe_2O_3, E172), zinc oxide (ZnO), and copper (Cu) used in the food industry (Moradi et al., 2022). For instance, titanium dioxide is used as a whitening agent for enhancing visual appeal and is approved by the Food and Drug Administration (FDA) and European Food Safety Authority (EFSA) (Musial et al., 2020). Besides TiO_2, the mixture of color additives may contain SiO_2 and Al_2O_3, but the FDA strictly prohibits using carbon black as a food coloring additive. NanoSiO_2 is used as an anticaking agent for maintaining the flow properties of powdered products and as a carrier of flavors and fragrances in food products. It has been registered within the European Union as a food additive, as per Directive E551 (Ameta et al., 2020). With the sustained release, it functions as a carrier of antioxidants like gallic acid and mesoporous poly(tannic acid).

Mesoporous silica nanomaterials were explored to functionalize caffeic acid and rutin with superior antioxidant profiles (Khalil et al., 2020). FDA approves nano antimicrobial agents such as silver-containing zeolites in food contact materials. They catalyze ROS generation in bacterial cells that cause cell death (Hoseinzadeh et al., 2017). Various nano-processed food products are commercially available in the markets of the United States, China, Australia, and Japan—enriched fruit juices, teas, nutritional drinks, shakes, and nanoencapsulation foods (Nile et al., 2020).

2.3.2 Nano delivery of functional components

Nano-delivery systems work to deliver functional components to the desired site to benefit consumers in general and targeted groups. They protect the components from degrading and control the release rate of an active ingredient under specific environmental conditions. Nanodispersions and nanocapsules are ideal for delivering functional elements because they can effectively perform all these tasks (Akhavan-Mahdavi et al., 2022). Nanoemulsions, solid lipid nanoparticles, nanoliposomes, inorganic carriers, casein micelles, protein nanoparticles, protein fibrils, polysaccharide nanoparticles, and protein-polysaccharide nanoconjugates are among the different types of nanomaterials used for the delivery purpose (Singh, 2016). Recently, a research group developed novel glycosylated zein nanoparticles to encapsulate zein with 10 times improved solubility. In this nanostructure, zein protein nanoparticles were glucosamine glycosylated by transglutaminase with improved entrapment efficiency and better antioxidant activity (Chang et al., 2022). Another protein-based nanomaterial, casein micelle used for the nano delivery of olive leaf extract in a sustained and controlled manner. It helps preserve the phenolic content, prevent antioxidant activity loss, and mask the bitter taste (Rikhtehgaran et al., 2021). An interesting study based on the co-delivery of curcumin and berberine using zein-chitosan complex nanoparticles, demonstrated that zein nanoparticles form the core structure with chitosan polysaccharide as a coating material that helped to improve solubility, storage stability, and anticancer functionality (Ghobadi-Oghaz et al., 2022).

A similar study was done based on nano complexes between chitosan and succinylated pea protein for curcumin nanoencapsulation by electrostatic binding process. Chitosan restored the hydrophobicity decreased by the succinylation process of pea protein, thereby, improving stability, solubility, and in-vitro bioaccessibility (Okagu et al., 2021). Additionally, the complex formed between xanthan gum and bovine serum albumin nanoparticles for binding of curcumin. The nanocomplex helps protect the curcumin from degradation at neutral pH and retains its properties (Papagiannopoulos and Sklapani, 2021). Nanocarriers such as nanogels, nanofibers, nanosponges, nanoparticle clustering, trojan horse nanoparticles, and intelligent nanomaterials are among the recent nanomaterials explored for delivery purposes (Montes et al., 2019). A critical study utilized nanogels made up of chitosan-stearic acid to improve the oxidative stability of sunflower pickering emulsion (Atarian et al., 2019). Fahami and Fathi (2018) developed mucilage nanofibers for carrying vitamin A, with a loading capacity of 29.51% with enhanced thermal stability.

2.3.3 Nanofiltration

Nanofiltration is an advanced approach operating at lower pressures than conventional membrane processes with unique selectivity making it a sustainable strategy. The broad applications of concentrating, fractionating, and purification reduce the amount of dissolved matter, colored compounds, and organic substances. Food industries employ this technology in fruit juice, beverages, dairy, sugar, lactic acid, and vegetables. Polyphenolic compounds responsible for browning and haziness are retained in conventional membrane filtration with MWCO range 1000–5000 Da but removed in the NF process. There are various examples of NF in fruit juice concentrations of apple, pear, grape must, blackcurrant, seabuckthorn berry, and bergamot. For example,

NF removed beetroot juice color, retained its flavor, and removed bacterial species from milk without boiling (Nile et al., 2020).

Red wine, coffee extraction processes, concentrating juices, and recovering bioactive compounds to produce functional fortified foods with improved sensory properties done with the help of NF (Nath et al., 2018). There is a continuously growing demand for nonalcoholic drinks, realizing the importance of NF as they can reduce the alcohol content 8–10 times while maintaining the drink's flavor. Additionally, low temperature during filtration minimizes the thermal impact on drinks. Whey protein concentration, protein hydrolysate fractionation, and effluent treatments in the dairy industry utilize NF technology. It is an excellent alternative to electrodialysis by reducing costs, waste disposal problems, simultaneous concentration, and demineralization of whey (Zhang and Feng, 2022; Castro-Muñoz and Gontarek, 2020).

2.3.4 Food packaging

The need for fresh food increases continuously due to the growing population, urbanization, and changing lifestyles. The global market size of the food packaging sector was USD 303.26 billion in 2019. Food wastage occurs due to excess buying, limited shelf life, big packets, confusion over labeling, improper meal planning, and inefficient storage space (FAO, 2021). The need for fresh foods raised the demand for high-performance packaging materials. The packaging materials with enhanced barrier properties, high mechanical strength, and smart/intelligent packaging for better food preservation. Nanotechnology has been revolutionized for improving active, smart/intelligent packaging with enhanced mechanical strength, physiological, physical, chemical, antimicrobial properties, and barrier performance (oxygen, CO_2, and moisture) (Chausali et al., 2022). Various nanomaterials (organic, inorganic, and a combination of these two) provide improved, active, bio-based, and intelligent packaging of foods (Ashfaq et al., 2022). Organic (natural products such as protein, carbohydrate, and fat), inorganic (metal and metal oxide), and a combination of these two (nanoclay) nanomaterials incorporation improved the packaging materials. Active packaging intentionally releases and absorbs the compounds from food during storage.

Intelligent packaging materials help target-specific sensing attributes such as freshness, temperature indicators, time-temperature integrators, food integrity, fruit ripeness, tracking, tracing, gas leakage, and microbial contaminants. In contrast, bio packaging employs biodegradable and sustainable material in place of plastic to preserve food (Cheng et al., 2022; Sanchez-Garcia et al., 2010). The various nanomaterials used for food packaging include nanocellulose, nanostarch (NS), protein nanoparticles, carbon nanotubes (CNTs), silver nanoparticles, zinc nanoparticles, titanium nanoparticles, and nanoclay (Onyeaka et al., 2022).

2.3.4.1 Nanocellulose

The breakdown of cellulose fibers prepared the nanocellulose. It is a biodegradable, renewable, abundant, and sustainable nanofiller (biopolymer) that produces a low carbon footprint. Three types of cellulose, such as cellulose nanocrystals (CNC), cellulose nanofibrils (CNF), and bacterial nanocellulose (BNC), are generally used in packaging. They assist in the controlled release of antimicrobials (Ahankari et al., 2021). A study by Mugwagwa and Chimphango (2022) reported that incorporating nanocellulose into the pectin/hemicellulose biocomposite films enhanced the physico-chemical properties and helped slow antioxidant release and has excellent potential in the active packaging of fatty foods. Another good research work based on the coating of nanocellulose and nanochitin on polypropylene films showed reduced bacterial adhesion, good thermal recyclability, high oxygen barrier, increased transparency, and potential as an advanced food packaging material (Nguyen et al., 2021).

2.3.4.2 Nanostarch

Nanostarch is a modified starch with improved properties such as nano quantum size effect, surface interfacial effect, and macroscopic quantum tunneling effect. The main types of NS used are starch nanocrystals (SNCs) and starch nanoparticles (SNPs) used in pickering emulsion stabilizers, film enhancers, active ingredients in packaging, and delivery of functional components (Wang and Zhang, 2021). The preparation meth-ods include acid hydrolysis, ball milling, high-pressure homogenization (top–down approaches), nanoprecipitation, micro emulsification, and recrystallization (bottom–up processes). The above preparation methods affect the physicochemical properties, mainly crystalline structure and morphology. For instance, Wang et al. (2021) reported the development of eco-friendly, pH-responsive SNPs, based on superhydrophobic coating for liquid food residue reduction and freshness monitoring. Mango kernels are utilized to source both starch and SNCs for nanobiocomposites films. The films at 5% SNC showed a higher 90% tensile strength, 120% elastic modulus, and 15% lower water vapor transmission rate (Oliveira et al., 2018).

2.3.4.3 Protein nanoparticles

Proteins are among the most abundant used biomaterials in food industries. They have excellent nutritional value, biodegradability, biocompatibility, safety status, and molecular properties (Martins et al., 2018). In a recent study, incorporating zein nanoparticles into modified cellulose films increased the film elasticity at lower concentrations, yet decreased elasticity at higher concentrations (Gilbert et al., 2017). The research group utilized zein nanoparticles to fabricate whey protein-based film with an increased tensile strength at 50% filler content (Oymaci and Altinkaya, 2016).

2.3.4.4 Carbon nanotubes

The rolling mechanism makes graphene sheets' CNTs into nanocylinders. The classification of CNTs into single-walled CNTs (SWCNTs) and multiwalled CNTs (MWCNTs) is done based on structural arrangements. They show exceptional ther-momechanical properties, unique crystallization structure, and antimicrobial activity (Azizi-Lalabadi et al., 2020). Recently, Alves et al. (2021) studied the effect of dif-ferent ionic surfactants (sodium dodecyl sulfate, cetyltrimethylammonium bromide, and sodium cholate) on the dispersibility of MWCNT in the starch film, with sodium cholate showing the most prominent effect on dispersibility, tensile strength, and young modulus.

2.3.4.5 Silver nanoparticles

Silver nanoparticles are the most commonly used nanofillers in food packaging, giving superior physicochemical, optical, thermal, biological, and antimicrobial properties. Recently, the interest in chemical synthesis methods shifted to green synthesis because of advantages such as cost-effectiveness, biocompatibility, sustainability, and large-scale production (Kumar et al., 2021).

For example, soluble soybean polysaccharide (SSPS) films fabricated in in-situ, generated silver nanoparticles, enhancing the UV-barrier, thermal, and antibacterial properties due to the reducing effect on silver ions (Liu et al., 2022). In another recent work, nanocomposite films were fabricated utilizing grape seed extract and chitosan with potent antimicrobial, antioxidant, and extended shelf life of grapes (Zhao et al., 2022).

2.3.4.6 Zinc oxide nanoparticles

Zinc oxide nanoparticles are the most affordable and safer nanomaterials for food packaging than silver nanoparticles. They are nontoxic, antimicrobial active, high aspect ratio, and barrier against the light. One interesting study is based on developing multifunctional polyvinyl alcohol and starch-based nanocomposite film utilizing ZnO-NPs as compatibilizers. The resulting film showed improved UV shielding, antimicrobial effect, compatibility, mechanical, and water barrier properties, and was highly optically transparent (Hu et al., 2022). One more work uses tragacanth, gelatin, and ZnO-NPs with improved mechanical stress, barrier, thermal properties, microstructure, and excellent antimicrobial activity against bacteria (Shahvalizadeh et al., 2021).

2.3.4.7 Titanium dioxide nanoparticles

Titanium dioxide (TiO_2) is present in the form of brookite, rutile, and anatase minerals. Its unique characteristics include a high refraction index, opaqueness, antibacterial activity, photocatalysis, and transmittance. It is used for coloring in foods, nano reinforcement, antimicrobial activity, protection from UV radiations, and prevention of food oxidation (Zhang and Rhim, 2022). For instance, TiO_2 nanoparticles are used to fabricate whey protein films, and improve the thermomechanical, physical, barrier, optical, microstructural, and chemical properties (Alizadeh Sani et al., 2017, 2022).

2.3.4.8 Nanoclay

Clay is the safe, biocompatible, simple origin, and economic nanomaterials most commonly used for food packaging. They are mineral silicates with layered structural units that form crystalline clay systems by stacking. The types of nanoclays such as halloysite, montmorillonite, bentonite, kaolinite, sepiolite, and laponite are used for food packaging and coatings (Nath et al., 2022). They also enhance mechanical, physical, and barrier properties (Tiwari et al., 2021). For instance, composite films fabricated with halloysite with concentration variation (0.5–2.0 mg) showed improved composite packaging film (Yang et al., 2020). Scientists studied the fabrication of κ-carrageenan films incorporated with bentonite at a concentration of (5%–15%) to improve the water-resistance properties (Dogaru et al., 2020). Further, researchers reported the gelatin/laponite nanocomposite films with more excellent thermal stability without affecting the barrier properties (López-Angulo et al., 2020).

2.3.5 Food monitoring

Food monitoring is a system of repeated measurement, evaluation, and analysis of unwanted substance levels, including heavy metals, plant-protection products, and other contaminants in and on food. Depending upon the food involved, the analysis includes plant protection product residues (fungicides, herbicides, insecticides, etc.), and the heavy metals include cadmium, lead, mercury, nitrite, nitrate, and mycotoxins (ochratoxin A and aflatoxins). As the demand for food is increasing tremendously in the food sector, the concern about health benefits and food quality is rising rapidly (Moguel et al., 2019). Moreover, the researchers are finding a way to improve the food standard by monitoring the undesirable substances and contaminants in food. Nanotechnology has been considered one of the relevant technologies that increasingly revolutionized the food industry (Singh et al., 2017). However, the desire for nanoscale materials has been increasing in the food sector as it offers complete food resolution from food processing and monitoring to packaging (Roselli et al., 2003; Sawai, 2003). Nanomaterials show advanced properties such as nontoxicity and stability and can be utilized as food additives, anticaking agents, fillers for enhancing the durability and mechanical strength and of the packaging material in food processing (Ezhilarasi et al., 2013).

Moreover, the most challenging feature of pathogen detection is the desire to mark zero-tolerance; that is, the concern that no feasible pathogens be permitted in the specific food item. Thus, a method must be delicate enough to monitor an individual pathogen in a sample to obtain zero-tolerance level detection. As a result, nanotechnology emerged as a novel technology that offers simpler and faster means of monitoring pathogens at much lower detection limits than conventional methods.

Nowadays, globally, all agricultural products are susceptible to agro-based chemicals, including fungicides and insecticides. The continually increasing utilization of chemicals in the agriculture system led to adverse effects on human health, primarily due to food safety problems. Gold nanoparticles (AuNPs) have been broadly utilized in environmental monitoring, chemical energy, biomedicine, electronic devices, sensing analysis, catalytic and food safety screening due to their narrow size distributions, good biological affinity, and unique optical properties (Rasheed and Sandhyarani, 2017; Wang et al., 2017; Lee et al., 2010). These exclusive properties are often utilized in monitoring toxic chemicals such as biotoxins, heavy metals, pesticide residues, melamine detection in milk, and banned preservatives in food as illustrated in Fig. 2.3.

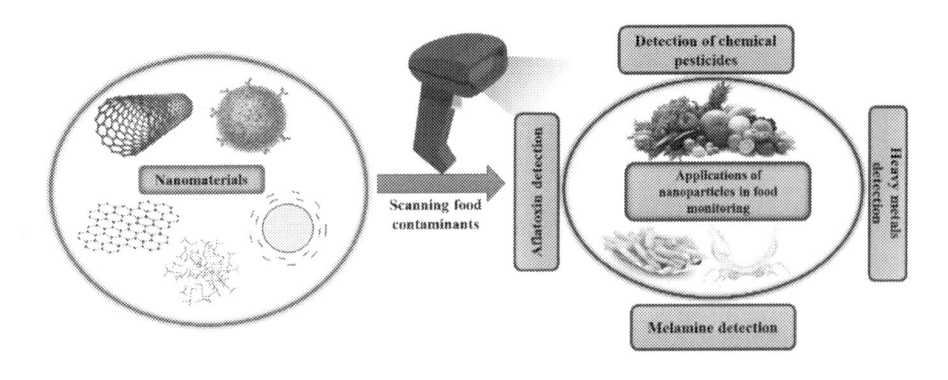

FIGURE 2.3

Represents the applications of nanomaterials in food monitoring.

Calorimetric sensors based on AuNPs possessing highly sensitive, rapid, and simple nature have been broadly applied in rapid testing and real-time monitoring of food safety and quality (Shi et al., 2015; Chen et al., 2015; Magar et al., 2021; Lee et al., 2010; Samanta et al., 2012; Liu et al., 2018). Zhang et al. (2022) prepared the flexible surface enhancing Raman scattering (SERS) chip via cross-linking method using aggregated gold nanoparticles (a-AuNPs) synthesized by Ca^{2+} mediated assembly, further dispersed in a polyvinyl alcohol solution. A study reported that the acquired hydrogel SERS chip exhibits numerous advantages, including good repeatability, robust anti-interference ability, high sensitivity, and long-term stability. These promising advantages of prepared hydrogel SERS chips can be utilized to monitor organophosphate pesticides such as phosmet and triazophos in an orange sample. Sun et al. (2012) reported an electrochemical immunosensor exhibiting properties such as high stability, good reproducibility, and high sensitivity based on AuNPs and Prussian blue-multiwalled CNTs-chitosan (PB-MWCNTs-CTs) nanocomposite film can be utilized in monitoring the carbofuran pesticide in food. Another study reported that the SERS method based on octanethiol-functionalized bimetallic core-shell nanoparticles (Oct/Au@AgNPs) shows promising efficiency in detecting ziram, an agricultural fungicide in apple and pear fruits (Hussain et al., 2020). One study reported that an electrochemical aptasensor which is labelled-free was developed to detect aflatoxin M1 in milk samples by utilizing AuNPs-based pencil graphite electrode (PGE) and reduced graphene oxide (rGO). Herein, the PGE is used as a working electrode due to its eco-friendly nature and low price (Ahmadi et al., 2022). An aptasensor belongs to a class of biosensors where the biological recognition element is a DNA or RNA aptamer, whereas aptamers are short single-stranded oligonucleotides (DNA or RNA) (Cheng et al., 2020). Aptamers are used as electrochemical biosensors due to their excellent advantages such as cost-effective, high affinity with their analyte, unique 3D structure, recyclable synthesis, and ease of labeling and modification (Li et al., 2020; Song et al., 2020). Furthermore, unfortunately, milk adulteration is a serious issue worldwide. Melamine is a nitrogen-rich chemical compound reported as another adulterant found in milk that increases the protein content artificially in milk and milk products. Kumar et al. (2016) investigated a simple and sensitive colorimetric method for melamine detection in milk utilizing unmodified silver nanoparticles (AgNPs). Moreover, quantum dots (QDs) are used as labeling materials for biosensors and optimal imaging in food safety conditions. Zhang et al. (2011) reported that the water-soluble CdTe QDs as fluorescence probes were utilized to determine the melamine adulterants in milk products.

Nanobiosensors also emerged as a powerful tool in monitoring harmful toxic contaminants such as heavy metal ions, pesticides, pathogens detection, detection of mycotoxins, etc. A nanobiosensor is usually a biosensor synthesized by using nanomaterials. Researchers are utilizing various nanocomposites and nanomaterials to improve food's shelf life and sensitivity (Thakur and Ragavan, 2013). Moreover, integrating biosensors with numerous nanomaterials like nanorods, thin films, and nanofibers in the examination methods for monitoring the impurities in food has enhanced the monitoring sensitivity and high portability. However, in food microbiology, nanosensors and nanobiosensors are utilized to monitor the microbes in food materials or processing plants, evaluate available food components, and inform distributors and consumers about the safety status of food (Mathivanan, 2021). One study reported that due to ample functional groups on surface area and good electrical conductivity, CNTs could be utilized to develop biosensors to monitor foodborne contagious bacteria such as *S. aureus* in fresh meat (Huang et al., 2016). Magnetic nanoparticles and TiO_2 nanocrystals are used to monitor the *Salmonella* in milk (Joo et al., 2012). Moreover, AuNPs-based biosensors are utilized to monitor the different pesticide molecules including dimethoate, chlorpyrifos, paraoxon, carbofuran, and carbaryl (Cui et al., 2018). Another study reported, that a stable and highly reproducible biosensor based on chitosan-TiO_2 graphene nanocomposite has recently been developed for monitoring organophosphate pesticides in cabbage (Ellman et al., 1961). On the other hand, colorimetric-based AuNP sensors were utilized to monitor the heavy metal ions such as Ag^+, Hg^{2+}, Cd^{2+}, Cr^{3+}, Pb^{2+}, Al^{3+}, and Fe^{3+} in food (Mathivanan, 2021).

2.4 **Cytotoxicity aspects of nanomaterials**

Cytotoxicity is one of the significant issues that occur in every consumer mindset. As the nanoparticles exhibit numerous applications in the food sector from food packaging to food monitoring, there are also some cytotoxicity aspects of nanomaterials, and this concern must be addressed. Nanoparticles are more mobile, more toxic, and more reactive. It has been investigated that there is an efficient possibility that nanoparticles can generate free radicals which lead to possible fatality, tumors, and DNA mutation, as well as a result of enhanced oxidative stress (Cushen et al., 2012). Moreover, some nanoparticles show negative impacts on tissues, such as early signs of tumor formation, oxidative stress, and inflammation. Nanotoxicity is mainly contemplated through reactive oxygen species (ROS) overproduction, which further results in the cell's futile to sustain normal physiological redox-regulated functions. Thus, nanotoxicity can cause unregulated cell signaling, DNA damage, cytotoxicity, cancer initiation, apoptosis, and alternation in cell motility (Vishwakarma et al., 2010; Wang et al., 2007).

Furthermore, unique properties of nanomaterials and nanostructures such as large surface area, small size, solubility, and high activity show harmful adverse side effects on the environment and health. Solubility serves as necessary in nanoparticle toxicity. For example, insoluble titanium oxide nanoparticles show less toxicity than soluble titanium oxide nanoparticles (Oberdörster, 2000). It has been reported that soluble nanoparticles are recognized as more carcinogenic as compared to insoluble nanoparticles (Salnikow and Kasprzak, 2005). Recent evidence shows that neutral or negative nanoparticles are less toxic than positively charged nanoparticles (Badawy et al., 2011). Hence, according to reports severity and types of injuries that can affect multiple cells and tissues are mainly determined by the calibrated exposure level of nanoparticles (Oberdörster et al., 2005). Nanoparticles that are utilized in food processing, packaging, fertilizers, and pesticides may also enter workers' respiratory tracts, which could be harmful and negatively impact health. Consequently, nanotoxicology plays a significant role in providing roadmaps and clear guidelines for minimizing risks in the excellent use of nanoparticles (Amini et al., 2014; Badawy et al., 2011).

2.5 Conclusion and prospects

Nanotechnology has gained popularity in food science and technology over the past years. The fundamental mystery of nanotechnology and the food industry is that the components of food products and the intercellular level of human cells can easily interact with the nanoparticles, assimilating nano-scale materials in all the zones of the food sector. Moreover, nanobiotechnology has shown multiple applications in the field of food processing, packaging, and monitoring. Nanomaterials have shown positive results in preserving food from unwanted gases, moisture, microbes, lipids, off-flavors, and odors. Moreover, nanomaterials have shown significant applications in food monitoring, such as detecting pesticides, fertilizers, heavy metals, and mycotoxins in food. As nanotechnology is covering new pathways day by day but still carries some of the challenges and cytological aspects that need to be resolved to ease consumers' concerns. However, numerous nanomaterials are still at the stage of development to find applications in the food sector. There is a need to create room to explore more nanomaterials in the food sector. Still, many researchers are trying to design biomarkers to help monitor the specific microorganisms responsible for contaminating processed food material.

References

Ahankari, S.S., Subhedar, A.R., Bhadauria, S.S., Dufresne, A., 2021. Nanocellulose in food packaging: a review. Carbohydr. Polym. 255, 117479. https://doi.org/10.1016/j.carbpol.2020.117479.

Ahmadi, S.F., Hojjatoleslamy, M., Kiani, H., Molavi, H., 2022. Monitoring of aflatoxin M1 in milk using a novel electrochemical aptasensor based on reduced graphene oxide and gold nanoparticles. Food Chem. 373, 131321. https://doi.org/10.1016/j.foodchem.2021.131321.

Akhavan-Mahdavi, S., Sadeghi, R., Esfanjani, A.F., Hedayati, S., Shaddel, R., Dima, C., Malekjani, N., Boostani, S., Jafari, S.M., 2022. Nanodelivery systems for d-limonene; techniques and applications. Food Chem. 384, 132479. https://doi.org/10.1016/j.foodchem.2022.132479.

Alizadeh Sani, M., Ehsani, A., Hashemi, M., 2017. Whey protein isolate/cellulose nanofibre/TiO$_2$ nanoparticle/rosemary essential oil nanocomposite film: its effect on microbial and sensory quality of lamb meat and growth of common foodborne pathogenic bacteria during refrigeration. Int. J. Food Microbiol. 251, 8–14. https://doi.org/10.1016/j.ijfoodmicro.2017.03.018.

Alizadeh Sani, M., Maleki, M., Eghbaljoo-Gharehgheshlaghi, H., Khezerlou, A., Mohammadian, E., Liu, Q., Jafari, S.M., 2022. Titanium dioxide nanoparticles as multifunctional surface-active materials for smart/active nanocomposite packaging films. Adv. Colloid Interface Sci. 300, 102593. https://doi.org/10.1016/j.cis.2021.102593.

Alves, Z., Abreu, B., Ferreira, N.M., Marques, E.F., Nunes, C., Ferreira, P., 2021. Enhancing the dispersibility of multiwalled carbon nanotubes within starch-based films by the use of ionic surfactants. Carbohydr. Polym. 273, 118531. https://doi.org/10.1016/j.carbpol.2021.118531.

Ameta, S.K., Rai, A.K., Hiran, D., Ameta, R., Ameta, S.C., 2020. Use of nanomaterials in food science. In: Biogenic Nano-Particles and Their Use in Agro-Ecosystems. Springer, Singapore, pp. 457–488. https://doi.org/10.1007/978-981-15-2985-6_24.

Amini, S.M., Gilaki, M., Karchani, M., 2014. Safety of nanotechnology in food industries. Electron. Physician 6 (4), 962–968. https://doi.org/10.14661/2014.962-968.

Ashfaq, A., Khursheed, N., Fatima, S., Anjum, Z., Younis, K., 2022. Application of nanotechnology in food packaging: pros and cons. J. Agric. Food Res. 7, 100270. https://doi.org/10.1016/j.jafr.2022.100270.

Atarian, M., Rajaei, A., Tabatabaei, M., Mohsenifar, A., Bodaghi, H., 2019. Formulation of Pickering sunflower oil-in-water emulsion stabilized by chitosan-stearic acid nanogel and studying its oxidative stability. Carbohydr. Polym. 210, 47–55. https://doi.org/10.1016/j.carbpol.2019.01.008.

Azizi-Lalabadi, M., Hashemi, H., Feng, J., Jafari, S.M., 2020. Carbon nanomaterials against pathogens; the antimicrobial activity of carbon nanotubes, graphene/graphene oxide, fullerenes, and their nanocomposites. Adv. Colloid Interface Sci. 284, 102250. https://doi.org/10.1016/j.cis.2020.102250.

Castro-Muñoz, R., Gontarek, E., 2020. Nanofiltration in the food industry. In: Handbook of Food Nanotechnology: Applications and Approaches. Academic Press, pp. 73–106. https://doi.org/10.1016/B978-0-12-815866-1.00003-0.

Chang, Y., Jiao, Y., Li, D.-J., Liu, X.-L., Han, H., 2022. Glycosylated zein as a novel nanodelivery vehicle for lutein. Food Chem. 376, 131927. https://doi.org/10.1016/j.foodchem.2021.131927.

Chausali, N., Saxena, J., Prasad, R., 2022. Recent trends in nanotechnology applications of bio-based packaging. J. Agric. Food Res. 7, 100257. https://doi.org/10.1016/j.jafr.2021.100257.

Chen, M., Shen, X., Liu, P., Wei, Y., Meng, Y., Zheng, G., Diao, G., 2015. β-cyclodextrin polymer as a linker to fabricate ternary nanocomposites AuNPs/PATP-β-CDP/RGO and their electrochemical application. Carbohydr. Polym. 119, 26–34. https://doi.org/10.1016/j.carbpol.2014.11.022.

Cheng, H., Xu, H., McClements, D.J., Chen, L., Jiao, A., Tian, Y., Miao, M., Jin, Z., 2022. Recent advances in intelligent food packaging materials: principles, preparation and applications. Food Chem. 375, 131738. https://doi.org/10.1016/j.foodchem.2021.131738.

Cheng, L., Xu, C., Cui, H., Liao, F., Hong, N., Ma, G., Xiong, J., Fan, H., 2020. A sensitive homogenous aptasensor based on tetraferrocene labeling for thrombin detection. Anal. Chim. Acta 1111, 1–7. https://doi.org/10.1016/j.aca.2020.03.017.

Cui, H.-F., Wu, W.-W., Li, M.-M., Song, X., Lv, Y., Zhang, T.-T., 2018. A highly stable acetylcholinesterase biosensor based on chitosan-TiO$_2$-graphene nanocomposites for detection of organophosphate pesticides. Biosens. Bioelectron. 99, 223–229. https://doi.org/10.1016/j.bios.2017.07.068.

Cushen, M., Kerry, J., Morris, M., Cruz-Romero, M., Cummins, E., 2012. Nanotechnologies in the food industry: recent developments, risks and regulation. Trends Food Sci. Technol. 24 (1), 30–46. https://doi.org/10.1016/j.tifs.2011.10.006.

Dogaru, B.I., Simionescu, B., Popescu, M.C., 2020. Synthesis and characterization of κ-carrageenan bio-nanocomposite films reinforced with bentonite nanoclay. Int. J. Biol. Macromol. 154, 9–17. https://doi.org/10.1016/j.ijbiomac.2020.03.088.

Du, J., Hu, Z., Yu, Z., Li, H., Pan, J., Zhao, D., Bai, Y., 2019. Antibacterial activity of a novel *Forsythia suspensa* fruit mediated green silver nanoparticles against food-borne pathogens and mechanisms investigation. Mater. Sci. Eng. C 102, 247–253. https://doi.org/10.1016/j.msec.2019.04.031.

El Badawy, A.M., Silva, R.G., Morris, B., Scheckel, K.G., Suidan, M.T., Tolaymat, T.M., 2011. Surface charge-dependent toxicity of silver nanoparticles. Environ. Sci. Technol. 45 (1), 283–287. https://doi.org/10.1021/es1034188.

Ellman, G.L., Courtney, K.D, Andres, V., Featherstone, R.M., 1961. A new and rapid colorimetric determination of acetylcholinesterase activity. Biochem. Pharmacol. 7 (2), 88–95. https://doi.org/10.1016/0006-2952(61)90145-9.

Ezhilarasi, P.N., Karthik, P., Chhanwal, N., Anandharamakrishnan, C., 2013. Nanoencapsulation techniques for food bioactive components: a review. Food Bioprocess Technol. 6 (3), 628–647. https://doi.org/10.1007/s11947-012-0944-0.

Fahàmi, A., Fathi, M., 2018. Development of cress seed mucilage/PVA nanofibers as a novel carrier for vitamin A delivery. Food Hydrocoll. 81, 31–38. https://doi.org/10.1016/j.foodhyd.2018.02.008.

FAO, 2021. The State of Food and Agriculture 2021. FAO. https://doi.org/10.4060/cb4476en.

Ghobadi-Oghaz, N., Asoodeh, A., Mohammadi, M., 2022. Fabrication, characterization and in vitro cell exposure study of zein-chitosan nanoparticles for co-delivery of curcumin and berberine. Int. J. Biol. Macromol. 204, 576–586. https://doi.org/10.1016/j.ijbiomac.2022.02.041.

Gilbert, J., Cheng, C.J., Jones, O.G., 2017. Vapor barrier properties and mechanical behaviors of composite hydroxypropyl methylcelluose/zein nanoparticle films. Food Biophys. 13, 25–36. https://doi.org/10.1007/S11483-017-9508-1.

Harke, S.S., Patil, R.V., Dar, M.A., Pandit, S.R., Pawar, K.D., 2022. Functionalization of biogenic silver nanoparticles with vitamin B12 for the detection of iron in food samples. Food Chem. Adv. 1, 100017. https://doi.org/10.1016/j.focha.2022.100017.

Hoseinzadeh, E., Makhdoumi, P., Taha, P., Hossini, H., Stelling, J., Kamal, M.A., Ashraf, G.Md., 2017. A review on nano-antimicrobials: metal nanoparticles, methods and mechanisms. Curr. Drug Metab. 18 (2), 120–128. https://doi.org/10.2174/1389200217666161201111146.

Hu, W., Zou, Z., Li, H., Zhang, Z., Yu, J., Tang, Q., 2022. Fabrication of highly transparent and multifunctional polyvinyl alcohol/starch based nanocomposite films using zinc oxide nanoparticles as compatibilizers. Int. J. Biol. Macromol. 204, 284–292. https://doi.org/10.1016/j.ijbiomac.2022.02.020.

Huang, H., Liu, M., Wang, X., Zhang, W., Yang, D.-P., Cui, L., Wang, X., 2016. Label-free 3D Ag nanoflower-based electrochemical immunosensor for the detection of escherichia coli O157:H7 pathogens. Nanoscale Res. Lett. 11 (1), 507. https://doi.org/10.1186/s11671-016-1711-3.

Hussain, N., Pu, H., Hussain, A., Sun, D.-W., 2020. Rapid detection of ziram residues in apple and pear fruits by SERS based on octanethiol functionalized bimetallic core-shell nanoparticles. Spectrochim. Acta Part A 236, 118357. https://doi.org/10.1016/j.saa.2020.118357.

Joo, J., Yim, C., Kwon, D., Lee, J., Shin, H.H., Cha, H.J., Jeon, S., 2012. A facile and sensitive detection of pathogenic bacteria using magnetic nanoparticles and optical nanocrystal probes. Analyst 137 (16), 3609. https://doi.org/10.1039/c2an35369e.

Khalil, I., Yehye, W.A., Etxeberria, A.E., Alhadi, A.A., Dezfooli, S.M., Julkapli, N.B.M., Basirun, W.J., Seyfoddin, A., 2020. Nanoantioxidants: recent trends in antioxidant delivery applications. Antioxidants 9 (1), 24. https://doi.org/10.3390/antiox9010024.

Kumar, N., Kumar, H., Mann, B., Seth, R., 2016. Colorimetric determination of melamine in milk using unmodified silver nanoparticles. Spectrochim. Acta Part A 156, 89–97. https://doi.org/10.1016/j.saa.2015.11.028.

Kumar, S., Basumatary, I.B., Sudhani, H.P.K., Bajpai, V.K., Chen, L., Shukla, S., Mukherjee, A., 2021. Plant extract mediated silver nanoparticles and their applications as antimicrobials and in sustainable food packaging: a state-of-the-art review. Trends Food Sci. Technol. 112, 651–666. https://doi.org/10.1016/j.tifs.2021.04.031.

Lee, O.-S., Prytkova, T.R., Schatz, G.C., 2010. Using DNA to link gold nanoparticles, polymers, and molecules: a theoretical perspective. J. Phys. Chem. Lett. 1 (12), 1781–1788. https://doi.org/10.1021/jz100435a.

Li, Y., Liu, D., Zhu, C., Shen, X., Liu, Y., You, T., 2020. Sensitivity programmable ratiometric electrochemical aptasensor based on signal engineering for the detection of aflatoxin B1 in peanut. J. Hazard. Mater. 387, 122001. https://doi.org/10.1016/j.jhazmat.2019.122001.

Liu, G., Lu, M., Huang, X., Li, T., Xu, D., 2018. Application of gold-nanoparticle colorimetric sensing to rapid food safety screening. Sensors 18 (12), 4166. https://doi.org/10.3390/s18124166.

Liu, J., Ma, Z., Liu, Y., Zheng, X., Pei, Y., Tang, K., 2022. Soluble soybean polysaccharide films containing in-situ generated silver nanoparticles for antibacterial food packaging applications. Food Packag. Shelf Life 31, 100800. https://doi.org/10.1016/j.fpsl.2021.100800.

López-Angulo, D., Bittante, A.M.Q.B., Luciano, C.G., Ayala-Valencia, G., Flaker, C.H.C., Djabourov, M., do Amaral Sobral, P.J., 2020. Effect of laponite® on the structure, thermal stability and barrier properties of nanocomposite gelatin films, Food Biosci. 35, 100596.

Magar, H.S., Hassan, R.Y.A., Mulchandani, A., 2021. Electrochemical impedance spectroscopy (EIS): principles, construction, and biosensing applications. Sensors 21 (19), 6578. https://doi.org/10.3390/s21196578.

Martins, J.T., Bourbon, A.I., Pinheiro, A.C., Fasolin, L.H., Vicente, A.A., 2018. Protein-based structures for food applications: from macro to nanoscale. Front. Sustain. Food Syst. 2, 77. https://doi.org/10.3389/fsufs.2018.00077/bibtex.

Mateus, A., Torres, J., Marimon-Bolivar, W., Pulgarín, L., 2021. Implementation of magnetic bentonite in food industry wastewater treatment for reuse in agricultural irrigation. Water Resour. Ind. 26, 100154. https://doi.org/10.1016/j.wri.2021.100154.

Mathivanan, S., 2021. Perspectives of nano-materials and nanobiosensors in food safety and agriculture. In: Novel Nanomaterials. IntechOpen. https://doi.org/10.5772/intechopen. 95345.

McClements, D.J., DeLoid, G., Pyrgiotakis, G., Shatkin, J.A., Xiao, H., Demokritou, P., 2016. The role of the food matrix and gastrointestinal tract in the assessment of biological properties of ingested engineered nanomaterials (IENMs): state of the science and knowledge gaps. NanoImpact 3–4, 47–57. https://doi.org/10.1016/j.impact.2016.10.002.

Moguel, E., Berrocal, J., García-Alonso, J., 2019. Systematic literature review of food-intake monitoring in an aging population. Sensors 19 (15), 3265. https://doi.org/10.3390/ s19153265.

Montes, C., Villaseñor, M.J, Ríos, Á., 2019. Analytical control of nanodelivery lipid-based systems for encapsulation of nutraceuticals: achievements and challenges. Trends Food Sci. Technol. 90, 47–62. https://doi.org/10.1016/j.tifs.2019.06.001.

Moradi, M., Razavi, R., Omer, A.K., Farhangfar, A., McClements, D.J., 2022. Interactions between nanoparticle-based food additives and other food ingredients: a review of current knowledge. Trends Food Sci. Technol. 120, 75–87. https://doi.org/10.1016/ j.tifs.2022.01.012.

Morris, V.J., 2011. Emerging roles of engineered nanomaterials in the food industry. Trends Biotechnol. 29 (10), 509–516. https://doi.org/10.1016/j.tibtech.2011.04.010.

Mugwagwa, L.R., Chimphango, A.F.A., 2022. Physicochemical properties and potential application of hemicellulose/pectin/nanocellulose biocomposites as active packaging for fatty foods. Food Packag. Shelf Life 31, 100795. https://doi.org/10.1016/j.fpsl.2021.100795.

Musial, J., Krakowiak, R., Mlynarczyk, D.T., Goslinski, T., Stanisz, B.J., 2020. Titanium dioxide nanoparticles in food and personal care products: what do we know about their safety? Nanomaterials 10 (6), 1–23. https://doi.org/10.3390/nano10061110.

Nath, D., Santhosh, R., Pal, K., Sarkar, P., 2022. Nanoclay-based active food packaging systems: a review. Food Packag. Shelf Life 31, 100803. https://doi.org/10.1016/j.fpsl. 2021.100803.

Nath, K., Dave, H.K., Patel, T.M., 2018. Revisiting the recent applications of nanofiltration in food processing industries: progress and prognosis. Trends Food Sci. Technol. 73, 12–24. https://doi.org/10.1016/j.tifs.2018.01.001.

Neme, K., Nafady, A., Uddin, S., Tola, Y.B., 2021. Application of nanotechnology in agriculture, postharvest loss reduction and food processing: food security implication and challenges. Heliyon 7 (12), e08539. https://doi.org/10.1016/j.heliyon.2021.E08539.

Nguyen, H.-L., Tran, T.H., Hao, L.T., Jeon, H., Koo, J.M., Shin, G., Hwang, D.S., Hwang, S.Y., Park, J., Oh, D.X., 2021. Biorenewable, transparent, and oxygen/moisture barrier nanocellulose/nanochitin-based coating on polypropylene for food packaging applications. Carbohydr. Polym. 271, 118421. https://doi.org/10.1016/j.carbpol.2021.118421.

Nile, S.H., Baskar, V., Selvaraj, D., Nile, A., Xiao, J., Kai, G., 2020. Nanotechnologies in food science: applications, recent trends, and future perspectives. Nano-Micro Lett. 12 (1), 45. https://doi.org/10.1007/S40820-020-0383-9.

Oberdörster, G., Oberdörster, E., Oberdörster, J., 2005. Nanotoxicology: an emerging discipline evolving from studies of ultrafine particles. Environ. Health Perspect. 113 (7), 823–839. https://doi.org/10.1289/ehp.7339.

Oberdörster, G., 2000. Pulmonary effects of inhaled ultrafine particles. Int. Arch. Occup. Environ. Health 74 (1), 1–8. https://doi.org/10.1007/s004200000185.

Okagu, O.D., Jin, J., Udenigwe, C.C., 2021. Impact of succinylation on pea protein-curcumin interaction, polyelectrolyte complexation with chitosan, and gastrointestinal release of curcumin in loaded-biopolymer nano-complexes. J. Mol. Liq. 325, 115248. https://doi.org/10.1016/j.molliq.2020.115248.

Oliveira, A.V., da Silva, A.P.M., Barros, M.O., Filho, M.M.S, Rosa, M.F., Azeredo, H.M.C., 2018. Nanocomposite films from mango kernel or corn starch with starch nanocrystals. Starch: Stärke 70 (11–12), 1800028. https://doi.org/10.1002/star.201800028.

Onyeaka, H., Passaretti, P., Miri, T., Al-Sharify, Z.T., 2022. The safety of nanomaterials in food production and packaging. Curr. Res. Food Sci. 5, 763–774. https://doi.org/10.1016/j.crfs.2022.04.005.

Oymaci, P., Altinkaya, S.A., 2016. Improvement of barrier and mechanical properties of whey protein isolate based food packaging films by incorporation of zein nanoparticles as a novel bionanocomposite. Food Hydrocoll. 54, 1–9. https://doi.org/10.1016/j.foodhyd.2015.08.030.

Papagiannopoulos, A., Sklapani, A., 2021. Xanthan-based polysaccharide/protein nanoparticles: preparation, characterization, encapsulation and stabilization of curcumin. Carbohydr. Polym. Technol. Appl. 2, 100075. https://doi.org/10.1016/j.carpta.2021.100075.

Rasheed, P.A., Sandhyarani, N., 2017. Electrochemical DNA sensors based on the use of gold nanoparticles: a review on recent developments. Microchim. Acta 184 (4), 981–1000. https://doi.org/10.1007/s00604-017-2143-1.

Rathee, S., Melaku, E.T., Ojha, A., 2021. Bionanomaterials utility in food industry and its challenges. In: Bionanomaterials for Environmental and Agricultural Applications. IOP Publishing, pp. 10-1–10-19. https://doi.org/10.1088/978-0-7503-3863-9CH10.

Rikhtehgaran, S., Katouzian, I., Jafari, S.M., Kiani, H., Maiorova, L.A., Takbirgou, H., 2021. Casein-based nanodelivery of olive leaf phenolics: preparation, characterization and release study. Food Struct. 30, 100227. https://doi.org/10.1016/j.foostr.2021.100227.

Rodríguez-Félix, F., López-Cota, A.G., Moreno-Vásquez, M.J., Graciano-Verdugo, A.Z., Quintero-Reyes, I.E., Del-Toro-Sánchez, C.L., Tapia-Hernández, J.A., 2021. Sustainable-green synthesis of silver nanoparticles using safflower (*Carthamus Tinctorius* L.) waste extract and its antibacterial activity. Heliyon 7 (4), e06923. https://doi.org/10.1016/j.heliyon.2021.E06923.

Roselli, M., Finamore, A., Garaguso, I., Britti, M.S., Mengheri, E., 2003. Zinc oxide protects cultured enterocytes from the damage induced by escherichia coli. J. Nutr. 133 (12), 4077–4082. https://doi.org/10.1093/jn/133.12.4077.

Sajid, M., 2022. Nanomaterials: types, properties, recent advances, and toxicity concerns. Curr. Opin. Environ. Sci. Health 25, 100319. https://doi.org/10.1016/j.coesh.2021.100319.

Salnikow, K., Kasprzak, K.S., 2005. Ascorbate depletion: a critical step in nickel carcinogenesis? Environ. Health Perspect. 113 (5), 577–584. https://doi.org/10.1289/ehp.7605.

Samanta, P.K., Periyasamy, G., Manna, A.K., Pati, S.K., 2012. Computational studies on structural and optical properties of single-stranded DNA encapsulated silver/gold clusters. J. Mater. Chem. 22 (14), 6774. https://doi.org/10.1039/c2jm16068d.

Sanchez-Garcia, M.D., Lopez-Rubio, A., Lagaron, J.M., 2010. Natural micro and nanobio-composites with enhanced barrier properties and novel functionalities for food biopackaging applications. Trends Food Sci. Technol. 21 (11), 528–536. https://doi.org/10.1016/j.tifs.2010.07.008.

Sawai, J., 2003. Quantitative evaluation of antibacterial activities of metallic oxide powders (ZnO, MgO and CaO) by conductimetric assay. J. Microbiol. Methods 54 (2), 177–182. https://doi.org/10.1016/S0167-7012(03)00037-X.

Shahvalizadeh, R., Ahmadi, R., Davandeh, I., Pezeshki, A., Moslemi, S.A.S., Karimi, S., Rahimi, M., Hamishehkar, H., Mohammadi., M., 2021. Antimicrobial bio-nanocomposite films based on gelatin, tragacanth, and zinc oxide nanoparticles: microstructural, mechanical, thermo-physical, and barrier properties. Food Chem. 354, 129492. https://doi.org/10.1016/j.foodchem.2021.129492.

Shi, J., Chan, C., Pang, Y., Ye, W., Tian, F., Lyu, J., Zhang, Yu, Yang, M., 2015. A fluorescence resonance energy transfer (FRET) biosensor based on graphene quantum dots (GQDs) and gold nanoparticles (AuNPs) for the detection of *Meca gene* sequence of *Staphylococcus aureus*. Biosens. Bioelectron. 67, 595–600. https://doi.org/10.1016/j.bios.2014.09.059.

Singh, H., 2016. Nanotechnology applications in functional foods; opportunities and challenges. Prev. Nutr. Food Sci. 21 (1), 1. https://doi.org/10.3746/pnf.2016.21.1.1.

Singh, T., Shukla, S., Kumar, P., Wahla, V., Bajpai, V.K., Rather, I.A., 2017. Application of nanotechnology in food science: perception and overview. Front. Microbiol. 8, 1501. https://doi.org/10.3389/fmicb.2017.01501.

Song, J., Huang, M., Jiang, N., Zheng, S., Mu, T., Meng, L., Liu, Y., Liu, J., Chen, G., 2020. Ultrasensitive detection of amoxicillin by TiO_2-g-C3N4@AuNPs impedimetric aptasensor: fabrication, optimization, and mechanism. J. Hazard. Mater. 391, 122024. https://doi.org/10.1016/j.jhazmat.2020.122024.

Sun, X., Du, S., Wang, X., 2012. Amperometric immunosensor for carbofuran detection based on gold nanoparticles and PB-MWCNTs-CTS composite film. Eur. Food Res. Technol. 235 (3), 469–477. https://doi.org/10.1007/s00217-012-1774-z.

Sun, J., Jiang, H., Wu, H., Tong, C., Pang, J., Wu, C., 2020. Multifunctional bionanocomposite films based on konjac glucomannan/chitosan with nano-ZnO and mulberry anthocyanin extract for active food packaging. Food Hydrocoll. 107, 105942. https://doi.org/10.1016/j.foodhyd.2020.105942.

Thakur, M.S., Ragavan, K.V., 2013. Biosensors in food processing. J. Food Sci. Technol. 50 (4), 625–641. https://doi.org/10.1007/s13197-012-0783-z.

Tiwari, K., Singh, R., Negi, P., Dani, R., Rawat, A., 2021. Application of nanomaterials in food packaging industry: a review. Mater. Today Proc. 46, 10652–10655. https://doi.org/10.1016/j.matpr.2021.01.385.

Vishwakarma, V., Sekhar Samal, S., Manoharan, N., 2010. Safety and risk associated with nanoparticles: a review. J. Miner. Mater. Charact. Eng. 9 (5), 455–459. https://doi.org/10.4236/jmmce.2010.95031.

Wang, F., Chang, R., Ma, R., Qiu, H., Tian, Y., 2021. Eco-friendly and PH-responsive nano-starch-based superhydrophobic coatings for liquid-food residue reduction and freshness monitoring. ACS Sustain. Chem. Eng. 9 (30), 10142–10153. https://doi.org/10.1021/acssuschemeng.1c02090.

Wang, J., Zhou, G., Chen, C., Yu, H., Wang, T., Ma, Y., Jia, G., Gao, Y., Li, B., Sun, J., 2007. Acute toxicity and biodistribution of different sized titanium dioxide particles in mice after oral administration. Toxicol. Lett. 168 (2), 176–185. https://doi.org/10.1016/j.toxlet.2006.12.001.

Wang, P., Lin, Z., Su, X., Tang, Z., 2017. Application of Au based nanomaterials in analytical science. Nano Today 12, 64–97. https://doi.org/10.1016/j.nantod.2016.12.009.

Wang, Y., Zhang, G., 2021. The preparation of modified nano-starch and its application in food industry. Food Res. Int. 140, 110009. https://doi.org/10.1016/j.foodres.2020.110009.

Yang, X., Zhang, Y., Zheng, D., Yue, J., Liu, M., 2020. Nano-Biocomposite Films Fabricated from Cellulose Fibers and Halloysite Nanotubes. Elsevier.

Yin, H.-Y., Li, Y.-T., Tsai, W.-C., Dai, H.-Y., Wen, H.-W., 2022. An immunochromatographic assay utilizing magnetic nanoparticles to detect major peanut allergen Ara h 1 in processed foods. Food Chem. 375, 131844. https://doi.org/10.1016/j.foodchem.2021.131844.

Youssef, A.M., El-Sayed, S.M., El-Sayed, H.S., Salama, H.H., Dufresne, A., 2016. Enhancement of Egyptian soft white cheese shelf life using a novel chitosan/carboxymethyl cellulose/zinc oxide bionanocomposite film. Elsevier.

Zhang, M., Ping, H., Cao, X., Li, H., Guan, F., Sun, C., Liu, J., 2011. Rapid determination of melamine in milk using water-soluble CdTe quantum dots as fluorescence probes. Food Addit. Contam. Part A 29, 1–12. https://doi.org/10.1080/19440049.2011.643459.

Zhang, B., Feng, X., 2022. Assessment of pervaporative concentration of dairy solutions vs ultrafiltration, nanofiltration and reverse osmosis. Sep. Purif. Technol. 292, 120990. https://doi.org/10.1016/j.seppur.2022.120990.

Zhang, J., Zhu, X., Chen, M., Chen, T., Liu, Z., Huang, J., Fu, F., Lin, Z., Dong, Y., 2022. Hybridizing aggregated gold nanoparticles with a hydrogel to prepare a flexible SERS chip for detecting organophosphorus pesticides. Analyst 147 (12), 2802–2808. https://doi.org/10.1039/d2an00541g.

Zhang, W., Rhim, J.W., 2022. Titanium dioxide (TiO_2) for the manufacture of multifunctional active food packaging films. Food Packag. Shelf Life 31, 100806. https://doi.org/10.1016/j.fpsl.2021.100806.

Zhao, X., Tian, R., Zhou, J., Liu, Y., 2022. Multifunctional chitosan/grape seed extract/silver nanoparticle composite for food packaging application. Int. J. Biol. Macromol. 207, 152–160. https://doi.org/10.1016/j.ijbiomac.2022.02.180.

Physicochemical characteristics and properties of nanobiocomposites for food packing and processing

3

Elaine Gabutin Mission[a] **and Andrew Lambert M. Tampoc**[b]

[a] *PressTech, Instituto de Bioeconomia de la Universidad de Valladolid, Valladolid University, Valladolid, Spain,* [b] *School of Chemical, Biological and Materials Engineering and Sciences, Mapua University, Manila, Philippines*

3.1 Introduction

A primary approach toward food preservation has been achieved through the development of food packaging materials capable of securing food from light, oxygen, and moisture thereby preserving its quality, keeping nutrients and flavor intact, avoidance of contamination and convenience during handling and transport, and avoidance of growth of microorganisms to extend shelf life (Qasim et al., 2021). The food packaging also contains labels and information such as food type, shelf-life duration, storage information, and nutrient contents to communicate with the handler, transporter, or consumer. Food packaging can be categorized into two packaging categories. The first one known as primary packaging is the one that is directly in contact with the food. Typical examples which include aluminum foils, glass bottles or jars, packaging trays, cling wraps, paper liners, or plastic bags. The secondary packaging normally acts as a second layer of protection and for ease of handling. Typical examples include boxes and packaging trays. Furthermore, tertiary packaging would be necessary for the combination, segregation, labeling, and transport of aggregated packed food products. Typical examples include shipping boxes, crates, and cases (Geueke et al., 2018).

While conventional food packaging materials encompass an extensive range of materials that include aluminum, plastics, cartons, paper glass, metals, and films, plastic remains to be the most used since it is lightweight and takes up less space than other materials. The most popular plastic types for packaging materials originate from major plastic families: polyethylene terephthalate (PET), polyvinylchloride (PVC), polyethylene (PE), polypropylene (PP), polystyrene (PS), and polyamide (PA). In 2018, the global plastic market reached 360 million tons that accounting for thermoplastics, polyurethanes, thermosets, elastomers, adhesives, coatings, and sealants, and PP-fibers and approximately 40% of these are utilized for packaging purposes due to the increasing demands of the society (Shamsuyeva and Endres, 2021). Most of the plastic packaging materials are sourced from petroleum and are intended for single use. In Europe for instance, it was estimated that recycling of plastic packaging materials only accounted for 14% in 2017 (Antonopoulos et al., 2021). Therefore, an estimated value of roughly 5–12 million tons of plastics would reach the oceans whereas more than 30% of the plastic waste would go to landfills. Obsolete means for the handling of postconsumer plastic wastes largely consist of unsustainable incineration and landfilling which induce human health and ecological hazards. While recycling has been adopted in order to minimize the inappropriate disposal of packaging materials and their subsequent accumulation in the environment, not everything can be recycled due to the heterogeneity and complexities in their structure (most packaging materials are a combination or composite of pure materials) and due to the level of contamination of organic and microbial substances post their utilization. The current usage of packaging applications as nonbiodegradable and non-renewable materials caused excessive concerns about environmental contamination. Thus, proper management of waste materials in such a context becomes highly crucial. As such, more states and governments introduced laws and regulations that ban single-use plastics (Jiang et al., 2020). Hence, aside from effective food preservation qualities, food packaging producers are also challenged to respond to the environmental footprint associated with the disposal of the used packaging materials. In this light, biopolymers have received widespread attention in developing biodegradable packaging materials. Biopolymers have a huge potential as substitute fossil-fuel-derived starting materials for the manufacture of biodegradable packaging materials to address the waste disposal problem since they are abundant and renewable.

Biopolymers are nature-derived polymers and macromolecules positioned in the cells of living organisms. They comprise repeated monomeric units that could be covalently bonded, assembling bigger molecules in a chain-like formation. While the prefix bio-signifies the biological origin of biopolymers, it is also commonly associated with biodegradability. Being biodegradable indicates that microorganisms are able to degrade them and convert them into harmless by-products, such as water and carbon dioxide under favorable environmental conditions. The most common type of biopolymers that have found utilization in food packaging includes starch, cellulose, chitosan, and agar which are derived from carbohydrates as well as gelatin, gluten, alginate, whey protein, and collagen which are derived from protein. However, due to the hydrophilic nature of these polymers, they are not a good option for food packaging that requires water resistance. This has led to the exploration of synthetic biopolymers which include polylactic acid (PLA), polycaprolactone (PCL), polyglycolic acid (PGA), polyvinyl alcohol (PVA), and polybutylene succinate (PBS) which have the potential to establish a renewable and sustainable industry. It has been demonstrated in various research investigations that synthetic biopolymers have enhanced properties over carbohydrate-based biopolymers such as better durability, enhanced clarity, and improved flexibility, more polished appearance, and higher tensile strength.

Nowadays, more biopolymer types have been studied and introduced into the market. Biopolymers can be categorized into four groups depending on the origin of the biopolymers, which are enumerated as follows:

- Natural biopolymers extracted from biomass and agricultural resources. Typical examples include lipids-based (acetoglycerides, fatty acids, monoglycerides, and phospholipids), carbohydrates-based (cellulose, starch, pectin, chitosan, and alginate) and proteins-based biopolymers (animal-derived proteins, such as collagen, whey protein, casein, milk protein, keratin, gums, carrageenan, and plant-derived protein, such as soy protein and wheat gluten). The carbohydrate-based polymers bear the D-glycopyranoside as their building block or repeating unit.
- Synthetic biopolymers from microbial production or fermentation (e.g., polyhydroxy-alkanoates [PHA], poly-3-hydroxybutyrate-co-3-hydroxyvalerate [PHBV], polyhydroxy-butyrate [PHB], and exopolysaccharides [EPS]).
- Synthetic biopolymers are conventionally and chemically synthesized from biomass (e.g., PLA).
- Synthetic biopolymers conventionally and chemically synthesized from petroleum products (e.g., PCL and polybutylene succinate adipate [PBAT]).

Among the biopolymers, starch and its derivatives found widespread utilization in food packaging, primarily due to starch being edible (Zhao et al., 2008; Sadeghizadeh-Yazdi et al., 2019; Mironescu et al., 2021). More importantly, starch was shown to be completely degradable, favoring the trend for environment-friendly packaging. However, materials developed with native or unprocessed starch suffer from poor mechanical properties and moisture sensitivity, which requires plasticizers and fillers to improve. Alginate is also another abundant biopolymer that has good film-forming properties but poor water barrier characteristics (Cheikh et al., 2020). Proteins are linear heteropolymers and are comprised of amino acids joined together by amide linkage called polypeptides. They also possess good film-forming properties due to the functional groups in the amino acids. These functional properties of the amino acids and consequently the protein as a whole are responsible for its inherent chemical reactivity and lower inertia that limits their utilization as plastic material. Furthermore, the presence of hydrophilic functionalities makes them susceptible to microbial spoilage (Zubair and Ullah, 2020). Aside from some weaknesses in terms of properties of bio-based polymers compared with petroleum-derived materials, they also have major issues such as difficulty in processing, which makes them largely inaccessible despite their abundance, and in turn, makes them more expensive.

Thus, alternative materials in the form of synthetically produced biopolymers based on biomass gained grounding. Amongst them, PLA was widely studied. PLA is a polymer usually produced from lactic acid in a process known as ring-opening polymerization. PLA is known for its biobased nature as the lactic acid raw material is synthesized via the fermentation of carbohydrates from plant resources such as soybean and corn (Jem and Tan, 2020). Similar to naturally derived biopolymers, PLA is more expensive than petroleum-based materials and also has sub-optimal mechanical and physical properties. It is suggested that PLA can be co-polymerized, physically blended, or sandwich laminated with PGA in order to enhance its properties. Acrylonitrile butadiene styrene (ABS) is also proposed to aid the need for polymers which may be great alternatives for their petroleum-based counterparts such as PET. However, certain limitations inhibit the use of these biobased polymers, including degradation at varying conditions, mechanical properties fluctuations upon storage, microbial growth, gas permeability, and high hydrophilicity (Reddy et al., 2013).

As discussed, while biopolymers possess the necessary properties that could achieve the transition from fossil-based plastics to packaging materials, the critical concerns include the poor characteristics they exhibits such as weak mechanical properties, thermal instability, and lack of water permeability. Current biopolymers are also brittle, sensitive to heat, and have low resistance to prolonged process operations. This caused researchers in the field of materials science to strive for the improvement of biopolymers. An essential area formed by these efforts is the utilization of nanocomposites. Studies have established that nanocomposites can enhance the mechanical and barrier properties of biopolymers. The product which is a composite of these two is called a nanobiocomposite. Nanobiocomposite is a multiphase material comprising two or more constituents which are continuous phase or matrix particularly biopolymer and discontinuous nanodimensional phase or nanofiller (<100 nm). The nano-sized fillers play a structural role in which they act as reinforcement to improve the mechanical and barrier properties of the matrix. The matrix tension is transferred to the nanofillers through the boundary between them. These properties can be enhanced by adding reinforcing nano-sized compounds or fillers to form composites. This article reviews the properties and physicochemical characteristics of a variety types of biopolymers and corresponding ones used to form nanobiocomposite materials.

3.2 Nanobiocomposites for food packaging and processing

The increasing concern for sustainable development has greatly impacted the current fields of science and technology. Recently, novel hybrid bio-derived materials, known as nanobiocomposites, which can be obtained from natural feedstocks or partially from petroleum feedstocks have been given considerable attention in the research community. Nanobiocomposites are constituted by a naturally occurring biobased polymer and in combination with inorganic partitions, which must bear at least one dimension in the range of 1–100 nm (Puiggalí and Katsarava, 2017). These advanced materials are deemed promising because green polymers with favorable properties are discovered to have an optimal design through the utilization of nano-supports in ecologically friendly polymers.

Nanobiocomposites may either be synthetic or naturally occurring. Natural processes paved the way for the formation of these materials, which essentially have incredible structural characteristics and functionalities. These processes are collectively identified as biomineralization, leading to the impeccable structure of organic and inorganic components from the nano- to macro-scale (Xu et al., 2007). Nacre is among the most popular and most researched types of naturally occurring nanobiocomposites due to its highly organized aragonite platelets connected by proteins and polysaccharide structures. The biomolecules such as proteins behave like an adhesive ligament patching up aragonites with inorganic layers apart by 30–600 nm. This leads to the material having a brick-like structure with excellent mechanical properties. Dentine and enamels in teeth, bones, ivory, shells, and pearls are common examples of natural specimens that contain nacre (Su and Cui, 1999).

Nanobiocomposites are known to have remarkable impacts in the fields of packaging, electronics, and biomedical sectors. In the food packaging industry, nanobiocomposite films with outstanding barrier properties are currently being explored. Even in electronic systems, nanobiocomposites are greenlighted, and it is anticipated that significant research is in development so that global reduction of e-waste can be foreseen. Nanobiocomposite carbon nanotubes (CNTs) have been proven to exhibit enhanced electrical conductivity. They are also being developed for their practical application as electromagnetic inference shielding materials (Reddy and Rhim, 2014). In the automobile industry, cellulose nanofibers are utilized in the development of durable load-bearing parts, automotive parts, and building blocks in vehicles. Furthermore, they are used in blades for vacuum cleaners, power tool housings, and mobile phone covers. Biopolymers such as PLA and PET glycol nanofibers offer remarkable biocompatibility (Arora et al., 2018). The prospect of chemical modifications of the composite materials is also explored as coating materials as well as fabric or cloth for the textile industry. An interesting fact about the rise of nanobiocomposites is that material scientists could replicate these natural processes to develop biomimetic and bioinspired materials. In practice, biomineralization processes producing natural nanobiocomposites were mimicked to create synthetic nanobiocomposites with more desirable properties (Xu et al., 2007). There are also a lot of subtypes of materials under synthetic nanobiocomposites, which may further increase due to the substantial attention being provided by researchers to the field. An example is the biopolymer clay nanocomposites which are formed through the intermolecular interactions of components of clay and natural polymers, including polysaccharides, chitosan, cellulose, starch, polyesters, and proteins. The particular interest in the advancement of synthetic nanobiocomposites that imitate the outstanding attributes of biogenic minerals such as nacre also pulled the creation of biopolymer-carbonate and -phosphate composites. The development of these nanobiocomposites is supplemental for clinical applications such as in the regeneration of functional bones and dental prostheses (Palin et al., 2005). Another fascinating type of synthetic nanobiocomposites is biopolymer-silica nanocomposites where biomolecules are applied in precipitating silica nanoparticles from sodium silicate solutions. This causes the entrapment of biological complement in a silica matrix (Schröder et al., 2006). Other widely studied synthetic biocomposites include biopolymer-CNTs and biopolymer-layered double hydroxide nanocomposites.

The biocompatible functionality of nanobiocomposites also has a leading role in biomedical purposes such as regenerative medicine and controlled drug release (Yokoyama et al., 2005). Current efforts in the field are related to the progress of biodegradable nanocomposites as bioresorbable scaffolds in tissue engineering. At the same time, their processing with reduced dimensions is favorable in drug delivery applications. There are also enzyme-based types of nanocomposites suitable for use in biosensors for the enhanced stability of the biological element brought by the protective influence of the inorganic matrix (Liu and Lin, 2006).

As medical biomaterials, nanobiocomposites' critical aspect is biocompatibility. This specific attribute provides their suitability in functioning internally in the human body so that they take their purpose into the desired effect without becoming detrimental to the biological components (Zafar et al., 2016). Nanobiocomposites have become increasingly relevant for use in the biomedical field for this reason. Specific applications in this field include tissue engineering, drug delivery, gene therapy, and cosmetics. Inherent abundance, flexibility, innate nontoxicity, and ecological soundness promote nanobiocomposites as a suitable candidate for expansive medical applications (Mousa et al., 2016).

Another notable application for nanobiocomposite materials is in polymeric materials research. Researchers found that polymers supported by functional nanostructures render electrical, electromagnetic, optical, and magnetic features. This results in the development of a wide array of electrical and electronic equipment. These are exemplified by light-emitting diodes, solar cells, display panels, and biomedical devices made more sustainable by their nanobiocomposite constituents (Kim et al., 2003). It was discovered that the high ratio of the material's surface area to volume and the heightened surface reactivity of the nanoscale antimicrobial agents resulted in greater efficiency in incapacitating microorganisms than their micro- or macro-scale equivalents. These features made it attractive for nanobiocomposites to be integrated into biomedical applications and the green polymer industry.

To sum up, nanobiocomposites correspond to a type of advanced materials that became a staple in the growth of environmentally suitable nanocomposites with applications in numerous fields of science and technology. Despite the expansive development, nanobiocomposites research is currently known to be in the embryonic phase. Thus, more investigations are required to fully understand its potential on the commercial and industrial scale. In particular, the selection and development of alternative biopolymers and inorganic counterparts can further be expanded for the acquisition of numerous functionalities. With this considered, it is foreseen that there are spectacular opportunities for these novel materials. This is due to the significant amount of potential building elements which can be assembled to create novel nanobiocomposites. Researchers have cited that the future trends of nanobiocomposites must be dedicated to the instrumental use of biocompatibility and biodegradability, and functional and structural properties to unlock more applications and possibilities for these futuristic materials (Darder et al., 2007). Recently, nanobiocomposites which are the desirable modification of these polymers, were proven to enable the acquisition of properties that make them a

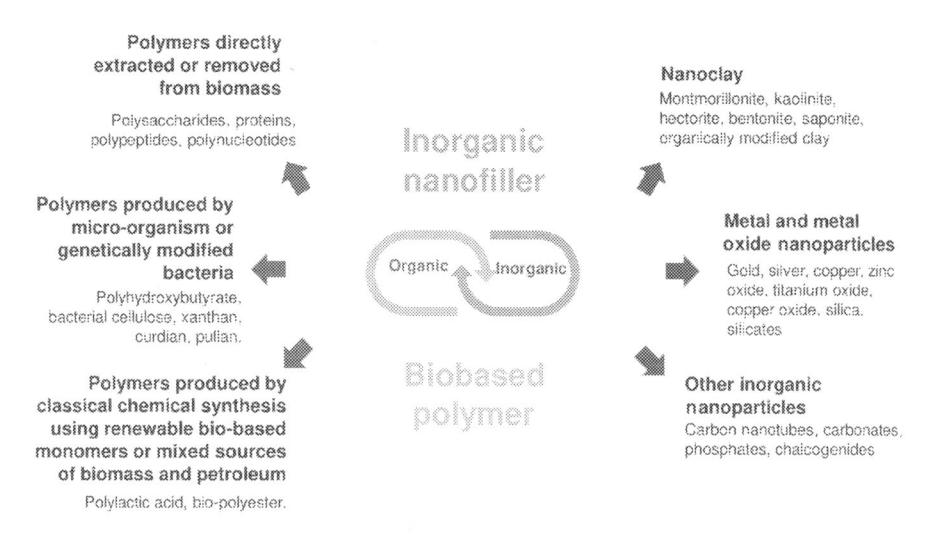

FIGURE 3.1

Common organic and inorganic constituents of nanobiocomposites.

better option for optimizing biopolymer-based packaging materials (Sorrentino et al., 2007).

3.2.1 Physicochemical characteristics of nanobiocomposites

Continuous research in nanobiocomposites has expanded the variations of its types to fulfill different applications that may be too extensive for an introductory discussion paper. For that matter, this section is only aimed to discuss the physicochemical characteristics of several nanobiocomposites designed by recent studies with specific biopolymers and inorganic nanofillers. Fig. 3.1 illustrates the common organic and inorganic constituents of nanobiocomposites. Amongst natural biopolymers, chitosan (1–4 linked 2-amino-deoxy-β-D-glucan) has attracted much attention since it is nontoxic for human beings and generally recognized as safe (GRAS) by the Food and Drug Administration.

3.2.1.1 Nanobiocomposites based on chitosan

Chitosan is a popular sugar acquired from the skeletons of shellfish. It is recognized for its myriad of chemical properties, including viscosity, solubility in organic and inorganic solvents, mucoadhesivity, polyelectrolyte behavior, chelation, and poly-oxysalt formation. Experimental studies have found that the unbranched and linear form structure of chitosan is the reason for the substance's notable viscosity. The viscosity can further be altered through the process of deacetylation, which is a distinctive characteristic because chitosan itself has a high degree of deacetylation. It is also known for having a low content of crystalline regions (Ghosh and Ali, 2012). These characteristics make chitosan-based nanobiocomposites an appealing contender for biomedical applications such as in pharmacy and biotechnology. However, there are also other exciting applications where they are known to have immense potential (Azmana et al., 2021).

In a study on the plasticized chitosan nanocomposites with $AgNO_3$ and Al_2O_3, it was observed that the developed nanobiocomposite from the solution casting method had the optimal DC conductivity and dielectric constant along with the lowest electric modulus. These properties are suggestive of the material's potential for use as an electrolyte and separator in a double-layer electrochemical capacitor (Hadi et al., 2020). With a similar preparation technique, graphene oxide and chitosan-infused nanobiocomposites can be produced with enhanced antimicrobial action against pathogenic microbes. Thus, the biopolymer modification makes it more suitable for antibacterial applications, water purification, and dye removal (Khalil et al., 2020). Another chitosan-based nanocomposite that can be prepared from the solution casting technique is chitosan/clay/glycerol nanobiocomposite film which is promoted as an innovative material for alternative food packaging. This is due to its extreme strength and stiffness, improved thermal stability and comparatively, the highest water resistance among all the chitosan-based nanobiocomposite variants studied (Kusmono and Abdurrahim, 2019). Aside from solution casting, there are also other types of preparation techniques for chitosan-based nanobiocomposites. This includes microwave synthesis, phase-inversion, graft polymerization, sol–gel polymerization, and others that may either be fully established or in the development phase. These preparation techniques also play an essential role in acquiring varying physicochemical properties, making the nanocomposite material desirable for diverse fields such as wastewater treatment, corrosion inhibitor, flame retardant, and, most importantly, biomedical applications (Azmana et al., 2021).

3.2.1.2 Nanobiocomposites based on protein

Proteins are among the most crucial naturally occurring polymers that can be equipped as biorenewable feedstock in the development of ecological bioplastics especially for food packaging (Silva et al., 2015). Protein molecules are linear hetero biopolymers with distinctive three-dimensional system constructs as they are composed of varying amino acid types. The existence of numerous functional groups in the amino acids of protein chains suggests an exceptional potential for the growth of proteins-based bioplastics as compared to carbohydrates and lipids. The main disadvantages of protein biopolymers are strong reactivity and lower inertia. These characteristics limit their applicability in polymer manufacturing. Furthermore, protein molecules consist of many hydrophilic sites which makes them vulnerable to microbial spoilage. Still, protein-based films are gaining significant attention from researchers because they are easily formed into films or coatings, have low cost, and can undergo natural decomposition (Hu et al., 2012).

To develop novel polymer materials with favorable qualities, modifications of molecular interactions of protein are used for film formation. Physical, chemical, and enzymatic treatment of proteins or fusing them with aquaphobic raw material or other polymers assists in the improvement of the mechanical and barrier properties of the protein-based films. Cross-linking of protein structures has a vital part in the fabrication of films or coatings with improved mechanical strength and water barrier functionalities of the resulting polymers since the enhanced structure has better strength (Cinelli et al., 2014). The challenge in this method is that cross-linking proteins would decrease the biodegradability of the films. Cross-linking of protein molecules leads to the development of strong intermolecular forces, tighter molecular packings, and reduction of polymer chain mobility (Avena-Bustillos and Krochta, 1993).

The conceptualization of nanobiocomposites further increased the attractiveness of protein-based polymer materials. Proteins-derived nanobiocomposites are commonly processed either by the dry methods or the wet methods which are also applied in other types of nanobiocomposites. Primarily, there are three phases in the production of protein-based molecules. This is initiated by the rupture of weak interactions existing among polypeptide chains in proteins' native state. This is followed by the polypeptide chain structural arrangement and orientation. Lastly, solvents are removed along with disruptive agents followed by the formation of a new three-dimensional polymer network that is stabilized by stronger connections and interactions (Cuq et al., 1996). Proteins are known to possess a high glass transition temperature that is typically higher than their degradation temperature. Hence, the utilization of plasticizers in the development of proteins-based films is typically essential, especially in attaining the desirable physiochemical properties such as material flexibility. Plasticizers make it easier for proteins to be processed thermally in nanobiocomposites formation. Water is hailed as the most fitting natural plasticizer. Other plasticizers usually provided in this application are glycerol, sucrose, sorbitol, phospholipids, and other organic substance. Introduction of nanofillers within the protein matrix leads to the further enhancement of protein-based nanobiocomposites particularly in terms of mechanical and barrier properties. Nanofillers mainly utilized in nanobiocomposites are organic nanoparticles such as nanocellulose (nanocrystals or fibers) and zein. There are also other inorganic nanofillers such as layered nanoclays, zinc oxide, silver (Kanmani and Rhim, 2014), and carbon nanotubes. The nanofillers' dispersion is usually correlated to the optimal properties observed on the nanocomposites. Nanofillers are prone to self-agglomeration because of the presence of hydrophilic groups on their surfaces which results in the loss of properties of the final product.

Protein-based nanobiocomposites are commonly studied because of their promising opportunities in food packaging applications. In a study on corn zein nanocomposites, it was inferred that it can be considered a great alternative to synthetic polymer barriers on PP films. The integration of organomodified montmorillonite via solution intercalation into the corn zein matrix enhanced the oxygen and water vapor barrier of coated PP films. With these findings, the researchers determined that additional studies in the field can lead to the acquisition of protein-based novel material that can be applied as edible film packaging for meat and meat products (Ozcalik and Tihminlioglu, 2013).

3.2.1.3 Nanobiocomposites based on cellulose

Cellulose nanostructures are known in the field of advanced materials research as components of systems or materials utilized in a variety of applications which include food packaging. They are essentially integrated as reinforcement phases in nanobiocomposites in the form of cellulose nanocrystals or nanofibrils. In some cases, cellulose nanostructures are treated as matrices. Bacterial cellulose is a widely studied specimen as it is a naturally nanostructured membrane that can be grown in a medium along with other biopolymers. This leads to the production of bottom–up built nanobiocomposites. They can also be impregnated with other components or disintegrated into nanocrystals or nanofibrils (Azeredo et al., 2017).

In the field of food packaging, nanobiocomposites based on cellulose nanostructures are developed as active food packaging systems. Active food packaging does not only protect food products but also has the added function of actively improving food quality and stability. Commonly, active food packaging systems are designed to release active compounds (such as antimicrobial agents) onto food surfaces (De Azeredo, 2013). A widely studied variant of active food packaging based on cellulose nanocomposites are those that contain silver nanoparticles. This is because of their antimicrobial efficiency, thermal stability, and excellent processability (Kumar and Münstedt, 2005). The only problem with these types of nanobiocomposites is their potential toxicity (Xiu et al., 2012). In a study by Fortunati et al. (2012), PLA films with crystallized nanocellulose and silver nanoparticles have been developed through melt extrusion. The silver nanoparticles demonstrated the sought-after antimicrobial character against gram-positive and gram-negative bacteria samples. Also, the added surfactant boosted the tensile properties of the film. General migration tests exposed that the surfactant-treated crystallized nanocellulose exceeded the specific migration levels allowed by the legislation when tested with 10% ethanol as the polar stimulant.

3.2.1.4 Nanobiocomposites based on hydroxyapatite

Hydroxyapatite is considered the fundamental inorganic component of bones and teeth. For that matter, it has been widely examined in the fields of bone tissue engineering applications. Furthermore, nanobiocomposites made from synthetic hydroxyapatite and collagen have wonderful opportunities to be utilized in mimicking and substituting skeletal bones (Turon et al., 2017). However, nanobiocomposites based on hydroxyapatite are versatile and can be utilized for a variety of other purposes such as in food packaging. This is especially suggestive for applications wherein the desirable natural properties of skeletal bones can be utilized.

Hydroxyapatite in its pure form comes with its inherent limitations which include difficulty to be molded into a shape, adverse mobility, and poor load-bearing properties. Nonetheless, these limitations can be conquered by integrating hydroxyapatite with other biopolymers. Nanobiocomposites also known as multicomponent materials can be produced with varying properties to target specific applications. The only concern in such strategies can be the typical phase separation problems (Nikpour et al., 2012). Coating or grafting hydroxyapatite nanoparticles in compatible polymer or copolymer matrices are great methods of developing nanobiocomposites. An example of novel nanobiocomposites based on hydroxyapatite is lactic acid polymers grafter on the surface of nanohydroxyapatite particles. The resulting material is formed from the reaction of the surface hydroxyls of hydroxyapatite particles and then combined in a poly(L-lactide-co-glycolide) matrix. The yielded nanobiocomposite was observed to have enhanced properties tensile strength and dispersability with respect to ungrafted systems (Turon et al., 2017).

As fundamentally an innovative biomedical material, the inherent properties of nanobiocomposites can also be promising when viewed from the perspective of food packaging applications. Moreover, a main topic in the field is also in the development of a novel material that can be safely placed within the human body. In addition, a lot of effort in the research is attributed to antimicrobial properties which is also a main case for studies developing nanobiocomposites for food packaging applications. Initially, the incorporation of antibiotics is implemented by pioneering researchers to achieve such properties, but other strategies are proposed because of the increasing resistance of microorganisms towards antibiotics (Zilberman and Elsner, 2008). Like other research areas of nanobiocomposites, silver nanoparticles are also the widely studied components for nanobiocomposites based on hydroxyapatite. Silver nanoparticles are determined to have exceptional antimicrobial activity but are technically limited due to their sensitivity to aggregation, which in turn reduces antimicrobial activity. This characteristic can be overcome by the manufacture of materials with silver nanoparticles immobilized on their surface (Xiong et al., 2016).

Apart from chitosan, proteins, collagen, cellulose, and hydroxyapatite other polysaccharide varieties were also utilized to produce nanobiocomposites. According to a study by Fabra and co-workers, sepiolite and palygorskite fibrous clays can serve as reinforcing fillers to optimize the mechanical and barrier characteristics of neutral and negatively charged polysaccharides matrices. The -OH and -COOH groups of the polysaccharide molecules exhibit strong intermolecular interactions with the surface of the fibrous clays, which makes the resulting nanobiocomposite highly stable. This observed compatibility led to the attainment of excellent mechanical properties increased stability, and reduction of water absorption. Furthermore, the enhanced barrier is determined to be resistant to ultraviolet (UV) light. Thus, the resulting material has a great opportunity in the food packaging sector. Fibrous clays altered with the hydrophobic protein zein also reduced the hydrophilic nature of the clays, which brought more unique properties to the novel biomaterial. The alginate/zein-clay systems are seen to develop self-sustaining films that further hinder water passage. The zein in the clay fiber structures is critical in acquiring the alginate films' water vapor barrier and gas permeation properties. This considerably decreases the water uptake and gas permeability even in highly humid environments (Lagarón et al., 2016).

Moreover, nanobiocomposite films are popular as packaging materials because they are recognized as cost-efficient, sustainable, and renewable. They are further considered favorable in the food packaging industry because of their enhanced antimicrobial characteristics. Currently, the comprehensively analyzed for these applications are nanobiocomposites with biopolymers such as starch, cellulose, PLA, PBS, and their derivatives. For this application, it is distinguished that the most capable nanoscale inorganic fillers are layered silicates and nanoclays (Zafar et al., 2016).

The ever-increasing demand for sustainable materials such as polymers truly fueled the craze for nanobiocomposites research. This is because even if biopolymers have already existed for a considerable amount of time, their primitive chemical and physical properties cannot compete with their petroleum-based counterparts. However, the areas currently being explored in the physicochemical studies of nanobiocomposites such as barrier properties, thermal stability, and mechanical strength have increasingly motivated environmentally friendly materials that are not just utilized for packaging applications but also in the manufacture of other industrial and commercial products.

A variety of techniques are being utilized in the characterization of nanobiocomposite properties as well as their presence and incorporation into the packaging films. These techniques include: morphology analysis via scanning electron microscopy (SEM) or tunneling electron microscopy (TEM); topography by atomic force microscopy (AFM); zeta potential for surface charges; Fourier transform infrared spectroscopy for the incidcation of surface functional groups; X-ray diffraction for crystallite sizes and degree of crystallinity; BET for porosity and surface area; themogravimetric analysis for thermal stability; rheological assessment; mechanical properties including tensile strength, texture, opacity, water vapor permeability; and moisture resistance and water contact angle (Ni et al., 2021).

3.3 Properties of nanobiocomposites

Given the extensive functional properties of nanobiocomposites, these innovative materials also fulfill a myriad of applications. For example, biopolymers such as PLA and its derivatives are improved by the addition of organically modified clay which promotes degradability as well as thermomechanical and gas barrier properties (Chivrac et al., 2009). In principle, the properties of nanobiocomposites are much better than native biopolymers resulting in improved bio-based packaging, smart packaging, and active packaging. Some examples of the advances in packaging that were developed for particular food packaging applications are illustrated in Fig. 3.2. Meanwhile, Table 3.1 summarizes representative nanobiocomposites including their components, properties and applications.

3.3.1 Improved bio-based packaging

Biopolymers as food packaging materials encounter major hurdles in terms of properties including poor mechanical, thermal and barrier properties. Thus, they need further processing or incorporation of other substances or chemicals to improve their properties. Material scientists made significant efforts in the areas of nanobiocomposites research because of the materials' admirable qualities in terms of functionality and structural identity, which reinforces their applicability as vital elements

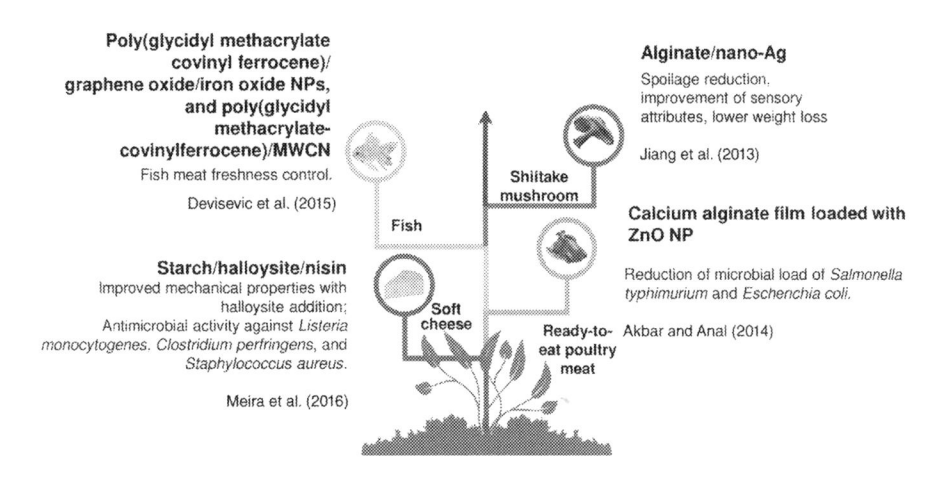

Poly(glycidyl methacrylate covinyl ferrocene)/graphene oxide/iron oxide NPs, and poly(glycidyl methacrylate-covinylferrocene)/MWCN
Fish meat freshness control.
Devisevic et al. (2015)

Starch/halloysite/nisin
Improved mechanical properties with halloysite addition; Antimicrobial activity against *Listeria monocytogenes, Clostridium perfringens*, and *Staphylococcus aureus*.
Meira et al. (2016)

Fish

Shiitake mushroom

Soft cheese

Ready-to-eat poultry meat

Alginate/nano-Ag
Spoilage reduction. improvement of sensory attributes, lower weight loss
Jiang et al. (2013)

Calcium alginate film loaded with ZnO NP
Reduction of microbial load of *Salmonella typhimurium* and *Eschenchia coli.*
Akbar and Anal (2014)

FIGURE 3.2

Sample food packaging applications of developed nanobiocomposites.

of innovative materials such as heterogeneous catalysts and optical, magnetic, and electrochemical devices (Puiggalí and Katsarava, 2017). For that matter, the different properties attributed to nanobiocomposites can then be classified or identified based on applications that can be further extended and diversified.

In the food packaging industry, the fundamental goal is to assure quality and safety of the food during the phases of transportation, storage, and handling. Furthermore, the packaging material must enable the shelf-life extension by preventing microbial, light, and chemical contamination (Wihodo and Moraru, 2013). To achieve these, the material must have optimum physicochemical conditions and be resistant to damages upon being in the phases, as mentioned earlier. Thus, this section is attributed to discussing past studies on nanobiocomposites and how the materials' enhanced properties can become substantial in the field of sustainable food packaging.

Excellent mechanical properties are yet another significant quality that nanobiocomposites possess. In a study concerning protein-derived nanobiocomposites, these properties are recognized to be the result of the strong intermolecular interactions and high energy bonds obtained during the novel material's processing (Zubair and Ullah, 2020). Plasticizers can also contribute to property optimization because they enhance the polymer chains' flexibility, which in turn improves elongation and lowers tensile strength (Vieira et al., 2011). Cross-linking agents can further increase the extensibility and stability of the film. Humid environments affect the films' mechanical strength, which is why nanobiocomposites with lower hydrophilic character are more desirable to reduce moisture absorption.

Other than the highlighted features of nanobiocomposites discussed above, other active packaging applications are desirable for these materials. Thus, future research in this field would heavily focus on color-containing films, light absorption systems,

Table 3.1 Nanobiocomposites: constitutions, properties and applications.

Biobased constituent	Composition	Properties	Application	References
Lipids	Triglycerides + montmorillonite nanoclay; 3, 5, and 10 wt.% based on nanoclay	Increased thermal stability (weight loss decreased from 98.2% to 91% at 600°C); flame retardancy (completely burning the nanocomposite polymer took 53 sec which is higher compared to the neat homopolymer at 30 sec); 10 wt.% nanoclay in triglyceride film	Thermally stable food packaging applications	Safder et al. (2020)
Crystallized nanocellulose (CNC)	CNC + agar; 1, 3, 5, and 10 wt.% based on agar	50–60 nm 40% improvement in tensile modulus and 25% improvement in tensile strength (5 wt.% CNC in agar film); 25% reduction in water vapor permeability (5 wt.% CNC in agar film)	Food packaging	Reddy and Rhim (2014)
Starch	70 wt.% ZnO nanoparticles + 30 wt.% pea starch (35% amylose, 65% pectin)	Tensile yield strength and Young's modulus increased from 3.94 MPa to 10.80 MPa and from 49.80 MPa to 137.00 MPa, respectively; water vapor permeability decreased from 4.76×10^{-10} g/m/s/Pa to 2.18×10^{-10} g/m/s/Pa	Food packaging with UV Shielding	Ma et al. (2009)
Chitosan	Chitosan + ZnO; 3, 5, and 7 wt.% based on ZnO	14.6 nm Tensile yield strength and Young's modulus decreased from 2811 MPa to 1603 MPa and from 82 MPa to 69 MPa, respectively (7 wt.% ZnO in chitosan film); high antimicrobial activity against *Brevibacterium lactofermentum* and *Escherichia coli* (7 wt.% ZnO in chitosan film)	Antimicrobial packaging	Boura-Theodoridou et al. (2020)

(continued on next page)

Alginate	Alginate + ZnO-rGO; 25, 30, 40, and 50 wt.% based on ZnO-rGO	Tensile yield strength and Young's modulus decreased from 46.3 MPa to less than 20 MPa and from 2.5 GPa to 1.25 GPa, respectively (50 wt.% ZnO-rGO in alginate film); water vapor permeability increased from 6.8×10^{-11} g/m/s/Pa to almost 2.00×10^{-10} g/m/s/Pa (50 wt.% ZnO-rGO in alginate film); antioxidant activity (38% ABTS inhibition in 0.5 hr) and antimicrobial activity against *Escherichia coli* and *Staphylococcus aureus* (50 wt.% ZnO-rGO in alginate film); high electrical conductivity (0.1 s/m)	Packaged food sterilization at low temperature	Alves et al. (2021)
Protein	Corn zein + polypropylene coatings + organically modified layered silicate nanoclay Cloisite 10A (OMMT); 1, 3, 5, and 7.5 wt.% based on OMMT	Reduction in oxygen permeability of up to 4 times; reduction of water vapor permeability by 30% (5 wt.% OMMT content in 5.9 μm corn zein coating)	Edible film packaging for meat and meat products	Ozcalik and Tihminlioglu (2013)
Polyhydroxyalkanoate (PHA)	Polyhydroxyalkanoate (PHA) + tailor-made long alkyl chain quaternary salt (LAQ) functionalized graphene oxide (GO-g-LAQ); 1, 3, 5, and 7 wt.% based on GO-g-LAQ	86% reduction in oxygen permeation; tensile strength and storage modulus at room temperature of the PHA films increased by 60% and 140% (5 wt.% GO-g-LAQ); 99.9% antibacterial activity against *Escherichia coli* and *S. aureus*	Heat-resistant food packaging with antimicrobial property	Xu et al. (2020)

(continued on next page)

Table 3.1 Nanobiocomposites: constitutions, properties and applications—cont'd

Biobased constituent	Composition	Properties	Application	References
Polyhydroxybutyrate (PHB)	Polyhydroxybutyrate (PHB)—thermoplastic starch + 3% OMMT + 3% eugenol	1565 MPa elastic modulus and 16.3 MPa tensile strength; 6.4 ± 2.5 cm^2 area of inhibition against Botritys cinerea; 92.9% DPPH radical scavenging activity	Food packaging material with antifungal and antioxidant activity	Garrido-Miranda et al. (2018)
Polylactic acid (PLA)	Polylactic acid (PLA) + halloysite nanotubes (HNTs); 1.5, 3.0, 4.5, and 6.0 wt.% based on HNTs	Young's modulus is approximately 4500 MPa and tensile strength is almost 40 MPa (3% HNT in PLA film); water vapor permeability decreased from 1.4×10^{-5} g/m/d/Pa to less than 1.2×10^{-5} g/m/d/Pa; oxygen permeability reduction from 3×10^{-13} cm^3 cm^{-2}/s/Pa to 2×10^{-13} cm^3 cm^{-2}/s/Pa	Shelf-life extension of packaged cherry tomatoes	Risyon et al. (2020)
Polyvinyl alcohol (PVA)	Polyvinyl alcohol (PVA) + boiled rice starch (BRS) + sAgNP	Young's modulus increased from 290 MPa to 682 MPa; tensile strength increased from 16.8 MPa to 26.5 MPa; strong antibacterial activity against Salmonella Typhimurium and S. aureus	Antimicrobial food packaging	Mathew et al. (2019)

nonfogging and nonstickfilms, microwave heat resistant, gas-permeable films, and insect-repellant packaging (Sothornvit and Krochta, 2001).

Nanobiocomposites in food packaging applications are specifically developed with the objective of attaining or improving the characteristics of the selected biobased polymers. In a study by Reddy and Rhim (2014), it was found that agar reinforced with crystallized nanocellulose has significant improvements in mechanical and barrier properties. This is reported to be caused by the strong interaction between the homogeneously dispersed nanocellulose and the biopolymer matrix. Similar improvements were discovered in the polymer based on plasticized starch when it was incorporated with ZnO nanoparticles. Due to the interactions between the inorganic nanofiller and the polymer, at below 4 wt.% filler loading level, nano-ZnO enhanced the Young's modulus, tensile yield strength, glass transition temperature and pasting viscosity of the nanocomposites. Furthermore, the developed material may be potentially applied to UV-shielding materials because its absorption peak in the UV-visible spectroscopic analysis demonstrated an apparent blue-shift phenomenon due to the quantum confinement effect. At elevated nano-ZnO content, UV-absorbance and absorbance peak intensity increased (Ma et al., 2009).

Nanobiocomposites' properties can further be altered by the fusion of nanofillers such as methylcyclopentadienyl manganese tricarbonyl, silver, zinc oxides, and other inorganic solids to the biopolymers to attain superior properties such as enhanced mechanical, thermal, and barrier characteristics (Zafar et al., 2016). There are also instances when these materials are supplemental in counteracting pollution. This is observed when a modification of biopolymer clay nanocomposites was implemented, and the results show how the manufactured material can be used for pollutant removal. An example is the prospect of uniting the adsorbent properties and ion exchange capacity in chitosan clay nanobiocomposites to recover heavy metal ions and azo dyes (Ruiz-Hitzky et al., 2010).

A less explored yet still very promising quality of nanobiocomposites are thermal stability and fire retardancy. In lipid-derived hybrid nanobiocomposites from spent hens, these are the highlighted characteristics. According to Safder et al. (2020), uniform dispersion in the nanobiocomposites normally decreases the flammability. This may not sound as astounding as the other properties that are excessively studied but the ability of these materials to withstand higher temperatures would contribute largely to the packaging's applicability for high-temperature environments. Increased thermal stability is also associated with better structural integrity which in the field of food packaging may be applied to microwavable goods and ready-to-consume food items.

The breadth of the relevant fields envisioned for nanobiocomposites is highly defined by their purpose as functional materials. This proposes that future research trends must be led by the opportunity to develop electrical, optical, and magnetic properties that can allow a modern generation of advanced devices or materials coupled with biodegradability and biocompatibility characteristics (Reddy et al., 2013). Nanobiocomposites are also promising as membranes (Bhat and Aminabhavi, 2006), superabsorbents (Qiu et al., 2005), or inks for water-sensitive inkjet printing (in het Panhuis et al., 2007). This has been reported of pervaporation processes involving alginate-montmorillonite systems. The hydrophilic nature of the inorganic substance establishes a favored pathway of water separating from organic solvents, which has been determined when dehydration activity increases as the clay content is of much higher concentration in the composite membrane (Bhat and Aminabhavi, 2006). Therefore, this implies that the use of nanobiocomposites is preferable in developing superabsorbent materials.

3.3.2 **Active packaging**

The nanobiocomposite can be an active food packaging whereby the food packaging can interact with food in some ways by releasing beneficial compounds such as antimicrobial agent, and antioxidant agent, or by eliminating some unfavorable elements such as oxygen or water vapor.

Among the optimal properties of nanobiocomposites, the most desirable must be their antimicrobial properties. Nanobiocomposites have recognized efficiency in terms of antimicrobial properties due to their high surface area to volume ratio as well as the greater surface reactivity of the nanoscaled antimicrobial agents. Most of these nanobiocomposites are synthesized with platinum, gold, silver, and copper, or metal oxides including magnesium oxide, zinc oxide, and titanium dioxide. Metals and metal oxides are utilized in photocatalysis and have undergone UV radiation so as to synthesize highly reactive oxygen species. This resulted in the antimicrobial activity of the fabricated material (Rhim et al., 2009). The nanoparticles can then be incorporated into biopolymers such as agar, alginate, carrageenan, chitosan, and starch to produce nanobiocomposite with high antimicrobial activity (Shankar and Rhim, 2016). For example, in the work of Sanuja and co-workers (2015), nano zinc oxide at concentrations between 0.1% to 0.5% was combined with neem essential oil in a chitosan nanobiocomposite film. The nano zinc oxide was expected to enhance the mechanical and barrier properties of the film, whereas the essential oil enhanced and prolonged the antibacterial properties. Crystallite sizes for nano zinc oxide and chitosan as derived from the Scherrer equation were found to be 27 nm and 70 nm, respectively. Overall, all tested properties improved in the presence of the nanoparticles, with the best conditions obtained in the presence of 0.5% nanozinc oxide. The tensile strength reached 51 MPa, elongation percentage increased to 12.3% versus 7% in the control, moisture solubility was 30.1% versus 60.3% in the control, and swelling property also decreased from 24.6% to 15.4%. Antibacterial properties were also observed against *Escherichia coli* and the film was tested for carrot packaging.

In one approach, starch films were incorporated with citric acid as a cross-linking and plasticizing agent, cellulose fiber as reinforcement filler, and sunflower-based phenolic extracts to impart antioxidant properties (Menzel, 2020). The citric acid was added to decrease the moisture sensitivity and improve film integrity while cellulose fibers would increase film strength and decrease water vapor permeability. Meanwhile, glycerol and sorbitol, both bio-based materials were used as plasticizing agents, which led to better puncture resistance with increasing plasticizer concentration from 10% to 50% starch basis but with an increase in the percentage of elongation. The potato starch film could be useful in developing edible candy wrapper film and edible coatings for fresh fruit and vegetables (Ballesteros-Mártinez et al., 2020).

Aside from mechanical and barrier properties, much of nanobiocomposites research are directed toward the acquisition of antibacterial properties. Mathew et al. (2019) reported a one-step preparation process of PVA/boiled rice starch blend film fabricated with in situ generated silver nanoparticles. In this study, they were able to correlate the presence of silver nanoparticles to the film's antibacterial activity against *Salmonella typhimurium* and *Staphylococcus aureus*. The most widely accepted mechanism of this activity proposes the interaction of positively charged silver ions with negatively charged sulfur or phosphorous-containing macromolecules in bacteria resulting in structural changes and disruption of metabolic activities which leads to cellular death. Antioxidant and antifungal activity were also featured in a PHB nanocomposite with eugenol, starch, and organically modified montmorillonite (OMMT). The antioxidant activity of nanobiocomposites was found to be caused by eugenol which, even in dilute forms, showed elevated eliminations of free radicals. This capability is credited by researchers to eugenol reducing two or more radicals of DPPH because it dimerizes and reacts with the DPPH to form dehydrodieugenol, contributing to the radical scavenging activity. Eugenol is also determined as the component on which the antifungal activity of the film is dependent. Significant growth reductions against Botrytis cinerea were obtained with increasing eugenol concentrations in the nanobiocomposites, where the greater area of inhibition was seen (Garrido-Miranda et al., 2018).

Essentially, nanobiocomposites are favored because of their excellent mechanical properties, biodegradability, and biocompatibility, all of which are integral to the novel materials because of the biopolymer and inorganic solid constituents. The impeccable structure of the material also contributes to the attainment of these qualities. In recent studies, nanobiocomposite antimicrobial systems have been developed. For instance, Ni et al. (2021) developed a nanobiocomposite film which incorporated chitosan and graphitic carbon nitride, whose antibacterial properties can be self-activated by visible light, therefore allowing a more sustained antibacterial properties over time. The graphitic carbon nitride was sourced from urea and the nanobiocomposite film was synthesized by electrostatic self-assembly by a simple one-step casting method. The chitosan films originally had no visible light-responsive enhanced antibacterial activity, and the antibacterial activity did not lasting. With the incorporation of surprisingly, the graphitic carbon nitride, the antibacterial activity of the produced nanobiocomposite films was significantly enhanced under visible light irradiation for 15 minutes. The graphitic carbon nitride contained preferably at 30 wt.% in the film was responsible for improving the mechanical, thermal, and hydrophobic properties of the chitosan films and preservation of tangerines for up to 24 days.

Aside from antimicrobial properties, Abral et al. (2020) also highlighted the importance of producing good transparency films despite the incorporation of ginger nanofiber in the PVA nanobiocomposite film. As expected, increasing the amount of nanofiber in PVA increases significantly in tensile properties, water vapor impermeability, and moisture resistance. While antibacterial is imparted by ginger fiber, antifungal properties were not present. Using 0.4 g ginger nanofiber reinforced film, tensile strength (44.2 MPa) increased by 65.6%, the maximum film decomposition temperature increased by 7% to 349°C and moisture decreased at 6.1% when compared to the PVA control. Despite the care and interest in highly transparent films, it was still observed that the film slightly increased its opacity, although visual observation suggests that this decrease in transparency was acceptable. This finding also indicates that the desire to improve some of the properties of the nanobiocomposite may also affect or diminish other desirable properties and therefore, a balance between these properties must be considered.

3.3.3 Smart packaging

Current packaging has made more progress as a result of international lifestyles and consumer demands for real-time data. These improvements are focused on to achieve better food quality and safety anchored on international standards. Additionally, due to globalization, packaging requirements required extended shelf life. The nanobiocomposite also introduced smart/intelligent food packaging mechanisms that would provide value-added capabilities such as microbial contamination indicators or expiry date notifications and as well as mechanisms to register and convey information about the quality or safety of the food. In general, intelligent packaging incorporates a color indicator responsive to time/temperature variations, gas evolution, or pH changes.

Ezati and Rhim (2020) reported on a pH-responsive film based on chitosan and alizarin. Alizarin or 1,2-dihydroxyanthraquinone ($C_{14}H_8O_4$) is a red pigment obtained from madder (Rubia tinctorum) plant roots, which has been used as a staining agent in biological research as well as textile dye. It has been used as an acid-base indicator by switching colors from yellow to purple. pH-responsive packaging materials are important for monitoring the quality packaged foods since food spoilage by microbes releases volatile basic compounds that increase pH. In the presence of alizarin, the microbial decay will manifest in a color-changing film, which consumers can quickly understand since the visual inspection is first and foremost a technique used in assessing the quality of fresh produce. They also found that alizarin greatly increased the UV-blocking properties, thermal stability, and surface hydrophobicity of the chitosan film without affecting the tensile strength and water vapor barrier properties of the film. The chitosan/alizarin composite film showed distinctive antibacterial activity against *E. coli* and *Listeria monocytogenes* and strong antioxidant activity. The film showed a distinct color change from khaki to light brown, indicating the onset of fish spoilage. Meanwhile, Ge and co-workers (2020) introduced anthocyanin sourced from black in chitin/gelatin for a pH-responsive nanocomposite film tested in monitoring shrimp and hairtail freshness. Anthocyanins are a group of polyphenolic pigments found extensively in red cabbage, grape skin, and black soybean that also possess pH-sensitivity. The presence of the anthocyanin up to 0.1% slightly reduced tensile strength but improved oxygen and water barrier properties. Bearing high antioxidant properties somehow retarded microbial spoilage lengthening product shelf life but eventually, the spoilage was evident when the films changed color from grayish blue to brown. Anthocyanin obtained from *Lycium ruthenicum* Murr were also incorporated in cassava starch and tested in pork products (Qin et al., 2019). In the work of Roy and team (2021), they mixed carboxymethyl cellulose (CMC) with agar to resolve CMC's high hydrophilicity and poor mechanical properties. Then, they used a combination of anthocyanin and shikonin to impart pH-sensitivity, antibacterial and antioxidant activities in the produced films. The CMC/agar-based films were transparent, showed high UV-light barrier properties, potent antimicrobial and antioxidant activities and an increase in water vapor barrier and surface hydrophobicity. The findings revealed that blending the new colorants resulted to unique pH responsive color changing properties and excellent reactivity to vapors such as ammonia and acetic acid.

Aside from pH, the evolution of gases such as hydrogen sulfide due to meat spoilage when sulfur-containing amino acids degrades can also be detected colorimetrically. Zhai and team (2019) developed a colorimetric hydrogen sulfide sensor based on gellan gum-silver nanoparticles nanobiocomposite for monitoring chicken breast and silver carp spoilage. The sensor in the form of hydrogel was attached to the lid to avoid contacting the meats and enabled the analysis of hydrogen sulfide with a limit of detection (LOD) of 0.81 μM at pH7. The on gellan gum-silver nanoparticles nanobiocomposite exhibited yellow-to-colorless color changes when exposed to hydrogen sulfide and is based on the ability of silver to form silver sulfide (Ag_2S) with hydrogen sulfide.

3.4 **Concluding statements**

This review states the certain characteristics and applications of nanobiocomposites for food packaging. Numerous researchers highlighted the necessity of replacing petroleum-based polymers with more sustainable polymers. The development of nanobiocomposite materials for food packaging is important not only to reduce the environmental problem but also to improve the functions of the food packaging materials. With an extensive number of biopolymers and inorganic nanofiller materials accessible, nanobiocomposites can be finely tuned to optimize novel materials with the most desirable features and functionalities. Many biomaterials, including chitosan, cellulose, and proteins have been provided with excellent versatility to prosper in varying applications (Joseph et al., 2020).

Indeed, the field of nanobiocomposites possesses a sensation and opportunities-filled research area. Initially, biopolymers were considered the only alternative materials to synthetic materials. However, with the rise of nanobiocomposites, the scientific community sees a brighter future for sustainable development. The greatest advantage that nanobiocomposites provide is the significant decrease in greenhouse gas emissions. This is a crucial issue to address in large industries, especially in the automotive and packaging industry. Sizeable cutbacks in the production of petrochemical polymers can be accomplished through the strategic development of nanobiocomposites. It is worth noting though that a major hurdle associated with bio-based materials regardless of their nature, is their difficulty for processing which makes them more expensive than petroleum-derived materials. Thus, it is important to develop effective biomass fractionation processes in order to selectively capture target polymers for processing. Recently, biomass fractionation processes such based on supercritical fluids (Cantero et al., 2013; Cocero et al., 2018; Adamovic et al., 2021) and subcritical fluids coupled with carbocatalysis (Mission et al., 2017, 2019) are making significant progress in this research fields that are worth considering.

The majority of recent studies are concentrated on films since films can be shrunk, stretch, and wrapped around food. Therefore, it is imperative to use biologically derived fillers, additives, or colorant to ensure the safety of the packaging materials. High-pressure extraction processes with green solvents such as water and carbon dioxide are also very prominent in this field (Quitain et al., 2017; Pazo-Cepeda et al., 2020).

3.5 Future trends on nanobiocomposites for food packaging application

Future studies on novel nanobiocomposites must be directed toward the cost efficiency and yield of sustainable materials. Although there is extensive growth in the field of nanobiocomposites, the conversion of laboratory-scale production to the macroscale or industrial level demands continuous progress and modernization in the material science and manufacturing field (Arora et al., 2018). While there are challenges with regulations and limited databases for production, the recognition of nanobiocomposites on a commercial scale is predicted because of the intense pressure among businesses, scientists, and other professionals to deliver more sustainable growth in the industries along with a circular economy.

The significant drawback of nanobiocomposite in food packaging applications is the severe lack of regulation, especially for those with intimate contact with food. There are a lot of studies confirming edibility and biocompatibility (Jiménez et al., 2016), but there are still excessive concerns regarding potential toxicity. In an unfortunate scenario, there is a critical danger of the purchaser having exposure to the nanoparticles from food packaging that may have migrated into the foodstuffs (Han et al., 2011). This is especially worrying as limited scientific data is associated with nanoparticle migration from packaging materials to food. There are general gaps in understanding regarding the behavior, outcome, and impacts of nanosized material through the gastrointestinal route. The characteristics of nanomaterials are drastically distinct from their bulk material and molecular forms. Some oral exposure to inorganic particles in rodents is studied. However, conflicts in nanomaterials quantification trapped in the animal exists. Although accurate toxicity reports are made at high-dose oral exposure for several nanoparticles such as silver, gold, and titanium dioxide, low-dose oral exposure still exhibits low accuracy and lack of adequate information. The majority of the concerns are for insoluble, indigestible, and biopersistant nanoparticles used in the food area. Further research is necessary to assess the prospective impacts of nanobiocomposite materials on human health and ecological safety (Han et al., 2011).

Still, nanobiocomposite packaging materials are foreseen to have a very desirable future, especially since the products' versatility is being optimized with the increasing variety of formulations and physicochemical applications. Furthermore, they could be considered smart, active, and intelligent materials with biofunctional properties in the food packaging industry.

Acknowledgment

Dr. Mission wishes to acknowledge her Marie Skłodowska-Curie postdoctoral fellowship (MSCA grant agreement No 894088).

References

Abral, H., et al., 2020. Highly transparent and antimicrobial PVA based bionanocomposites reinforced by ginger nanofiber. Polym. Test. 81, 106186. https://doi.org/10.1016/j.polymertesting.2019.106186.

Adamovic, T., et al., 2021. A feasibility study on green biorefinery of high lignin content agro-food industry waste through supercritical water treatment. J. Cleaner Prod. 323, 1–11. ISSN 0959-6526. https://doi.org/10.1016/j.jclepro.2021.129110 .

Alves, Z., et al., 2021. Design of alginate-based bionanocomposites with electrical conductivity for active food packaging. Int. J. Mol. Sci. 22 (18). https://doi.org/10.3390/ijms22189943.

Antonopoulos, I., Faraca, G., Tonini, D., 2021. Recycling of post-consumer plastic packaging waste in EU: process efficiencies, material flows, and barriers. Waste Manage. 126, 694–705. https://doi.org/10.1016/j.wasman.2021.04.002.

Arora, B., Bhatia, R., Attri, P., 2018. Bionanocomposites: green materials for a sustainable future. In: New Polymer Nanocomposites for Environmental Remediation. Elsevier Inc. https://doi.org/10.1016/B978-0-12-811033-1.00027-5.

Avena-Bustillos, R.J., Krochta, J.M., 1993. Water vapor permeability of caseinate-based edible films as affected by pH, calcium crosslinking and lipid content. J. Food Sci. 58 (4), 904–907. https://doi.org/10.1111/j.1365-2621.1993.tb09388.x.

Azeredo, H.M.C., Rosa, M.F., Mattoso, L.H.C., 2017. Nanocellulose in bio-based food packaging applications. Ind. Crops Prod. 97, 664–671. https://doi.org/10.1016/j.indcrop.2016.03.013.

Azmana, M., et al., 2021. A review on chitosan and chitosan-based bionanocomposites: promising material for combatting global issues and its applications. Int. J. Biol. Macromol. 185, 832–848. https://doi.org/10.1016/j.ijbiomac.2021.07.023.

Ballesteros-Mártinez, L., Pérez-Cervera, C., Andrade-Pizarro, R., 2020. Effect of glycerol and sorbitol concentrations on mechanical, optical, and barrier properties of sweet potato starch film. NFS J. 20, 1–9. https://doi.org/10.1016/j.nfs.2020.06.002.

Bhat, S.D., Aminabhavi, T.M., 2006. Novel sodium alginate-Na^+MMT hybrid composite membranes for pervaporation dehydration of isopropanol, 1,4-dioxane and tetrahydrofuran. Sep. Purif. Technol. 51 (1), 85–94. https://doi.org/10.1016/j.seppur.2005.12.025.

Boura-Theodoridou, O., et al., 2020. Performance of ZnO/chitosan nanocomposite films for antimicrobial packaging applications as a function of NaOH treatment and glycerol/PVOH blending. Food Packag. Shelf Life 23, 100456. https://doi.org/10.1016/j.fpsl.2019.100456.

Cantero, D.A., Dolores Bermejo, M., José Cocero, M., 2013. High glucose selectivity in pressurized water hydrolysis of cellulose using ultra-fast reactors. Bioresour. Technol. 135, 697–703. https://doi.org/10.1016/j.biortech.2012.09.035.

Cheikh, D., et al., 2020. Alginate bionanocomposite films containing sepiolite modified with polyphenols from myrtle berries extract. Int. J. Biol. Macromol. 165, 2079–2088. https://doi.org/10.1016/j.ijbiomac.2020.10.052.

Chivrac, F., Pollet, E., Avérous, L., 2009. Progress in nano-biocomposites based on polysaccharides and nanoclays. Mater. Sci. Eng. R Rep. 67 (1), 1–17. https://doi.org/10.1016/j.mser.2009.09.002.

Cinelli, P., et al., 2014. Whey protein layer applied on biodegradable packaging film to improve barrier properties while maintaining biodegradability. Polym. Degrad. Stab. 108, 151–157. https://doi.org/10.1016/j.polymdegradstab.2014.07.007.

Cocero, M.J., et al., 2018. Understanding biomass fractionation in subcritical & supercritical water. J. Supercrit. Fluids 133, 550–565. https://doi.org/10.1016/j.supflu.2017.08.012.

Cuq, B., et al., 1996. Stability of myofibrillar protein-based biopackagings during storage. LWT – Food Sci. Technol. 29 (4), 344–348. https://doi.org/10.1006/fstl.1996.0052.

Darder, M., Aranda, P., Ruiz-Hitzky, E., 2007. Bionanocomposites: a new concept of ecological, bioinspired, and functional hybrid materials. Adv. Mater. 19 (10), 1309–1319. https://doi.org/10.1002/adma.200602328.

De Azeredo, H.M.C., 2013. Antimicrobial nanostructures in food packaging. Trends Food Sci. Technol. 30 (1), 56–69. https://doi.org/10.1016/j.tifs.2012.11.006.

Ezati, P., Rhim, J.W., 2020. pH-responsive chitosan-based film incorporated with alizarin for intelligent packaging applications. Food Hydrocoll. 102, 105629. https://doi.org/10.1016/j.foodhyd.2019.105629.

Fortunati, E., et al., 2012. Multifunctional bionanocomposite films of poly(lactic acid), cellulose nanocrystals and silver nanoparticles. Carbohydr. Polym. 87 (2), 1596–1605. https://doi.org/10.1016/j.carbpol.2011.09.066.

Garrido-Miranda, K.A., et al., 2018. Antioxidant and antifungal effects of eugenol incorporated in bionanocomposites of poly(3-hydroxybutyrate)-thermoplastic starch. LWT – Food Sci. Technol. 98, 260–267. https://doi.org/10.1016/j.lwt.2018.08.046.

Ge, Y., et al., 2020. Intelligent gelatin/oxidized chitin nanocrystals nanocomposite films containing black rice bran anthocyanins for fish freshness monitorings. Int. J. Biol. Macromol. 155, 1296–1306..https://doi.org/10.1016/j.ijbiomac.2019.11.101.

Geueke, B., Groh, K., Muncke, J., 2018. Food packaging in the circular economy: overview of chemical safety aspects for commonly used materials. J. Cleaner Prod. 193, 491–505. https://doi.org/10.1016/j.jclepro.2018.05.005.

Ghosh, A., Ali, M.A., 2012. Studies on physicochemical characteristics of chitosan derivatives with dicarboxylic acids. J. Mater. Sci. 47 (3), 1196–1204. https://doi.org/10.1007/s10853-011-5885-x.

Hadi, J.M., et al., 2020. Electrical, dielectric property and electrochemical performances of plasticized silver ion-conducting chitosan-based polymer nanocomposites. Membranes 10 (7), 1–22. https://doi.org/10.3390/membranes10070151.

Han, W., et al., 2011. Application and safety assessment for nano-composite materials in food packaging. Chin. Sci. Bull. 56 (12), 1216–1225. https://doi.org/10.1007/s11434-010-4326-6.

Hu, X., et al., 2012. Protein-based composite materials. Mater. Today 15 (5), 208–215. https://doi.org/10.1016/S1369-7021(12)70091-3.

in het Panhuis, M., et al., 2007. Inkjet printed water sensitive transparent films from natural gum-carbon nanotube composites. Soft Matter 3 (7), 840–843. https://doi.org/10.1039/b704368f.

Jem, K.J., Tan, B., 2020. The development and challenges of poly (lactic acid) and poly (glycolic acid). Adv. Ind. Eng. Polym. Res. 3 (2), 60–70. https://doi.org/10.1016/j.aiepr.2020.01.002.

Jiang, T., et al., 2020. Starch-based biodegradable materials: challenges and opportunities. Adv. Ind. Eng. Polym. Res. 3 (1), 8–18. https://doi.org/10.1016/j.aiepr.2019.11.003.

Jiménez, A., Vargas, M., Chiralt, A., 2016. Antimicrobial nanocomposites for food packaging applications: novel approaches. In: Novel Approaches of Nanotechnology in Food. Elsevier Inc. https://doi.org/10.1016/b978-0-12-804308-0.00011-x.

Joseph, B., et al., 2020. Bionanocomposites as industrial materials, current and future perspectives: a review. Emergent Mater. 3 (5), 711–725. https://doi.org/10.1007/s42247-020-00133-x.

Kanmani, P., Rhim, J.W., 2014. Physical, mechanical and antimicrobial properties of gelatin based active nanocomposite films containing AgNPs and nanoclay. Food Hydrocoll. 35, 644–652. https://doi.org/10.1016/j.foodhyd.2013.08.011.

Khalil, W.F., et al., 2020. Graphene oxide-based nanocomposites (GO-chitosan and GO-EDTA) for outstanding antimicrobial potential against some *Candida* species and pathogenic bacteria. Int. J. Biol. Macromol. 164, 1370–1383. https://doi.org/10.1016/j.ijbiomac.2020.07.205.

Kim, J.Y., et al., 2003. Electrical and optical studies of organic light emitting devices using SWCNTs-polymer nanocomposites. Opt. Mater. 21 (1–3), 147–151. https://doi.org/10.1016/S0925-3467(02)00127-1.

Kumar, R., Münstedt, H., 2005. Silver ion release from antimicrobial polyamide/silver composites. Biomaterials 26 (14), 2081–2088. https://doi.org/10.1016/j.biomaterials.2004.05.030.

Kusmono, Abdurrahim, I., 2019. Water sorption, antimicrobial activity, and thermal and mechanical properties of chitosan/clay/glycerol nanocomposite films. Heliyon 5 (8), e02342. https://doi.org/10.1016/j.heliyon.2019.e02342.

Lagarón, J.M., López-Rubio, A., José Fabra, M., 2016. Bio-based packaging. J. Appl. Polym. Sci. 133 (2), 42971. https://doi.org/10.1002/app.42971.

Liu, G., Lin, Y., 2006. Carbon nanotube-templated assembly of protein. J. Nanosci. Nanotechnol. 6 (4), 948–953. https://doi.org/10.1166/jnn.2006.133.

Ma, X., et al., 2009. Preparation and properties of glycerol plasticized-pea starch/zinc oxide-starch bionanocomposites. Carbohydr. Polym. 75 (3), 472–478. https://doi.org/10.1016/j.carbpol.2008.08.007.

Mathew, S., et al., 2019. One-step synthesis of eco-friendly boiled rice starch blended polyvinyl alcohol bionanocomposite films decorated with in situ generated silver nanoparticles for food packaging purpose. Int. J. Biol. Macromol. 139, 475–485. https://doi.org/10.1016/j.ijbiomac.2019.07.187.

Menzel, C., 2020. Improvement of starch films for food packaging through a three-principle approach: antioxidants, cross-linking and reinforcement. Carbohydr. Polym. 250, 116828. https://doi.org/10.1016/j.carbpol.2020.116828.

Mironescu, M., et al., 2021. Green design of novel starch-based packaging materials sustaining human and environmental health. Polymers 13 (8), 1–35. https://doi.org/10.3390/polym13081190.

Mission, E.G., et al., 2017. Synergizing graphene oxide with microwave irradiation for efficient cellulose depolymerization into glucose. Green Chem. 19 (16), 3831–3843. https://doi.org/10.1039/c7gc01691c.

Mission, E.G., et al., 2019. Carbocatalysed hydrolytic cleaving of the glycosidic bond in fucoidan under microwave irradiation. RSC Adv. 9 (52), 30325–30334. https://doi.org/10.1039/c9ra03594j.

Mousa, M.H., Dong, Y., Davies, I.J., 2016. Recent advances in bionanocomposites: preparation, properties, and applications. Int. J. Polym. Mater. Polym. Biomater. 65 (5), 225–254. https://doi.org/10.1080/00914037.2015.1103240.

Ni, Y., et al., 2021. Visible light responsive, self-activated bionanocomposite films with sustained antimicrobial activity for food packaging. Food Chem. 362, 130201. https://doi.org/10.1016/j.foodchem.2021.130201.

Nikpour, M.R., Rabiee, S.M., Jahanshahi, M., 2012. Synthesis and characterization of hydroxyapatite/chitosan nanocomposite materials for medical engineering applications. Compos. Part B: Eng. 43 (4), 1881–1886. https://doi.org/10.1016/j.compositesb.2012.01.056.

Ozcalik, O., Tihminlioglu, F., 2013. Barrier properties of corn zein nanocomposite coated polypropylene films for food packaging applications. J. Food Eng. 114 (4), 505–513. https://doi.org/10.1016/j.jfoodeng.2012.09.005.

Palin, E., Liu, H., Webster, T.J., 2005. Mimicking the nanofeatures of bone increases bone-forming cell adhesion and proliferation. Nanotechnology 16 (9), 1828–1835. https://doi.org/10.1088/0957-4484/16/9/069.

Pazo-Cepeda, V., et al., 2020. Valorization of wheat bran: ferulic acid recovery using pressurized aqueous ethanol solutions. Waste Biomass Valorization 11 (9), 4701–4710. https://doi.org/10.1007/s12649-019-00787-7.

Puiggalí, J., Katsarava, R., 2017. Bionanocomposites. In: Clay-Polymer Nanocomposites. Elsevier Inc. https://doi.org/10.1016/B978-0-323-46153-5.00007-0.

Qasim, U., et al., 2021. Renewable cellulosic nanocomposites for food packaging to avoid fossil fuel plastic pollution: a review. Environ. Chem. Lett. 19 (1), 613–641. https://doi.org/10.1007/s10311-020-01090-x.

Qin, Y., et al., 2019. Preparation and characterization of active and intelligent packaging films based on cassava starch and anthocyanins from *Lycium ruthenicum* Murr. Int. J. Biol. Macromol. 134, 80–90. https://doi.org/10.1016/j.ijbiomac.2019.05.029.

Qiu, H., Yu, J., Zhu, J., 2005. Polyacrylate/(chitosan modified montmorillonite) nanocomposite: water absorption and photostability. Polym. Polym. Compos. 13 (2), 167–172. https://doi.org/10.1177/096739110501300205.

Quitain, A.T., et al., 2017. Microwave-assisted pressurized hot water extraction of alkaloids. In: Water Extraction of Bioactive Compounds: From Plants to Drug Development. Elsevier, 2017, pp. 269–289, ISBN 9780128093801. https://doi.org/10.1016/B978-0-12-809380-1.00010-3.

Reddy, J.P., Rhim, J.W., 2014. Characterization of bionanocomposite films prepared with agar and paper-mulberry pulp nanocellulose. Carbohydr. Polym. 110, 480–488. https://doi.org/10.1016/j.carbpol.2014.04.056.

Reddy, M.M., et al., 2013. Biobased plastics and bionanocomposites: current status and future opportunities. Prog. Polym. Sci. 38 (10–11), 1653–1689. https://doi.org/10.1016/j.progpolymsci.2013.05.006.

Rhim, J.W., Hong, S.I., Ha, C.S., 2009. Tensile, water vapor barrier and antimicrobial properties of PLA/nanoclay composite films. LWT – Food Sci. Technol. 42 (2), 612–617. https://doi.org/10.1016/j.lwt.2008.02.015.

Risyon, N.P., et al., 2020. Characterization of polylactic acid/halloysite nanotubes bionanocomposite films for food packaging. Food Packag. Shelf Life 23, 100450. https://doi.org/10.1016/j.fpsl.2019.100450.

Roy, S., Kim, H.J., Rhim, J.W., 2021. Effect of blended colorants of anthocyanin and shikonin on carboxymethyl cellulose/agar-based smart packaging film. Int. J. Biol. Macromol. 183, 305–315. https://doi.org/10.1016/j.ijbiomac.2021.04.162.

Ruiz-Hitzky, E., et al., 2010. Hybrid materials based on clays for environmental and biomedical applications. J. Mater. Chem. 20 (42), 9306–9321. https://doi.org/10.1039/c0jm00432d.

Sadeghizadeh-Yazdi, J., et al., 2019. Application of edible and biodegradable starch-based films in food packaging: a systematic review and meta-analysis. Curr. Res. Nutr. Food Sci. 7 (3), 624–637. https://doi.org/10.12944/CRNFSJ.7.3.03.

Safder, M., Temelli, F., Ullah, A., 2020. Lipid-derived hybrid bionanocomposites from spent hens. Mater. Today Commun. 25, 101327. https://doi.org/10.1016/j.mtcomm.2020.101327.

Sanuja, S., Agalya, A., Umapathy, M.J., 2015. Synthesis and characterization of zinc oxide-neem oil-chitosan bionanocomposite for food packaging application. Int. J. Biol. Macromol. 74, 76–84. https://doi.org/10.1016/j.ijbiomac.2014.11.036.

Schröder, H.C., et al., 2006. Co-expression and functional interaction of silicatein with galectin: matrix-guided formation of siliceous spicules in the marine demosponge Suberites domuncula. J. Biol. Chem. 281 (17), 12001–12009. https://doi.org/10.1074/jbc.M512677200.

Shamsuyeva, M., Endres, H.J., 2021. Plastics in the context of the circular economy and sustainable plastics recycling: comprehensive review on research development, standardization and market. Compos. Part C: Open Access 6, 100168. https://doi.org/10.1016/j.jcomc.2021.100168.

Shankar, S., Rhim, J.W., 2016. Tocopherol-mediated synthesis of silver nanoparticles and preparation of antimicrobial PBAT/silver nanoparticles composite films. LWT – Food Sci. Technol. 72, 149–156. https://doi.org/10.1016/j.lwt.2016.04.054.

Silva, K.S., et al., 2015. Effects of edible coatings on convective drying and characteristics of the dried pineapple. Food Bioproc. Technol. 8 (7), 1465–1475. https://doi.org/10.1007/s11947-015-1495-y.

Sorrentino, A., Gorrasi, G., Vittoria, V., 2007. Potential perspectives of bio-nanocomposites for food packaging applications. Trends Food Sci. Technol. 18 (2), 84–95. https://doi.org/10.1016/j.tifs.2006.09.004.

Sothornvit, R., Krochta, J.M., 2001. Plasticizer effect on mechanical properties of β-lactoglobulin films. J. Food Eng. 50 (3), 149–155. https://doi.org/10.1016/S0260-8774(00)00237-5.

Su, X.W., Cui, F.Z., 1999. Hierarchical structure of ivory: from nanometer to centimeter. Mater. Sci. Eng. C 7 (1), 19–29. https://doi.org/10.1016/S0928-4931(98)00067-8.

Turon, P., et al., 2017. Biodegradable and biocompatible systems based on hydroxyapatite nanoparticles. Appl. Sci. 7 (1), 60. https://doi.org/10.3390/app7010060.

Vieira, M.G.A., et al., 2011. Natural-based plasticizers and biopolymer films: a review. Eur. Polym. J. 47 (3), 254–263. https://doi.org/10.1016/j.eurpolymj.2010.12.011.

Wihodo, M., Moraru, C.I., 2013. Physical and chemical methods used to enhance the structure and mechanical properties of protein films: a review. J. Food Eng. 114 (3), 292–302. https://doi.org/10.1016/j.jfoodeng.2012.08.021.

Xiong, Z.C., et al., 2016. One-step synthesis of silver nanoparticle-decorated hydroxyapatite nanowires for the construction of highly flexible free-standing paper with high antibacterial activity. Chem. Eur. J. 22 (32), 11224–11231. https://doi.org/10.1002/chem.201601438.

Xiu, Z.M., et al., 2012. Negligible particle-specific antibacterial activity of silver nanoparticles. Nano Lett. 12 (8), 4271–4275. https://doi.org/10.1021/nl301934w.

Xu, A.W., Ma, Y., Cölfen, H., 2007. Biomimetic mineralization. J. Mater. Chem. 17 (5), 415–449. https://doi.org/10.1039/b611918m.

Xu, P., et al., 2020. Multifunctional and robust polyhydroxyalkanoate nanocomposites with superior gas barrier, heat resistant and inherent antibacterial performances. Chem. Eng. J. 382 (May 2019), 122864. https://doi.org/10.1016/j.cej.2019.122864.

Yokoyama, A., et al., 2005. Biomimetic porous scaffolds with high elasticity made from mineralized collagen: an animal study. J. Biomed. Mater. Res.: Part B Appl. Biomater. 75 (2), 464–472. https://doi.org/10.1002/jbm.b.30331.

Zafar, R., et al., 2016. Polysaccharide based bionanocomposites, properties and applications: a review. Int. J. Biol. Macromol. 92, 1012–1024. https://doi.org/10.1016/j.ijbiomac.2016.07.102.

Zhai, X., et al., 2019. A colorimetric hydrogen sulfide sensor based on gellan gum-silver nanoparticles bionanocomposite for monitoring of meat spoilage in intelligent packaging. Food Chem. 290 (March), 135–143. https://doi.org/10.1016/j.foodchem.2019.03.138.

Zhao, R., Torley, P., Halley, P.J., 2008. Emerging biodegradable materials: starch- and protein-based bio-nanocomposites. J. Mater. Sci. 43 (9), 3058–3071. https://doi.org/10.1007/s10853-007-2434-8.

Zilberman, M., Elsner, J.J., 2008. Antibiotic-eluting medical devices for various applications. J. Controlled Release 130 (3), 202–215. https://doi.org/10.1016/j.jconrel.2008.05.020.

Zubair, M., Ullah, A., 2020. Recent advances in protein derived bionanocomposites for food packaging applications. Crit. Rev. Food Sci. Nutr. 60 (3), 406–434. https://doi.org/10.1080/10408398.2018.1534800.

Nanobiotechnology for the food industry: Current scenario, risk assessment, and management

4

Mariam M. Abady [a,b,c], **Sila Imtiaz** [d], **Sahar Imtiaz** [d] and **Dina Mostafa Mohammed** [c]

[a] *Organic Metrology Group, Division of Chemical and Medical Metrology, Korea Research Institute of Standard and Science, Daejeon, Yuseong-gu, Republic of Korea,* [b] *Department of Bio-Analytical Science, University of Science and Technology, Daejeon, Yuseong-gu, Republic of Korea,* [c] *Department of Nutrition and Food Science, National Research Centre, Dokki, Cairo, Egypt,* [d] *Department of Microbiology, University of Health Sciences, Lahore, Pakistan*

4.1 Introduction

The methodical management of substances on an atomic, molecular, and supramolecular scale is known as nanotechnology. Manipulation of atoms and molecules is also included in order to create structures with macroscale dimensions (Nogrady, 2021). Nowadays, this is referred to as molecular nanotechnology (Drexler, 1992). Nanotechnology may be described as the examination and application of objects crossing in dimensions beginning 1–100 nm. Nanomaterials are endowed with a multitude of unique characteristics and improved properties over the bulk materials from which they are formed which including shape, charge, surface structure, chemical composition, and variance from nanoparticles including quantum dots, nano-fibers and nanocrystals to carbon nanotubes (Abdel-Wahhab and Márquez, 2015; Hassan and Singh, 2014; Singh et al., 2014, 2018). The biological efficiency of nanomaterials' and their attributes has significantly increment due to acquired characteristics for instance surface area and surface energy elevation in addition to catalytic reactivity (Abdel-Mohsen et al., 2013). On the additional contrary, biotechnology has the ability for utilization of biological techniques as well as information to handle the processes to create novel products and services (Tisato et al., 2018). There are two different nanotechnological ways, which are applied to food: "bottom–up" and "top–down," both of which have gained a lot of traction in the development of valuable consumer and therapeutic products (Singh et al., 2010, 2012; Singh and Mehta, 2016).

Nanobiotechnology is a unique combination of both nanotechnology and biotechnology permitting the combination of traditional microtechnology and biologically molecular approach in real-world claims (Fakruddin et al., 2012; Singh et al., 2014). In the case of the food industry, the top–down approach focuses on the physical handling of food components, such as milling and grinding. On the other side, the bottom–up approach focuses on developing and expanding larger structures from atoms and molecules (Singh et al., 2010, 2012; Singh and Mehta, 2016; Yang et al., 2015).

Nanobiotechnology has revolutionized a variety of research and industrial disciplines, particularly the food industry. Food processing, development of functional food, food packaging, and food safety are just a few of the nanotechnology applications that have emerged as a consequence of the increasing need for nanoparticles in relevant disciplines of food knowledge besides food microbiology. Nanobiotechnology has a great potentiality for improving food items both within and outside of them. Actuality, nanobiotechnology is swiftly providing new challenges for the industry of food innovation at massive rapidity, but uncertainty in addition to health is correspondingly developing.

Nanotechnology's ultimate goal echoes the needs of the food industry: error-free and quick processing with low-cost components that are free of any toxic properties. Even still, novel ingredient processing formats suffer from a host of flaws and are a source of intense debate among researchers. Nanomaterials have been dubbed "wonders of modern medicine" for their exceptional virtues (Dwivedi et al., 2018; Singh et al., 2018).

On the nanoscale, nanofibrous scaffolds may presently be reliably produced using only a few processing processes (Dwivedi et al., 2018). Food technology is considered as being part of the industry areas where nanobiotechnology could encourage fulfilling a pivotal role in the development. Food additives (means nano-inside) in addition to food packaging (means nano-outside) are the two broad categories of food industry applications (Ravichandran, 2010). Nanotechnology could be used to create strong nanofibers and confined nanoparticles for food industry microbial resistance components. As a result, nanomaterials may be worth pondering in the context of food technology research and applications.

Moreover, functional bioactive substances have piqued the curiosity of scientists, consumers, and food makers in the 2000s. Vitamins, probiotics, bioactive peptides, antioxidants, and other functional bioactive substances are descriptions (Singh et al., 2018). Lipids, proteins, carbohydrates, and vitamins, which are considered the bulk of bioactive compounds are vulnerable to the high acidity of the stomach plus duodenum as well as enzyme activity.

To optimize the release of these compounds, nanoencapsulation techniques have been widely used (Singh et al., 2017, 2018). The bioactive complexes are encapsulated not just to make them resistant to unfavorable conditions, but rather to make them easier to incorporate inside food products, which is problematic in noncapsulated formulas due to the poor solubility in water (Singh et al., 2017). Furthermore, nanoencapsulation enhances the food products shelf life by slowing or preventing

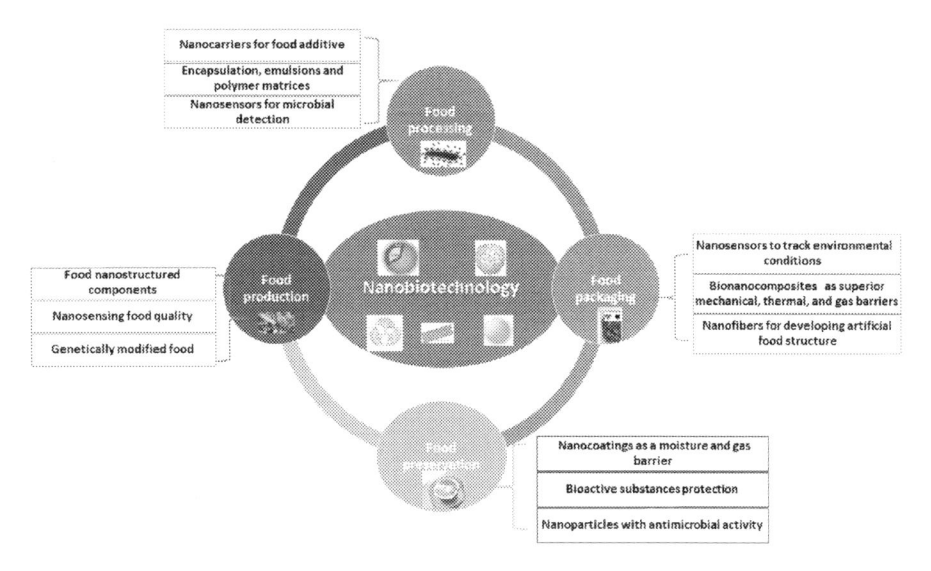

FIGURE 4.1

Systematic illustration of the application of nanobiotechnology in numerous parts of the food industry which could be summarized into four major field: food processing, food production, food packaging, and food preservation to improve food quality, safety, and shelf-life extension.

deterioration until the product arrives at the intended location (Gidwani and Singh, 2013).

Furthermore, nanocoating on several edible food materials might operate as a humidity plus gas exchange blockade even whereas supplying colors, flavors, enzymes, antioxidants as well and antibrowning mediators, in addition to extending the shelf lifespan of industrial foodstuffs even though the packaging has definitely been released (Bratovčić et al., 2015; Uttara et al., 2009). Currently, nanotechnology is primarily used in the food industry (Rashidi and Khosravi-Darani, 2011; Singh et al., 2018).

- Designing nanosensors and nanobiosensors to assure quality control and safety.
- Develop devices for targeted nutrient delivery using nutritional nanotherapy.
- Develop nanoencapsulation-based schemes for regulated relief of nutrients, antioxidants, proteins, and tastes (intelligent/smart systems).
- Develop an enzymatic nanoscale reactor for product creation and nutrition fortification with omega-3 fatty acids, haem, licopene, beta-caraton, phitosterols, and docosahexaenoic acid/eicosapentaenoic acid (DHA/EPA). Furthermore, the nanobiotechnology role in the food industry, which concludes the current scenario, risk assessment, and management as seen in Fig. 4.1 will be discussed in this chapter.

4.2 Role of nanobiotechnology in food production

This is the procedure of converting raw materials into food products for human intake, either in food processing manufacturing or at home. It uses scientific principles to guide its approach. A variety of herbal and animal foods are used to promote human healthiness including grains, spices, pulses, cereals, honey, nuts, milk, egg, poultry, vegetables, meat, fruits, and additional foods. There are numerous ways to produce food, including slicing vegetables or chopping, food fermentation, food curing, emulsification, food grinding and marinating, and the brewing industry. Cooking methods include boiling, broiling, steaming, grilling, mixing, in addition to frying. Pasteurization, processing of fruit liquor peeling, and skinning are two methods for removing the outer layers. Soft drink gasification, vacuum packs are used to preserve and package food. Some of the numerous methods of food production include cultivating, picking, crop management and production, maintaining, fermenting, pickling, as well as cooking, broiling, grilling, baking, and stewing, in addition to braising.

Food flavors, colors, besides textures may all be modified and changed at the nanoscale level because of nanotechnology could be based on many new and useful foods. The unique properties of nanomaterials improve sensory food quality by giving innovative texture, color, and appearance. Food nanostructured components and food nanosensing are two types of nanotechnological applications in food production as illustrated in Fig. 4.2. The category of food nanostructured ingredients includes additives that are carriers for the smart delivery of nutrients and antimicrobial agents, as well as food processing and packaging. Food nanosensing contributes to improved food quality and safety. Nanosensors/nanobiosensors are considered the peak prominent applications of nanobiotechnology in food quality observing in the processing industries in addition to bacterium identification. Furthermore, the nanosensors could be employed with pinpoint accuracy for the detection of the presence of fungus or insects in grain bulk storage locations. Scientists proposed nanobiotechnology practice simulations, either as a stand-alone technology or even as a complement to present technologies (Sastry and Rao, 2013).

The researchers were capable of changing the color of rice from purple to green. For instance, scientists have modified golden rice genetically with carbon nanofibers bearing external DNA through cellular injection (Torney et al., 2007). Now, industry has extraordinary capabilities to manipulate genes and even make new species by virtue of nanobiotechnology. This is due to the information that it permits foreign DNA and chemicals to be carried via nanofibers, nanoparticles as well and nanocapsules (Torney et al., 2007). Synthetic biology can similarly be used to make novel plant kinds (a novel division drawing on various techniques of nanotechnology, genetic engineering, and informatics (Serrano, 2007).

Many of the important concerns confronting the world's food supply today may be solved through nanotechnology. Food polymers and polymeric assemblages can be manipulated using nanotechnology to deliver tailored improvements in food quality and safety (Sekhon, 2014). Many fields in the dairy and food industries will be

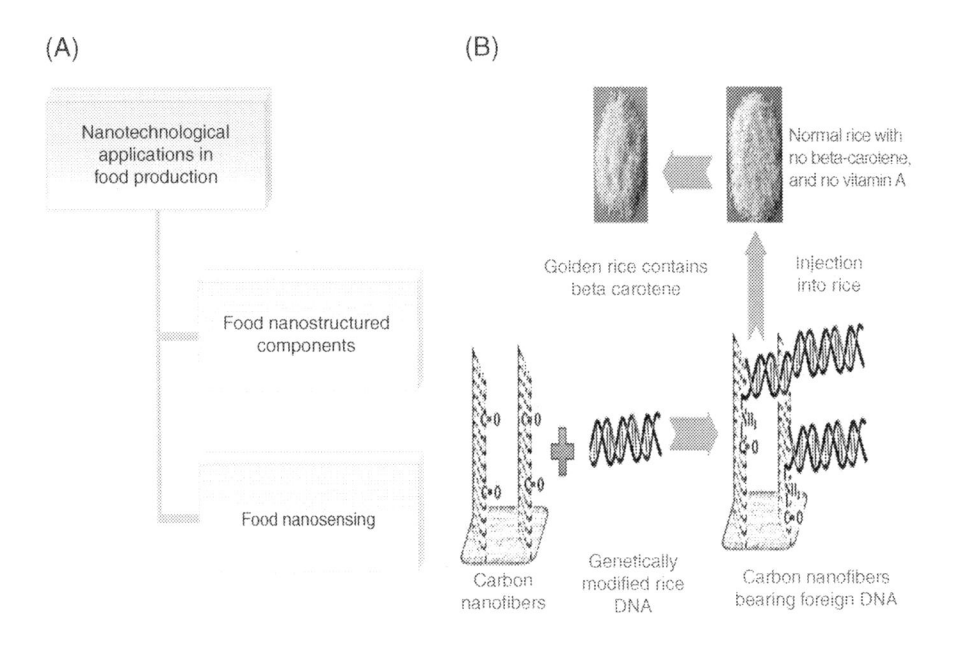

FIGURE 4.2

Role of nanotechnology in food production: (A) Nanotechnological applications in food; (B) steps in the production of golden rice with high nutritious value with beta-carotene by using carbon nanofibers, which act as a carrier for genetically modified DNA for beta carotene genes, and injection into rice for production golden rice rich in beta carotene.

replaced by nanobiotechnology, which has enormous application potential (Qureshi et al., 2012).

4.3 Role of nanobiotechnology in food processing

Food ingredients with nanostructured structures are indeed being developed, with the ability to improve texture, consistency, and flavor (Weiss et al., 2006). Additionally, nanotechnology is prolonging the shelf lifetime of several foodstuff products as well as decreasing food waste due to microbial infection (Pradhan et al., 2015). Additionally, nanocarriers are now being employed in food products as transfer methods for foodstuff additives without modification of the product's basic shape.

Because only submicron nanoparticles (not larger-size microparticles) are freely absorbed in some cell lines, particle size could have a direct influence on the prevalence of some bioactive chemicals to numerous locations inside the body (Ezhilarasi et al., 2013). The impactful qualities of the best delivery system are: (1) the ability to precisely deliver the active chemical to the region of interest, (2) the availability at a defined time and rate besides, and (3) efficacy in keeping dynamic compounds at optimum points for prolonged times (optimum storage condition). When

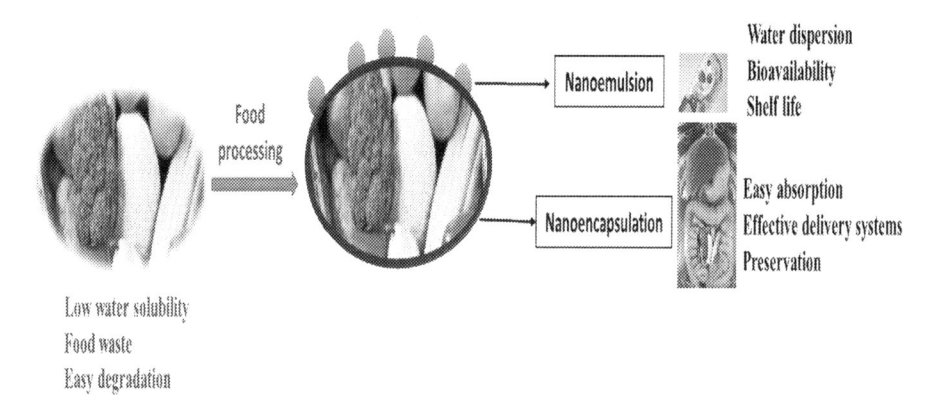

FIGURE 4.3

Incorporation of nanotechnology in food processing could improve food quality by increasing shelf life as well as facilitate its bioavailability and absorption in gastrointestinal tract. Such a target can be achieved by nanotechnology via nanoemulsion and/or nanocapsulation.

nanotechnology is employed to create emulsions, biopolymer matrices, encapsulation, simple solutions, and association colloids, efficient delivery systems with the aforementioned features are generated. Nano polymers are being used to substitute traditional materials in food packaging (Bratovčić et al., 2015).

The existence of contaminants, bacteria, and mycotoxins can be assessed through using nanosensors. Moreover, nanoencapsulations can be used for various applications. It can conceal tastes or odors, and control the interactions of active ingredients in the food matrix. Furthermore, it can control the active ingredient relief, and ensure availability at any certain rate and time, as well as protect active compounds from heat and humidity (Ubbink and Krüger, 2006), biological and/or chemical degradation, through processing, in addition to storage, and use to be compatible with further substances in the system (Weiss et al., 2006). In addition, those delivery systems can penetrate intensely into tissues, providing effective delivery of active compounds to specific regions through the body because of their tiny size (Lamprecht et al., 2004).

A range of delivery techniques with synthetic and natural polymer-based encapsulating has been developed to enhance the preservation and bioavailability of energetic food components. Nanotechnology deals with a number of preferences for improving nutrition quality and flavor (Nakagawa, 2014; Yang et al., 2015; Zhang et al., 2014). Nanoemulsions are increasingly being treated to distribute the bioactive chemical, which are lipid soluble since they can be completed with natural food ingredients and can be tailored to improve bioavailability and water dispersion (Ozturk et al., 2015).

Because of its subcellular size, nanoencapsulation provides a viable approach to enhancing the bioavailability of nutraceutical components when matched to bigger particles, which releases encapsulated substances extra gradually and over a longer time (Danquah-Amoah and Morya, 2017; Dekkers et al., 2011) as shown in Fig. 4.3. Encapsulation of these composites not only permits them to tolerate such harsh circumstances, but it also permits them to easily absorb into nutrition yields, which is problematic to attain in noncapsulated system due to little water solubility (Koo et al., 2005; Langer and Peppas, 2003; Yan and Gilbert, 2004).

4.4 **Role of nanobiotechnology in food packaging**

Food packaging is among the first commercialized uses of nanotechnology in the food industry (Singh et al., 2018). Food packaging has a significant impact on food safety and quality through storage, transit, and delivery to customers, especially for fruits and vegetables. There are several types of food packaging depending on nanotechnology as illustrated in Table 4.1.

A variety of functional nanostructural materials may be utilized as constructing elements to develop new productions and enhance the functionality of foods. Several of these structures have been characterized, as well as their current and potential submissions in the food region (Bajpai et al., 2018). Incorporating inorganic particles including clay into the biopolymer matrix can make the packaging material biodegradable, and can even be regulated by utilizing surfactants to modify layered silicates (Bratovčić et al., 2015). Nanofibers are ideal materials for developing artificial food structure matrixes and environmentally responsive food packaging. Moreover, nanofibers are now being manufactured from food-grade ingredients, and their application shall be expected to rise in the near future. Nanofibers manufactured by using the electrospinning technique are considered better technologies compared to current ones for the production of nanostructured materials due to their shape (Ravichandran, 2010). Electrospinning which is being used to encapsulate functional components creates nanofibers and innovative structures from synthetic and natural polymers, allowing them to be applied in a varied range of applications concluding new food components, food additives as well as novel packaging, food sensors, and additives encapsulation (Da Silva, 2012; Nikmaram et al., 2017).

Integrating nanomaterials to enhance packaging characteristics (gas barrier and flexibility properties); smart or intelligent food packaging integrating nanosensors for signaling and sensing of biochemical and microbial variations; active packaging to integrate nanoparticles with oxygen scavenging or antimicrobial properties; relief of antioxidants, antimicrobials, flavors, enzymes, and nutraceuticals are some of the nanobiotechnology applications for food contact materials.

Various food-grade antibacterial agents and polysaccharides are used in the packaging of edible food (Dutta et al., 2009; Joerger, 2007; Leceta et al., 2013; Rhim et al., 2013).

The present and expected market for nano-enabled products in the food sector is dominated by food packaging applications as presented in Fig. 4.4. Food nanoparticles have essential elements which are nontoxic and resistant to heat and pressure providing enhanced taste and consistency (Sawai, 2003). Organic nanoparticles containing natural polymers were used for active packaging with biodegradable properties, antibacterial, and antioxidant characteristics, as well as inorganic nanoparticles (metal

Table 4.1 The commercial applications of nanotechnology in food packaging.

Food packaging technology	Types	Roles	Mechanism	Examples
Active packaging	Antioxidants	Keep food from oxidizing	Release of antioxidant compounds into the meal as well as scavenging unfavorable molecules like oxygen	Eugenol-loaded chitosan nanoparticles
	Antimicrobials	Extend the shelf life of packaged foods	Inhibiting the growth of spoiling germs	Potassium sorbate Nisin Layered silicate fillers (montmorillonite silver nanoparticles)
Intelligent packaging	Time-temperature indicators	Estimate the proper temperature using time	Temperature and time-dependent migration of a dye through a porous material producing color change.	Fluorescent dye particles attached to bacteria antibodies
	Gas indicators	Monitor the composition of gases inside a package	A chemical or enzymatic reaction causes a change in the color of the indicator	TiO_2 nanoparticles SnO_2 nanoparticles
	Thermochromic inks	Indicate whether a food item is hot or cold to the consumer	Changes color at high temperatures	Polymer/layered silicate nanocomposites

(continued on next page)

Biodegradable coatings and films	Natural waxy	Protective coatings	Physical barrier to carbon dioxide, oxygen, and moisture movement	Hydrocolloids (polysaccharides and proteins)
				Lipids (waxes and resins)
				Synthetic polymers
	Thin layers of edible component	Gas barrier capabilities	Improve the appearance, delay ripening, prevent water loss, and lengthen shelf life	Cellulose (carboxymethyl cellulose, hydroxypropyl cellulose, and hydroxypropyl methylcellulose)
				Chitosan, a cationic and antibacterial polysaccharide
	Antibacterial compounds	Enhance the fruit's microbiological stability during storage	Providing effective antimicrobial activity	Benzoates
				Sorbates
				Short-chain organic acids
				Fungicides (benomyl, thiabendazole, iprodione, and captan)

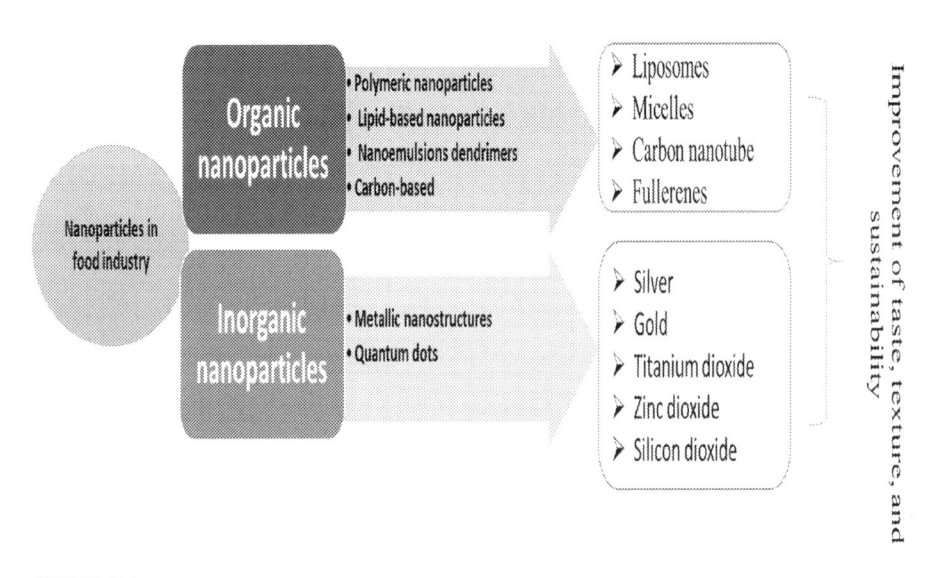

FIGURE 4.4

Summary of two types of nanoparticles that are used in the food industry including organic and inorganic nanoparticles to improve food quality and safety.

types and quantum dots). Most of these polymers are consisted of lipid-based nanoparticles, polysaccharides, and protein, which can be substituted with numerous packaging materials to preserve food. These biological nanoparticles improve texture, taste, and sustainability owing to the unique features (Suppakul et al., 2003). There are many of broadly used as biodegradable nanoparticle include" polylactic acid to encapsulate micronutrients such as vitamin, iron and protein; polyethylene glycol with functional elements to release in response of specific environmental conditions (Thangavel and Thiruvengadam, 2014). One of the most popular polymer nanoparticles is chitosan with unique properties including biodegradability, biofunctionality, antibacterial activity, and nontoxicity to extend shelf lifetime, in active packaging and additional applications such as cosmetics, pharmaceutics, wastewater treatment, and drug delivery.

Different processes can be used to produce chitosan nanoparticles such as molecular self-assembly, emulsion droplet coalescence, emulsion cross-linking, template polymerization, and ionotropic gelation (Assa et al., 2017; Malmiri et al., 2012).

In the case of ionic gelation, it is a gentle and simple process for chitosan nanoparticles production in aqueous media, without any modification on the chitosan surface, which requires a basic operating medium and counter ions (Kunjachan et al., 2014; Vaezifar et al., 2013). Tripolyphosphate (TPP) can be used as a nontoxic counter polyanion with negatively charged connect to protonated amine groups of chitosan through electrostatic interactions forming cross-linked ionic networks (Martins et al., 2012; Saleh et al., 2021) as summarized in Fig. 4.4.

On the other side, metal nanoparticles such as silver (Ag) and gold, and its oxides demonstrate microbial inactivation as revealed in Table 4.2. These nanoparticles

Table 4.2 Classification of various types' nanocarriers.

Types	Polymeric nanoparticles	Liposomes	Dendrimers	Carbon-based nanocarriers	Hydrogel nanoparticles	Quantum dots	Nanoemulsions
Shape							
Characteristics	Biocompatible and biodegradable polymers	Concentric lipid-bilayer comprised of an aqueous core surrounded by surfactant	Monodispersed macromolecular complexes repeatedly branched around an internal core	Tubular structures in the shape of a graphene sheet into a cylinder or enclosed at both tops producing buckyball shape	Three-dimensional polymer networks absorbing huge volumes of water or other biological fluid	Nanocrystals of inorganic fluorescent semiconductor atoms (2–10 nm)	Droplets of the oil and water phases (10–100 nm)
Examples	• Natural polymers (agarose, sodium alginate, chitosan, collagen, and fibrin) • Synthetic polymers (polylactic acid, polyglycolic acid, and polyamino acid)	• Multilamellar vesicles • Oligolamellar vesicles • Unilamellar vesicles	• Chitin • Melamine • Polyamidoamine • Poly L-glutamic acid • Polyethyleneimine • Polyethylene glycol • Polypropyleneimine	• Single-walled nanotubes • Multiwalled nanotubes • Fullerenes	• Alginate • Chitosan • Polyvinyl alcohol • Polyethylene oxide • Polyvinyl pyrrolidone • Poly-N-isopropylacrylamide	• Semiconducting core material (cadmium selenide) • Aqueous zinc sulfide shell	• Micellar system • Reversed micellar system

(continued on next page)

Table 4.2 Classification of various types' nanocarriers—cont'd

Types	Polymeric nanoparticles	Liposomes	Dendrimers	Carbon-based nanocarriers	Hydrogel nanoparticles	Quantum dots	Nanoemulsions
Advantages	Excellent carrier for controlled release of core materials and suitable for integration with biomaterials	Improvement of the solubility and stability of core materials	Loading of Core materials in either interior or conjugated to free surface clusters to boost targeted delivery	Target-oriented Ligand-attached Solvent-dispersed Surfactant-grafted	Crosslinks in the polymer networks provided by covalent bonds, hydrogen bonds, dipole-dipole interactions, van der Waals interactions, and physical entanglements	• Emission of light from the ultraviolet to infrared wavelength. For detection at the subcellular level • Stable and inert delivery vessel	Stable and suitable for integration into optically transparent goods such as fortified soft drinks and sauces

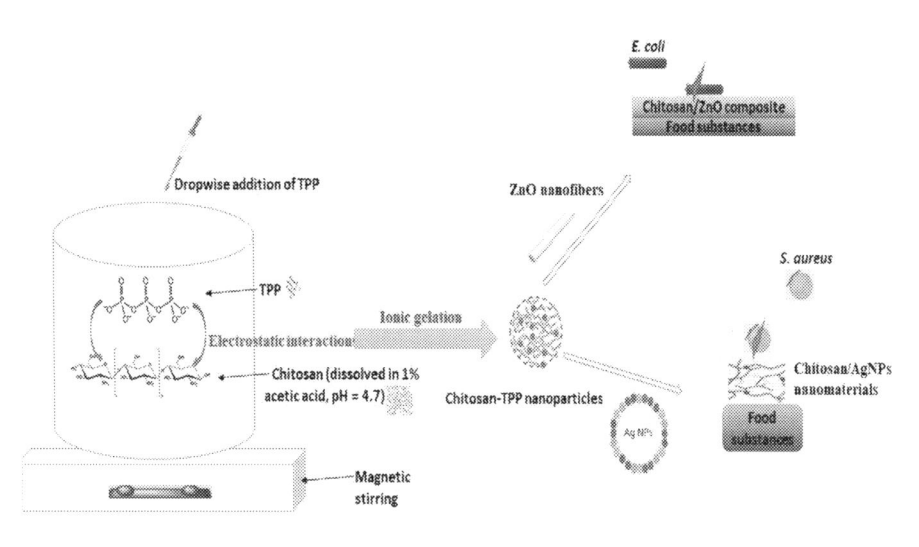

FIGURE 4.5

Scheme of ionic gelation of chitosan in acetic acid through cross-linking with TPP (*left*) and addition of some nanoparticles (NPs) to chitosan films improving its antibacterial activity (*right*). *AgNPs*, silver nanoparticles; *TPP*, tripolyphosphate.

can extend the shelf lifespan of vegetables and fruits, for example, food packaging sheets with Ag nanoparticles can absorb and break down toxic chemicals such as ethylene. Moreover, Ag nanoparticles with Food and Drug Administration (FDA)-approved concentrations have rarely side effects (Ansari et al., 2013). In addition, carbon black and silica oxide nanoparticles have been used in Europe for some food contact (Higashisaka et al., 2015). Furthermore, the combination of chitosan films and inorganic nanoparticles can improve antibacterial activity. For instance, the addition of Ag nanoparticles to chitosan decrease the inhibitory concentrations of *Staphylococcus aureus* to 500 times lesser than bulk chitosan (Ali et al., 2011). Another example is the combination of ZnO nanofibers to chitosan creating a composite, which can fight *Candida albicans* and *Escherichia coli* (Wang et al., 2012) as shortened in Fig. 4.5.

Recently, ultraviolet (UV)-activated colorimetric oxygen indicator which practices nanoparticles of TiO_2 in polymer encapsulating medium for photosensitization of methylene blue reduction by triethanolamine (Lee et al., 2002). By exposure to UV light, it becomes colorless till oxygen exposure which returns to blue color which accelerated the recovery rate of color (Mills and Hazafy, 2009).

4.5 Role of nanobiotechnology in food preservation

Food yields are stored in an inert, low-oxygen environment to avoid microbial growth plus spoilage. As a result, the material employed should be impervious to humidity and gas exchange even while carrying colors, flavors, enzymes, antioxidants, and antibrowning agents, along with lengthening shelf-lifetime of manufactured foodstuffs even if they have been released (Singh et al., 2017; Weiss et al., 2006).

4.5.1 Encapsulation technique

By employing the qualities of the interfacial layer surrounding them, encapsulating functional components within droplets can typically slow down chemical breakdown processes. The word nanoencapsulation refers to the use of nanometer-scale encapsulation techniques such as films, layers, and coatings, or simply microdispersion. The nanometer-scale encapsulation layer forms a defensive coating on the food or flavor ingredients/molecules. The active substance is frequently found in a nano or molecular form. The main advantage is that homogeneousness leads to improved efficiency of the encapsulation along with chemical and physical characteristics. This approach can be used to protect bioactive substances (such as vitamins, proteins, antioxidants, and carbohydrates) in the development of efficient meals to improve functioning and stability (Sekhon, 2010).

Food packaging materials generated from nanotechnology are currently the most widely used type of nanotechnology in the nutrition industry (Duncan, 2011). These uses conclude integrating nanomaterials to enhance packaging features (gas barrier capabilities, flexibility, and temperature/moisture stability); "intelligent/smart" food packaging with nanosensors that could check and report on the nutrition state; integrating nanoparticles with oxygen scavenging or antimicrobial properties in addition to biodegradable composites of polymer nanomaterial.

4.5.2 Production technique of nanocapsule

Nanocapsules can be manufactured using together "top–down" and "bottom–up" processes. The prior method requires energy for achieving nanonization, whereas the latter method employs physicochemical regulation to construct the nanocapsule by aggregating indeed atoms, ions, molecules, or monomers as shown in Fig. 4.6. Emulsification, solvent extraction, high pressure, and homogenization are all part of the top–down approach. Nanoprecipitation, coacervation, supercritical fluid, and inclusion complexation, are examples of bottom–up techniques.

4.6 Current scenario

The future of global food systems can be divided into four scenarios:

1. The wealthiest survive: In an ecosphere of resource-intensive intake and isolated marketplaces, there is a lethargic worldwide economy and a visible divide among the "haves" as well as "have-nots."
2. Unchecked consumption: This is a world of high gross domestic product (GDP) development at a tremendous cost to the environment, due to higher market connections and resource-intensive consumerism.

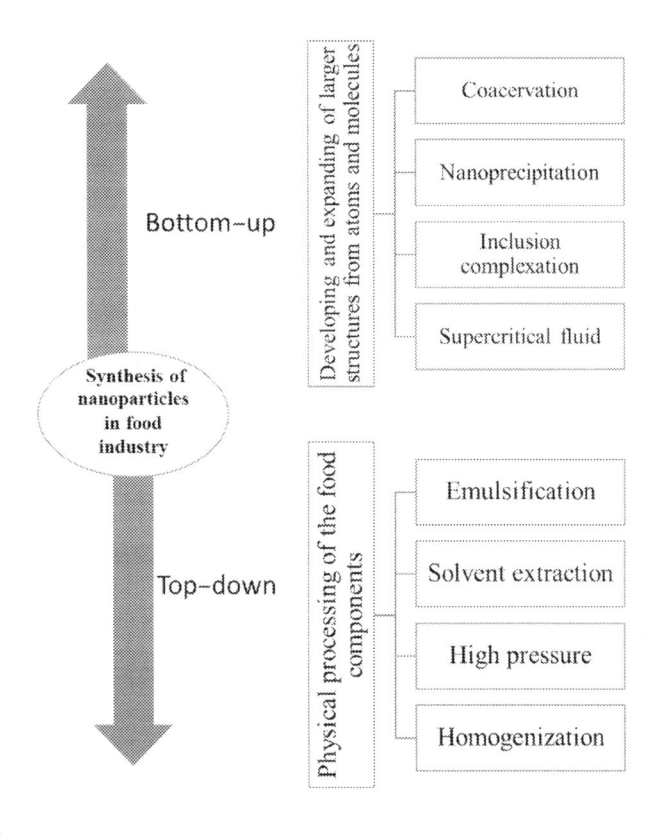

FIGURE 4.6

Preparation of nanomaterials in the food industry is done by two main processes:
bottom–up and top–down through various techniques.

3. Open-source sustainability: International cooperation and innovation have expanded because of the future combining highly networked markets and resource-efficient consumption, while others may be left behind.
4. The new global is local: In an ecosphere of scrappy local markets and resource-effective consumption, wealthy countries rely on native goods, whilst import-dependent areas become malnutrition hotspots.

4.6.1 The implications: today's decisions will figure out tomorrow's world

All of these four scenarios can be found in our world today, and a few of them could materialize by 2030. They show that to meet human demands within planetary constraints in 2030, today's food systems will need to undergo a dramatic shift.

Additional information is provided by the scenarios:

1. Consumption will determine whether or not the world's health and sustainability are preserved. The scenarios emphasize incentivizing, allowing, and motivating customers to eat additional resource-effective diets in their own circumstances.
2. A major rethink of food production methods has been needed to place nutritious and sustainable food in every bowl. A transformation like this would put a higher emphasis on the quality of agricultural production rather than just the quantity.

Climate change will have an impact on the hypothetical outcomes and is a serious hazard. Climate change and environmental degradation may jeopardize food systems' long-term productivity, jeopardizing societal stability and economic well-being.

4.6.2 How to build scenarios

We made various decisions in defining this question, including:

1. Geographic scope: This analysis' geographic scope is worldwide in order to present a large viewpoint for international spectators. This allows for a holistic, systems-level view while also limiting specificity by recognizing that are taken by stakeholders in accordance with the more detailed circumstances of their organization or country. Action leaders from Africa, Asia, and Latin America provided regional and stakeholder perspectives.
2. Goals for the future: Nutrition and sustainability are highlighted since they are two of the most pressing issues confronting the world's food systems. The World Health Organization (WHO) defines nutrition as "a sufficient, well-balanced diet with appropriate exercise," while a maintainable food scheme "transports food and nutrition safety for all in such means that the economic, social, and environmental foundations to create food safety and nutrition for upcoming generations are not jeopardized."
3. Timeframe: 2030 year was selected to arrange in line with the schedule for the Sustainable Development Goals. A mid-term timeline is proposed to emphasize the significance of leaders' food-system actions.

4.7 Risk assessment

Although nanoscale configurations in the food industry are unlikely to have a straight impact on human health, the nanoscale properties could have inevitable side effects. Nanoscale edible coatings have developed as a viable option for preserving food quality, extending storage life, and preventing microbial decomposition. Inhalation, ingestion, and dermal penetration are all ways for nanoparticles from manufactured nanoscale materials to reach the body. New data and international measurement are needed for accurate assessment of safety in food materials resulting from nanotechnology (WHO, 2008).

Table 4.3 The toxic effect of some nanoparticles frequently used in the food.

Nanoparticle	Toxicity	Mechanism	References
ZnO nanoparticles	Genotoxic potential in epidermal cells	Promotion of sensory changes to foods	Sharma et al. (2009)
Titanium oxide	Cell membrane disruption	Promotion of lipid oxidation in cell membranes	de Azeredo (2013)
		Rancidity due to lipid oxidation in the food	
Silver nanoparticles	Toxic to human neuronal cells and lung epithelial cells	The migration of silver ions with toxic levels in consumer goods	Ahamed et al. (2010)

The quality and direction of nanomaterial entrance inside the body are critical determinant for their harmfulness, along with the concentration of nanoparticle exposure, accompanied by sensitivity and the genetic basis of the organism. When the oral route of transmission was investigated, only rather high dosages of nanosilver or nano-TiO_2 were shown to cause toxicity (Aschberger et al., 2011).

Risk is the possibility that a person will be affected when there is an exposure to a threat. Nanoparticles have their harmful impacts on the atmosphere and human health, which are the major fears related to business. Due to their bioavailability and increased biochemical reactivity, small-scale particles have shown greater toxicity. Nanoparticles are stored in different body organs like the skin, liver, kidney, lung, etc. and cause different side effects (Bumbudsanpharoke and Ko, 2015). Risk assessment is vital to maintain the quality of food besides health standards. Moreover, proper risk assessment of the nanonutrition product consumption is highly required beforehand commercialization. Importantly, the assessment of risk is considered as a scientific process to estimate a risk and understand the influencing issues. New nanotechnology-based materials may create threats of environmental contamination or even harmful effects on human health. In addition, some nanomaterials enter the human body. Nanostructures are frequently used in the field of food packaging but there is restricted scientific information on the movement of nanostructures from packing ingredients to the food (de Azeredo, 2013). Some examples of nanomaterials toxicity are shown in Table 4.2. The probability of nanomaterial passage through the body barriers (barriers of blood-brain, blood-milk, and placenta in addition to other cellular barriers) is considered as extra serious matter for the risk assessment. Such blood-brain barrier permeability, which represents a strict defense mechanism to the brain, is extremely limited to molecules that are either are small soluble (<500 Da), lipophilic, or actively transported particles. Nevertheless, a suggestion exists that some nanoparticles might be permeable to the brain as shown in Table 4.3 (Bouwmeester et al., 2009).

Risk assessment principles are necessary for the administration of safety measures for nanomaterials used in the food industry (Hwang et al., 2012). Prophylactic standards are practical in case of highly threat of food to human health and atmosphere, and at that point, constraints are applied for hazard control. According to Greenpeaces Leggetts "Do not emit a substance unless you have proof it will do no harm to the environment" (Goklany, 2001).

On the other side, microbiological risk assessment (MRA) is an efficient method to conclude and reduce risk or hazards associated with public health (Voysey and Brown, 2000). There are two methods for evaluation of risk assessment, e.g., qualitative (includes information in a categorical manner) and quantitative (mathematical analyses of numerical data which is further divided into two categories, e.g., point estimate and probabilistic). Risk assessment is a scientific examination containing four defined steps: first, hazard identification; second, exposure assessment; third, hazard characterization; and last, risk characterization.

4.7.1 **Hazard identification**

The term hazard refers to a pathogen and its toxins that are capable of causing human illness. It is the qualitative determination of risk. In the hazard identification step, it is significant to know the agent that is present in food and responsible for serious health effects. There are three categories of hazards including biological hazards (bacteria, parasites, viruses, etc.), chemical hazards (pesticides, antibiotics, etc.) in addition to physical ones (glass, etc.).

4.7.2 **Assessment exposure**

The assessment of exposure involves estimation qualitatively and/or quantitatively of the possible consumption of the agent by the food (WHO, 2010). As illustrated in Fig. 4.7, the disruption potential of nanoparticle to the epithelia barrier of the gastrointestinal tract (GIT) after its passage either across the cells or by endocytosis. It means the chance of an individual exposure to a microbial hazard as well as the ingested bacterial numbers. Properties of microorganism and its poisonous effects on human health is also important. For viral and parasitic lives, contamination frequency and concentration in addition to distribution and inactivation steps play an essential role in exposure assessment. For bacterial growth and inactivation within food is also a major concern in risk assessment. There are various factors that an assessor should know, e.g., characteristics of pathogenic agent, biology of food, degree of sanitation, processing methods, packaging, and food storage. Furthermore, the entrance of a pathogen into food could be done due to deprived processing and worse food handling through an unhygienic person. Microbiologist, food scientist, food processor, health experts, and nutritionist are risk assessors. However, there is huge data on the food industry that is ready to read and copy paste but most of the information are not acknowledged. In risk assessment, constructing models are very important part. In qualitative assessment exposure pathways are described while in relationships of quantitative assessment among components of assessment are argued (Lammerding and Fazil, 2000).

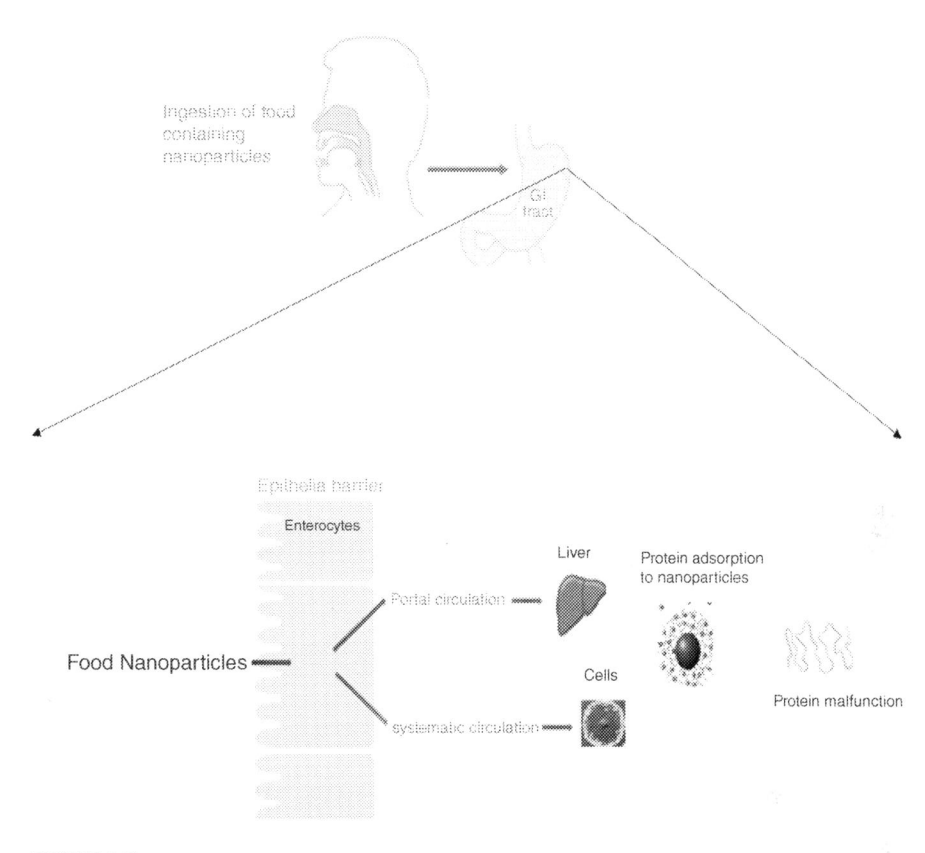

FIGURE 4.7

The risks involved in the consumption of a nanoscale food product. Potential contamination with bacteria or disruption potential to epithelia cells of the gastrointestinal tract could increase microbial exposure in the human body which may lead to the inactivation of important protein and enzymes affecting metabolism.

4.7.3 Hazard characterization

The aim of hazard characterizations is to make logic of the current information and demonstrate what they mean to the common person. This type of risk assessment is occasionally given inadequate devotion to health risk evaluations. In risk assessment, identification and characterization of the environment and amount of human health risks is the essential step in the analytical process (Williams and Paustenbach, 2002).

4.7.4 Risk characterization

It involves the integration of data collected from prior steps to know the extent of risk to a population or specific consumer. Risk assessments are done by various applications on a computer (Gerba et al., 1996).

4.8 Risk assessment principle

The risk assessment principle is necessary because it has three reasons: First, the probability of highly exposure of consumers to nanomaterials with a minor size which exhibit harmfulness. Second, the implication produced by the biological action of nanomaterials does not demonstrate the risk of nanomaterials. Third, the community has little knowledge about the care of nanomaterial (Sonkaria et al., 2012).

4.8.1 Tools for assessment of risk in food industry

To access various risk in food industry, many food companies use different tools, e.g., diagnostic tools, selection tools, and improvement tools. These tools can be used for auditing purposes and provide strength to the food safety and management system (Jacxsens et al., 2011). For MRA, there are various tools available to access the potential risk to the food products. These are: characterize dose-response associations, model growth, survival, and eventually death of pathogens, develop a typical food chain, and incorporate information from MRA steps (Nauta et al., 2012).

The Centers for Disease Control and Prevention (CDC) has developed a novel tool for the reduction of food safety risk from production to consumption of food. Food safety info sheets are introduced for changing the practices of food handlers (Chapman et al., 2010). Meta-analysis is also a valuable tool for the assessment of risk in food areas. It includes organized evaluation, withdrawal of information, and assessment of conclusion and its variability (Gonzales-Barron and Butler, 2011). Online survey feedback form is furthermore broadly used risk assessor in the food industry. The vulnerability assessment tools of food fraud are as well applied in the food industry.

4.8.2 Types of risk assessments in food industry

- Threat assessment (TACCP)
- Hazard assessment of food (HACCP)
- Vulnerability assessment (VACCP)
- Raw materials
- Food fraud
- Safety and health of work area
- Environmental
- Business operational

  Excess use of nanomaterials in food industry causes various risk to community:

- Pulmonary infections
- Change in body metabolic rate
- Accumulation of these materials in diverse tissues as well as organs (such as lung, skin, kidney, liver, brain, spleen, reproductive, and vascular tissues)
- Cellular changes (shape, size, etc.)
- Autoimmune diseases (Hwang et al., 2012)

For quantitative simulations of microbial risk assessment, assessment of exposure necessitates pathogen occurrence, and density, besides the distribution in living animals and foods, growth and decline factors, as well as consumption (Coleman and Marks, 1999). However, qualitative assessments have been required widely in risk assessments of foodstuffs projected for human intake, particularly those of animal origin (Wooldridge, 2007). There are certain methods that are able to detect food allergens are enzyme-linked immunosorbent assay (ELISA), polymerase chain reaction (PCR), and real-time-PCR (Sancho and Mills, 2010).

4.8.3 Risk communication

Collaborating the outcomes of the risk analysis procedure aids numerous determinations. It offers the community with the results of technical evaluation of the identification of food risk plus risk calculation to the community groups such as infants or the elders. Communication offers information to both the private and public sectors with essential for avoiding, and dropping food risks to acceptable safety levels. In addition, it affords adequate information for the population. It is considered as the final section of the risk analysis procedure.

Technological risk and the process of explaining risks to the public have become major community problems. The National Center Risk Communication Team for Food Protection and Defense (NCFPD) established a list of eleven best practices for investors to implement for actual risk communication in the food field.

These are prompt response, common network, community partnership, public alarm, exposed, truthful, and accessible media, self-preservation, compassion, and update plan (McEntire and Boateng, 2012).

4.9 Challenges and limitations

At this time, there is no internationally known definition for "nanobiotechnology." Still, no confirmation is that those nutrients or their contact materials resulting from nanotechnology are harmless to the public. Methods for the safety of nanomaterial vary from region to region. Novel information about the safety of products using nanotechnology is needed. Knowledge about the nanomaterial presence in foodstuffs depends on industry information, producers, and marketing organizations. Korea Food and Drug Administration (KFDA) began an examination to make a "Strategic Action Plan" to assess the safety and management of nano risk related to foodstuffs, cosmetics, and medicinal devices by nano-scale supplies. Above 60% of food scientist said that there should be precautionary measures for the safety of food products to protect public health issues besides at that moment regulatory agency should develop management procedure related to risk assessment (Hwang et al., 2012).

On the other side, important limits of the principal development of risk assessment for nanoproducts safety and management still exist. The first thing is the deficiency of definition, and suitable information to evaluate possible hazards to public health.

The second issue is that most of the nanomaterial yields are gradually developing. The last one is a noteworthy information breach in considerate of the risk assessment. So there should be increased research on nanobiotechnology products (Hwang et al., 2012; Kandlikar et al., 2007).

Another limitation is that a monitoring agency has to regulate the planned needs; in addition, a risk assessor has to determine evaluated needs. There is the absence of a common measure of risk that would allow a comparison of various types of risks associated with exposures to the different ingredients of food. The Global Burden of Disease Study involved the use of the disability-adjusted life-year (DALY) as a communal measure of various fatal and nonfatal diseases. There is some possibility that current regulations can accommodate this type of management strategy for all modules of food substances (Stamm, 2011).

4.10 Risk management

In the past "risk," word was used for the protection of food. First, the threats are recognized and then the severity and possibility of each threat are evaluated. By increasing the worldwide trade, new demands on food safety and excellence could be raised. As food is our basic need so, our health level depends on good quality products to avoid any kind of miss happen. To provide safe and healthy food many countries works on different food safety programs to minimize food-related risks (Samimi, 2020). Presently, no specific regulations are for nanotechnology usage in food. FDA controls on a product-by-product basis. FDA has precise products through particulate food in the range of nano-size. Recommendations by the Royal Society of the U.K. government to measure the possible nanotechnology influence, comprised a demand for nanoparticles identification inside ingredient lists for consumers to make informed decisions (Rashidi and Khosravi-Darani, 2011).

4.10.1 Risk management decisions

Risk management means the management of techniques and methods related to risk, which is the fundamental component of the control system (Bozorgian et al., 2020). Risk management in an organization is highly essential because a society or group cannot attain its future objectives without it so, the leader who receipts the charge for risk in an association must be capable of achieving its goals (Samimi, 2020).

Depending upon the type of threats, different techniques are used. Warranty of foodstuffs related to its safety is one of the major tasks in developing states to keep the health in good state. Many improvements have been made in foods care systems in current eras (Yuryevich et al., 2014). In the study of Smithson and Simkins (2005) on collected works of risk management and stable worth, it can be concluded a limited evidence about the increment of risk management firm value.

Theoretically, the actual management of threats in food connected to bacterio-logical risks is a difficult method. In recent times, the method of risk examination

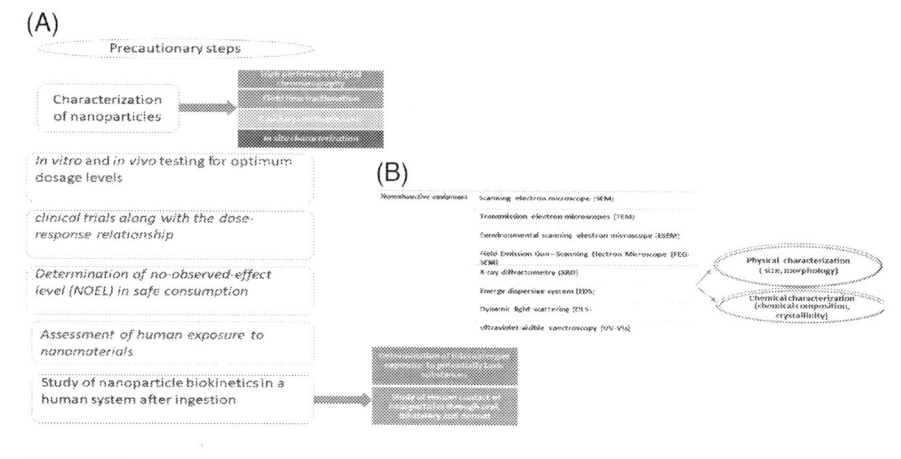

FIGURE 4.8

Measurement and characterization of nanoparticles: (A) Precautionary measures for nanoparticle before human ingestion. (B) Nonexhaustive list of equipment required to characterize nanoparticle. Both steps are essential for the management of potential risk from food exposure.

includes (A) risk assessment, (B) risk management, and (C) communication of risk components to control the bacteriological hazard to assist the consumers' health. In microbial risk management practice, an accurate defense is a basic model and an image of the national food safety goals fixed by the skilled establishments of the country (Samimi, 2020). Additionally, the characterization of nanonparticle before ingestion is one of the most precautionary measures.

The precautionary steps as well as analytical techniques and equipment are widely used to characterize nanoparticle as illustrated in Fig. 4.8.

4.10.2 Steps for selection of risk management options

Risk analysis and management should be done constantly, patently, and with complete evidence. If a novel data or study is found, then proper analysis and evaluation should be needed.

4.10.3 Steps for effective risk management process

Governments should develop certain rules and regulations to make an effective food safety system for public health. Food can become polluted at any step of its production, etc. Here are some steps to effectively manage the food risk (WHO):

- Maintain acceptable food systems to respond and achieve food safety risks
- Communication between community health, animal health and farming
- Incorporate food safety into food strategies
- Pondered universally and deed in the vicinity to assure food safety

4.10.4 Risk management policies

4.10.4.1 Price increases

In the past 8 years, food prices have been raised globally. Food and Agriculture Organization (FAO) international food price index has been increased. Increased food prices also lead to malnourishment in poor people.

4.10.4.2 Adverse weather

Thrilling climate affects the availability of foods, diminishes contact to food, and disturbs the food features. Food will soil if not stored accurately, which can lead to food-borne sickness.

4.10.4.3 Infrastructure issues

The global food system focuses on its manufacturing, handling, delivery, and also infrastructure issues of food (Neethirajan and Jayas, 2011). Policymakers maintain satisfactory food systems and infrastructures (e.g., laboratories) (WHO, 2010).

4.10.4.4 Damage to a supplier location impacting the food industry

Healthy and safe food is vital to support and encourage life. Safe food deliveries support state thrifts, occupation, and travel. As the world's inhabitant's breed, the request for food as well as chances and challenges for food including its production, safety, and delivery, are also increased. These trials are tackled by food producers and handlers to ensure the food safety (WHO).

4.10.4.5 Brand innovation

Nanotechnology helps the food industries to provide a good quality product that is designed for a specific markets. In order to track or follow up on any product, nanotechnology offers obscure nano barcodes of the company or brand on the food product directly in addition to batch information. This nano barcode technology suggests food safety by permitting the trademark holders to observe their stock without sharing information with suppliers and vendors. By allocating different codes to all food products aids the brands in tracing food lots (Neethirajan and Jayas, 2011).

4.10.4.6 Product quality

Maintenance of quality assurance of food is of great importance as the users request nontoxic and healthy food and the government enforces certain rules to assure food safety. Devices or some sort of discovery schemes for speedy finding of decay of product, for quality control are possible through nanotechnology (Neethirajan and Jayas, 2011).

4.10.5 Practices to control the risk

Although nanotechnology has advantages in the food industry but still its protection problems cannot be ignored. Researchers that worked on safety issues of foodstuffs also highlight the likelihood of nanoparticles transmitting from the packing material into the foodstuff and their effect on consumer's well-being (Jain et al., 2016). Antimicrobial foodstuff wrapping works to lessen bacterial growth, to expand the shelf life of foodstuffs, and to sustain food features for a longer time. There is a variety of materials with antimicrobial agents in food storage material such as alcohol, bacteriocins and metals (Suppakul et al., 2003).

4.10.6 Mitigate the risk of nanotechnology

The first phase is the establishment of a risk outline. This outline aids the managers to evaluate the risks of food safety. It acts as an appropriate model for arranging food safety risks on a priority basis and checks whether the problem can be resolved (Samimi et al., 2019). Regardless of several studies on risk management, its appliance and its relationship with organizational performance investigators have shown less care for the subject and have not studied how to perform risk management in numerous administrations. Careful examination of raw materials and control of all authorizations, etc. are the means to control physical threats in food (Samimi et al., 2019).

4.11 Conclusion and perspective

Biotechnology and nanotechnology are two speedily emerging technologies. Nanobiotechnology is an innovative technology. By the usage of its devices and techniques for better cultivation, food processing, food packaging, and production results in the development of nutritional food quality and safety. The existing level of nanotechnology applications in food industries has massive alterations because of the specific properties of nanomaterials. The nanotechnology applications in the industry of food are extensively discussed in the current chapter together with various preparation protocols and types of nanoparticles to improve the nutritional value and constancy of nutritious ingredients. On the other hand, this promising technology could lead to indefinite and occasionally hazardous effects on ecological units. The essential part is providing a pure guiding principle for the estimation and elimination of risks in the optimal custom of nanomaterials. Additionally, the introduction of nanotechnology in food products into the market entails education for the proper use of it as well as an understanding of its complications. As a final point, nanotechnology supports the alteration of the prevailing food systems and processing to confirm food safety, produce a healthy food culture, and improve the dietary worth of the food.

Acknowledgment

Worth mentioning is the help, cooperation, and encouragement established between all contributed authors.

References

Abdel-Mohsen, A., Hrdina, R., Burgert, L., Abdel-Rahman, R.M., Hašová, M., Šmejkalová, D., Kolář, M., Pekar, M., Aly, A., 2013. Antibacterial activity and cell viability of hyaluronan fiber with silver nanoparticles. Carbohydr. Polym. 92 (2), 1177–1187.

Abdel-Wahhab, M.A., Márquez, F., 2015. Nanomaterials in biomedicine. Soft Nanosci. Lett. 5 (3). United States, ISSN : 2160-0600, doi:10.4236/snl.2015.53006. https://www.osti.gov/biblio/1335956.

Ahamed, M., Alsalhi, M.S., Siddiqui, M.K.J., 2010. Silver nanoparticle applications and human health. Clin. Chim. Acta 411 (23–24), 1841–1848. https://doi.org/10.1016/j.cca.2010.08.016.

Ali, S.W., Rajendran, S., Joshi, M., 2011. Synthesis and characterization of chitosan and silver loaded chitosan nanoparticles for bioactive polyester. Carbohydr. Polym. 83 (2), 438–446.

Ansari, M.A., Anurag, A., Fatima, Z., Hameed, S., 2013. Natural phenolic compounds: a potential antifungal agent. Combat. Sci. Technol. Educ. 1, 1189–1195.

Aschberger, K., Micheletti, C., Sokull-Klüttgen, B., Christensen, F.M., 2011. Analysis of currently available data for characterising the risk of engineered nanomaterials to the environment and human health—lessons learned from four case studies. Environ. Int. 37 (6), 1143–1156.

Assa, F., Jafarizadeh-Malmiri, H., Ajamein, H., Vaghari, H., Anarjan, N., Ahmadi, O., Berenjian, A., 2017. Chitosan magnetic nanoparticles for drug delivery systems. Crit. Rev. Biotechnol. 37 (4), 492–509.

Bajpai, V.K., Kamle, M., Shukla, S., Mahato, D.K., Chandra, P., Hwang, S.K., Kumar, P., Huh, Y.S., Han, Y.-K., 2018. Prospects of using nanotechnology for food preservation, safety, and security. J. Food Drug Anal. 26 (4), 1201–1214.

Bouwmeester, H., Dekkers, S., Noordam, M.Y., Hagens, W.I., Bulder, A.S., De Heer, C., Ten Voorde, S.E., Wijnhoven, S.W., Marvin, H.J., Sips, A.J., 2009. Review of health safety aspects of nanotechnologies in food production. Regul. Toxicol. Pharm. 53 (1), 52–62.

Bozorgian, A., Zarinabadi, S., Samimi, A., 2020. Optimization of well production by designing a core pipe in one of the southwest oil wells of Iran. J. Chem. Rev. 2 (2), 122–129.

Bratovčić, A., Odobašić, A., Ćatić, S., Šestan, I., 2015. Application of polymer nanocomposite materials in food packaging. Croat. J. Food Sci. Technol. 7 (2), 86–94.

Bumbudsanpharoke, N., Ko, S., 2015. Nano-food packaging: an overview of market, migration research, and safety regulations. J. Food Sci. 80 (5), R910–R923.

Chapman, B., Eversley, T., Fillion, K., MacLaurin, T., Powell, D., 2010. Assessment of food safety practices of food service food handlers (risk assessment data): testing a communication intervention (evaluation of tools). J. Food Prot. 73 (6), 1101–1107.

Coleman, M., Marks, H., 1999. Qualitative and quantitative risk assessment. Food Control 10 (4-5), 289–297.

Danquah-Amoah, A., Morya, S., 2017. Application of nanotechnology in bioengineering industry and its potential hazards to human health and the environment. Pharm. Innov. 6 (7), 49.

Da Silva, L.J., 2012. Functional nanofibers in food processing. In: Functional Nanofibers and Their Applications. Elsevier, pp. 262–304.

de Azeredo, H.M., 2013. Antimicrobial nanostructures in food packaging. Trends Food Sci. Technol. 30 (1), 56–69.

Dekkers, S., Krystek, P., Peters, R.J., Lankveld, D.P., Bokkers, B.G., van Hoeven-Arentzen, P.H., Bouwmeester, H., Oomen, A.G., 2011. Presence and risks of nanosilica in food products. Nanotoxicology 5 (3), 393–405.

Drexler, K., 1992. Nanosystems: Molecular Machinery, Manufacturing, and Computation. Wiley, New York.

Duncan, T.V., 2011. Applications of nanotechnology in food packaging and food safety: barrier materials, antimicrobials and sensors. J. Colloid Interface Sci. 363 (1), 1–24.

Dutta, P., Tripathi, S., Mehrotra, G., Dutta, J., 2009. Perspectives for chitosan based antimicrobial films in food applications. Food Chem. 114 (4), 1173–1182.

Dwivedi, C., Pandey, I., Pandey, H., Patil, S., Mishra, S.B., Pandey, A.C., Zamboni, P., Ramteke, P.W., Singh, A.V., 2018. In vivo diabetic wound healing with nanofibrous scaffolds modified with gentamicin and recombinant human epidermal growth factor. J. Biomed. Mater. Res. Part A 106 (3), 641–651.

Ezhilarasi, P., Karthik, P., Chhanwal, N., Anandharamakrishnan, C., 2013. Nanoencapsulation techniques for food bioactive components: a review. Food Bioprocess Technol. 6 (3), 628–647.

Fakruddin, M., Hossain, Z., Afroz, H., 2012. Prospects and applications of nanobiotechnology: a medical perspective. J. Nanobiotechnol. 10 (1), 1–8.

Gerba, C.P., Rose, J.B., Haas, C.N., 1996. Sensitive populations: who is at the greatest risk? Int. J. Food Microbiol. 30 (1-2), 113–123.

Gidwani, M., Singh, A.V., 2013. Nanoparticle enabled drug delivery across the blood brain barrier: in vivo and in vitro models, opportunities and challenges. Curr. Pharm. Biotechnol. 14 (14), 1201–1212.

Goklany, I.M., 2001. The Precautionary Principle: A Critical Appraisal of Environmental Risk Assessment. Cato Institute.

Gonzales-Barron, U., Butler, F., 2011. The use of meta-analytical tools in risk assessment for food safety. Food Microbiol. 28 (4), 823–827.

Hassan, S., Singh, A.V., 2014. Biophysicochemical perspective of nanoparticle compatibility: a critically ignored parameter in nanomedicine. J. Nanosci. Nanotechnol. 14 (1), 402–414.

Higashisaka, K., Yoshioka, Y., Tsutsumi, Y., 2015. Applications and safety of nanomaterials used in the food industry. Food Safety 3 (2), 39–47.

Hwang, M., Lee, E.J., Kweon, S.Y., Park, M.S., Jeong, J.Y., Um, J.H., Kim, S.A., Han, B.S., Lee, K.H., Yoon, H.J., 2012. Risk assessment principle for engineered nanotechnology in food and drug. Toxicol. Res. 28 (2), 73–79.

Jacxsens, L., Luning, P., Marcelis, W., van Boekel, T., Rovira, J., Oses, S., Kousta, M., Drosinos, E., Jasson, V., Uyttendaele, M., 2011. Tools for the performance assessment and improvement of food safety management systems. Trends Food Sci. Technol. 22, S80–S89.

Jain, S., White, M., Radivojac, P., 2016. Estimating the class prior and posterior from noisy positives and unlabeled data. Adv. Neural Info. Process. Syst. 29, 2693–2701.

Joerger, R.D., 2007. Antimicrobial films for food applications: a quantitative analysis of their effectiveness. Packag. Technol. Sci. Int. J. 20 (4), 231–273.

Kandlikar, M., Ramachandran, G., Maynard, A., Murdock, B., Toscano, W.A., 2007. Health risk assessment for nanoparticles: a case for using expert judgment. J. Nanopart. Res. 9 (1), 137–156.

Koo, O.M., Rubinstein, I., Onyuksel, H., 2005. Role of nanotechnology in targeted drug delivery and imaging: a concise review. Nanomed. Nanotechnol. Biol. Med. 1 (3), 193–212.

Kunjachan, S., Jose, S., Lammers, T., 2014. Understanding the mechanism of ionic gelation for synthesis of chitosan nanoparticles using qualitative techniques. Asian J. Pharmaceut. 4 (2). https://doi.org/10.22377/ajp.v4i2.220.

Lammerding, A.M., Fazil, A., 2000. Hazard identification and exposure assessment for microbial food safety risk assessment. Int. J. Food Microbiol. 58 (3), 147–157.

Lamprecht, A., Saumet, J.-L., Roux, J., Benoit, J.-P., 2004. Lipid nanocarriers as drug delivery system for ibuprofen in pain treatment. Int. J. Pharm. 278 (2), 407–414.

Langer, R., Peppas, N.A., 2003. Advances in biomaterials, drug delivery, and bionanotechnology. AIChE J. 49 (12), 2990–3006.

Leceta, I., Guerrero, P., Cabezudo, S., de la Caba, K., 2013. Environmental assessment of chitosan-based films. J. Cleaner Prod. 41, 312–318.

Lee, S.-W., Mao, C., Flynn, C.E., Belcher, A.M., 2002. Ordering of quantum dots using genetically engineered viruses. Science 296 (5569), 892–895.

Malmiri, H.J., Jahanian, M.A.G., Berenjian, A., 2012. Potential applications of chitosan nanoparticles as novel support in enzyme immobilization. Am. J. Biochem. Biotechnol. 8 (4), 203–219.

Martins, A.F., de Oliveira, D.M., Pereira, A.G., Rubira, A.F., Muniz, E.C., 2012. Chitosan/TPP microparticles obtained by microemulsion method applied in controlled release of heparin. Int. J. Biol. Macromol. 51 (5), 1127–1133.

McEntire, J., Boateng, A., 2012. Industry challenge to best practice risk communication. J. Food Sci. 77 (4), R111–R117.

Mills, A., Hazafy, D., 2009. Nanocrystalline SnO_2-based, UVB-activated, colourimetric oxygen indicator. Sens. Actuat. B 136 (2), 344–349.

Nakagawa, K., 2014. Nano-and microencapsulation of flavor in food systems. In: Nano- and Microencapsulation for Foods. Wiley Online Library, pp. 249–271.

Nauta, M., Lindqvist, R., Zwietering, M., 2012. Tools for microbiological risk assessment. ILSI, Europe, p. 40, ISBN: 9789078637349.

Neethirajan, S., Jayas, D.S., 2011. Nanotechnology for the food and bioprocessing industries. Food Bioprocess Technol. 4 (1), 39–47.

Nikmaram, N., Roohinejad, S., Hashemi, S., Koubaa, M., Barba, F.J., Abbaspourrad, A., Greiner, R., 2017. Emulsion-based systems for fabrication of electrospun nanofibers: food, pharmaceutical and biomedical applications. RSC Adv. 7 (46), 28951–28964.

Nogrady, B., 2021. How nanotechnology can flick the immunity switch. Nature 595 (7865), 18–19.

Ozturk, B., Argin, S., Ozilgen, M., McClements, D.J., 2015. Formation and stabilization of nanoemulsion-based vitamin E delivery systems using natural biopolymers: whey protein isolate and gum arabic. Food Chem. 188, 256–263.

Pradhan, N., Singh, S., Ojha, N., Shrivastava, A., Barla, A., Rai, V., Bose, S., 2015. Facets of nanotechnology as seen in food processing, packaging, and preservation industry. Biomed. Res. Int. 2015. https://doi.org/10.1155/2015/365672.

Qureshi, M., Karthikeyan, S., Karthikeyan, P., Khan, P., Uprit, S., Mishra, U., 2012. Application of nanotechnology in food and dairy processing: an overview. Pak. J. Food Sci. 22 (1), 23–31.

Rashidi, L., Khosravi-Darani, K., 2011. The applications of nanotechnology in food industry. Crit. Rev. Food Sci. Nutr. 51 (8), 723–730.

Ravichandran, R., 2010. Nanotechnology applications in food and food processing: innovative green approaches, opportunities and uncertainties for global market. Int. J. Green Nanotechnol. Phys. Chem. 1 (2), P72–P96.

Rhim, J.-W., Park, H.-M., Ha, C.-S., 2013. Bio-nanocomposites for food packaging applications. Prog. Polym. Sci. 38 (10-11), 1629–1652.

Saleh, S.R., Abady, M.M., Nofal, M., Yassa, N.W., Abdel-Latif, M.S., Nounou, M.I., Ghareeb, D.A., Abdel-Monaem, N, 2021. Berberine nanoencapsulation attenuates hallmarks of scoplomine induced Alzheimer's-like disease in rats. Curr. Rev. Clin. Exp. Pharmacol. 16 (2), 139–154.

Samimi, A., 2020. Risk management in information technology. Prog. Chem. Biochem. Res. 3 (2), 130–134.

Samimi, A., Rajeev, P., Bagheri, A., Nazari, A., Sanjayan, J., Amosoltani, A., Esfahani, M.T., Zarinabadi, S., 2019. Use of data mining in the corrosion classification of pipelines in Naphtha Hydro-Threating Unit (NHT). Pipeline Sci. Technol. 3 (1), 14–21.

Sancho, A., Mills, E., 2010. Proteomic approaches for qualitative and quantitative characterisation of food allergens. Regul. Toxicol. Pharm. 58 (3), S42–S46.

Sastry, R.K., Rao, N., 2013. Emerging technologies for enhancing Indian agriculture-case of nanobiotechnology. Asian Biotechnol. Dev. Rev. 15 (1), 1–19.

Sawai, J., 2003. Quantitative evaluation of antibacterial activities of metallic oxide powders (ZnO, MgO and CaO) by conductimetric assay. J. Microbiol. Methods 54 (2), 177–182.

Sekhon, B.S., 2010. Food nanotechnology: an overview. Nanotechnol. Sci. Appl. 3, 1–15.

Sekhon, B.S., 2014. Nanotechnology in agri-food production: an overview. Nanotechnol. Sci. Appl. 7, 31.

Serrano, L., 2007. Synthetic biology: promises and challenges. Mol. Syst. Biol. 3 (1), 158.

Sharma, V., Shukla, R.K., Saxena, N., Parmar, D., Das, M., Dhawan, A., 2009. DNA damaging potential of zinc oxide nanoparticles in human epidermal cells. Toxicol. Lett. 185 (3), 211–218.

Singh, A., Ferri, M., Tamplenizza, M., Borghi, F., Divitini, G., Ducati, C., Lenardi, C., Piazzoni, C., Merlini, M., Podestà, A., 2012. Bottom-up engineering of the surface roughness of nanostructured cubic zirconia to control cell adhesion. Nanotechnology 23 (47), 475101.

Singh, A.V., Gemmati, D., Anurag, K., Ishan, P., Vatsala, M., Vimal, K., Timotheus, J., Joachim, B., 2018. Nanobiomaterials for vascular biology and wound management: a review. Veins Lymph. 7 (1). https://doi.org/10.4081/vl.2018.7196.

Singh, A.V., Jahnke, T., Wang, S., Xiao, Y., Alapan, Y., Kharratian, S., Onbasli, M.C., Kozielski, K., David, H., Richter, G., 2018. Anisotropic gold nanostructures: optimization via in silico modeling for hyperthermia. ACS Appl. Nano Mater. 1 (11), 6205–6216.

Singh, A.V., Maheshwari, S., Giovanni, D., Naikmasur, V.G., Rai, A., Aditi, A., Gade, W., Vyas, V., Gemmati, D., Zeri, G., 2010. Nanoengineering approaches to design advanced dental materials for clinical applications. J. Bionanosci. 4 (1-2), 53–65.

Singh, A.V., Mehta, K.K., 2016. Top-down versus bottom-up nanoengineering routes to design advanced oropharmacological products. Curr. Pharm. Des. 22 (11), 1534–1545.

Singh, A.V., Mehta, K.K., Worley, K., Dordick, J.S., Kane, R.S., Wan, L.Q., 2014. Carbon nanotube-induced loss of multicellular chirality on micropatterned substrate is mediated by oxidative stress. ACS Nano 8 (3), 2196–2205.

Singh, T., Shukla, S., Kumar, P., Wahla, V., Bajpai, V.K., Rather, I.A., 2017. Application of nanotechnology in food science: perception and overview. Front. Microbio. 8, 1501.

Smithson, C., Simkins, B.J., 2005. Does risk management add value? A survey of the evidence. J. Appl. Corporate Finance 17 (3), 8–17.

Sonkaria, S., Ahn, S.-H., Khare, V., 2012. Nanotechnology and its impact on food and nutrition: a review. Recent Patents Food Nutr. Agric. 4 (1), 8–18.

Stamm, H., 2011. Nanomaterials should be defined. Nature 476 (7361), 399.

Suppakul, P., Miltz, J., Sonneveld, K., Bigger, S.W., 2003. Active packaging technologies with an emphasis on antimicrobial packaging and its applications. J. Food Sci. 68 (2), 408–420.

Thangavel, G., Thiruvengadam, S., 2014. Nanotechnology in food industry: a review. Int. J. Chem. Tech. Res. 6 (9), 4096–4101.

Tisato, V., Zuliani, G., Vigliano, M., Longo, G., Franchini, E., Secchiero, P., Zauli, G., Paraboschi, E.M., Vikram Singh, A., Serino, M.L., 2018. Gene-gene interactions among coding genes of iron-homeostasis proteins and APOE-alleles in cognitive impairment diseases. PLoS One 13 (3), e0193867.

Torney, F., Trewyn, B.G., Lin, V.S.-Y., Wang, K., 2007. Mesoporous silica nanoparticles deliver DNA and chemicals into plants. Nat. Nanotechnol. 2 (5), 295–300.

Ubbink, J., Krüger, J., 2006. Physical approaches for the delivery of active ingredients in foods. Trends Food Sci. Technol. 17 (5), 244–254.

Uttara, B., Singh, A.V., Zamboni, P., Mahajan, R., 2009. Oxidative stress and neurodegenerative diseases: a review of upstream and downstream antioxidant therapeutic options. Curr. Neuropharmacol. 7 (1), 65–74.

Vaezifar, S., Razavi, S., Golozar, M.A., Karbasi, S., Morshed, M., Kamali, M., 2013. Effects of some parameters on particle size distribution of chitosan nanoparticles prepared by ionic gelation method. J. Cluster Sci. 24 (3), 891–903.

Voysey, P., Brown, M., 2000. Microbiological risk assessment: a new approach to food safety control. Int. J. Food Microbiol. 58 (3), 173–179.

Wang, Y., Zhang, Q., Zhang, C.-l., Li, P., 2012. Characterisation and cooperative antimicrobial properties of chitosan/nano-ZnO composite nanofibrous membranes. Food Chem. 132 (1), 419–427.

Weiss, J., Takhistov, P., McClements, D.J., 2006. Functional materials in food nanotechnology. J. Food Sci. 71 (9), R107–R116.

WHO, 2008. The International Food Safety Authorities Network (INFOSAN). http://www.who.int./foodsafety/fs_management/infosan/en/.

WHO, 2010. FAO/WHO Expert Meeting on the Application of Nanotechnologies in the Food and Agriculture Sectors: Potential Food Safety Implications: Meeting Report. World Health Organization.

Williams, P.R., Paustenbach, D.J., 2002. Risk characterization: principles and practice. J. Toxicol. Environ. Health Part B: Critic. Rev. 5 (4), 337–406.

Wooldridge, M., 2007. Qualitative risk assessment. In: Microbial Risk Analysis of Foods. Wiley, pp. 1–28.

Yan, S.S., Gilbert, J.M., 2004. Antimicrobial drug delivery in food animals and microbial food safety concerns: an overview of in vitro and in vivo factors potentially affecting the animal gut microflora. Adv. Drug Deliv. Rev. 56 (10), 1497–1521.

Yang, R., Zhou, Z., Sun, G., Gao, Y., Xu, J., Strappe, P., Blanchard, C., Cheng, Y., Ding, X., 2015. Synthesis of homogeneous protein-stabilized rutin nanodispersions by reversible assembly of soybean (Glycine max) seed ferritin. RSC Adv. 5 (40), 31533–31540.

Yuryevich, D.A., Yakovlevich, K.M., Mikhaylovich, K.P., 2014. Optimization of finances into regional energy. Econ. Reg. 2014 (2), 248–254. https://doi.org/10.17059/2014-2-24.

Zhang, T., Lv, C., Chen, L., Bai, G., Zhao, G., Xu, C., 2014. Encapsulation of anthocyanin molecules within a ferritin nanocage increases their stability and cell uptake efficiency. Food Res. Int. 62, 183–192.

Nanobiotechnological utility for the removal of food contaminants: Physicobiochemical

5

Xiaoyi Liu[a], K. M. Faridul Hasan[b] and Shaofeng Wei[a]

[a] *School of Public Health, The Key Laboratory of Environmental Pollution Monitoring and Disease Control, Ministry of Education, Guizhou Medical University, Guiyang, China,* [b] *Simonyi Károly Faculty of Engineering, University of Sopron, Sopron, Hungary*

5.1 Introduction

As one of the main sources of food safety problems, food contaminants seriously threaten people's health, it means that the food itself does not contain toxic or harmful substances. However, due to human factors or environmental influences, food may be contaminated by various toxic and harmful substances at all stages from planting, and growth, to consumption, resulting in a decline in the hygiene quality and nutritional value of the food. These substances that cause food contamination are called food contaminants in general. At present, according to different sources, people habitually divide food contaminants into three categories: biological contaminants (Multari et al., 2013; Lim and Kim, 2016), chemical contaminants (Schrenk, 2004; Hird et al., 2014), and physical contaminants (Nie et al., 2011; Carbery et al., 2018). Biological contaminants mainly include microorganisms, parasites, insects, and viruses; chemical contaminants mainly include pesticides, veterinary drugs, and toxic metals; and physical pollution mainly includes radioactive pollution and the addition of unnecessary sand and grass seeds to food (see Fig. 5.1 for details).

At present, the discovery of efficient and economical technology to remove food contaminants is one of the hot issues that the food science community is paying attention to. In this case, the rapid development of nanobiotechnology has provided a good theoretical basis and technical support for the innovation of food contaminant removal methods, and has brought revolutionary contributions to materials science and food science (Bhople et al., 2016; Verma, 2017; Kaur et al., 2020). Nanobiotechnology is mainly used to develop and use materials and systems by manipulating matter at the nano-level (that is, at the level of atoms, molecules, and supramolecular structures) (Morrow et al., 2007). It is a cutting-edge technology in the international biological field, it is a technology that combines nanotechnology and biology and

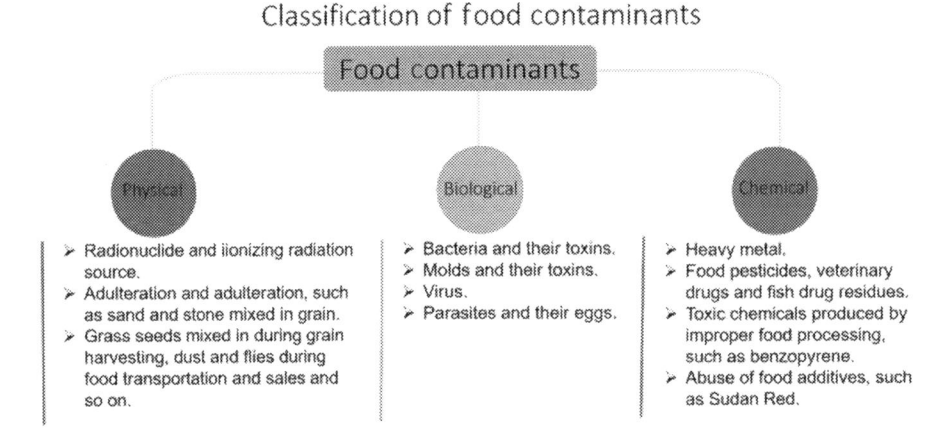

Classification of food contaminants

Food contaminants

Physical
- Radionuclide and iionizing radiation source.
- Adulteration and adulteration, such as sand and stone mixed in grain.
- Grass seeds mixed in during grain harvesting, dust and flies during food transportation and sales and so on.

Biological
- Bacteria and their toxins.
- Molds and their toxins.
- Virus.
- Parasites and their eggs.

Chemical
- Heavy metal.
- Food pesticides, veterinary drugs and fish drug residues.
- Toxic chemicals produced by improper food processing, such as benzopyrene.
- Abuse of food additives, such as Sudan Red.

FIGURE 5.1

Classification of different food contaminants.

is mainly used to study life phenomena (Jafarizadeh-Malmiri et al., 2019a). The development of nanobiotechnology depends to a large extent on the development of nanomaterials science. Nanomaterials are divided into nanoparticles, nanofilms, and nanosolids. Studies have shown that the influence of nanomaterials on nanobiotechnology has far-reaching significance (Sobha et al., 2010; Crean Neé Lynam et al., 2011). Different nanomaterials target food contaminants and play different influential roles. For example, nanofilms can selectively permeate oxygen and enzymes when used as packaging materials to inhibit the growth of microorganisms (Jafarzadeh et al., 2019). Silicate nanoparticles, as well as metal nanoparticles, can remove pathogens from food by reducing oxygen flow and moisture leakage in food packaging containers (Horner et al., 2006). Therefore, this chapter begins with nanomaterials to describe nanobiotechnology and its application in the inhibition or removal of food contaminants from physical and biochemical perspectives.

5.2 **Nanomaterials**

5.2.1 **Nanoparticle**

Nanoparticle is a typical representative element of the nanosystem, which belongs to the range of ultrafine particles and has a diameter from 1 nm to 1000 nm (Chandrasekaran et al., 2011; de Morais et al., 2014; He et al., 2019; Hasan et al., 2021a). So far, nanoparticles can be divided into two types according to their composition, namely organic and inorganic nanoparticles (Moghtaderi and Salehi-Abargouei, 2018). Organic nanoparticles can be divided into the following three types, protein-based, carbohydrate-based, and lipid-based. They are widely used in food science (Shin et al., 2015). Inorganic nanoparticles are mainly composed of metals including nanoparticles into metallic nanoparticles, magnetic nanoparticles, quantum dots, and silver and silica nanoparticles (He and Hwang, 2016; Lee et al., 2021). Due to the tiny size of nanoparticles, quantum size effect, and other reasons, these factors all determine the final physical and biochemical properties of nanoparticles, resulting in new properties that are different from conventional solids (Ge et al., 2014). For example, when the size is reduced to tens of nanometers, the original metal conductor will become an insulator, and the original insulator will become a conductor due to its greatly reduced resistance (Stark et al., 2015). Generally speaking, the physical properties of conventional solids are relatively stable under certain conditions, while the magic number effect appears in their properties in the nanometer state (Narouz et al., 2019). From the perspective of technical application, the effect of nanoparticles makes it have huge application prospects in food packaging and preservation, fuel, magnetic recording, radar wave invisibility, and gas-sensing (Zheng et al., 2015).

5.2.2 Nanofilm

Nanofilm refers to a film composed of nanoparticles or a single-layer or multilayer film with a thickness in the nanometer range. Because it has about 50% of the interface components, the nanofilm shows a difference from crystalline and amorphous substances' physicobiochemical properties (Zheng et al., 2013). For example, nanofilm has the advantages of good thermal stability and high doping effect. In view of these characteristics, nanofilm can play an important role in electronic devices such as piezoresistive sensors and opto-electromagnetic devices, based on the performance of nano-membrane, nano-membrane has been widely used in the manufacturing of bio-pharmaceuticals, production electronic component packaging material network cable products, and so on (Huang et al., 2008; Takahashi et al., 2012; Zheng et al., 2013; Wan et al., 2021). In the food field, nanofilms can also be used to make antibacterial food packaging materials (Jafarzadeh et al., 2019).

5.3 Utilization of nanobiotechnology

The research fields of nanobiotechnology mainly focus on nano-related biological materials and biological devices, which are mainly reflected in: (1) The application of nanomaterials in biology, including nanoinorganic biological materials, nanopolymer biological materials, and nanocomposite biological materials; (2) Apply nanotechnology to observe biomolecules, further manipulate biomolecules, and make nanobiological devices (Jingming, 2007). Based on the research scope of nanobiotechnology, nowadays, nanobiotechnology is mainly used in the food field and other fields such as military, marine transportation, and so on.

5.3.1 Nanobiotechnology utilization in food sector

Nanobiotechnology is a relatively new field in the food field. However, some of the merits and demerits are summarized in Table 5.1. It has revolutionized the food

Table 5.1 Merits and demerits of nanomaterials for food technology.

Nanomaterials	Name and application	Merits	Demerits	References
Metallic nanoparticles	Au nanoparticles: (1) Detection of aflatoxins (2) Detection of oxytetracycline in milk.	Simple and cost-effective	Nano-Au sol has poor stability and is easy to form agglomerates	Ahmadi et al. (2022), Blidar et al. (2022), Mechouche et al. (2022)
	Ag nanoparticles: resistant to Streptomyces aureus	Strong pertinence	Colloidal silver accumulates in the body to produce silver deposits	Lugani et al. (2021), Mechouche et al. (2022)
	Copper nanoclusters: detection of glutathione	High sensitivity	Lively and unstable	Li et al. (2022)
Magnetic nanoparticles	Coated magnetic nanoparticles: capture and detect hepatitis A virus from solid food	Efficient and sensitive	Magnetic nanoparticles lose their stable magnetic order when they are smaller than a certain size	Assa et al. (2016), Wu et al. (2022)

(continued on next page)

Quantum dots	Nitrogen-sulfur co-doped carbon dots: nitrogen and Sulfur synthesize carbon dots, and develop fluorescent probes to detect glutathione in vegetables	Simple	The synthesis conditions are harsh, the water solubility is not good, it is easy to aggregate and has toxic side effects to the organism, and it is not easy to surface modification	Lugani et al. (2021), Sun et al. (2022)
Silica nanoparticles	Silica nanoparticle film: the film containing silica nanoparticles can be used as a new type of antioxidant food packaging composite film	The mechanical properties, antioxidant activity, water vapor barrier properties, and UV barrier capabilities have been significantly improved	Absorb antibiotic protein	Dong et al. (2022)
Chitosan nanoparticles	Chitosan nanoparticles: has broad-spectrum antibacterial, antifungal activity, and antiviral	Wide range of antibacterial		Beyth et al. (2015), Lugani et al. (2021)

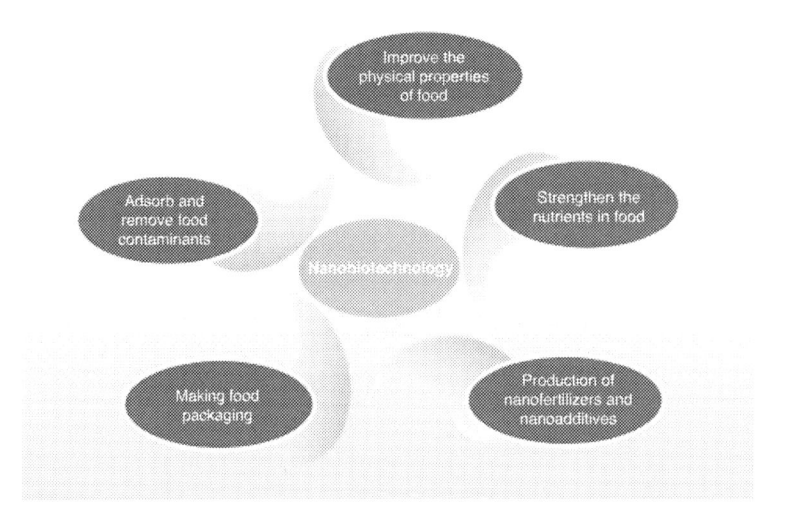

FIGURE 5.2

Utilization of nanobiotechnology in the food sector.

industry in the process of food processing, packaging, storage, and the development of innovative products (Jafarizadeh-Malmiri et al., 2019b; Cheba, 2020). For different nanoparticles, including gold, titanium dioxide, quantum dots, nanorods, etc., due to their properties: (1) their antibacterial and antipathogenic (Patra and Baek, 2016); (2) nanoparticles can interact with microbial cell membranes through electrostatic action (Nowak et al., 2016); and (3) nanoparticles can be accumulated in cytoplasm and other organelles (Angélique et al., 2017). These characteristics make nanomaterials play an important role in the food field. For example, nanobiotechnology can use nanoparticles for food supplements. Controlled release (Ogończyk et al., 2011), antibacterial, extends the shelf life of food, thereby improving the stability of food flavor and texture (Sugumar et al., 2016; Hosseini and Jafari, 2020). Studies have shown that nanobiotechnology can track and detect pathogens through nanotracers and nanosensors, and act on it (Ali et al., 2020). In addition, nanobiotechnology is also used in food processing and delivery, keeping them fresh, nanofertilizers, nanoadditives, smart packaging, and so on (Özer et al., 2014; Brandelli et al., 2016; Chaudhry et al., 2018; Lopes and Brandelli, 2018), all require the nanobiotechnology. Therefore, nanobiotechnology can be used in the development of food industry resources to create and increase global wealth and market value. The nanobiotechnology utilization in the food sector is shown in Fig. 5.2.

5.3.2 Nanobiotechnology utilization in other sectors

According to recent research (Fakruddin et al., 2012), nanobiotechnology also involves physics, quantum science, materials science, and comprehensive disciplines such as medicine have a wide range of applications and clear industrialization prospects in the field of medicine and health (Dash et al., 2020). For instance, the

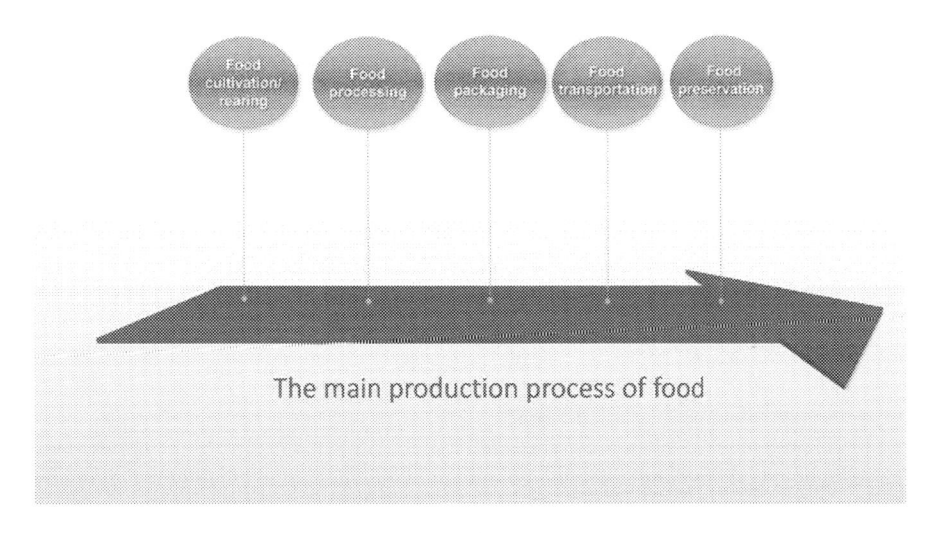

The main production process of food

FIGURE 5.3

Different food production processes.

advanced biocomposite materials produced by the application of nanobiotechnology have been used in various fields such as automobiles, biomedicine, aerospace and navigation, and national defense (Abdelrahman et al., 2020; Hasan, K.F. et al., 2021; Hasan et al., 2021b; Hasan et al., 2021c; Hasan, K.M.F. et al., 2021). Dinnyes et al. (2020) use nanobiotechnology for β-cell and islet transplantation to treat type 1 diabetes, the results show nanobiotechnology can perform high-resolution imaging of stem cell-derived beta cell function to test advanced therapeutics. Research by Linxuan Che and others found that nanomanganese dioxide-modified activated carbon can trigger the electron transfer pathway to change and promote the anaerobic treatment of printing and dyeing wastewater (Che et al., 2022). Lorena García-Hevia and other studies have found that magnetic lipid nanocarriers can synergistically control the heat release of chemotherapeutic drugs through magnetic ablation, and at the same time, noninvasively monitor the treatment and diagnosis of melanoma through magnetic resonance imaging (MRI). In addition, Nxele et al. (2022) found that the coupling of Au nanoparticles and alkynyl Co (II) phthalocyanine has a synergistic effect on the detection of prostate-specific antigens.

5.4 Nanobiotechnology for food contaminants removal

China is a big country in the production and consumption of animals and plants food. The production process of food includes planting/feeding of food raw materials (animals and plants), processing, packaging, transportation, and storage, the main production process of food is shown in Fig. 5.3. Among them, each link may be exposed to food contaminants and become contaminated. Numerous studies have shown that different forms of nanobiotechnology products, such as nanosensors, nanocapsules, nanotubes, and so on, have been used in food processing, packaging, and preservation (Duncan, 2011; Bajpai et al., 2018; Lugani et al., 2021).

In terms of combating food contaminants, nanobiotechnology has a certain role in food planting, harvesting, processing, packaging, and preservation. For example, Xiang and other studies have shown that iron sulfide nanoparticles can reduce Hg (II) methylation and MeHg bioavailability, reduce Hg pollution, and thereby, reduce the risk of exposure to aquatic organisms such as fish (Xiang et al., 2022). Sardar et al. (2022) confirmed that selenium nanoparticles reduce the absorption of cadmium by coriander under cadmium toxic conditions and regulate its nutritional homeostasis and antioxidant system. Moradi et al. (2021) found that C-dots synthesized by adding nanoparticles into packaging materials can be used as antibacterial agents to inhibit the growth of disease-causing and spoilage microorganisms in food.

5.4.1 Removal of contaminants in the food cultivation/rearing process

In recent years, organic (synthetic) pesticides and chemical fertilizers have been widely used in food cultivation, resulting in food pesticide residues, soil pollution, and plant food contamination (Fang, 2006). In addition, veterinary drug residues caused by the use of veterinary drugs in animal breeding are also one of the main sources of animal food contamination. In this case, nanobiotechnology has become a valid choice in the removal of pollutants generated in the process of food cultivation/rearing.

5.4.1.1 Application of nanobiotechnology in pesticides

In the process of plant cultivation, pests are a major risk factor that is unavoidable. Common pesticides often have shortcomings such as long half-life and easy residue, and nanoinsecticides can overcome these shortcomings. Nanoinsecticide is a product produced by nanobiotechnology, it not only has a good effect on pests but is also easily absorbed by plants, which helps to control agricultural pests and improve crop safety and yield (Jing et al., 2020). In addition, metallic nanoparticles are considered to be one of the most promising nanoparticles in the field of biomedicine and agricultural research (Hussain et al., 2016). Among metallic nanomaterials, nanosilver (Ag NPs) is the most intensively studied. Studies have confirmed that nanosilver synthesized from biological sources has insecticidal activity (Liguo, 2019; Xiaojun et al., 2021). For nonmetallic nanoparticles, they are widely used in nanopesticides, nanoherbicides, and nanofertilizers. For example, silicon nanoparticles and silica nanoparticles, which are used as nanopesticides, can be directly applied to the field and play the role of pesticides (Ziaee and Ganji, 2016; Rastogi et al., 2019).

In the nanometer range, the smaller size enhances the penetration of plant cell walls and cuticles, thereby increasing the uptake of organic matter. For example, pesticide microcapsule formulations can effectively improve the utilization rate of pesticides and reduce residues and pollution through their advantages of controlled release, attenuation, and resistance to photolysis (Anqi et al., 2018). Nano pesticide-controlled release agents can effectively reduce pesticide residues in the environment with less side effects (Wang et al., 2021). Organic pesticides are effectively degraded under the combined action of semiconductor nanoparticles and photocatalytic systems (Hongyun et al., 2017). Through the combination of existing agricultural active ingredients and chitosan nanocarriers, chitosan-based agricultural nanochemicals can be used as effective fungicides for plant pathogens and pests (Maluin and Hussein, 2020).

5.4.1.2 Application of nanobiotechnology in veterinary drugs

Intracellular infectious diseases are one of the global diseases that endanger the health and life of animals and humans. Biodegradable nanoparticles have become effective carriers for targeted intracellular delivery of antibiotics due to their unique advantages such as surface effect, quantum size effect, and small size effect, which can be used to treat infectious diseases in livestock and poultry cells (Shuyu et al., 2015).

5.4.1.3 Application of nanobiotechnology in the remediation of soil heavy metal and metalloid pollution

During the process of plant growth, various nutrients are basically supplied by the soil, and plants which can enrich heavy metals in the soil through their roots. Therefore, the problem of heavy metals in soil has always been one of the urgent problems to be solved. Nanobiotechnology provides the possibility to solve the problem of heavy metal pollution in soil. Some studies have indicated that nanozeolite can adsorb and fix a large amount of Cd^{2+}, reducing the adsorption of heavy metal cadmium by plants. At the same time, potassium, calcium, magnesium, silicon, and other elements in zeolite can improve soil nutrients and promote plant growth. Nanohydroxyapatite can significantly improve the removal rate of metals such as lead, nickel, and zinc in soil, reduce the absorption of heavy metals by crops (Weili et al., 2018), and improve the soil microbial community (Chuanbao et al., 2010). In addition, nano-metals and their oxides also have a good remediation effect on heavy metal-contaminated soils. Nano-iron and its oxides are also important passivators for the remediation of heavy metal pollution in soils (Huang et al., 2020), and have good remediation effects on chromium-contaminated soils (Wei et al., 2019).

5.5 Nanobiotechnology removes pollutants in the food industry chain

5.5.1 Remove contaminants from food processing

It is difficult to avoid contact with bacteria during food processing. It is also very important to find a way to remove bacteria exposed during food processing. Studies have shown that silver nanoparticles and silver colloids can inhibit the bacterial composition of the microbiome of wheat and rye cereal crops (Suvorov et al., 2017). Through further experiments, the specific bacteriostatic effect of silver nanoparticles and the effective destruction of bacterial contamination in food were further demonstrated. Through nanobiotechnology, problems in food processing can be effectively solved to produce high-quality products. Furthermore, recent developments in nanotechnology have demonstrated that effective antimicrobial packaging strategies can significantly inhibit food contamination, thereby increasing the sustainability of food production.

5.5.2 Remove contaminants during food packaging and transportation

Food packaging is an important means of preventing food from being contaminated during transportation. Healthy shipping protects and preserves those foods that produce any unacceptable changes in quality until they reach the consumer. Mycotoxin contamination is one of the most common types of contamination during food transportation (Janssen et al., 2014). The growth of mold and the production of mycotoxins can be prevented and inhibited by controlling environmental conditions. Nanoparticles synthesized based on green models are one of the best-emerging technologies in food packaging. Nanoparticles can improve the mechanical strength, water and oxygen barrier properties, and light-shielding properties of packaging. Nanoparticles are stable, beneficial, and environment-friendly (Jafarzadeh and Jafari, 2021). The applications of nanomaterials in food packaging can be divided into six categories: nanocomposites, nanocoatings, surface fungicides, active packaging, smart packaging, and bioplastics (Tan et al., 2019; Kumar et al., 2021). Metal-based nanoparticles also have potential effects on a variety of pathogens. Antibacterial, antioxidant, and antifungal activities (Dos et al., 2020), copper nanoparticles (Cu-NPs), and copper oxide nanoparticles (CuO-NPs) are the two main active agent packaging materials used in food. CuO-NPs can significantly inhibit the growth of fungi, bacteria, and viruses in food packaging, and CuO-NPs have shown antibacterial activity against gram-positive and gram-negative bacteria in all biopackaging systems. Traditional freshness preservation methods and food packaging do not allow real-time monitoring of food quality during storage and transportation, but smart food packaging is able to use indicators and sensors to detect physiological changes in food (due to microbial and chemical degradation) during shelf life of these indicators information, such as the freshness of packaged products, is often provided by color changes and can be easily identified by food distributors and consumers alike (Alfei et al., 2020; Rodrigues et al., 2021).

5.5.3 Removal of contaminants during food preservation

Meat food and its products are rich in protein, lipids, and high in water content, and are the "natural culture medium" for the growth of microorganisms (Li and Tang, 2021). Nowadays, in order to prevent meat products from being contaminated by bacteria, the commonly used preservation method is cold chain storage. *Listeria monocytogenes* is one of the main pathogenic bacteria of frozen meat products. Xuejing's (2016) research shows that the stimulated release of cardamom essential oil nanoliposomes can be targeted to remove *L. monocytogenes*. Dan et al. (2021) confirmed that the composite nanopackaging film materials and coating materials have good antibacterial and bacteriostatic effects in animal food and military food and achieve the effect of preservation. For plant-based foods, the high spoilage rates of fruits and vegetables pose a major challenge to their storage, leading to their nutrient loss and degradation and the spread of microorganisms, as well as the spread of harmful microorganisms that pose a serious threat to human health. Research has shown that innovative technologies such as edible coatings and films and air-conditioned packaging are considered possible solutions to maintain the quality of produce during storage and shelf life (Xing et al., 2019). Das et al. (2021) showed that linalool-containing chitosan nanocomposites exhibited significant antifungal and antiaflatoxin activities in in vitro and in situ studies, indicating that they can be used as a new environmental-friendly ecological smart preservative in the food industry in practical application.

5.6 Worldwide concerns on nanoparticles in food sector

Nanobiotechnology has received extensive attention in the global food industry. Nowadays, many countries, led by the United States, have devoted themselves to nanobiotechnology research. The research and development institutions in the food field in the United States include the National Institutes of Health (NIH), the Center for Nanoscale Science and Technology (CNST), the National Nanotechnology Initiative (NNI), and the United States Department of Agriculture (USDA). Europe includes the Biotechnology and Biological Sciences Research Council (BBSRC), the European Commission (EC), and the Medical Research Council (MRC) in the field of nanobiotechnology. India's funding agencies for funding innovation in nanotechnology are Council for Scientific and Industrial Research (CSIR), Defense Research Development and Organization (DRDO), Indian Council for Medical Research (ICMR), and so on (McClements and Rao, 2011). Nowadays, the research contents of various countries include but are not limited to the following points:

- Nanobiotechnology to manufacture food.
- Nanofood packaging materials.
- Using nanotechnology to monitor food and nanofood labels.
- Nanofood additives.
- Application of nanotechnology to transport active substances.

5.7 Conclusion and future recommendations

Today, nanobiotechnology is one of the emerging fields in the food field. A big challenge for the food industry today is to find technologies to remove food contaminants and provide healthier and safer food for humans. Nanobiotechnology can not only improve the quality and value of food, but also develop new products for food that are resistant to toxin-producing fungi and mycotoxins, pesticides, reduce heavy metal bioaccumulation, and antipollutant food packaging products. Therefore, nanobiotechnology is another opportunity to overcome global problems and help to provide a basis for future research. However, nanobiotechnology is still in its early stages, and despite its advantages, its possible impact on human health is still not clearly understood. Therefore, the improvement and commercialization of nanomaterials require a very technical market development and education with a view to producing new, green, and safe nanoproducts for removing food contaminants.

References

Abdelrahman, M.S., Nassar, S.H., et al., 2020. Studies of polylactic acid and metal oxide nanoparticles-based composites for multifunctional textile prints. Coatings (Basel) 10 (1), 58.

Ahmadi, S.F., Hojjatoleslamy, M., et al., 2022. Monitoring of Aflatoxin M1 in milk using a novel electrochemicalaptasensorbased on reduced graphene oxide and gold nanoparticles. Food Chem. 373 (Pt A), 131321.

Alfei, S., Marengo, B., et al., 2020. Nanotechnology application in food packaging: a plethora of opportunities versus pending risks assessment and public concerns. Food Res. Int. 137, 109664.

Ali, R., Batool, T., Manzoor, B., et al., 2020. Nanobiotechnology-based drug delivery strategy as a potential weapon against multiple drug-resistant pathogens: sciencedirect. In: Antibiotics and Antimicrobial Resistance Genes in the Environment. Elsevier, pp. 350–368.

Angélique, D., Ducray, A., et al., 2017. Uptake of silica nanoparticles in the brain and effects on neuronal differentiation using different in vitro models: sciencedirect. Nanomed. Nanotechnol. Biol. Med. 3 (13), 1195–1204.

Assa, F., Jafarizadeh-Malmiri, H., et al., 2016. A biotechnological perspective on the application of iron oxide nanoparticles. Nano Res. 9 (8), 2203–2225.

Bajpai, V.K., Kamle, M., et al., 2018. Prospects of using nanotechnology for food preservation, safety, and security. J. Food Drug Anal. 26 (4), 1201–1214.

Beyth, N., Houri-Haddad, Y., et al., 2015. Alternative antimicrobial approach: nano-antimicrobial materials. Evid. Based Complement. Alternat. Med. 2015, 246012.

Bhople, S., Gaikwad, S., et al., 2016. Myxobacteria-mediated synthesis of silver nanoparticles and their impregnation in wrapping paper used for enhancing shelf life of apples. IET Nanobiotechnol. 10 (6), 389–394.

Blidar, A., Hosu, O., et al., 2022. Gold-based nanostructured platforms for oxytetracycline detection from milk by a "signal-on" aptasensing approach. Food Chem. 371, 131127.

Brandelli, A., Brum, L.F.W., et al., 2016. Nanobiotechnology Methods to Incorporate Bioactive Compounds in Food Packaging. Springer International Publishing, Cham, pp. 27–58.

C., M., A., R.C., et al., 2018. Nanotechnology research directions for societal needs in 2020. J. Nanopart. Res. 3 (13), 897–919.

Carbery, M., O'Connor, W., et al., 2018. Trophic transfer of microplastics and mixed contaminants in the marine food web and implications for human health. Environ. Int. 115, 400–409.

Chandrasekaran, A.R., Venugopal, J., et al., 2011. Fabrication of a nanofibrous scaffold with improved bioactivity for culture of human dermal fibroblasts for skin regeneration. Biomed. Mater. 6 (1), 015001.

Chaudhry, N., Dwivedi, S., et al., 2018. Bio-inspired nanomaterials in agriculture and food: current status, foreseen applications and challenges. Microb. Pathog. 123, 196–200.

Che, L., Xu, H., et al., 2022. Activated carbon modified with nano manganese dioxide triggered electron transport pathway changes for boosted anaerobic treatment of dyeing wastewater. Environ. Res. 203, 111944.

Cheba, B.A., 2020. Chitosan: properties, modifications and food nanobiotechnology. Procedia Manuf. 46, 652–658.

Crean Neé Lynam, C., Lahiff, E., et al., 2011. Polyaniline nanofibres as templates for the covalent immobilisation of biomolecules. Synth. Met. 161 (3-4), 285–292.

Das, S., Singh, V.K., et al., 2021. Fabrication, physico-chemical characterization, and bioactivity evaluation of chitosan-linalool composite nano-matrix as innovative controlled release delivery system for food preservation. Int. J. Biol. Macromol. 188, 751–763.

Dash, D., Panik, R., Sahu, A., 2020. Role of nanobiotechnology in drug discovery. Development and Molecular Diagnostic.

de Morais, M.G., Martins, V.G., et al., 2014. Biological applications of nanobiotechnology. J. Nanosci. Nanotechnol. 14 (1), 1007.

Dinnyes, A., Schnur, A., et al., 2020. Integration of nano- and biotechnology for beta-cell and islet transplantation in type-1 diabetes treatment. Cell Prolif. 53 (5), e12785.

Dong, W., Su, J., et al., 2022. Characterization and antioxidant properties of chitosan film incorporated with modified silica nanoparticles as an active food packaging. Food Chem. 373, 131414.

Dos, S.C., Ingle, A.P., et al., 2020. The emerging role of metallic nanoparticles in food. Appl. Microbiol. Biotechnol. 104 (6), 2373–2383.

Du, C., Haiyan, Z., et al., 2010. Remediation of heavy metal contaminated soil by nano-hydroxyapatite and its effect on microbial community structure. Jiangsu Agric. J. 4 (26), 5.

Duncan, T.V., 2011. Applications of nanotechnology in food packaging and food safety: barrier materials, antimicrobials and sensors. J. Colloid Interface Sci. 363 (1), 1–24.

Fakruddin, Md., Hossain, Z., et al., 2012. Retraction—prospects and applications of nanobiotechnology: a medical perspective. J. Nanobiotechnol. 1 (10), 31.

Fang, C., 2006. Food safety during planting and breeding. Chin. J. Food Hyg. (3) 256–258.

Ge, L., Li, Q., et al., 2014. Nanosilver particles in medical applications: synthesis, performance, and toxicity. Int. J. Nanomed. 9, 2399–2407.

Hasan, K.F., Horváth, P.G., Alpár, T., 2021. Silk protein and its nanocomposites. In: Biopolymeric Nanomaterials. Elsevier, Amsterdam, Netherlands, pp. 309–323.

Hasan, K.M.F., Horváth, P.G., et al., 2021a. Coloration of woven glass fabric using biosynthesized silver nanoparticles from Fraxinus excelsior tree flower. Inorg. Chem. Commun. 126, 108477.

Hasan, K.M.F., Horváth, P.G., et al., 2021b. Colorful and facile in situ nanosilver coating on sisal/cotton interwoven fabrics mediated from European larch heartwood. Sci. Rep. 11 (1), 22397.

Hasan, K.M.F., Horváth, P.G., et al., 2021c. Potential fabric-reinforced composites: a comprehensive review. J. Mater. Sci. 56 (26), 14381–14415.

Hasan, K.M.F., Liu, X., et al., 2021. Nanosilver coating on hemp/cotton blended woven fabrics mediated from mammoth pine bark with improved coloration and mechanical properties. J. Text. Inst. 113 (12), 2641–2650.

He, X., Deng, H., et al., 2019. The current application of nanotechnology in food and agriculture. J. Food Drug Anal. 27 (1), 1–21.

He, X., Hwang, H., 2016. Nanotechnology in food science: functionality, applicability, and safety assessment. J. Food Drug Anal. 24 (4), 671–681.

Hird, S.J., Lau, B.P.Y., et al., 2014. Liquid chromatography-mass spectrometry for the determination of chemical contaminants in food. TrAC Trends Anal. Chem. 59, 59–72.

Horner, S.R., Mace, C.R., et al., 2006. A proteomic biosensor for enteropathogenic *E. coli*. Biosens. Bioelectron. 21 (8), 1659–1663.

Hosseini, H., Jafari, S.M., 2020. Introducing nano/microencapsulated bioactive ingredients for extending the shelf-life of food products. Adv. Colloid Interface Sci. 282, 102210.

Huang, H., Pierstorff, E., et al., 2008. Protein-mediated assembly of nanodiamond hydrogels into a biocompatible and biofunctional multilayer nanofilm. ACS Nano 2 (2), 203–212.

Huang, K., Yingjie, S., et al., 2020. Research progress on preparation and remediation of hexavalent chromium polluted soil by biochar-loaded nano-zero valent iron. Environ. Eng. 11 (38), 9.

Hussain, I., Singh, N.B., et al., 2016. Green synthesis of nanoparticles and its potential application. Biotechnol. Lett. 38 (4), 545–560.

Jafarizadeh-Malmiri, H., Sayyar, Z., et al., 2019a. Nanobiotechnology in Food Packaging. Springer, Cham, pp. 69–79.

Jafarizadeh-Malmiri, H., Sayyar, Z., et al., 2019b. Nanobiotechnology in Food: Concepts, Applications and Perspectives. Springer, Cham.

Jafarzadeh, S., Rhim, J.W., et al., 2019. Application of antimicrobial active packaging film made of semolina flour, nano zinc oxide and nano-kaolin to maintain the quality of low-moisture mozzarella cheese during low-temperature storage. J. Sci. Food Agric. 99 (6), 2716–2725.

Jafarzadeh, S., Jafari, S.M., 2021. Impact of metal nanoparticles on the mechanical, barrier, optical and thermal properties of biodegradable food packaging materials. Crit. Rev. Food Sci. Nutr. 61 (16), 2640–2658.

Janssen, S., Pankoke, I., et al., 2014. Two underestimated threats in food transportation: mould and acceleration. Philos. Trans. A Math. Phys. Eng. Sci. 372 (2017), 20130312.

Jingming, T., 2007. Research progress of nano biotechnology. Chem. Technol. Devel. 7 (36), 4.

Kaur, P., Choudhary, R., et al., 2020. Polymer—metal nanocomplexes based delivery system: a boon for agriculture revolution. Curr. Top. Med. Chem. 20 (11), 1009–1028.

Kumar, A., Choudhary, A., et al., 2021. Metal-based nanoparticles, sensors, and their multifaceted application in food packaging. J. Nanobiotechnol. 19 (1), 256.

Lee, K.Y., Pham, X.H., et al., 2021. Introduction of nanobiotechnology. Adv. Exp. Med. Biol. 1309, 1–22.

Li, D., Shunhai, Q., et al., 2021. Application of nanomaterials in marine food preservation. Chin. J. Navigational Med. Hyperbaric Med. 3 (28), 5.

Li, J., Liang, G., et al., 2020. Research progress on absorption and transport of nano-pesticides in plants. Chin. Bull Bot. 4 (55), 16.

Li, H., Tang, R., 2021. Application of gelatin composite coating in pork quality preservation during storage and mechanism of gelatin composite coating on pork flavor. Gels 8 (1), 21.

Li, W., Xin, L., et al., 2019. Experimental study on remediation of cadmium-contaminated paddy soil by bio-nanomaterials for rice cultivation. Southern Agric. 32 (13), 4.

Li, L., Fu, M., et al., 2022. Sensitive detection of glutathione through inhibiting quenching of copper nanoclusters fluorescence. Spectrochim. Acta A Mol. Biomol. Spectrosc. 267 (Pt 1), 120563.

Liguo, W., 2019. Response Surface Method to Optimize the Removal of 2,4-Dichlorophenol from Groundwater by Modified Nano-Zero Valent Iron. Lanzhou Jiaotong University.

Lim, M.C., Kim, Y.R., 2016. Analytical applications of nanomaterials in monitoring biological and chemical contaminants in food. J. Microbiol. Biotechnol. 26 (9), 1505–1516.

Lopes, N.A., Brandelli, A., 2018. Nanostructures for delivery of natural antimicrobials in food. Crit. Rev. Food Sci. Nutr. 13 (58), 2202–2212.

Lugani, Y., Sooch, B.S., Singh, P., Kumar, S., 2021. 8 - Nanobiotechnology applications in food sector and future innovations. In: Ray, R.C., (Ed.), Microbial Biotechnology in Food and Health: Applied Biotechnology Reviews. Academic Press, pp. 197–225, ISBN 9780128198131, https://doi.org/10.1016/B978-0-12-819813-1.00008-6.

Maluin, F.N., Hussein, M.Z., 2020. Chitosan-based agronanochemicals as a sustainable alternative in crop protection. Molecules 25 (7), 1611.

McClements, D.J., Rao, J., 2011. Food-grade nanoemulsions: formulation, fabrication, properties, performance, biological fate, and potential toxicity. Crit. Rev. Food Sci. Nutr. 51 (4), 285–330.

Mechouche, M.S., Merouane, F., et al., 2022. Biosynthesis, characterization, and evaluation of antibacterial and photocatalytic methylene blue dye degradation activities of silver nanoparticles from Streptomyces Tuirus strain. Environ. Res. 204, 112360.

Moghtaderi, F., Salehi-Abargouei, A., 2018. Nanotechnology in food industries: application and safety. J. Environ. Health Sustain. Devel. 3 (3), 551–553.

Moradi, M., Molaei, R., et al., 2021. Carbon dots synthesized from microorganisms and food by-products: active and smart food packaging applications. Crit. Rev. Food Sci. Nutr. 63 (14), 1–17.

Morrow, K.J., Bawa, R., et al., 2007. Recent advances in basic and clinical nanomedicine. Med. Clin. North Am. 91 (5), 805–843.

Multari, R.A., Cremers, D.A., et al., 2013. Detection of biological contaminants on foods and food surfaces using laser-induced breakdown spectroscopy (LIBS). J. Agric. Food Chem. 61 (36), 8687–8694.

Narouz, M.R., Osten, K.M., et al., 2019. N-heterocyclic carbene-functionalized magic-number gold nanoclusters. Nat. Chem. 11 (5), 419–425.

Nie, J., Duan, X., et al., 2011. The pollution levels of PAHs in Chinese food. Epidemiology 22, S90.

Nowak, A., Szade, J., et al., 2016. Physicochemical and antibacterial characterization of ionocity Ag/Cu powder nanoparticles. Mater. Charact. 117, 9–16.

Nxele, S.R., Nkhahle, R., et al., 2022. The synergistic effects of coupling Au nanoparticles with an alkynyl Co(II) phthalocyanine on the detection of prostate specific antigen. Talanta 237, 122948.

Ogończyk, D., Siek, M., et al., 2011. Microfluidic formulation of pectin microbeads for encapsulation and controlled release of nanoparticles. Biomicrofluidics 5 (1), 13405.

Özer, E.A., Özcan, M., et al., 2014. Nanotechnology in Food and Agriculture Industry. Springer, New York, pp. 477–497.

Patra, J.K., Baek, K.H., 2016. Biosynthesis of silver nanoparticles using aqueous extract of silky hairs of corn and investigation of its antibacterial and anticandidal synergistic activity and antioxidant potential. IET Nanobiotechnol. 10 (5), 326–333.

Rastogi, A., Tripathi, D.K., et al., 2019. Application of silicon nanoparticles in agriculture. 3 Biotech 9 (3), 1–11.

Rodrigues, C., Souza, V., et al., 2021. Bio-based sensors for smart food packaging-current applications and future trends. Sensors (Basel) 21 (6), 2148.

Sardar, R., Ahmed, S., et al., 2022. Selenium nanoparticles reduced cadmium uptake, regulated nutritional homeostasis and antioxidative system in coriandrum sativum grown in cadmium toxic conditions. Chemosphere 287 (Pt 3), 132332.

Schrenk, D., 2004. Chemical food contaminants. Bundesgesundheitsblatt Gesundheitsforsch. Gesundheitsschutz 47 (9), 841–847.

Shin, G.H., Kim, J.T., et al., 2015. Recent developments in nanoformulations of lipophilic functional foods. Trends Food Sci. Technol. 46 (1), 144–157.

Sobha, K., Surendranath, K., et al., 2010. Emerging trends in nanobiotechnology. Biotechnol. Mol. Biol. Rev. 1 (5), 1–12.

Stark, W.J., Stoessel, P.R., et al., 2015. Industrial applications of nanoparticles. Chem. Soc. Rev. 44 (16), 5793–5805.

Sugumar, S., Ghosh, V., et al., 2016. Essential oil-based nanoemulsion formation by low- and high-energy methods and their application in food preservation against food spoilage microorganisms. In: Essential Oils in Food Preservation, Flavor and Safety. Academic Press, pp. 93–100.

Sun, X., Wang, C., et al., 2022. The facile synthesis of nitrogen and sulfur co-doped carbon dots for developing a powerful "on-off-on" fluorescence probe to detect glutathione in vegetables. Food Chem. 372, 131142.

Suvorov, O.A., Volozhaninova, S.Y., et al., 2017. Antibacterial effect of colloidal solutions of silver nanoparticles on microorganisms of cereal crops. Foods Raw Mater. 5 (1), 100–107.

Takahashi, H., Tsukahara, H., et al., 2012. Liquid resin composition for electronic components and electronic component device: JP20160000194[P]. JP2016074918A.

Tan, R., Lv, Z., et al., 2019. Theoretical study of the adsorption characteristics and the environmental influence of ornidazole on the surface of photocatalyst TiO_2. Sci. Rep. 9 (1), 1–8.

Verma, M.L., 2017. Nanobiotechnology advances in enzymatic biosensors for the agri-food industry. Environ. Chem. Lett. 15 (4), 555–560.

Wan, H., He, G.Q., et al., 2021. Numerical study and experimental verification on spray cooling with nanoencapsulated phase-change material slurry (NPCMS). Int. Commun. Heat Mass Transf. 123, 105187.

Wang, A., Haixin, C., et al., 2018. Research progress on nano-encapsulated formulations of pesticides. China Agric. Sci. Technol. Herald 2 (20), 9.

Wu, R., Meng, B., et al., 2022. Efficient capturing and sensitive detection of hepatitis A virus from solid foods (green onion, strawberry, and mussel) using protamine-coated iron oxide (Fe_3O_4) magnetic nanoparticles and real-time RT-PCR. Food Microbiol. 102, 103921.

Xiang, Y., Zhu, A., et al., 2022. Decreased bioavailability of both inorganic mercury and methylmercury in anaerobic sediments by sorption on iron sulfide nanoparticles. J. Hazard. Mater. 424 (Pt B), 127399.

Xiao, W., Guangfu, Y., et al., 2021. A nano-pesticide controlled release agent and its preparation method: CN202011520383. 7[P]. CN112450222A.

Xie, S., Yuanhu, P., et al., 2015. Huazhong Agricultural University. Research progress of biodegradable nano-targeted intracellular delivery of antibacterial drugs. In: The 11th

Member Congress and 13th Academic Discussion of Veterinary Pharmacology and Toxicology Branch of China Animal Husbandry and Veterinary Association, Proceedings of the Fifth Academic Symposium of the Association and the Veterinary Toxicology Professional Committee of the Chinese Society of Toxicology. Chinese Society of Animal Husbandry and Veterinary Medicine.

Xie, X., Guiying, Y., et al., 2021. Research progress on green synthesis, insecticidal activity and action mechanism of silver nanoparticles. Chin. J. Pestic. 6 (23), 9.

Xing, Y., Li, W., et al., 2019. Antimicrobial nanoparticles incorporated in edible coatings and films for the preservation of fruits and vegetables. Molecules 24 (9), 1695.

Xuejing, Z., 2016. Preparation of Stimulated Release Cardamom Essential Oil Nanoliposomes and Its Application in Meat Food. Jiangsu University.

Zhang, W., Haiyun, J., et al., 2018. Application of nanomaterials in soil heavy metal pollution treatment. Chem. Eng. Technol. 2 (8), 10.

Zhao, H., Shan, L., et al., 2017. Research progress on photocatalytic degradation mechanism of organic pesticides. Chin. J. Agron. (8) 8.

Zheng, S., Zhu, Y., et al., 2013. Fiber humidity sensors with high sensitivity and selectivity based on interior nanofilm-coated photonic crystal fiber long-period gratings. Sens. Actuators B Chem. 176, 264–274.

Zheng, Y., Tang, Y., et al., 2015. FePt nanoparticles as a potential X-ray activated chemotherapy agent for HeLa cells. Int. J. Nanomed. 10, 6435–6444.

Ziaee, M., Ganji, Z., 2016. Insecticidal efficacy of silica nanoparticles against *Rhyzopertha dominica* F. and *Tribolium confusum* Jacquelin du Val. J. Plant Prot. Res. 56 (3), 250–256.

Emerging issues in the food processing

6

Kamana Singh[a], Vineeta Kashyap[a] and Addanki P. Kumar[b]

[a]*Department of Biochemistry, Deshbandhu College, University of Delhi, New Delhi, India,* [b]*The University of Texas Health Science Center at San Antonio, TX, United States*

6.1 Introduction

Food processing entails a range of techniques for converting raw materials into edible products. This can include traditional methods such as washing, cutting, freezing, fermenting, curing, treatment, smoking, and packing, as well as more modern methods such as pasteurization, high-pressure processing (HPP) (Kim et al., 2008), cold plasma (Sonawane et al., 2020), ionizing radiations (Indiarto et al., 2020), ultra-heat treatment, or modified atmosphere packaging, among others.

At first, food technology was used to suit military demands. Food Technology grew in the 20th century as a result of world wars, space exploration, and increasing consumer demand for a range of foodstuffs. Packaged food like Maggi, instant soups, and beverages were made especially for working women. In addition, the production of safe and nutritionally balanced food was the major concern of the food industry. When people's eating habits and preferences changed, they began incorporating foods/preparations from various places and countries into their diets. Their desire to have seasonal food available for the whole year pushed the food technologists to toil harder to create food that was fresh, nutritious, delicious, minimally processed, and safe. In the 21st century, food scientists are tasked with developing not only healthy food but also tasked to create a variety of functional food that is fortified with special ingredients for the specific needs of the consumers. The use of new emerging technologies in food processing enabled the food industry to accomplish this goal and sustain the food supply as well as provide the food of their choice. This rapidly growing and developing industry has helped disadvantaged countries improve their food security while also creating job possibilities at all levels.

6.1.1 Do we really require processed food?

Foods degrade physically, chemically, and biologically. Food degradation causes rotting, off-flavor development, texture deterioration, discoloration, and nutritional content loss in varying degrees, reducing aesthetic appeal and rendering it unfit/unsafe for consumption. Pests, insect infestations, microbial infestation, pesticides, poor processing and/or storage temperatures, excessive exposure to light and other

Nanobiotechnology for Food Processing and Packaging. DOI: https://doi.org/10.1016/B978-0-323-91749-0.00009-5

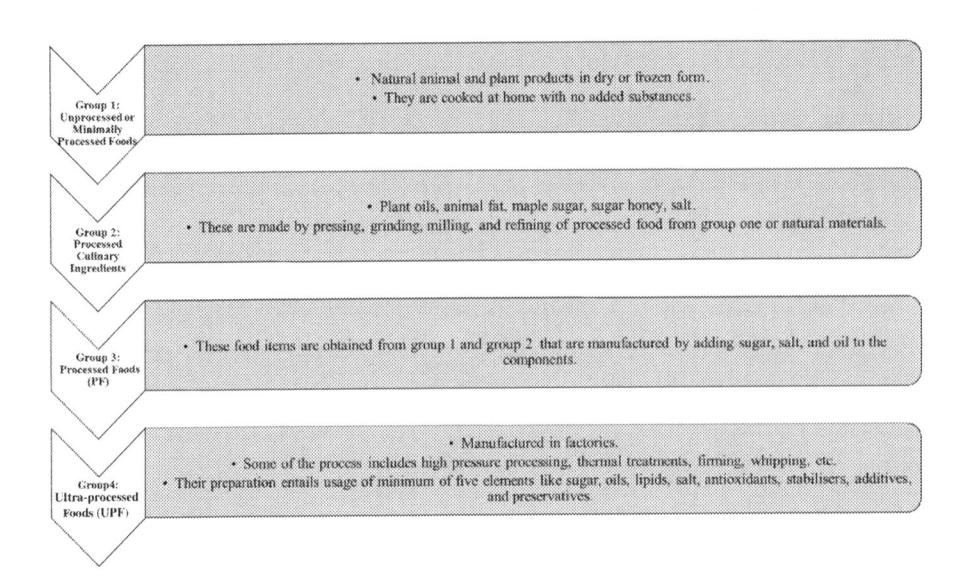

FIGURE 6.1

"NOVA food classification system" divides food into four groups based on how they are processed and to what extent they are being processed.

radiations, oxygen, and moisture can all contribute to food deterioration or spoilage. Foods can also be spoiled by the denaturation of naturally occurring enzymes. Furthermore, several constituents of plant and animal-based meals undergo physical and chemical changes quickly after harvesting or slaughtering, compromising food quality. As a result, to keep food edible and safe, food processing and preservation are required.

Different food preservation methods have been employed to protect food from spoilage after it is harvested. Some of the traditional methods include sun drying, controlled baking, salting/pickling, fermentation, roasting, candying, smoking, and certain spices that can be used as preservatives. Despite the emergence of new technology at the beginning of the Industrial Revolution, these tried-and-true procedures are still in use.

Food processing incorporates regulatory-compliant manufacturing processes to develop different classes of food based on the principles of food technology, biotechnology, and microbiology while preserving the sensory qualities of the food.

Professor Carlos Monteiro and his colleagues developed the "NOVA food categorization system", which separates foods into four groups based on processing quantity and function rather than nutrients (Monteiro et al., 2019) (Fig. 6.1).

6.1.2 **NOVA food classification system**

Group 1: Unprocessed or minimally processed foods

These include plant (seeds, fruits, leaves, stems, roots) and animal edible parts (muscles, eggs, milk). These include naturally occurring vegetables, fruits, grains, and nuts.

These foods are cooked at home by using processes like boiling, chilling, pasteurizing, roasting, crushing, grounding, fermenting, frying, or freezing, for example, yogurt obtained from the fermentation of milk.

Group 2: Processed culinary ingredients
These are made by pressing, refining, grinding, milling, and spray drying processed food (PF) from group one or natural materials. To improve the flavor of food, they are used in cooking and seasoning, for example, vinegar, butter, salt, milk fat, crushed spices, and sugar.

Group 3: Processed foods
This group includes PFs manufactured by adding sugar, salt, and oil to the components. The amount of sugar and salt in PFs has an impact on how damaging they are. Only a few examples are fruits preserved in sugar syrup (murabba), vegetables preserved in saltwater or oil (pickles), basic cheese made from milk, and canned fruits and vegetables.

Group 4: Ultra-processed foods
The least healthy of the lot are ultra-processed foods (UPF), which contain little to no real food. These are manufactured in factories and sold as packaged food or ready-to-eat meals that may be consumed at any time and in any location. A few examples of UPF include packaged food items like bread, biscuits, baby milk, soups, ready-to-cook meals, juices, beverages, sweet and salty snacks, yogurts, ice-creams, sauces, cakes, and protein bars. The process includes bulking firming and antibulking, IOR, whipping, HPP, carbonation, and defoaming. UPF contains a minimum of five ingredients such as sugar, oils, lipids, salt, antioxidants, stabilizers, preservatives, inverted sugar, hydrolyzed proteins, casein, hydrogenated oil, lactose, whey, and gluten which are dangerous for health. UPFs are made extremely attractive by adding many food additives, food dyes, and flavors.

There is a tremendous demand and interest of consumers in high-quality food with natural flavors and tastes, and functional foods that are designed to be rich in specific ingredients like vitamins, minerals, folic acid, etc. to be healthy and fit. This lays challenges for the food industries to supply various types of functional food required for special needs and for this lots of effort has been made to upgrade the thermal and nonthermal technologies involved in the processing of food.

6.2 Adverse effects of food processing technologies on food

6.2.1 Shortcomings of thermal treatment of food

Various commercial and home cooking methods involve the application of thermal treatment to generate certain food characteristics. Boiling, frying, steaming, baking, stewing, and roasting are some of the methods used in conventional, microwave, and steam ovens. Toasting, drying, sterilization, canning, coffee roasting, and pasteurization are all examples of conventional transformation processes that use heat. Heat breaks down cell walls, allowing bound phenolics to spread throughout the plant, and increase their bioavailability. The antioxidant properties of phenolics are known to protect humans from degenerative or cardiovascular diseases. They are, however, more prone to oxidation, and some may be thermostable to varying degrees. Thermal treatment of food on the one hand can make the food easier to digest in the gastrointestinal (GI) tract by modifying the carbohydrates or proteins present in them (Cartus and Schrenk, 2017). On the other hand, high temperatures can modify the contents of the food and can result in the production of some by-products that may be carcinogenic or may change the texture and appearance of the food. Thermal treatments can also speed up the oxidation of some of the lipids and proteins present in the food by increasing free radical production and decreasing some of the antioxidant substances (Santé-Lhoutellier et al., 2008).

Furthermore, the heat treatment method influences the extent and the formation of different types of food toxicants. Thermal treatment of the food includes the cooking of food in a traditional oven at around 250°C temperature (Calabrò and Magazù, 2012) or involves boiling, steaming, and blanching of food in hot water. These methods have been associated with significant weight loss and loss of dietary nutrients by leaching. In contrast to classical heating by conduction or convection, the use of electromagnetic radiation in microwaves results in the cooking of food more quickly as the microwaves are absorbed only by the food instead of heating the walls of the whole oven. These microwaves are converted to heat in the outer layers of the food and so are unable to cook the food from inside (Fan et al., 2018), thereby maintaining their nutritional value by retaining all the vitamins and minerals. Several reports suggest microwave heating as an alternative to conventional thermal cooking methods like blanching to preserve nutritional integrity. One of the studies confirms that the amount of vitamin C, proteins, iron, phosphorus, and ashes was significantly reduced in broccoli which was blanched with hot water in comparison to fresh broccoli, but these contents were substantially retained in microwave-blanched broccoli (Patricia et al., 2011).

Different cooking methods have different influences on the chemical composition of food, for example, microwave-prepared rice has more fat, protein, and ash levels (8.49%, 2.45%, and 1.42%, respectively) than rice cooked by regular boiling and steaming processes. When compared to normal cooking, unsoaked rice (14%–24%) and presoaked rice (12%–33%) cooked in the controlled temperature settings of microwave ovens result in much lesser use of energy (Daomukda et al., 2011). Arab et al. (2010) discovered that different cooking procedures, such as heating food for 90 minutes on a hot plate, for 5 minutes in a microwave on high power, and frying food at 170°C using corn oil for 1 minute, leads to the change in the contents of fat and ash levels, as well as changes the composition of the chickpea flour before and after cooking. Megahey et al. (2005) compared the baking of cake in a microwave oven (250 W) over the convection oven (200°C) and observed the difference in cooking conditions on the texture of the cake. When compared to convection-baked cake, the microwave-baked cake had higher springiness, moisture content, and lower stiffness as textural qualities. All these studies suggest that microwave heating of food is a safe alternative over conventional oven, and controlled heating of food at 500 W using the microwave, uniformly cooks the food in a fast manner, retains the moisture content, uses less energy, and preserves all the natural ingredients in the food. However, microwave's electromagnetic radiations can lead to the oxidation of lipids present in animal fats and vegetable oils (Oueslati et al., 2010), and may produce free radicals which may be further oxidized to hydroperoxides and various other secondary oxidation products that may be toxic to human health. Different heating methods can sometimes hamper the release of carbohydrates and influence their digestion in the GI tract. During the thermal treatment, the Maillard reaction which is a chemical reaction between the sugars and the amino acids in the food, and oxidation of lipids in food can result in the production of around 800 compounds. 50 out of 800 compounds being discovered were found to be toxic possessing mutagenic and carcinogenic properties, a threat to human health.

A study by Sugimura et al. (1977) discovered the first carcinogenic substances known as heterocyclic aromatic amines (HAA) produced from high-temperature thermal processing of cooked protein-rich food products like cooked meat, fish, grilled foods, seafood, and tobacco smoke condensate, as well as diesel exhaust. Another food mutagen produced, as a result of the Maillard reaction during the thermal processing of food, was Acrylamide (AA, $H_2C = CH–CO–NH_2$) (Zamani et al., 2017). Further research found that the major source of AA is the roasting, frying, or baking of foods high in carbohydrates and proteins (Kwolek-Mirek et al., 2011; Tareke et al., 2000). When food items like morning cereals, coffee, potato chips, and cereal grains which are part of the daily diet of consumers are made in industry by different thermal methods, acrylamide is produced which may damage the DNA, impair the liver, and destroy the neurons in mammalian systems (El-Assouli, 2009).

Along with acrylamide, another toxicant 5-Hydroxymethylfurfural (HMF, $C_6H_6O_3$) is produced upon heating of food items like milk, bakery goods, coffee, honey, juices, baby formulas containing hexose at 150°C (Capuano and Fogliano, 2011).

Furan (C_4H_4O) is a cyclic dienyl ether with a low molecular weight of 68.07 kDa, formed upon degradation of amino acids, carbohydrates, and ascorbic acid and oxidation of polyunsaturated fats and carotenoids (Yaylayan, 2006) upon heating at 150°C. It is a heat pollutant which has a low odor threshold that influences the sensory qualities of thermally PF and beverages. Furan has been associated with causing cancer in both animals and humans (Byrns et al., 2006).

Polycyclic aromatic hydrocarbons (PAH) are a broad family of chemical molecules that are produced by incomplete combustion of food components at temperatures above 120°C and are considered widespread environmental and food pollutants (Sampaio et al., 2021). Some PAHs like phenanthrene found in the bark of *Vismia cayennensis* trees are synthesized autogenously in plants (Krauss et al., 2005), whereas some may be formed in plants during cooking as a result of the decomposition of organic materials such as lipids, proteins, and steroids. This thermal food toxicant constitutes the largest class of carcinogens known worldwide owing to their capacity to induce gene mutation and cancer (EFSA, 2008).

Nitrosamines (NA) are one of the most potent carcinogen and mutagenic compounds known to cause cancer in the colorectum, gastric, and esophagus (Lee et al., 2006). In general, people are exposed to NAs mostly through different thermally processed consumables above 130°C, such as cured meat, sausage, fish, beer, drinking water, and so on. The nitrosation reaction of a nitrosating reagent formed from either nitrites or nitrogen oxide with the N-containing compounds has proven their creation (Yurchenko and Mölder, 2005). Thus, the use of nitrites and nitrates as preservatives is highly regulated in the meat business.

Acrolein (C_3H_4O) is generated during thermal food processing of food containing carbohydrates usually at temperatures above 150°C. It is also found in fruits, vegetables, seafood, and wines. Acrolein forms adducts upon reaction with primary amines and secondary amines, amino acids, nucleic acids, thiol group of cysteine residue as well as other cellular components (Cai et al., 2009) and leads to depletion of glutathione which is a powerful antioxidant in the cell. It crosslinks with many peptides, proteins, and DNA (LoPachin et al., 2009) and was found to be carcinogenic to many cell lines reducing their viability (Yoshida et al., 2009).

Chloropropanols are foodborne pollutants generated from the thermal processing of edible oils like rapeseed meals and gluten in maize at temperatures exceeding 150°C (Wenzl et al., 2007). During edible oil extraction, chloropropanols and their isomers are produced on acid hydrolysis of leftover lipids and glycerol linked with proteinaceous by-products and are formed in hydrolyzed vegetable protein (HVP) (Velísek et al., 1978). 3-monochloropropane-1,2-diol (MCPD) is generally the most common chloropropanols formed in acid-HVPs with smaller levels of 2-MCPD, 1,3-DCP, 2,3-DCP, and 3 chloropropan1-1. Several studies reported mutagenic effects of MCPD in vitro (Robjohns et al., 2003) and some suggested their toxic effect on the kidneys of the experimental animals, but so far there are no in vivo clinical studies showing mutagenic activity of chloropropanols in humans reported, although negative findings from a bone marrow micronucleus assay in rats have been reported. Although there are no epidemiological studies reported in humans, humans are more exposed to the risk of 2-chloropropane-1,3-diol(MCPD) released from the action of lipases on the esters present in these processed vegetable oils as these oils are consumed on a regular basis by them for preparation of a variety of foodstuff (Seefelder et al., 2008).

It's worth noting that the thermal processing of food and edible oils in a single batch can result in the generation of many different food-related toxicants. These food toxicants are a big threat to human health as they can alter the structure of DNA and cross-link with various amino acids/peptides and lipids to form toxic adducts in tissues and organs in the human body. Food scientists need to explore and study the metabolic pathways of these toxicants inside the human body as sometimes the xenobiotics that are produced from the thermal PFs are more harmful than the substance itself. Nonetheless, decreasing carcinogens in food while maintaining the quality of food remains a significant challenge for the food processing industries.

6.2.2 **Shortcomings of nonthermal treatment of food**

Thermal treatment of food has been primarily used for many years to prevent spoilage of food and the growth of microorganisms but under some circumstances, the application of heat has adversely affected the physio-chemical properties of food (Chacha et al., 2021; López-Gámez et al., 2021). To overcome the drawbacks of thermal treatment, food technologists alternatively developed many nonthermal technologies that do not influence the nutritional content and the quality of food. (Zhong et al., 2019). Nonthermal food processing simply refers to techniques of microbial inactivation of food items that do not include the use of direct heat (Bhattacharjee et al., 2019; Hernández-Hernández et al., 2019; Troy et al., 2016). Such new technologies like pulsed light (PL), HPP (Chemat et al., 2017; Hernández-Hernández et al., 2019), ozone technology (O'Donnell et al., 2012), nonthermal plasma (NTP), ultrasonic technology (Dong et al., 2021; Rahaman et al., 2016), are paired with hurdle technology to replace thermal food processing processes, are increasingly regarded as emergent, unique, or new food processing methods. Physical processes (ultraviolet radiation pulse electric field (PEF), PL, ultrasound (US), HPP, and ionizing radiation) and chemical processes (cold plasma and ozone treatment) are the two primary categories of technologies used in food processing.

6.2.3 **Impact of physical treatments**

6.2.3.1 *Pulse electric fields*

The downsides of PEF include high start-up expenses, which are one of the key barriers to using PEF on a broad scale. The many types of equipment necessary for its initial setup are quite expensive, ranging in price from 250,000 to 2,000,000 USD (Pal, 2017). As a result, in order to attain entire industrial applications, the method and equipment capacity must be improved (Shahbaz et al., 2018). There might also be a failure to process the food effectively as sometimes the spores of bacteria acquire resistance to the PEF, this might pose a danger to the health of humans (Bahrami et al., 2020). PEF has been regarded as ineffective in treating solid foods as compared to liquid meals (Priyadarshini et al., 2019). Furthermore, scaling up PEF has been difficult due to the technology's sophistication and the fact that just a few research on its many industrial features are now accessible (Arshad et al., 2020).

6.2.3.2 *High-pressure processing*

High-pressure technology is a nonthermal technique available for food processing that can help the food industry to produce high-quality convenience food which is devoid of any additives and preservatives. This food preservation technique for killing microorganisms by high pressure was discovered in 1899 and has been used successfully in the plastic and chemical industries for decades, but it was not until the late 1980s, that the industries started using the technique commercially for food processing. High isostatic pressure in HPP sterilizes food at a pressure of 6000 bar to inactivate food-borne microorganisms (Kim et al., 2008). Hite (1899) used high pressure in his lab to preserve milk, fruits, and vegetables (Hite et al., 1914). HPP is deemed unsuitable for use in dried and porous products. This is because, while such goods must be kept dry, the usage of water is required during the HPP process (Huang et al., 2017). Furthermore, HPP-treated goods should always be refrigerated after treatment. This is because pressure (600 MPa) treatment of food along with treatment of food with temperature (90°C–120°C) are more efficient than pressure alone in inactivating vegetative cells of microorganisms. Although this appears to make the HPP approach efficient, it may turn out to be inefficient and time-consuming (Huang et al., 2017) HPP-treateded food needs to be packed in a packaging material like plastic that can be compressed to around 15%, so this could be one of the limitations as it precludes the use of all the different types of packaging material available in market.

High-pressure processing has also been shown to have a negative impact on milk and dairy products, such as diminishing casein micelle size, increasing free fatty acid levels, and modifying the natural properties of whey (Liepa et al., 2016). Like the other nonthermal food processing technology, HPP is limited by large beginning charges. High equipment expenses (even exceeding $7000) should be addressed, particularly in the case of PEF technology. This is likely to be mitigated by the lower-than-expected operational expenses (Verma et al., 2020). Furthermore, the facility's complexity and size would require suitable talent and space to run properly.

6.2.3.3 Ultrasound technology

Ultrasound technology is one of the newly emerging technologies used in the food industry. The high power (low frequency) US may bring about physico-chemical changes in the liquid food products by the free radicals produced in the cavitation process that is used in various processes such as drying, freezing, and emulsification. These, in turn, cause lipid oxidation, protein denaturation, and ascorbic acid degradation, all of which have a negative impact on sensory properties (Bhargava et al., 2021; Hernández-Hernández et al., 2019). Furthermore, high-power ultrasonic wave (20 kHz) when combined with moderate pressure keeping the temperature at room temperature (manosonication) was found to increase its inactivation efficiency and lead to modest inactivation of certain microorganisms, particularly *Listeria monocytogenes* (Chemat et al., 2017). There appears to be one major limitation to its use at the industry level. Magnavita and Fileni (1994) have shown that exposure of medical workers to contact the US for a long time has led to the development of some neural changes in the exposed organ. US waves pose a risk to the health of the operators/workers working in the US technology industry. Therefore, it is important to optimize the power and time and use US technology in a safe and controlled manner in order to avoid any negative effects on the treated foodstuff.

6.2.3.4 Pulsed light

Similar to UV technology, PL requires a huge amount of capital to produce a profitable investment (Bhavya and Umesh Hebbar, 2017). PL has been demonstrated to be unsuitable for use in foods that are opaque and irregular in shape since they can serve as possible breeding grounds for bacteria. Furthermore, prolonged PL treatment might have a "heating effect" on food goods, reducing the efficacy of bacterial elimination (Bhavya and Umesh Hebbar, 2017).

6.2.3.5 Ultraviolet radiation

The harmful effect of high doses of UV radiation on human health is a major reason for consumers to be apprehensive about the use of UV radiation for the treatment of food and has been a major setback for UV irradiation for a long time. Unfortunately, many individuals are still suspicious, particularly about whether UV-treated goods are safe (Hernández-Hernández et al., 2019; Khan et al., 2018). Consumers appear to be concerned that UV radiation may result in radioactive elements in meals, exposing them to major health risks (Bahrami et al., 2020). Indeed, if humans are exposed to UV-C radiation, they may be in danger. Ultraviolet-C (UV-C) radiation can cause serious skin burns and eye damage (photokeratitis). This UV-C restriction implies that persistent exposure to materials or processing surfaces may alter their chemistry and behavior. Furthermore, UV radiation may promote isomerization and oxidation of lycopene (Bhattacharjee et al., 2019) or protein cross-linking or oxidation of lipids, especially at high radiation concentrations and contact periods. Another issue is that the action of UV-C radiation on liquid meals is influenced by turbidity. After the food items are packed, they need to be transported to the irradiation plants for treatment, which consumes a lot of time and labor, so this could be one of the drawbacks of not being extensive application in the food industry (Bahrami et al., 2020). UV radiation, due to its poor penetration capability, would be useless when applied to foodstuffs having indefinite shape and structure. However, combining UV light with other nonthermal processes would enhance this. Furthermore, the large money required for set-up can be one of the limiting factors in reaching the full practicality of the UV radiation process (Hernández-Hernández et al., 2019).

6.2.3.6 Ionizing irradiation

The dose of IOR used for food processing should neither be too high as it can affect the sensory qualities of the food nor below the optimal recommended dose that may leave the food still contaminated with the viruses. This means that the duration, intensity, and dose of the degree of irradiation required to be applied for the treatment of food is dependent on factors like the texture and density of the food products as well as the load patterns (Indiarto and Qonit, 2020). High doses of ionizing irradiations can produce free radicals like hydroperoxides upon oxidation of lipids in food that can spoil its flavor and odor (Indiarto et al., 2020). Irradiation can either cause mutations in microorganisms residing in the food leading to more pathogenic strains than the native strain or can make these microorganisms resistant to radiation and thus, alter the chemical structure of the food by producing toxins like Aflatoxins produced following irradiation of foodstuff like wheat, corn, sorgum, potatoes which is public health hazard leading to liver cancer. These harmful radiations may pose a risk to the health of processors and workers as they are constantly being exposed to these radiations and can lead to chromosomal abnormalities, various types of cancers, or even lead to death in some cases. Furthermore, another disadvantage of IOR technology is the huge capital associated with the set-up of ionizing radiation facilities in the food industry.

6.2.4 Impact of chemical treatments

6.2.4.1 Ozone treatment

A high initial cost is required for the set-up of a commercial ozone treatment plant and its maintenance (O'Donnell et al., 2012). Ozone is corrosive if employed over 4 ppm (Ölmez and Kretzschmar, 2009), which makes it necessary for adequate monitoring, especially in indoor applications. Because ozone has a short half-life, is very reactive, and unstable gas, so it is generally generated on-site and cannot be stored, which makes it difficult to analyze its efficacy in such a short time. High investments are required for setting up of industrial ozone treatment plant on a commercial scale. Trained skilled operators are also required to operate the ozone equipment. The accessible ozone concentrations, such as those found in home (as well as certain commercial) varieties, are normally fixed (Okpala, 2017a, 2017b, 2018), and while this might cause issues, it also necessitates extreme attention while handling. Another disadvantage is the way ozone interacts with organic debris, which may restrict either the antibacterial effect or the treatment's effectiveness (O'Donnell et al., 2012).

6.2.4.2 Cold plasma (or non-thermal atmospheric pressure plasma)

Non-thermal atmospheric pressure plasma (NTAPP) is used in the food industry to increase the viability and function of beneficial microorganisms that can promote the expression and secretion of microbial enzymes. To produce NTAPP, high energy is applied to the gas that can produce many reactive oxygen and nitrogen species like ozone, superoxide, nitric oxide, etc. that can lead to the oxidation of lipids and cause the breakdown of naturally occurring antioxidants changing in the flavor and taste of the food (Sonawane et al., 2020). When used alone, foods exposed to cold plasma are not able to remove significantly the bacterial cells or fungal spores, so pressure and/or temperature must be utilized in conjunction with it. One of the major drawbacks of this nonthermal technology is the large financial expenditure necessary to set up plasma-producing equipment, as well as the difficulties connected with understanding plasma chemistry (Atik and Gumus, 2021; Roobab et al., 2018). Furthermore, there is a lack of optimal process and product characteristics in a cold plasma, which limits the successful scale-up of the process from lab or pilot to the industrialized stage to assure cost-effectiveness (Hernández-Hernández et al., 2019; Ölmez and Kretzschmar, 2009; Priyadarshini et al., 2019). Cold plasma may hasten the oxidation of lipids in dairy products, compromising their sensory qualities (Coutinho et al., 2019). There are challenges, particularly when treating meals with no specific shape and structure, owing to the plasma effect's inability to penetrate the food matrix. Because of their widespread distribution, certain bacteria may go untreated (Mandal et al., 2018). Other difficulties include the fact that cold plasma technology is still in its infancy (laboratory or pilot scale). The entire technicalities involved in the procedure of NTAPP for food items are not clear and it remains a challenge with regard to harmful pathogenic bacteria, so therefore, a complete knowledge of its usage and the parameters controlling the microbial species need to be carefully understood. There are only a few research reports which suggest its beneficial effect on the microbial cells and their enzyme production or its impact on a matrix of food and product quality (Asaithambi et al., 2021).

6.2.5 Impact of fermentative treatments

Fermented foods (FFs) are an essential component in the diet of Asian people. The global market for fermented goods and components is expected to reach $28.4 billion by 2022. Different FF have different concentrations and different genera and species of naturally occurring microorganisms that decompose carbohydrates in the absence or presence of oxygen to produce a variety of end-products that increase the nutritional value of the food (Sivamaruthi et al., 2018a). Microbes isolated from FF have been observed to contain bacteriocin, enzyme and neurotransmitter synthesis, and probiotic activity (Sivamaruthi et al., 2018b). The production procedure of FF also differs depending on geographical area and resource availability. Even while they share the same basic ingredient, each fermented product is made using a distinct method. Milk, for example, is the primary component in Ayran, Koumiss, Kurut, Torba, and Kefir, all of which are produced using various ways (Kabak and Dobson, 2011). The fermentation process benefits food goods in a variety of ways, including flavor, taste, shelf life, and so on. Bioactive metabolites such as conjugated linoleic acids, vitamins, exopolysaccharides, peptides, neurotransmitters, and oligosaccharides can be released by fermenting bacteria (Beermann and Hartung, 2013). On the other side, during fermentation, certain undesired harmful pollutants and microbial metabolites are released into the food basis. The most common hazardous chemicals in FF are fungal and bacterial toxins, cyanogenic glycosides, and biogenic amines (BA) (Sivamaruthi et al., 2018a). Excess levels of unwanted hazardous chemicals in the FF over the allowed threshold might cause major health problems in customers and possibly death. Pathogenic bacteria (*Listeria spp., Brucella spp., Clostridium botulinum, Staphylococcus aureus, Salmonella spp.,* and *Streptococcus spp.,*) have been found in FF, particularly fermented milk products, causing significant outbreaks. If harmful incidents occur often, the quality of the FF becomes doubtful, affecting the whole FF industry. As a result, several precautions and changes in the traditional methods of production of FF in the food processing sectors should be implemented to obtain a high-quality product with minimal levels of pathogenic bacteria or toxins in food items.

Mycotoxins are the aggregate term for all fungal poisons. FF was found to contain many mycotoxins (ochratoxin A, aflatoxins, citrinin, deoxynivalenol, and zearalenone).

The presence of fungal toxins released from *Aspergillus spp., Rhizopus spp., Penicillium spp.,* and *Mucor spp.*, were observed in the grains, barley, and maize used in beer production. Some of the commercial and local beverages contained aflatoxins, ochratoxin A, and zearalenone, respectively (Odhav and Naicker, 2002).

Ochratoxin A contamination of 5 µg of toxin/kg of sample was detected in food samples like maize, sesame, fermented cassava, sorghum, millet, and flakes of Nigeria origin which is considered dangerous to human health according to European Union standards. Ochratoxin A is a known carcinogen and nephrotoxin (Beermann and Hartung, 2013). Ochratoxin A was found to be present in wine while alternariol and alternariol methyl were observed in cider and cork stoppers when liquid chromatography-tandem mass spectrometry was used to look for mycotoxins. The findings revealed that the packaging of FFs required additional attention (Scussel et al., 2013). Alternariol is a mutagen (Brugger et al., 2006), a genotoxic (Fehr et al., 2009), cytotoxic, teratogenic, and fetotoxic (Weidenborner, 2001) substance. *Rhizopus spp.* commonly used for fermentation in FF in Asian region was found to produce rhizoxins in immunocompromised people while some may associate endosymbiotically with another bacteria *Burkholderia spp.* and create more toxic compounds in the FF.

Pathogens such as *Proteus mirabilis, Alcaligenes faecalis, Staphylococcus sciuri* subsp., and *Bacillus anthracis* were found in many of the Nigerian condiments and their raw materials. Aflatoxin contamination was also prevalent in fermented milk products and ogiri was found to be carcinogenic to various immune and liver cells (Kensler et al., 2011; Roze et al., 2013). Some of the Iranian fermented dairy products like doogh, kashk, and yogurt were found to be contaminated with Shiga toxin-producing *Escherichia coli* (STEC) (Ivbade et al., 2014).

Another important harmful component in FF is BA. FF products have been shown to have significant levels of BA that are produced by decarboxylation of amino acids caused by microorganisms, posing major health risks to humans. The presence of BA in stored or PFs indicates the quality of food. The most frequent BA in FF are cadaverine, tyramine, putrescine, and phenylethylamine histamine, especially, histamine is responsible for various food poisonings and epidemics. Histamine toxicity in fish products is caused by unsanitary production and storage techniques. BA can be produced by a variety of *Streptococcus spp., Lactobacillus spp., Bacillus spp., Staphylococcus spp., Bacillus subtilis strains,* and *Enterococci spp.* in FF such as wine sausages, douchi, miso, and natto (Doeun et al., 2017).

Thus, through advanced new technologies and automation in the food industry, the toxicity of FFs with BA, pathogenic microorganisms, bacterial toxins, and mycotoxins are some of the most significant spoilers of food items and should be taken care of for preparing good quality food.

6.3 **Health hazards associated with processed food**

6.3.1 **Impact of ultra-processed food on human health**

Ultra-processed foods comprise a variety of food products like packaged salty, sugary, or fatty snacks to beverages, ready-to-eat meals, organic food, plant-based dairy or meat substitutes, and certain functional foods like milk and meat obtained from plant or gluten-free items (Fardet and Rock, 2019). UPF therefore includes both cheap foods with minimal or almost no nutritional value and good quality healthy functional foods that are enriched with specific ingredients for specific purposes. Thus, 40% of total energy intake comes from consumption of these UPFs that form a major part of a vegetarian diet (Drewnowski, 2021). In 10 years, since its inception, NOVA's definition of UPFs has evolved dramatically and has been categorized based not on the basis of different types of components but lately, the classification of PF takes into account their number, origin, and category of food processing methods (primary, secondary, and tertiary) (Gibney, 2019). Due to the large demand for these convenient meals by the consumers, the industrialists were forced to produce cheap foods that were made in large numbers by the new emerging technologies in food processing. Due to the complex nature of these substances, especially the mixed dishes (Lorenzoni et al., 2021) manufactured on a large scale in the food industry, often the reviewers get confused in categorizing these products into a specific defined food group category (Bleiweiss-Sande et al., 2019) as proposed by NOVA. The over-refining of these products during their manufacturing can lead to a loss of wholeness as well as damage to the matrix of the food, (Fardet and Rock, 2019) resulting in a loss of sensory qualities of the products.

According to several reports, the consumption of UPF differs among people belonging to different countries and ethnic groups, age groups, gender, number of children per family, and the time required to prepare a particular meal depending on the food product, and individual's and country's economic status. UPF is consumed by people of developed and developing countries in a varied manner, as high-income people from developed countries can easily afford these products in contrast to people from low- or middle-income countries which are largely dependent on farming for supporting their families (Khandpur et al., 2020).

The food processing industries are driven largely by the need to enhance profits, which necessitates lowering costs throughout all stages of production. In certain circumstances, this may be accomplished by using less expensive alternative food items wherever and to the degree practicable; but in some cases, this technique can result in UPFs that are harmful to one's health. The food industries, for example, made cheap solid and spreadable margarines from vegetable oils which they used in baking and cooking food (Astrup et al., 2008; Stender et al., 2006). These margarines are cheaper and were projected by the industries to be healthier than butter that contains saturated fats, but later researchers reported the ill effects of trans-fat containing margarine on the health of humans as it increased bad cholesterol. The first warning signs appeared in 1993, indicating the association of large intake of food high in trans-fat to the development of cardiovascular diseases, as a result, the usage of these trans-fats in food products was strictly banned or limited in several countries (Astrup et al., 2008; Stender et al., 2006). Still, it is crucial to remember that the kind and quantity of food processing are not always related to the finished product's health potential. Indeed, several developing processing methods can increase the nutritional content of food items, for example, by reducing nutrient loss (Sammugam and Pasupuleti, 2019; Weaver et al., 2014).

UPFs constitute the major part of the diet of most high- and middle-income countries and account for more than 40% of dietary calories (Adams et al., 2020; Matos et al., 2021; Monteiro et al., 2019). These highly PFs are hyperpalatable, have fewer micronutrients and fibers, have high salt and sugar content, high energy density with low quality of protein and fat, and have a glycemic index (Martínez Steele et al., 2017; Rauber et al., 2018). According to several association studies, it was believed that it was the quantity of UPFs and not the quality of the food that has negative effect on health and is the cause of obesity and various noncommunicable diseases (Adams et al., 2020; Popkin, 2019).

The higher calorie intake on the UPF diet might be only due to the quicker eating rate (Hall et al., 2019), regardless of any impacts from processing, variations in food type, shape, or matrix (Hawton et al., 2018; Robinson et al., 2014). The majority of the trials in a prior meta-analysis that came to this result looked at the impact of consuming the same meal in the same form and matrix at different rates (Robinson et al., 2014). The impact was considerably stronger in tests where the eating rate was manipulated by verbal instructions as opposed to studies where the eating rate was manipulated by changing the food structure of the meal (Robinson et al., 2014).

UPFs have also been linked to contributing to a gut environment that favors microorganisms linked to inflammatory illness (Zinöcker and Lindseth, 2018). It is believed that the addition of additives during the processing of food alters the matrix of the food and softens the texture of the food (Juul et al., 2021), thereby, delaying the satiety feeling and leading to overconsumption of these foods. Natural foods which are rich in fibers and micronutrients are easily digested by the colon bacteria (Grundy et al., 2016) in contrast to UPFs which are acellular, difficult to digest by the gut microbiota, and lead to an inflammatory gut that is linked to cardiometabolic disorders (Zinöcker and Lindseth, 2018). Thus, UPFs have a negative influence on the composition as well as the amount of healthy gut microbiota and a positive influence on the growth of harmful bacteria that lead to diseases associated with inflammation (Lane et al., 2020). A study in the animal model confirms that pigs who were fed a diet high on fiber-containing extruded grains showed reduced diversity of bacteria present in their colon as compared to pigs who were fed a diet comprising natural whole grains (Moen et al., 2016). Further, the presence of high sugar/salt/fat contents in UPF makes them more palatable and results in low low-quality diet that leads to increased adiposity in children. There may also be the presence of endocrine-disrupting substances like bisphenol A (BPA) in the packaging materials used for UPF goods, that may affect biology or metabolism (Juul et al., 2021). These substances may lead to obesity, hypertension, inflammatory diseases, and the development of many CVDs (Rancière et al., 2015) but the action of BPA is unclear. Any negative impacts reported from UPFs, according to Gibney (2019), are attributed to dietary considerations rather than the method or the extent to which foods are processed (Gibney, 2019; Monteiro et al., 2019). According to the results of our lab's SWAP-MEAT crossover intervention trial, participants improved in numerous cardiovascular disease risk factors after 8 weeks of consuming alternative plant-based meat products versus organic animal meats (Crimarco et al., 2020). This was due to the consumption of less amount of saturated fatty acids and more dietary fibers that are present in plant-based meat products. In multiple prospective studies, higher intake of UPFs in children was linked to a faster increase in body weight and fat mass index, and also led to increased waist circumference in adolescence and early adulthood (Chang et al., 2021; Costa et al., 2019; Vedovato et al., 2021). Consumption of ultra-processed meals during childhood and adolescence is positively related to obesity, according to a 2018 comprehensive review (Leffa et al., 2020).

Though several reports suggest that the consumption of UPFs by children, adolescents, and adults is associated with an increase in body weight and may lead to the development of non-communicable disease (NCD). There has only been one randomized clinical research particularly examining the impact of UPF consumption (Hall et al., 2019), and thus, one cannot be sure that UPF is the main cause behind these negative health effects, therefore, more clinical research still needs to be done in this area to check negative health impacts of UPF consumption on weight gain and development of chronic diseases.

6.3.2 Effect of food processing on the structure and matrix of food

The quantity of a component released from the food matrix through the GI system and accessible for absorption is known as bioaccessibility. Bioaccessibility of bioactive substances is mostly determined by their amount and chemical structure, as well as matrix qualities and association with other natural components (like proteins, fibers, carbohydrates, and lipids) present in food during digestion (Rodríguez-Roque et al., 2014). Food preparation or the addition of milk or oil as an adjuvant might change these properties and impact the bioaccessibility of bioactive chemicals (Ribas-Agustí et al., 2018). Bioavailability refers to an amount of compound released from the food matrix upon its digestion, absorption, and metabolism that occur in the gut and liver and are released to the target tissues. Thus, there is an increasing demand for new novel food processing technologies to produce food products rich in bioactive compounds that can be easily digested and absorbed by the gut, their influence on the structure and metabolism of these bioactive compounds along with their association with the gut microbiome composition. For example, according to some studies, thermal treatments boost the bioaccessibility of bioactive substances, but high temperatures on the other hand, can cause carotenoids to degrade or isomerize, which can have negative impacts on sensory and nutritional quality. Individual carotenoids are absorbed in different ways in the human body based on their molecular structure and deposition form. In general, carotenoids with a linear form and comprising a greater number of double bonds are more prone to aggregation (Meléndez-Martínez et al., 2014), which explains why trans isomers are less bioavailable. By inducing isomerization, various processing processes can alter bioaccessibility. Zhang et al. (2019) found that the bioaccessibility of carotenoid in tomato juice was increased on the application of US for 20 minutes (800 W; 25 kHz) because of the conversion of trans-lycopene to cis-lycopene. In contrast, Knockaert et al. (2012) found that when HPP at 600 MPa was used to sterilize tomato puree at 117°C, compact aggregates formed, reducing carotenoid bioaccessibility and blocking isomerization to cis-lycopene (Colle et al., 2010). Following the application of HPP and US, fiber networks have been seen, which have been linked to decreased carotenoid bioaccessibility. From these observations, the authors theorized that processing altered the structure of phenols (hydroxylation, methylation, glycosylation, and so on), allowing them to be absorbed (Rodríguez-Roque et al., 2020, 2015). Additionally, during digestion, they may be degraded or conjugated, resulting in a reduction in bioaccessibility (Colle et al., 2010).

Nonetheless, several studies report that the type of food processing method applied and the matrix of the food, affects the bioaccessibility of naturally occurring antioxidants like phenolic compounds or carotenoids in the food. It has been suggested that nonthermal food processing technologies like PEF, US, and HPP improve the bioaccessibility of several phenolic compounds in liquid matrices as compared to thermal treatments. One of the most critical variables affecting bioactive chemical bioaccessibility is matrices' structural characteristics. The size and structural changes of the matrix have an impact on the release of some of the bioactive compounds from the food as well as their metabolism during digestion and the interaction of these compounds with protein/lipid, for example, the association of polyphenols with dietary fibers making their absorption difficult in GI tract. Therefore, more research into the effects of these technologies on fiber content, size and viscosity of particles, pectin properties, and microstructural characteristics is required in order to develop more novel strategies to improve bioaccessibility and to comprehend the mechanism behind these changes.

6.3.3 **Food additives, food colors, and food flavors**

Food additives are usually incorporated in PFs to improve the flavor, texture, or storage properties. They are categorized into three groups (1) Cosmetic additives that include flavors, flavor enhancers, sweeteners, and texture modifiers that make the food look more attractive, (2) Preservatives include antioxidants that extend the shelf life of food by preventing the growth of microorganisms or slow down the deterioration of fats and flavoring of food, and (3) Processing aids include food additives that aid during production. Color additives are often added to the food to replace the color lost during food processing or create an impression that a particular ingredient has been added to the PF to mislead the consumers in a plea to improve the quality of food. PFs containing these additives have poor nutritional value. These additives increase the fat, sugar, and starch content in food making them more palatable and attractive than they otherwise would be, but these additives usually have no nutritional value. For example, many consumers drink soft drinks having orange flavor thinking that these drinks might contain orange juice or vitamin C. Though food preservatives are helpful in preserving the food by preventing food poisoning but generally they have been used in excessive amounts during food processing by food industries. Some commonly used food additives like tartrazine, quinoline yellow, amaranth, sunset yellow, and brilliant blue can cause asthma, migraine, eczema, and hyperactivity in some patients while some additives such as sulphite, and caramel can destroy the existing vitamins in the food or may lead to diseases such as diabetes, coronary heart diseases, stroke, dental caries, and different types of cancer. Additives like nitrates present in meat products, canned meats, and cheese may reduce the ability of blood to carry oxygen. High-nitrate water used to make bottled milk for unwell babies reaches into the bloodstream where it produces methaemoglobin that has a very low capacity to bind oxygen leading to methaemoglobinaemia or blue baby syndrome. Thus, the level of nitrate content in baby foods is regulated in countries such as Austria and Switzerland. Nowadays, chemical preservatives and food irradiation technology are used along with the traditional techniques like drying, refrigeration, and fermentation used for food preservation. Large doses of X-rays or gamma rays used in food irradiation delay fruit ripening and promote the killing of insect pests and microorganisms infesting the food. However, one needs to be careful regarding the hazards of irradiation technology as the ionizing radiations should not make food radioactive and alter the chemical structure of the food by generating toxic-free radicals/chemicals that have an adverse effect on human health.

6.3.4 Diseases due to microbial contamination of food

Microbial contamination in food must be managed using techniques such as heat, chemical disinfectants, or ionizing radiations to ensure food safety. Microbial contaminants come in a variety of forms and can cause food poisoning as well as serious human and animal illnesses such as mycotoxicoses. Ingestion of toxins generated by molds growing on food such as grains, oilseeds, and other agricultural commodities during the postharvest period causes mycotoxicoses. Ergotism, which is caused by eating cereals infected with ergot (a fungus) of the genus *Claviceps* and can induce convulsions, hallucinations, miscarriages, or death, was the first recognized mycotoxicosis in humans. In the past, ergotism epidemics have impacted various European countries, including France.

Aflatoxicosis disease occurs in humans from the consumption of maize contaminated by aflatoxin released by *Aspergillus flavus*. These toxigenic fungi lead to the development of several diseases e.g., sweet clover poisoning, infertility problems due to lucerne, degnala disease, and some other diseases. Fats in irradiated food can develop rancid flavors and odors.

Indian childhood cirrhosis is one of the various forms of liver disorders that affect mostly Indian subcontinent laborers because of aflatoxin ingestion, which causes severe liver tissue death.

Cardiac beriberi is a condition caused by a vitamin B1 deficiency caused by excessive cooking or food processing. Neurological symptoms, muscular weaknesses, rhythm disturbances, low blood pressure, and the emergence of valve murmurs are all signs of the condition, which can lead to heart failure and death. Beriberi was a serious health issue in East Asian areas where polished rice (thiamine found mostly in the husk of rice) was the main source of nutrition. Many parts of Europe including France had suffered in the past with outbreaks of ergotism. Since then, it has been shown that in addition to vitamin B1 insufficiency (thiamine deficiency), a poisonous chemical called citreo virdin, which was discovered in rice that had developed a yellow color, causes clinical signs that are comparable to beriberi in many laboratory animals.

Thus, food spoilage due to microbial contamination remains a major issue to be tackled hampering the growth of the food industry. And the deadly diseases caused by microorganisms, yeasts, and molds by consumption of contaminated foods have a great impact on human health. Food spoilage is thus an economic concern that, despite contemporary food technology and a variety of preservation measures, is still not well controlled. When microbiological issues are under control, the chemical reaction caused by oxidation determines the food's quality. As a result, along with the preservation of food, new methods should be developed for the packaging of the food. One such popular packaging method is the modified atmosphere packaging which keeps the food fresh for a long time and reduces its wastage.

The possible relationship between food technology and biotechnology is another issue that should not be overlooked. Biotechnology's application of food will be fantastic in many ways. Gene manipulation of microbes, plants, and animals, for example, will make a range of food additives available at reasonable rates. Plants with microbial genes can produce enzymes that are required for FF and other health-promoting compounds in large quantities. With such vast potential capabilities, food scientists must apply new findings in the field of bio and information technology, in order to enhance existing food processing technologies and make them better by further inventing more new innovative technologies for promoting the optimal health of mankind.

6.4 Factors affecting the food processing industry

In a pan India survey conducted across a set of 250 companies by The Federation of Indian Chambers of Commerce & Industry (FICCI) to figure out the major bottlenecks and challenges (*"food-processing-bottlenecks-study.pdf"*) that gripples the growth of food processing industries; they identify four major issues:

1. **Insufficient infrastructure facilities**
   The food industry lacks sufficient warehousing and infrastructure facilities. The improper logistics infrastructure and connectivity (road and rail) lead to mismanagement of the food supply chain incurring huge losses for the food industries.

   Lack of government support for providing investment in the set-up and adoption of new emerging food processing technologies hampers the growth of the sector.

2. **Absence of a comprehensive national policy related to the processing of food**
   India desperately requires a comprehensive food processing policy that gives tax benefits for setting up food industries in the backward regions of India, so that there is growth in the region and provides employment to the local people. The policy must be broad and holistic, including various administrative, legislative,

and promotional initiatives. The study found that the lack of a clear national policy on the food processing sector is the second most important reason impeding the industry's expansion. According to the people, policy should be developed via extensive conversations among all stakeholders throughout the full value chain and should encourage sustainable agri-business and agro-industry models based on varied agro-climates and regions.

3. **Existence of different food policies and safety laws between the center and the states of India**

   The government of India (GoI) has made various food laws to regulate the proper functioning of food industries. These policies were developed in order to achieve comprehensive food sufficiency, safety, and quality. As a result, rather than a single comprehensive enactment, India's food sector is controlled by a variety of distinct statutes. This piecemeal approach has resulted in regulatory incoherence and inconsistency in the food industry. Furthermore, at both the federal and state levels, the multitude of ministries and administering agencies has resulted in a complicated regulatory framework that is not effectively coordinated, adding to the food industry's burden.

4. **Inadequately trained manpower**

   According to the FICCI study, a shortage of well-qualified workforce is a key impediment to the sector's growth. Technical know-how and assistance are severely lacking at every stage of the value chain. The productivity of food industries is retarded severely by the huge investments involved in the manufacturing of food products, high wastage, and lack of availability of trained and skilled manpower. So, to fully realize the sector's development potential, present obstacles must be adequately handled, and actions must be taken to eliminate bottlenecks impeding growth.

6.5 Conclusion

Today, consumers are becoming more inclined towards consuming functional foods as part of their daily diets as they are beneficial for human health. The FF contains bioactive nutrients that help people to overcome their health deficiencies in order to fight various diseases. To meet the needs of people of different age groups, who have different eating habits and nutrition, tremendous advancements in the technologies for effective food processing, packaging, and preservation will profoundly lead to the growth of food technology in the future.

Though innovative technologies in the processing of food have proven to be beneficial to mankind, they still have certain shortcomings. PF containing high levels of salt, sugar, or other food additives is associated with the development of obesity and other health disorders. Food industries should manufacture nutritionally balanced quality food hygienically in an eco-friendly environment. But consumers should be aware about the ill effects of excessive consumption of PF. Therefore, food technologists must educate consumers about the nutritional value of processed and functional foods.

To meet the growing demand for food and to provide diverse varieties of functional foods, old and new technologies should be used in a coordinated manner by the food processing industries.

Another problem which we face today is that people of developed countries have surplus food and high purchasing power as compared to the people of the developing countries, who face scarcity of food and have limited purchasing power/resources. So, it is a big challenge for the food industry to cater to such a diverse population while taking care of the nutritional value of the food. Further, on one hand, the industries have to do mass production of food to maintain the food supply that could quench the hunger of people of developing countries, on the other hand, in the years to come, food technologists should be able to use the knowledge of biotechnology to manipulate the microbial genes that can produce more thermostable enzymes and improve the existing technologies that will benefit the mankind and promote health.

References

Adams, J., Hofman, K., Moubarac, J.-C., Thow, A.M., 2020. Public health response to ultra-processed food and drinks. BMJ 369, m2391. https://doi.org/10.1136/bmj.m2391.

Arab, E.A.A., Helmy, I.M.F., Bareh, G.F., 2010. Nutritional evaluation and functional properties of chickpea (*Cicer arietinum* L.) flour and the improvement of spaghetti produced from its. J. Am. Sci. 66, 1055–1072.

Arshad, R.N., Abdul-Malek, Z., Munir, A., Buntat, Z., Ahmad, M., Jusoh, Y., Bekhit, A., Qazalbash, U., Manzoor, M., 2020. Electrical systems for pulsed electric field applications in the food industry: an engineering perspective. Trends Food Sci. Technol. 104, 1–13. https://doi.org/10.1016/j.tifs.2020.07.008.

Asaithambi, N., Singh, S.K., Singha, P., 2021. Current status of non-thermal processing of probiotic foods: a review. J. Food Eng. 303, 110567. https://doi.org/10.1016/j.jfoodeng.2021.110567.

Astrup, A., Dyerberg, J., Selleck, M., Stender, S., 2008. Nutrition transition and its relationship to the development of obesity and related chronic diseases. Obes. Rev. Off. J. Int. Assoc. Study Obes. 9 (1), 48–52. https://doi.org/10.1111/j.1467-789X.2007.00438.x.

Atik, A., Gumus, T., 2021. The effect of different doses of UV-C treatment on microbiological quality of bovine milk. LWT 136, 110322. https://doi.org/10.1016/j.lwt.2020.110322.

Bahrami, A., Moaddabdoost Baboli, Z., Schimmel, K., Jafari, S.M., Williams, L., 2020. Efficiency of novel processing technologies for the control of *Listeria monocytogenes* in food products. Trends Food Sci. Technol. 96, 61–78. https://doi.org/10.1016/j.tifs.2019.12.009.

Beermann, C., Hartung, J., 2013. Physiological properties of milk ingredients released by fermentation. Food Funct. (2). https://doi.org/10.1039/c2fo30153a.

Bhargava, N., Mor, R.S., Kumar, K., Sharanagat, V.S., 2021. Advances in application of ultrasound in food processing: a review. Ultrason. Sonochem. 70, 105293. https://doi.org/10.1016/j.ultsonch.2020.105293.

Bhattacharjee, C., Saxena, V.K., Dutta, S., 2019. Novel thermal and non-thermal processing of watermelon juice. Trends Food Sci. Technol. 93, 234–243. https://doi.org/10.1016/j.tifs.2019.09.015.

Bhavya, M.L., Umesh Hebbar, H., 2017. Pulsed light processing of foods for microbial safety. Food Qual. Saf. 1, 187–202. https://doi.org/10.1093/fqsafe/fyx017.

Bleiweiss-Sande, R., Chui, K., Evans, E.W., Goldberg, J., Amin, S., Sacheck, J., 2019. Robustness of food processing classification systems. Nutrients 11, E1344. https://doi.org/10.3390/nu11061344.

Brugger, E.-M., Wagner, J., Schumacher, D.M., Koch, K., Podlech, J., Metzler, M., Lehmann, L., 2006. Mutagenicity of the mycotoxin alternariol in cultured mammalian cells. Toxicol. Lett. 164, 221–230. https://doi.org/10.1016/j.toxlet.2006.01.001.

Byrns, M.C., Vu, C.C., Neidigh, J.W., Abad, J.-L., Jones, R.A., Peterson, L.A., 2006. Detection of DNA adducts derived from the reactive metabolite of furan, *cis-* 2-butene-1,4-dial. Chem. Res. Toxicol. 19, 414–420. https://doi.org/10.1021/tx050302k.

Cai, J., Bhatnagar, A., Pierce, W.M., 2009. Protein modification by acrolein: formation and stability of cysteine adducts. Chem. Res. Toxicol. 22, 708–716. https://doi.org/10.1021/tx800465m.

Calabrò, E., Magazù, S., 2012. Comparison between conventional convective heating and microwave heating: an FTIR spectroscopy study of the effects of microwave oven cooking of bovine breast meat. J. Electromagn. Anal. Appl. 4, 433–439. https://doi.org/10.4236/jemaa.2012.411060.

Capuano, E., Fogliano, V., 2011. Acrylamide and 5-hydroxymethylfurfural (HMF): a review on metabolism, toxicity, occurrence in food and mitigation strategies. LWT Food Sci. Technol. 44, 793–810. https://doi.org/10.1016/j.lwt.2010.11.002.

Cartus, A., Schrenk, D., 2017. Current methods in risk assessment of genotoxic chemicals. Food Chem. Toxicol. 106, 574–582. https://doi.org/10.1016/j.fct.2016.09.012.

Chacha, J.S., Zhang, L., Ofoedu, C.E., Suleiman, R.A., Dotto, J.M., Roobab, U., Agunbiade, A.O., Duguma, H.T., Mkojera, B.T., Hossaini, S.M., Rasaq, W.A., Shorstkii, I., Okpala, C.O.R., Korzeniowska, M., Guiné, R.P.F., 2021. Revisiting non-thermal food processing and preservation methods—action mechanisms, pros and cons: a technological update (2016–2021). Foods 10, 1430. https://doi.org/10.3390/foods10061430.

Chang, K., Khandpur, N., Neri, D., Touvier, M., Huybrechts, I., Millett, C., Vamos, E.P., 2021. Association between childhood consumption of ultraprocessed food and adiposity trajectories in the Avon longitudinal study of parents and children birth cohort. JAMA Pediatr. 175, e211573. https://doi.org/10.1001/jamapediatrics.2021.1573.

Chemat, F., Rombaut, N., Sicaire, A.-G., Meullemiestre, A., Fabiano-Tixier, A.-S., Abert-Vian, M., 2017. Ultrasound assisted extraction of food and natural products—mechanisms, techniques, combinations, protocols and applications: a review. Ultrason. Sonochem. 34, 540–560. https://doi.org/10.1016/j.ultsonch.2016.06.035.

Colle, I., Buggenhout, S.V., Loey, A.V., Hendrickx, M., 2010. High pressure homogenization followed by thermal processing of tomato pulp: influence on microstructure and lycopene in vitro bioaccessibility. Food Res. Int. 43 (8), 2193–2200.

Costa, C.S., Rauber, F., Leffa, P.S., Sangalli, C.N., Campagnolo, P.D.B., Vitolo, M.R., 2019. Ultra-processed food consumption and its effects on anthropometric and glucose profile: a longitudinal study during childhood. Nutr. Metab. Cardiovasc. Dis. NMCD 29, 177–184. https://doi.org/10.1016/j.numecd.2018.11.003.

Coutinho, N.M., Silveira, M.R., Fernandes, L.M., Moraes, J., Pimentel, T.C., Freitas, M.Q., Silva, M.C., Raices, R.S.L., Ranadheera, C.S., Borges, F.O., Neto, R.P.C., Tavares, M.I.B., Fernandes, F.A.N., Fonteles, T.V., Nazzaro, F., Rodrigues, S., Cruz, A.G., 2019. Processing chocolate milk drink by low-pressure cold plasma technology. Food Chem. 278, 276–283. https://doi.org/10.1016/j.foodchem.2018.11.061.

Crimarco, A., Springfield, S., Petlura, C., Streaty, T., Cunanan, K., Lee, J., Fielding-Singh, P., Carter, M.M., Topf, M.A., Wastyk, H.C., Sonnenburg, E.D., Sonnenburg, J.L., Gardner, C.D., 2020. A randomized crossover trial on the effect of plant-based compared with animal-based meat on trimethylamine-N-oxide and cardiovascular disease risk factors in generally healthy adults: study with appetizing plantfood-meat eating alternative trial (SWAP-MEAT). Am. J. Clin. Nutr. 112, 1188–1199. https://doi.org/10.1093/ajcn/nqaa203.

Daomukda, N., Moongngarm, A., Payakapol, L., Noisuwan, A., 2011. Effect of cooking methods on physicochemical properties of brown rice 4, 4.

Doeun, D., Davaatseren, M., Chung, M.-S., 2017. Biogenic amines in foods. Food Sci. Biotechnol. 26, 1463–1474. https://doi.org/10.1007/s10068-017-0239-3.

Dong, X., Wang, J., Raghavan, V., 2021. Critical reviews and recent advances of novel non-thermal processing techniques on the modification of food allergens. Crit. Rev. Food Sci. Nutr. 61, 196–210. https://doi.org/10.1080/10408398.2020.1722942.

Drewnowski, A., 2021. Perspective: identifying ultra-processed plant-based milk alternatives in the USDA branded food products database. Adv. Nutr. 12 (6), 2068–2075. https://doi.org/10.1093/advances/nmab089.

EFSA, 2008. Polycyclic aromatic hydrocarbons in food [1] – scientific opinion of the panel on contaminants in the food chain. EFSA. https://www.efsa.europa.eu/en/efsajournal/pub/724. (Accessed 15 January 2022).

El-Assouli, S.M., 2009. Acrylamide in selected foods and genotoxicity of their extracts. J. Egypt. Public Health Assoc. 84, 371–392.

Fan, D., Li, L., Zhang, N., Zhao, Y., Cheng, K.-W., Yan, B., Wang, Q., Zhao, J., Wang, M., Zhang, H., 2018. A comparison of mutagenic PhIP and beneficial 8-C-(E-phenylethenyl)quercetin and 6-C-(E-phenylethenyl)quercetin formation under microwave and conventional heating. Food Funct. 9, 3853–3859. https://doi.org/10.1039/c8fo00542g.

Fardet, A., Rock, E., 2019. Ultra-processed foods: a new holistic paradigm? Trends Food Sci. Technol. 93, 174–184. https://doi.org/10.1016/j.tifs.2019.09.016.

Fehr, M., Pahlke, G., Fritz, J., Christensen, M.O., Boege, F., Altemöller, M., Podlech, J., Marko, D., 2009. Alternariol acts as a topoisomerase poison, preferentially affecting the IIalpha isoform. Mol. Nutr. Food Res. 53, 441–451. https://doi.org/10.1002/mnfr.200700379.

Gibney, M.J., 2019. Ultra-processed foods: definitions and policy issues. Curr. Dev. Nutr. 3, nzy077. https://doi.org/10.1093/cdn/nzy077.

Grundy, M.M.-L., Lapsley, K., Ellis, P.R., 2016. A review of the impact of processing on nutrient bioaccessibility and digestion of almonds. Int. J. Food Sci. Technol. 51, 1937–1946. https://doi.org/10.1111/ijfs.13192.

Hall, K.D., Ayuketah, A., Brychta, R., Cai, H., Cassimatis, T., Chen, K.Y., Chung, S.T., Costa, E., Courville, A., Darcey, V., Fletcher, L.A., Forde, C.G., Gharib, A.M., Guo, J., Howard, R., Joseph, P.V., McGehee, S., Ouwerkerk, R., Raisinger, K., Rozga, I., Stagliano, M., Walter, M., Walter, P.J., Yang, S., Zhou, M., 2019. Ultra-processed diets cause excess calorie intake and weight gain: an inpatient randomized controlled trial of ad libitum food intake. Cell Metab. 30, 67–77. https://doi.org/10.1016/j.cmet.2019.05.008.

Hawton, K., Ferriday, D., Rogers, P., Toner, P., Brooks, J., Holly, J., Biernacka, K., Hamilton-Shield, J., Hinton, E., 2018. Slow down: behavioural and physiological effects of reducing eating rate. Nutrients 11, E50. https://doi.org/10.3390/nu11010050.

Hernández-Hernández, H.M., Moreno-Vilet, L., Villanueva-Rodríguez, S.J., 2019. Current status of emerging food processing technologies in Latin America: novel non-thermal processing. Innov. Food Sci. Emerg. Technol. 58, 102233. https://doi.org/10.1016/j.ifset.2019.102233.

Hite, B.H., 1899. The effect of pressure in the preservation of milk: a preliminary report. West Virginia Agricultural and Forestry Experiment Station Bulletins, 58. https://doi.org/10.33915/agnic.58.

Hite, B.H., Giddings, N.J., Weakley, C.E., 1914. The effect of pressure on certain microorganisms encountered in the preservation of fruits and vegetables. West Virginia Agricultural and Forestry Experiment Station Bulletins, 146. https://doi.org/10.33915/agnic.146.

Huang, H.-W., Wu, S.-J., Lu, J.-K., Shyu, Y.-T., Wang, C.-Y., 2017. Current status and future trends of high-pressure processing in food industry. Food Control 72, 1–8. https://doi.org/10.1016/j.foodcont.2016.07.019.

Indiarto, R., Pratama, A., Theodora, H., Sari, T., 2020. Food irradiation technology: a review of the uses and their capabilities. Int. J. Eng. Trends Technol. 68, 91–98. https://doi.org/10.14445/22315381/IJETT-V68I12P216.

Indiarto, R., Qonit, M.A.H., 2020. A review of irradiation technologies on food and agricultural products 9, 4.

Ivbade, A., Ojo, O.E., Dipeolu, M.A., 2014. Shiga toxin-producing *Escherichia coli* O157:H7 in milk and milk products in Ogun state. Nigeria. Vet. Ital. 50, 185–191. https://doi.org/10.12834/VetIt.129.2187.1.

Juul, F., Vaidean, G., Parekh, N., 2021. Ultra-processed foods and cardiovascular diseases: potential mechanisms of action. Adv. Nutr. 12, 1673–1680. https://doi.org/10.1093/advances/nmab049.

Kabak, B., Dobson, A.D.W., 2011. An introduction to the traditional fermented foods and beverages of Turkey. Crit. Rev. Food Sci. Nutr. 51, 248–260. https://doi.org/10.1080/10408390903569640.

Kensler, T.W., Roebuck, B.D., Wogan, G.N., Groopman, J.D., 2011. Aflatoxin: a 50-year odyssey of mechanistic and translational toxicology. Toxicol. Sci. 120 (Suppl 1), S28–S48. https://doi.org/10.1093/toxsci/kfq283.

Khan, M., Ahmad, K., Hassan, S., Imran, M., Ahmad, N., Xu, C., 2018. Effect of novel technologies on polyphenols during food processing. Innov. Food Sci. Emerg. Technol. 45, 361–381.

Khandpur, N., Cediel, G., Obando, D.A., Jaime, P.C., Parra, D.C., 2020. Sociodemographic factors associated with the consumption of ultra-processed foods in Colombia. Rev. Saude Publica 54, 19. https://doi.org/10.11606/s1518-8787.2020054001176.

Kim, H.Y., Kim, S.H., Choi, M.J., Min, S.G., Kwak, H.S., 2008. The effect of high pressure–low temperature treatment on physicochemical properties in milk. J. Dairy Sci. 91, 4176–4182. https://doi.org/10.3168/jds.2007-0883.

Knockaert, G., Pulissery, S.K., Colle, I., Van Buggenhout, S., Hendrickx, M., Loey, A.V., 2012. Lycopene degradation, isomerization and in vitro bioaccessibility in high pressure homogenized tomato puree containing oil: effect of additional thermal and high pressure processing. Food Chem. 135, 1290–1297. https://doi.org/10.1016/j.foodchem.2012.05.065.

Krauss, M., Wilcke, W., Martius, C., Bandeira, A.G., Garcia, M.V.B., Amelung, W., 2005. Atmospheric versus biological sources of polycyclic aromatic hydrocarbons (PAHs) in a tropical rain forest environment. Environ. Pollut. Barking Essex 135, 143–154. https://doi.org/10.1016/j.envpol.2004.09.012.

Kwolek-Mirek, M., Zadrag-Tecza, R., Bednarska, S., Bartosz, G., 2011. Yeast Saccharomyces cerevisiae devoid of Cu,Zn-superoxide dismutase as a cellular model to study acrylamide toxicity. Toxicol. Vitro Int. J. Publ. Assoc. BIBRA 25, 573–579. https://doi.org/10.1016/j.tiv.2010.12.007.

Lane, M., Howland, G., West, M., Hockey, M., Marx, W., Loughman, A., O'Hely, M., Jacka, F., Rocks, T., 2020. The effect of ultra-processed very low-energy diets on gut microbiota and metabolic outcomes in individuals with obesity: a systematic literature review. Obes. Res. Clin. Pract. 14, 197–204. https://doi.org/10.1016/j.orcp.2020.04.006.

Lee, S., Munerol, B., Pollard, S., Youdim, K., Pannala, A., Kuhnle, G., Debnam, E., Rice-Evans, C., Spencer, J., 2006. The reaction of flavanols with nitrous acid protects against N-nitrosamine formation and leads to the formation of nitroso derivatives which inhibit cancer cell growth. Free Radic. Biol. Med. 40, 323–334. https://doi.org/10.1016/j.freeradbiomed.2005.08.031.

Leffa, P.S., Hoffman, D.J., Rauber, F., Sangalli, C.N., Valmórbida, J.L., Vitolo, M.R., 2020. Longitudinal associations between ultra-processed foods and blood lipids in childhood. Br. J. Nutr. 124, 341–348. https://doi.org/10.1017/S0007114520001233.

Liepa, M., Zagorska, J., Galoburda, R., 2016. High-pressure processing as novel technology in dairy industry: a review. In: Research for Rural Development. International Scientific Conference Proceedings (Latvia). Latvia University of Agriculture.

LoPachin, R.M., Gavin, T., Petersen, D.R., Barber, D.S., 2009. Molecular mechanisms of 4-Hydroxy-2-nonenal and acrolein toxicity: nucleophilic targets and adduct formation. Chem. Res. Toxicol. 22, 1499–1508. https://doi.org/10.1021/tx900147g.

López-Gámez, G., Elez-Martínez, P., Martín-Belloso, O., Soliva-Fortuny, R., 2021. Recent advances toward the application of non-thermal technologies in food processing: an insight on the bioaccessibility of health-related constituents in plant-based products. Foods 10, 1538. https://doi.org/10.3390/foods10071538.

Lorenzoni, G., Benedetto, R.D., Ocagli, H., Gregori, D., Silano, M., 2021. A validation study of NOVA classification for ultra-processed food on the USDA food and nutrient database. Curr. Dev. Nutr. 5, 594. https://doi.org/10.1093/cdn/nzab044_025.

Magnavita, N., Fileni, A., 1994. Occupational risk caused by ultrasound in medicine. Radiol. Med. (Torino) 88, 107–111.

Mandal, R., Singh, A., Pratap Singh, A., 2018. Recent developments in cold plasma decontamination technology in the food industry. Trends Food Sci. Technol. 80, 93–103. https://doi.org/10.1016/j.tifs.2018.07.014.

Martínez Steele, E., Popkin, B.M., Swinburn, B., Monteiro, C.A., 2017. The share of ultra-processed foods and the overall nutritional quality of diets in the US: evidence from a nationally representative cross-sectional study. Popul. Health Metr. 15, 6. https://doi.org/10.1186/s12963-017-0119-3.

Matos, R.A., Adams, M., Sabaté, J., 2021. Review: the consumption of ultra-processed foods and non-communicable diseases in Latin America. Front. Nutr. 8, 622714. https://doi.org/10.3389/fnut.2021.622714.

Megahey, E.K., McMinn, W.A.M., Magee, T.R.A., 2005. Experimental study of microwave baking of madeira cake batter. Food Bioprod. Process. 83, 277–287. https://doi.org/10.1205/fbp.05033.

Meléndez-Martínez, A.J., Paulino, M., Stinco, C.M., Mapelli-Brahm, P., Wang, X.-D., 2014. Study of the time-course of cis/trans (Z/E) isomerization of lycopene, phytoene, and phytofluene from tomato. J. Agric. Food Chem. 62, 12399–12406. https://doi.org/10.1021/jf5041965.

Moen, B., Berget, I., Rud, I., Hole, A.S., Kjos, N.P., Sahlstrøm, S., 2016. Extrusion of barley and oat influence the fecal microbiota and SCFA profile of growing pigs. Food Funct. 7, 1024–1032. https://doi.org/10.1039/c5fo01452b.

Monteiro, C.A., Cannon, G., Levy, R.B., Moubarac, J.-C., Louzada, M.L., Rauber, F., Khandpur, N., Cediel, G., Neri, D., Martinez-Steele, E., Baraldi, L.G., Jaime, P.C., 2019. Ultra-processed foods: what they are and how to identify them. Public Health Nutr. 22, 936–941. https://doi.org/10.1017/S1368980018003762.

Odhav, B., Naicker, V., 2002. Mycotoxins in South African traditionally brewed beers. Food Addit. Contam. 19, 55–61. https://doi.org/10.1080/02652030110053426.

O'Donnell, C., Tiwari, B.K., Cullen, P.J., Rice, R.G., 2012. Ozone in Food Processing. John Wiley & Sons.

Okpala, C., 2017a. Fish processing by ozone treatment: is further investigation of domestic applications needful? Chem. Eng. Trans. 57, 1813–1818. https://doi.org/10.3303/CET1757303.

Okpala, C., 2017b. Ozone delivery on food materials incorporating some bio-based processes: a succinct synopsis. Adv. Mater. Proc. 2, 469–478. https://doi.org/10.5185/amp.2017/802.

Okpala, C.O.R., 2018. Changes in some biochemical and microbiological properties of ozone-processed shrimps: effects of increased ozone discharge combined with iced storage. J. Food Nutr. Res. 1–9.

Ölmez, H., Kretzschmar, U., 2009. Potential alternative disinfection methods for organic fresh-cut industry for minimizing water consumption and environmental impact. LWT Food Sci. Technol. 42, 686–693. https://doi.org/10.1016/j.lwt.2008.08.001.

Oueslati, I., Taamalli, W., Haddada, F.M., Zarrouk, M., 2010. Microwave heating effects on the chemical composition and the antioxidant capacity of tataouine virgin olive oils from Tunisia. J. Food Prot. 73, 1891–1901. https://doi.org/10.4315/0362-028x-73.10.1891.

Pal, M., 2017. Pulsed electric field processing: an emerging technology for food preservation. J. Exp. Food Chem. 3 (2), 1–2. https://doi.org/10.4172/2472-0542.1000126.

Patricia, C., Bibiana, D., José, P., 2011. Evaluation of microwave technology in blanching of broccoli (Brassica oleracea L. var Botrytis) as a substitute for conventional blanching. Procedia Food Sci. 1, 426–432. https://doi.org/10.1016/j.profoo.2011.09.066.

Popkin, B., 2019. Ultra-processed Foods' Impacts on Health: Document No. 34, 2030 - Food, Agriculture and Rural Development in 2030 in Latin America and the Caribbean. FAO, Santiago, Chile.

Priyadarshini, A., Rajauria, G., O'Donnell, C.P., Tiwari, B.K., 2019. Emerging food processing technologies and factors impacting their industrial adoption. Crit. Rev. Food Sci. Nutr. 59, 3082–3101. https://doi.org/10.1080/10408398.2018.1483890.

Rahaman, T., Vasiljevic, T., Ramchandran, L., 2016. Effect of processing on conformational changes of food proteins related to allergenicity. Trends Food Sci. Technol. 49, 24–34. https://doi.org/10.1016/j.tifs.2016.01.001.

Rancière, F., Lyons, J.G., Loh, V.H.Y., Botton, J., Galloway, T., Wang, T., Shaw, J.E., Magliano, D.J., 2015. Bisphenol A and the risk of cardiometabolic disorders: a systematic review with meta-analysis of the epidemiological evidence. Environ. Health Glob. Access Sci. Source 14, 46. https://doi.org/10.1186/s12940-015-0036-5.

Rauber, F., da Costa Louzada, M.L., Steele, E.M., Millett, C., Monteiro, C.A., Levy, R.B., 2018. Ultra-processed food consumption and chronic non-communicable diseases-related dietary nutrient profile in the UK (2008–2014). Nutrients 10, E587. https://doi.org/10.3390/nu10050587.

Ribas-Agustí, A., Martín-Belloso, O., Soliva-Fortuny, R., Elez-Martínez, P., 2018. Food processing strategies to enhance phenolic compounds bioaccessibility and bioavailability in plant-based foods. Crit. Rev. Food Sci. Nutr. 58, 2531–2548. https://doi.org/10.1080/10408398.2017.1331200.

Robinson, E., Almiron-Roig, E., Rutters, F., de Graaf, C., Forde, C.G., Tudur Smith, C., Nolan, S.J., Jebb, S.A., 2014. A systematic review and meta-analysis examining the effect of eating rate on energy intake and hunger. Am. J. Clin. Nutr. 100, 123–151. https://doi.org/10.3945/ajcn.113.081745.

Robjohns, S., Marshall, R., Fellows, M., Kowalczyk, G., 2003. In vivo genotoxicity studies with 3-monochloropropan-1,2-diol. Mutagenesis 18, 401–404. https://doi.org/10.1093/mutage/geg017.

Rodríguez-Roque, M.J., de Ancos, B., Sánchez-Moreno, C., Cano, M.P., Elez-Martínez, P., Martín-Belloso, O., 2015. Impact of food matrix and processing on the in vitro bioaccessibility of vitamin C, phenolic compounds, and hydrophilic antioxidant activity from fruit juice-based beverages. J. Funct. Foods 14, 33–43. https://doi.org/10.1016/j.jff.2015.01.020.

Rodríguez-Roque, M.J., De Ancos, B., Sánchez-Vega, R., Sánchez-Moreno, C., Elez-Martínez, P., Martín-Belloso, O., 2020. In vitro bioaccessibility of isoflavones from a soymilk-based beverage as affected by thermal and non-thermal processing. Innov. Food Sci. Emerg. Technol. 66, 102504. https://doi.org/10.1016/j.ifset.2020.102504.

Rodríguez-Roque, M.J., Rojas-Graü, M.A., Elez-Martínez, P., Martín-Belloso, O., 2014. *In vitro* bioaccessibility of health-related compounds from a blended fruit juice–soymilk beverage: influence of the food matrix. J. Funct. Foods Complete 7, 161–169. https://doi.org/10.1016/j.jff.2014.01.023.

Roobab, U., Aadil, R.M., Madni, G.M., Bekhit, A.E.-D., 2018. The impact of nonthermal technologies on the microbiological quality of juices: a review. Compr. Rev. Food Sci. Food Saf. 17, 437–457. https://doi.org/10.1111/1541-4337.12336.

Roze, L.V., Hong, S.-Y., Linz, J.E., 2013. Aflatoxin biosynthesis: current frontiers. Annu. Rev. Food Sci. Technol. 4, 293–311. https://doi.org/10.1146/annurev-food-083012-123702.

Sammugam, L., Pasupuleti, V.R., 2019. Balanced diets in food systems: emerging trends and challenges for human health. Crit. Rev. Food Sci. Nutr. 59, 2746–2759. https://doi.org/10.1080/10408398.2018.1468729.

Sampaio, G.R., Guizellini, G.M., da Silva, S.A., de Almeida, A.P., Pinaffi-Langley, A.C.C., Rogero, M.M., de Camargo, A.C., Torres, E.A.F.S., 2021. Polycyclic Aromatic hydrocarbons in foods: biological effects, legislation, occurrence, analytical methods, and strategies to reduce their formation. Int. J. Mol. Sci. 22, 6010. https://doi.org/10.3390/ijms22116010.

Santé-Lhoutellier, V., Astruc, T., Marinova, P., Greve, E., Gatellier, P., 2008. Effect of meat cooking on physicochemical state and in vitro digestibility of myofibrillar proteins. J. Agric. Food Chem. 56, 1488–1494. https://doi.org/10.1021/jf072999g.

Scussel, V.M., Scholten, J.M., Rensen, P.M., Spanjer, M.C., Giordano, B.N.E., Savi, G.D., 2013. Multitoxin evaluation in fermented beverages and cork stoppers by liquid chromatography–tandem mass spectrometry. Int. J. Food Sci. Technol. 48, 96–102. https://doi.org/10.1111/j.1365-2621.2012.03163.x.

Seefelder, W., Varga, N., Studer, A., Williamson, G., Scanlan, F.P., Stadler, R.H., 2008. Esters of 3-chloro-1,2-propanediol (3-MCPD) in vegetable oils: significance in the formation of 3-MCPD. Food Addit. Contam. Part Chem. Anal. Control Expo. Risk Assess. 25, 391–400. https://doi.org/10.1080/02652030701385241.

Shahbaz, H.M., Kim, J.U., Kim, S.H., Park, J., 2018. Advances in nonthermal processing technologies for enhanced microbiological safety and quality of fresh fruit and juice products. In: Food Processing for Increased Quality and Consumption. Elsevier, pp. 179–217. https://doi.org/10.1016/B978-0-12-811447-6.00007-2.

Sivamaruthi, B.S., Kesika, P., Chaiyasut, C., 2018a. Toxins in fermented foods—prevalence and preventions: a mini review. Toxins 11, E4. https://doi.org/10.3390/toxins11010004.

Sivamaruthi, B.S., Kesika, P., Chaiyasut, C., 2018b. Thai fermented foods as a versatile source of bioactive microorganisms: a comprehensive review. Sci. Pharm. 86, E37. https://doi.org/10.3390/scipharm86030037.

Sonawane, S.K., T, M., Patil, S., 2020. Non-thermal plasma: an advanced technology for food industry. Food Sci. Technol. Int. Cienc. Tecnol. Los Aliment. Int. 26, 727–740. https://doi.org/10.1177/1082013220929474.

Stender, S., Dyerberg, J., Astrup, A., 2006. High levels of industrially produced trans fat in popular fast foods. N. Engl. J. Med. 354, 1650–1652. https://doi.org/10.1056/NEJMc052959.

Sugimura, T., Kawachi, T., Nagao, M., Yahagi, T., Seino, Y., Okamoto, T., Shudo, K., Kosuge, T., Tsuji, K., Wakabayashi, K., Iitaka, Y., Itai, A., 1977. Mutagenic principle(s) in tryptophan and phenylalanine pyrolysis products. Proc. Japan Acad. 53, 58–61. https://doi.org/10.2183/pjab1945.53.58.

Tareke, E., Rydberg, P., Karlsson, P., Eriksson, S., Törnqvist, M., 2000. Acrylamide: a cooking carcinogen? Chem. Res. Toxicol. 13, 517–522. https://doi.org/10.1021/tx9901938.

Troy, D.J., Ojha, K.S., Kerry, J.P., Tiwari, B.K., 2016. Sustainable and consumer-friendly emerging technologies for application within the meat industry: an overview. Meat Sci. 120, 2–9. https://doi.org/10.1016/j.meatsci.2016.04.002.

Vedovato, G.M., Vilela, S., Severo, M., Rodrigues, S., Lopes, C., Oliveira, A., 2021. Ultra-processed food consumption, appetitive traits and BMI in children: a prospective study. Br. J. Nutr. 125, 1427–1436. https://doi.org/10.1017/S0007114520003712.

Velísek, J., Davídek, J., Hajslová, J., Kubelka, V., Janícek, G., Mánková, B., 1978. Chlorohydrins in protein hydrolysates. Z. Lebensm. Unters. Forsch. 167, 241–244. https://doi.org/10.1007/BF01135595.

Verma, D.K., Thakur, M., Kumar, J., Srivastav, P.P., Al-Hilphy, A.R.S., Patel, A.R., Suleria, H.A.R., 2020. High pressure processing (HPP): fundamental concepts, emerging scope, and food application. In: Emerging Thermal and Nonthermal Technologies in Food Processing. Apple Academic Press, p. 33.

Weaver, C.M., Dwyer, J., Fulgoni III, V.L., King, J.C., Leveille, G.A., MacDonald, R.S., Ordovas, J., Schnakenberg, D., 2014. Processed foods: contributions to nutrition. Am. J. Clin. Nutr. 99, 1525–1542. https://doi.org/10.3945/ajcn.114.089284.

Weidenborner, M., 2001. Encyclopedia of food mycotoxins: buy encyclopedia of food mycotoxins by weidenborner martin at low price in India. https://www.flipkart.com/encyclopedia-food-mycotoxins/p/itmfc6gxqashcr3e. (Accessed 15 January 2022).

Wenzl, T., Lachenmeier, D.W., Gökmen, V., 2007. Analysis of heat-induced contaminants (acrylamide, chloropropanols and furan) in carbohydrate-rich food. Anal. Bioanal. Chem. 389, 119–137. https://doi.org/10.1007/s00216-007-1459-9.

Yaylayan, V.A., 2006. Precursors, formation and determination of furan in food. J. Für Verbraucherschutz Leb. 1, 5–9. https://doi.org/10.1007/s00003-006-0003-8.

Yoshida, M., Tomitori, H., Machi, Y., Hagihara, M., Higashi, K., Goda, H., Ohya, T., Niitsu, M., Kashiwagi, K., Igarashi, K., 2009. Acrolein toxicity: comparison with reactive oxygen species. Biochem. Biophys. Res. Commun. 378, 313–318. https://doi.org/10.1016/j.bbrc.2008.11.054.

Yurchenko, S., Mölder, U., 2005. N-nitrosodimethylamine analysis in Estonian beer using positive-ion chemical ionization with gas chromatography mass spectrometry. Food Chem. 89, 455–463. https://doi.org/10.1016/j.foodchem.2004.05.034.

Zamani, E., Shokrzade, M., Fallah, M., Shaki, F., 2017. A review of acrylamide toxicity and its mechanism. Pharm. Biomed. Res. 3, 1–7. https://doi.org/10.18869/acadpub.pbr.3.1.1.

Zhang, W., Yu, Y., Xie, F., Gu, X., Wu, J., Wang, Z., 2019. High pressure homogenization versus ultrasound treatment of tomato juice: effects on stability and in vitro bioaccessibility of carotenoids. Lebensm.-Wiss. Ie Technol. 116, 108597.

Zhong, S., Vendrell-Pacheco, M., Heskitt, B., Chitchumroonchokchai, C., Failla, M., Sastry, S.K., Francis, D.M., Martin-Belloso, O., Elez-Martínez, P., Kopec, R.E., 2019. Novel processing technologies as compared to thermal treatment on the bioaccessibility and caco-2 cell uptake of carotenoids from tomato and kale-based juices. J. Agric. Food Chem. 67, 10185–10194. https://doi.org/10.1021/acs.jafc.9b03666.

Zinöcker, M.K., Lindseth, I.A., 2018. The western diet: microbiome-host interaction and its role in metabolic disease. Nutrients 10, 365. https://doi.org/10.3390/nu10030365.

Challenges of nanobiotechnology in food processing

Kantrol Kumar Sahu[a], Ramakant Joshi[b], Wasim Akram[b], Ganesh Kumar[c], Krishna Yadav[d], Madhulika Pradhan[e] and Sunita Minz[f]

[a]*Institute of Pharmaceutical Research, GLA University, Mathura, Uttar Pradesh, India,* [b]*School of Studies in Pharmaceutical Sciences, Jiwaji University, Gwalior, Madhya Pradesh, India,* [c]*Sita Ram Kashyap College of Pharmacy, Rahaud, Chhattisgarh, India,* [d]*Raipur Institute of Pharmaceutical Education and Research, Raipur, Chhattisgarh, India,* [e]*Gracious College of Pharmacy, Abhanpur, Chhattisgarh, India,* [f]*Department of Pharmacy, Indira Gandhi National Tribal University, Amarkantak, Madhya Pradesh, India*

7.1 Introduction

Food processing is one of the oldest, critical, and integral parts of the field of the food supply chain. It is the part of ancient civilization to be used by humans. In ancient times, the use of fire by humans to cook food is one such example of food processing. Processes like drying, milling, and extraction of oil are examples of primary processing while further converting these food materials into finished products to extend shelf life is secondary processing. Nowadays, food processing is a big profit business for companies and it generates opportunities for income and employment in society (Kim, 2013). Processed food products are an essential part of life now and they also contribute to our nutrition security (Augustin et al., 2016).

Nanoscience technology is fast-rising in the food industry as a novel area for research and development. Nanotechnology covers the atomic, molecular, and macro-molecular aspects of understanding and recovering food product quality. Nano-sized particles are more active biologically as it has a larger surface area per mass in comparison with similar chemistry having bigger particles (Oberdorster et al., 2005). Along with this prime property, it has other special characteristics like greater penetrability, reactivity, and quantum property. This makes it used in a lesser quantity with enhanced chemical and physical characteristics as compared to large-size particles. Thus, nanotechnology has the ability to produce revolutionary results for both science and people with multiple paybacks to the stakeholders in the food product chain (Lane and Kalil, 2005).

Nano-range nutrients are naturally present in food materials such as globular protein (10–100 nm) containing food sources. Other example includes milk protein casein which has a nano-size range or filamentous protein in meat (Prasad et al., 2017). Food nanosensing is also a way to attain security assessment of food products by enhancing the stability and mechanical power of the packing material. This can be achieved by the use of proper filling agents, antimicrobial agents, food additives, and anticaking agents (Ezhilarasi et al., 2013). Food products where stability and bioavailability are important concerns apply this type of nanotechnology (Duncan, 2011).

Nanotechnology makes easy accessibility of food products with newer manufactured products. It is one of the capable areas to boost the novel and special applications in food as well as agriculture (Sadeghi et al., 2017). Growing foodstuffs with increased productivity and declined postharvest expenses shows better results of applied nanotechnology and biotechnology in this field (Yadollahi et al., 2010). Some of the rising aspects of nanotechnology for food processing include nanoencapsulation of nutraceuticals, minimizing biological and chemical contaminants, solubilization, bioseparation of proteins, and smart delivery of nutrients (Ravichandran, 2010).

Food processing nanotechnology involves the strengthening of the bioactive elements to change their biological understanding and improve their adaptation against various chemical as well as environmental differences (Mozafari et al., 2008). It comprises of enhanced sensory features like color, odor, and consistency in food (Kalita and Baruah, 2019). It also amplifies the biological convenience and captivation of nutraceuticals along with drug delivery systems (Jafari and McClements, 2017). Nanotechnology is a profitable option for food industries as a newer food packing material with customized fencing, and mechanical as well as antimicrobial properties (Mustafa and Andreescu, 2020). Another advantage of this technology is sensors for trace assessment of food conditions in terms of storage, and transportation along with encapsulation of nutrient adjusters (Chaudhry et al., 2008).

In a broad way, nanotechnology has several applications in food industries (Fig. 7.1) and ensures the progress of novel and enhanced products in the most promising way. However, some studies showed that public health risks is connected with nanotechnology products which should be a major concern for various scientist to do further investigations (Kamarulzaman et al., 2020).

Thus, this technology provides visualized information and enables incremental and significant progress in food productivity and effectiveness (Scott, 2005). In this review, the fundamentals of nanobiotechnology, its importance, application, and challenges in food processing are discussed.

7.2 Challenges of nanobiotechnology in food processing

The advancements in food processing nanotechnology are more concerned with the manufacture of modified foodstuffs that improve the taste as well as the texture of food components, encapsulate food additives, regulate the release of tastes, and increase the nutritional component's bioavailability and make food more nutritive for customers. According to customer demands, food production has experienced a

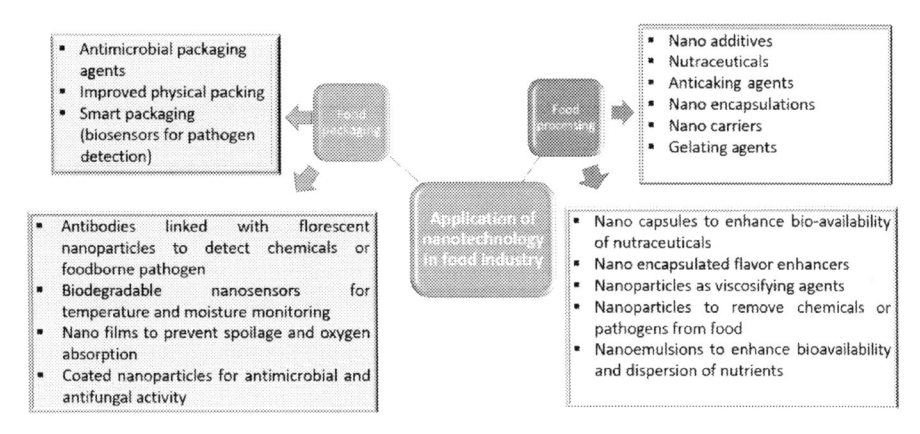

FIGURE 7.1

Applications of nanotechnology in the field of food industry.

number of modifications over time. The application of nanotechnology in the food business has evolved, but there are still significant obstacles to overcome in standings of environmental concerns, human health, and safety (Fig. 7.2).

7.2.1 Nanotoxicity of nanofood

There are worries regarding the accumulation of produced nanomaterial and their potential infiltration into the food chain as a result of the fast development of nanotechnology. The carbohydrates, proteins, and fat globules in milk and mayonnaise are just a few examples of traditional foods that contain a lot of nanoscale materials. However, using engineered nanomaterials in water as well as food products could be hazardous to the environment and human well-being (Gruère et al., 2011). Nanomaterials created by bioengineering are typically less than 100 nm in size, highly reproducible, and monodisperse. Nanomaterials have changed characteristics that are innovative and quite unlike those seen in bulk constituents due to the smaller molecular scale as well as because of altered interactions between molecules. These bioengineered compounds could have a wide range of harmful consequences. The disintegration of these nanostructures might result in a distinct hazardous impact that is challenging to anticipate. Numerous catalytic and oxidative processes using nanomaterials can lead to cytotoxicity in some cases. Because of their higher surface-to-volume ratios compared to their bulk equivalents, the consequent toxicity could be higher. Quantum dots (QDs), fullerenes, metallic, metallic-oxide nanoparticles as well and fibrous nanomaterials have all been discovered to have negative consequences that comprise chromosomal fragmentation, DNA strand breaks, and alterations in gene precision. Because of the physical size of an engineered nanostructure when compared to numerous biological molecules like proteins, structures like viruses, and antibodies, these substances may readily infiltrate tissues, cells, and organelles, simultaneously in

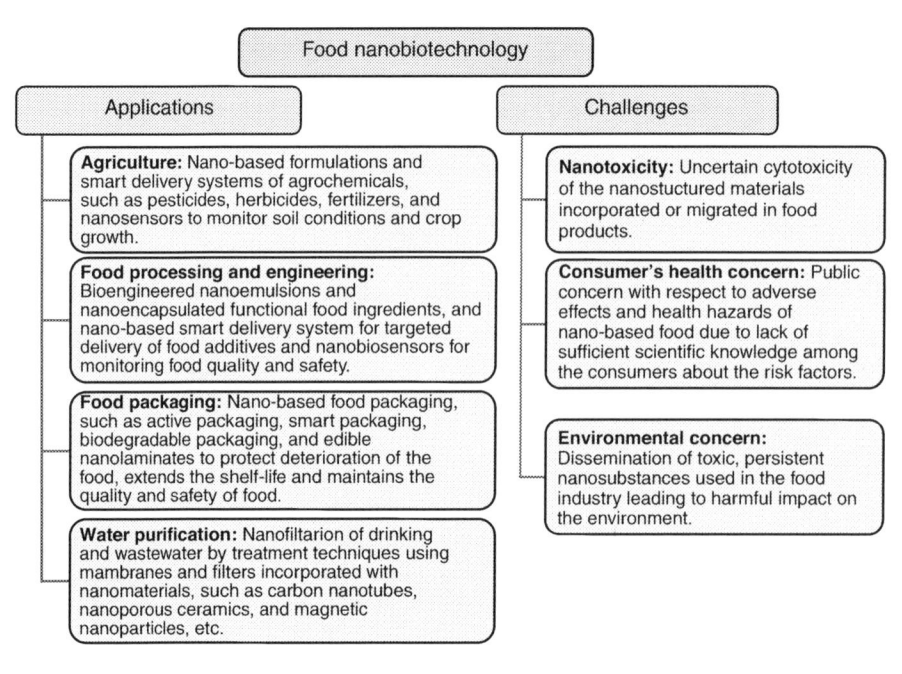

FIGURE 7.2

Applications and challenges of nanobiotechnology in food processing.

functional biomolecular structures such as DNA, RNA, and ribosomes. Through three different entry points—dermal exposure, inhalation, and ingestion—nanoparticles can enter the body and cause injury by moving between organs. Nanoparticles may enter the dermis and go to nearby lymph nodes through the lymph (Oberdorster et al., 2005). Skin oxidative injury may result from the photogeneration of hydroxyl radicals by nanomaterials like ZnO and TiO_2. It has been shown that TiO_2 nanoparticles may enter cells and even pass through the skin, generating free radicals that can harm the cells' internal structures. Carbon nanoparticles can also induce pulmonic swelling when exposed to them over an extended period of time, and their migration from the lungs into the blood vessel system would lead to more vascular illness (Nemmar et al., 2002). Aerosol nanoparticles can enter the lungs through the nasal cavity by being inhaled. The nanoparticles may build up in the lungs and cause chronic conditions such as oxidative stress, pulmonary inflammation, pneumonia, and pulmonary granuloma. The nanoparticles can also go through several channels and defensive systems to escape the respiratory tract's protective defenses. For the safety assessment of these materials, nanoparticles toward the inside the body through ingested routes are predominantly crucial for food items comprising designed nanomaterials. Nanomaterials can enter the body directly through food, drink, cosmetics, or medications, as well as indirectly through the mucociliary escalator in the respiratory system and end up in the digestive tract (Nel et al., 2005). According to studies, silica-based nanoparticles increase lipid peroxidation (LPO), release reactive oxygen species (ROS), and lower levels of cellular glutathione, which lead to oxidative stress and cytotoxicity both in vitro and in vivo. According to research by Peters et al. (2012), after in vitro ingestion of meals using silica as a food additive, the gut epithelium is almost certainly exposed to nanosized silica when consumed. When functionalized multiwalled carbon nanotubes (CNTs) were tested for cytotoxicity, it was found that the CNTs had a low to moderate level of toxicity. Regardless of nanotube functionalization, the in vivo research findings showed a wide-ranging lung toxicity accompanied by an inflammatory response, deprived of overt evidence of fibrosis and granuloma formation.

7.2.2 Consumers health and safety

Nanomaterials with characteristics not seen at the macroscale, such as CNTs, silver, silica, TiO_2, and ZnO-based nanoparticles may present unforeseen danger. The applications of nanoparticles involved in the food sector have grown significantly recently, which has prompted public concern about their impact and health risks. Consumers' growing worry is a result of a lack of scientific understanding of risk variables such as nanoparticle toxicity, bioaccumulation, exposure information, and ingestion dangers. The efforts to use nanotechnology in the poor world will undoubtedly be threatened by increased awareness of its hazards, notwithstanding its probable significance (Maclurcan, 2005). The practice of nanotechnology in different food processing methods is expanding quickly and transforming the sector. If properly utilized, nanomaterials including silver, gold, TiO_2, ZnO, QDs, dendrimers, and nanocomposites will have a positive effect. However, a buildup of harmful pollutants in foods and detrimental effects on human health might emerge from broad and intense usage of nanoparticles in food manufacturing operations. If left unchecked, the nanomaterials utilized in the creation of the nanosensors used in the arena to collect data may constitute a risk to the environment and eventually humans. Smart drug delivery systems and nanoadditives both have the potential to affect people's health. The likelihood of exposure is decreased by the fact that nanomaterials in nanocoatings remain locked inside the coating matrix which are unlikely to be released. To guarantee that the majority of the impacts are positive rather than negative, it is crucial to evaluate the safe usage of nanomaterials. It will be easier to examine the safe usage of the materials if you are aware of the risks and advantages of the nanomaterial to be employed. One should take into account the impact of nanomaterials on heterogeneity, shape, and complexes generated after the nanomaterial is absorbed in foodstuffs when conducting the analysis. As a result, there should be a well-defined methodology for sample collection, methods for consuming food, the form of processed food ingested, the quantity of food consumed as well and the movement of the food particles in the alimentary canal from which the sample of food is to be obtained for investigation (Szakal et al., 2014). The usage of nanomaterials may have hazardous and damaging impacts that go beyond what is anticipated, although research has also shown that not all nanoparticles created using various processing techniques would inevitably result in products with negative effects. By dispersing safe nanomaterials like TiO_2 and ZnO nanoparticles in the biocompatible dispersal medium using direct probe-type sonicators and indirect cup-type sonicators, Wu et al. (2013) showed specialized approaches to design the safest nanomaterials.

Nanoparticles' unknown and unexpected long-term behavior still exists. Chronic exposure to insoluble nanoparticles might cause accumulation in secondary target tissues, with unknown long-term effects. In order to understand and avoid any negative impacts of nanoparticle connection with numerous brain illnesses, there is another special worry over the potential migration of nanoparticles from the nose into the brain or to a developing fetus. Lack of research done on the risks posed by exposure to the existence of such tiny nanoparticles in the environment or living things. Analyzing the possible toxicity of nanocomposite materials and the security of their usage for the environment are urgently needed.

7.2.3 Biological adverse effects of nanomaterials used in food technology

Food technology frequently makes use of nanomaterials and it is added to the processed foods. These deliver nanoparticles to the consumer's digestive system. These experience physiological changes in turn. The interaction of these nanomaterials with the tissue, cellular organelles, and biomolecules is therefore quite plausible. Physiologically, biochemically, cytologically, histologically, or genetically, this interaction may be advantageous and/or detrimental. Some nanomaterials, including fullerenes, QDs, metallic and fibrous nanomaterials have negative consequences, including chromosomal fragmentation, DNA strand breaks, changes in gene precision, etc. Clinical toxicity has been reported, however, substantial research shows that nanoparticles might start harmful biological interactions that eventually result in toxicological effects and premature physiological disturbance.

In-vitro as well as in-vivo cytotoxicity caused by silica nanoparticles resulted in increased LPO, ROS, and decreased levels of cellular glutathione. Research done on cells treated with various nanomaterials with regard to cytotoxicity as well as inflammation, there is always a chance of various recognized and undiscovered risks. Both AuNPs and AgNPs have the ability to depolarize tubulin, a significant microtubule component, which would negatively impact the cell's cytoskeleton and cellular structure. It is currently unclear what mechanism nanomaterials and live biosystems interact through. The effectiveness of nanomaterials and in what manner they interact with biological molecules in the surrounding environment to produce a variety of end products is a very complicated issue. The modest modifications made to the nanomaterials by size, charge, surface characteristics, dispersion, agglomeration, and concentration make this complexity more difficult to manage (Kumar, 2015).

Nanomaterials have the capacity to induce apoptosis, autophagy, mitoptosis (death of mitochondria), aneugenicity (producing the presence of aberrant chromosomal numbers), and other biologically harmful processes. For silica, titanium dioxide, and silver, there is toxicological information available regarding its cytotoxicity, genotoxicity, biokinetics, and related to its repeated dose toxicity; however, there are few records pertaining to the usage of nanomaterials in agriculture and food feed applications. The list of laws and regulations in both European and non-European nations has embraced European Union (EU) legislative acts and created the standards that must be adhered to for the connected businesses (Peters et al., 2014).

7.2.4 Impact on environment and ecology

By offering solutions for high energy use, gas emissions, and pollution, nanotechnology in the food business has the possibility to support lessen the impact of humans on the environment on all fronts, including economic, social, and environmental advantages. Nanoparticles are brought into direct touch with consumers when food items with nanostructured materials are used, whether in the meal itself or as a component of the packaging. Applications of nanotechnology come with several concerns and implications for the environment. The potential environmental effects of nanotechnology are listed below (Zhang et al., 2011).

- The huge energy consumption brought on by the need to create nanoparticles.
- The spread of unsafe, long-lasting nanomaterials that have an effect on the environment.
- Lower rates of recycling and recovery.
- Other life cycle stages' effects on the environment are likewise unclear.
- More worries are being raised due to a lack of educated engineers and laborers.

Since nanoparticles have a wide choice of physical characteristics and activities, it is difficult to estimate the environmental concerns they represent. The size, shape as well as charge of these materials have an impact on their kinetic, i.e., absorption, distribution, metabolism, and excretion (ADME), and their hazardous characteristics too; as a result, even nanomaterials with identical chemical configuration but different size and shape may have quite varied toxicities. Therefore, particle size may alone be not a suitable way to distinguish between technologies and materials that are more or less harmful (Pragya et al., 2012). It is to be expected that nano-sized structures and their byproducts will reach the environment and can represent a serious hazard when the application of nanotechnology in the field of the food sector moves into large-scale manufacturing. Effective risk assessment practices must be done as soon as feasible to address the risk associated with nanofood technology.

The effects and control of food-related nanotechnology are of great concern to the group ETC (Action Group on Erosion, Technology, and Concentration). The ETC claimed that the fusion of nanotechnology and biotechnology has unidentified effects on human well-being, biodiversity, and the environment in a report from November 2004. Owing to a deficiency of a standardized framework for the usage and testing of nanotechnology, as well as several food items containing nanomaterials are being developed and sold without review. A worry for the safety of the environment and the general people has resulted from this. Even though nanotechnology is a young and developing field of research with still-uncertain chemical characteristics, there is now no scant proof that it has harmed humans or the environment.

There have been reports on the photocatalysis of pesticides like tetrachlorvinphos, pirimiphos-methyl, dichlorvos, and fenitrothion as well as herbicides like 2,4-dichlorophenoxyacetic acid. According to Bandala et al. (2012), the widely used organochlorine pesticide aldrin was degraded via solar photocatalysis. Water, mineral salts, and carbon dioxide—all of which escape into the atmosphere—are the byproducts of photocatalysis. These substances eventually improve the soil's fertility. The method of photocatalysis degradation has grown in favor of the treatment of wastewater. In their investigation of the photocatalysis-based purification technique, Peral et al. (1997) showed that photocatalysis may be utilized to clean, decontaminate, and deodorize air. According to Mills et al. (1997), the elimination of organics can eradicate cancer cells, bacteria, and viruses and were accomplished via semiconductor-sensitized photosynthetic as well as photocatalytic processes.

7.2.5 Emerging policies and regulations of nanomaterials in food

In terms of researching and regulating nanoparticles in food, Europe has assumed the lead. The European Food Safety Authority (EFSA) released "the first useful advice for evaluating nano uses in food and feed" in May 2011 (EFSA, 2011). The Authority's Scientific Committee produced this guideline, which is the first of its type to offer actionable advice for managing possible dangers associated with food-related uses of nanoscience and nanotechnologies. The recommendations encompass risk for pesticides, new foods, flavors, food contact materials, enzymes, and additives used in foods. The EFSA advice outlines factors for risk assessment that may be affected by the unique traits and features of engineered nanomaterials (ENM). The ENM guidance, which specifies the supplementary data required for the physical as well as chemical characterization of ENM in contrast with applications and summaries of various toxicity testing methods to be followed by applicants, is an important addition to existing supervision documents for constituents and products.

The European Union enacted rules governing the labeling of nanoparticles in food in July 2011. The new law merged two directives—2000/13/EC on food labeling, demonstration, and advertising and 90/496/EEC on food nutrition labeling—into a single piece of legislation. The new rule requires the labeling of all substances that meet the concept of "nanomaterials," which is contained in the statute's definition. Contrarily, the Food and Drug Administration (FDA) has not published any laws governing the application of nanomaterials in food, further, it was initially unclear if the FDA would even treat nanoparticles as a distinct chemical. The FDA announced in April 2012 that: still we are unaware of food ingredient and food contact substance (FCS) that has been purposefully engineered in the range of nanometer for which there are generally no sufficient safety data available to serve as the basis for the determination of the use a food ingredient or FCS is generally recognized as safe (GRAS). This was a significant step toward providing an answer to that question.

Because "the practice of nanotechnology can bring certain outcomes in features of numerous products and those are differed from those of conventionally-modified and manufactured foodstuffs," the FDA acknowledged that substances that are GRAS at the macro level are safe at the nano level. Furthermore, "nano-engineered food ingredients can drastically change bioavailability and it may raise new safety problems that can't be identified in their conventionally made equivalents" (FDA Guidance Documents, 2008, 2011, 2012).

7.2.6 The Food and Drug Administration notes

Data extrapolation from conventionally made food ingredients is often only possible on an individual basis. Safety evaluations must be based on information pertinent to the nanometer type of the foodstuffs when a food substance is prepared to comprise a particle size dispersal that has been pushed more fully into the nanoscale range. It may be required to use extra or alternative testing techniques to verify the safety of food ingredients where nano-engineered versions exhibit novel features. In order to serve as a jumping-off point for the nanotechnology debate, the FDA released draught advice for the industry in June 2011 titled "Considering Whether an FDA Regulated Product Involves the Application of Nanotechnology." According to the FDA's draught, examinations of the safety, efficacy, or potential effects on public health of such goods (those comprising nanomaterials) "shall reflect several unique features and behaviors' of nanomaterials." This was of an important advance since it puts the material's characteristics and behaviors above its size. The FDA also pledges to abide by the "Policy Principles for the U.S. Decision-Making Concerning Regulation and Oversight of Applications of Nanotechnology and Nanomaterials" in the same paper. However, there are concerns regarding the FDA's capability to carry out such analyses; the agency's capability to control the safety of nanomaterials in food is severely constrained by a lack of knowledge, resources, and legislative power in several crucial areas. The FDA acknowledged that some substances may not be safe at the nanoscale even if they are commonly regarded as safe at the macro level. In an issue brief on nanomaterials in food, entitled "Sliding Through the Cracks," the FDA has only officially acknowledged these two points: that nanomaterials are not GRAS and that their categorization is based on their characteristics, not on their size. The FDA's conclusion that nanomaterials are not GRAS suggests that at least one more layer of research is necessary to establish a nanomaterial's safety for use in food or food packaging, although it is still unclear how the FDA evaluates the toxicity exposure of nanomaterials in food items. The identification or determination of exposure, absorption levels, or behaviors of nanoparticles in the human body have not yet been made using any standardized criteria, assays, testing apparatus, or testing technique. Once that has been decided, a procedure should be established and followed to determine the safety of nanomaterials intended to enter the food chain. This procedure will take years (FDA, 2011).

Despite the absence of U.S. government regulations, some business executives have taken it upon themselves to formulate their own official statements about the use of nanomaterials in meals and food wrapping. Examples of current public corporate positions on nanomaterials in foods are described briefly.

7.2.6.1 McDonald's

The McDonald's corporation is attempting to comprehend how nanotechnology is used and how it may be used in food and packaging items. McDonald's does not presently support suppliers using nano-engineered materials in the manufacturing of any of our food, wrapping, or toys due to the current lack of clarity on the possible effects of such materials ("Nanotechnology," McDonald's).

7.2.6.2 Kraft foods

Scientifically speaking, nanotechnology is a young field. It includes regulating the form and size of buildings, electronics, and systems at incredibly tiny scales—between one and 100 nm. Additionally, this technique has enormous potential in a variety of fields. We do not currently employ nanotechnology. There are possibilities that this technology are safe, having product quality, enhanced nutritional value and sustainability must be understood, though, as a major food firm. Because of this, several research and development teams constantly keep an eye on the latest scientific findings and take into account future uses for nanotechnology in packaging materials. We are focusing in particular on packaging that uses less material, which lowers waste. Only usage that abides by legal standards and are regarded as safe by the scientific community would be taken into consideration. We also consider what our customers' thoughts and feelings are. If we ever plan to employ nanotechnology, and we will make sure the pertinent environmental as well as health and safety issues have been taken into account. This entails undergoing our own rigorous quality-control procedures and collaborating with our suppliers to ensure that the appropriate assessments have been carried out.

7.3 Future prospects

Nanotechnology is a speedily emerging field and has the potential to improve food systems. Nanotechnology has impacted all aspects of the food processing industry, including improvement of product shelf life, improved food storage as well as safety, and providing nutritional supplements in food. Several research are going on the application of nanoparticles to improve the quality of food, still there is a requirement of advance research to maintain the freshness and prevent the accumulation of toxins during storage. More research is required to discover the harmful properties of nanoparticles, so that it will not affect edible materials. It is already explored that nanotechnology improved the manufacturing process and enhanced preservation. In the future, an attempt should be focused on ensuring that nanotechnology is carefully used for better impact on health. Moreover, use of polymers in food industries is often used for flavor, texture, food color, and fragrance. But synthetic polymer should not be used as it is nonbiodegradable, so biodegradable natural polymer must be used, though this area still needs scientific validation.

7.4 Conclusion

Nanotechnology has brought a revolutionary effect on the field of the food industry. No doubt nanotechnologies are far better than conventional food processing technologies. It has exhaustive applicability in developing nano-formulations that maintain food color, flavor, and nutritional value to meet consumer preference as well as dietary requirements. Nanocarriers improve food products, color, taste, texture, and shelf-life and also give better health benefits. Therefore, it is concluded that nanotechnology is applied in the food processing industry for monitoring the unregulated usage of food additives to ensure the protection of food products. Nanotechnology has immense potential in postharvest food processing. Nanoscience-based technology has a very good impact on food quality, food safety as well as food packaging aspects.

References

Augustin, M.A., Riley, M., Stockmann, R., Bennett, L., Kahl, A., Lockett, T., et al., 2016. Role of food processing in food and nutrition security. Trends Food Sci. Technol. 56, 115–125.

Bandala, E.R., Gelover, S., Leal, M.T., Arancibia-Bulnes, C., Jimenez, A., Estrada, C.A., 2002. Solar photocatalytic degradation of Aldrin. Catal. Today 76 (2-4), 189–199.

Chaudhry, Q., Scotter, M., Blackburn, J., Ross, B., Boxall, A., Castle, L., Aitken, R., Watkins, R., 2008. Applications and implications of nanotechnologies for the food sector. Food Addit. Contam. 25, 241–258.

Duncan, T.V., 2011. Applications of nanotechnology in food packaging and food safety: barrier materials, antimicrobials and sensors. J. Colloid Interface Sci. 363, 1–24.

European Food Safety Authority, 2011. EFSA Publishes First Practical Guidance for Assessing Nano Applications in Food & Feed, News Release. https://www.efsa.europa.eu/en/press/news/sc110510.

Ezhilarasi, P., Karthik, P., Chhanwal, N., Anandharamakrishnan, C., 2013. Nanoencapsulation techniques for food bioactive components: a review. Food Bioprocess Technol. 6, 628–647.

FDA, 2008. Draft guidance for industry: considering whether an FDA-regulated product involves the application of nanotechnology; U.S. Office of Science and Technology, "Policy principles for the U.S. decision-making concerning regulation and oversight of applications of nanotechnology and nanomaterials." Assuring the safety of nanomaterials in food packaging: the regulatory process and key issues. Woodrow Wilson International Center for Scholars Project on Emerging Nanotechnologies and GMA, p. 45. https://www.fda.gov/regulatory-information/search-fda-guidance-documents/considering-whether-fda-regulated-product-involves-application-nanotechnology.

FDA, 2011. Draft guidance for industry: considering whether an FDA-regulated product involves the application of nanotechnology; U.S. Office of Science and Technology, Office of Management and Budget, Office of the United States Trade Representative, "Policy principles for the U.S. decision-making concerning regulation and oversight of applications of nanotechnology and nanomaterials" p. 4. https://www.fda.gov/regulatory-information/search-fda-guidance-documents/considering-whether-fda-regulated-product-involves-application-nanotechnology.

FDA, 2012. Draft guidance for industry: assessing the effects of significant manufacturing process changes, including emerging technologies, on the safety and regulatory status of food ingredients and food contact substances, including food ingredients that are color additives. https://www.fda.gov/regulatory-information/search-fda-guidance-documents/guidance-industry-assessing-effects-significant-manufacturing-process-changes-including-emerging.

Gruère, G., Narrod, C., Abbott, L., 2011. Agricultural, Food, and Water Nanotechnologies for the Poor. International Food Policy Research Institute, Washington, DC.

Jafari, S.M., McClements, D.J., 2017. Nanotechnology approaches for increasing nutrient bioavailability. Adv. Food Nutr. Res. 81, 1–30.

Kalita, D., Baruah, S., 2019. The impact of nanotechnology on food. In: Nanomaterials Applications for Environmental Matrices. Elsevier, pp. 369–379.

Kamarulzaman, N.A., Lee, K.E., Siow, K.S., Mokhtar, M., 2020. Public benefit and risk perceptions of nanotechnology development: psychological and sociological aspects. Technol. Soc. 62, 101329.

Kim, E., 2013. The amazing multimillion-year history of processed food. Sci. Am. 309, 50–55.

Kumar, L.Y., 2015. Role and adverse effects of nanomaterials in food technology. J. Toxicol. Health 2 (2), 245–255.

Lane, N., Kalil, T., 2005. The national nanotechnology initiative: present at the creation. Issues Sci. Technol. 21 (4), 49–54.

Maclurcan, D., 2005. Nanotechnology and developing countries–part 2: what realities? Online J. Nanotechnol. 1, 1–19.

Mills, A., Le Hunte, S., 1997. An overview of semiconductor photocatalysis. J. Photochem. Photobiol. A 108 (1), 1–35.

Mozafari, M.R., Johnson, C., Hatziantoniou, S., Demetzos, C., 2008. Nanoliposomes and their applications in food nanotechnology. J. Liposome Res. 18, 309–327.

Mustafa, F., Andreescu, S., 2020. Nanotechnology-based approaches for food sensing and packaging applications. RSC Adv. 10 (33), 19309–19336.

Nel, A., Xia, T., Madler, L., Li, N., 2005. Toxic potential of materials at the nanolevel. Science 311 (5761), 622–627.

Nemmar, A., Hoet, P.M., Vanquickenborne, B., Dinsdale, D., Thomeer, M., Hoylaerts, M.F., Vanbilloen, H., Mortelmans, L., Nemery, B., 2002. Passage of inhaled particles into the blood circulation in humans. Circulation 105 (4), 411–424.

Oberdorster, G., Oberdorster, E., Oberdorster, J., 2005. Nanotoxicology: an emerging discipline evolving from studies of ultrafine particles. Environ. Health Perspect. 113, 823–839.

Peral, J., Domènech, X., Ollis, D.F., 1997. Heterogeneous photocatalysis for purification, decontamination and deodorization of air. J. Chem. Technol. Biotechnol.: Int. Res. Process Environ. Clean Technol. 70 (2), 117–140.

Peters, R., Brandhoff, P., Weigal, S., Marvin, H., Bouwmeester, H., 2014. Inventory of nanotechnology applications in agriculture, feed and food Sector. EFSA Supporting Publication 621. https://efsa.onlinelibrary.wiley.com/doi/abs/10.2903/sp.efsa.2014.EN-621.

Peters, R., Kramer, E., Oomen, A.G., Herrera Rivera, Z.E., Oegema, G., Tromp, P.C., Fokkink, R., Rietveld, A., Marvin, H.J., Weigel, S., Peijnenburg, A.A., 2012. Presence of nano-sized silica during in vitro digestion of foods containing silica as a food additive. ACS Nano 6 (3), 2441–2451.

Pragya, R., Nandini, P., Bhavesh, P., 2012. Nanomaterials: a future concern. Int. J. Res. Chem. Environ. 2 (2), 1–7.

Prasad, R., Kumar, M., Kumar, V., 2017. Nanotechnology: An Agricultural Paradigm. Springer Nature, Singapore, pp. 371–376.

Ravichandran, R., 2010. Nanotechnology applications in food and food processing: innovative green approaches, opportunities and uncertainties for global market. Int. J. Green Nanotechnol. Phys. Chem. 20 (2), 72–96.

Sadeghi, R., Rodriguez, R.J., Yao, Y., Kokin, J.L., 2017. Advances in nanotechnology as they pertain to food and agriculture: benefits and risks. Annu. Rev. Food Sci. Technol. 8, 467–492.

Scott, N., 2005. Nanotechnology and animal health. Rev. Sci. Tech. Off. Int. Epiz. 24, 425–432.

Szakal, C., Roberts, S.M., Westerhoff, P., Bartholomaeus, A., Buck, N., Illuminato, I., Canady, R., Rogers, M., 2014. Measurement of nanomaterials in foods: integrative consideration of challenges and future prospects. ACS Nano 8 (4), 3128–3135.

Wu, W., Ichihara, G., Suzuki, Y., Izuoka, K., Oikawa-Tada, S., Chang, J., Sakai, K., Miyazawa, K., Porter, D., Castranova, V., Kawaguchi, M., 2013. Dispersion method for safety research on manufactured nanomaterials. Ind. Health 4 2012-0218.

Yadollahi, A., Arzani, K., Khoshghalb, H., 2010. The role of nanotechnology in horticultural crops postharvest management. In: Southeast Asia Symposium on Quality and Safety of Fresh and Fresh-Cut Produce, 875, pp. 49–56.

Zhang, B., Misak, H., Dhanasekaran, P.S., Kalla, D., Asmatulu, R., 2011. Environmental impacts of nanotechnology and its products. In: Proceedings of the 2011 Midwest Section Conference of the American Society for Engineering Education, 26, pp. 1–9.

Factors influencing food processing

Monika Bhattu[a], Noorkamal Kaur[b], Shikha Kapil Soni[c] and Meenakshi Verma[a]

[a] *University Centre for Research and Development, Chandigarh University, Mohali, Punjab, India,* [b] *Department of Food Processing Technology, Sri Guru Granth Sahib World University, Fatehgarh Sahib, Punjab, India,* [c] *University Institute of Biotechnology, Chandigarh University, Mohali, Punjab, India*

8.1 Introduction

Food is processed for a variety of reasons. Grain has been dried after harvesting since ancient times to extend its shelf life. Food was initially processed mainly to enhance nutritional quality, and consumer acceptability, and to maintain a steady supply. With the passing years, increased efficiency, connectivity, and more urbanization, consumer demands have become more diversified, and there has been a rising need for fresh, organic, safer, and healthier foods with an optimal shelf life. Consumers seek higher-quality meals with nutrient retention, often with specified functional qualities and taste/texture/consistency, that are shelf-stable and easy to package, store, and transport. This has prompted scientists to create methods and techniques for processing foods in such a way that the packaged foods fit the needs and expectations of customers. India has advanced from an agro-deficit to an agro-surplus country, necessitating the need for agricultural and horticulture output storage and processing. As a result, India's food sector has emerged as a significant manufacturer of processed foods, ranking sixth in terms of size and contributing about 6% of gross domestic product (GDP). Furthermore, changes in lifestyle, increased mobility, and globalization have raised demand for numerous sorts of items, necessitating the development of novel technologies. Simple diets focused mostly on basic foods like cereals are widely recognized to be low in specific nutrients, leading to deficiency problems. Consequently, food fortification is carried out by including the deficient supplement in the foodstuffs or sauces to ensure that basic nutritional requirements are satisfied. Iodized salt, folic acid added to wheat, and vitamins A and D added to milk and oils/fats are a few examples. Indeed, the Food Safety and Standards Authority of India (FSSAI) has established standards for staple foods such as salt, wheat flour, milk, and oats. In practice, the FSSAI has established criteria for the fortification of staple foods. Because of the increased incidences of health issues and public concern about being well, scientists have to modify/enhance the food nutrient value by attempting alternatives such as lowering the calorie content of processed foods through the use of

Nanobiotechnology for Food Processing and Packaging. DOI: https://doi.org/10.1016/B978-0-323-91749-0.00014-9

artificial sweeteners, the addition of protein instead of fats in ice-creams to give them a smooth texture and energy hike. In the present time, people are more into organic food, which definitely has increased the value of food processing and its technologies.

Previously, initial researches were carried out on food and its preservation which has been proven as a boon in this field. In 1810, the trend of canned food was brought by Nicolas Appert and later on, Louise Paster's concept of pasteurization put people's perception of food processing up to the next level back then. In the 20th century, the main purpose of food technology was to serve military needs. Moreover, world wars, space tours, and the rising population have led to the growth of food technology. In addition to employment, food technology was able to meet the nutritional value of processed food which is actually the main concern these days. In the 21st century, developing countries are able to meet this criterion at all levels with the help of advanced technologies.

Food processing, as previously said, is a discipline of production in which raw ingredients are turned into transitional delicacies or palatable goods by the use of modern science and technologies. Various methods are employed to transform bulky, perishable, and occasionally nonedible food resources into more usable, condensed, nonperishable, and pleasant meals or beverages. Food processing provides value to the finished product by improving storability, portability, palatability, and convenience. Professionals in the food processing industry must be familiar with the general properties of raw food products. Let us look at the necessity, principles, techniques, and modernization of food processing in a nutshell.

Food deterioration/degradation is related to food spoilage, off-flavor, off-texture, and less-nutrients which makes it unsafe to eat. Food spoilage is render by physical factors such as pests, insect's manifestation, improper processing and storage, temperature, moisture content, and oxygen level, chemical factors such as pesticides, and biological factors such as microorganisms and enzymes. In back time, traditional practices such as drying, stable fermentation, candying/salting/pickling/spicing, and roasting/smoking/baking were followed to secure food from spoilage. In addition to these old practices, new methods/science such as food science, food technology, food microbiology, food nutrition, sensory, and statistical analysis with the addition of good manufacturing practices have been incorporated so far. The concept of food processing is to increase food shelf life and to eliminate the risk of microbial and other contamination. Microbial and enzymatic activity is highly influenced by nutrient availability, pH, temperature, moisture, oxygen, and absence of antibiotics, etc.

The selection of raw materials is vital for quality processed food. Other factors involved are preharvest factors, harvesting methods, postharvest conditions, storage, and handling also define quality traits. Moreover, good knowledge of food properties is an important aspect for scientists and engineers to solve the problem of food processing, packaging, storage, marketing, and consumption (Paulus, 2000) (Table 8.1).

8.2 Processes involved in food processing

The conversion of agricultural products into substances with certain textural, sensory, and nutritional qualities using commercially viable methods is known as food

Table 8.1 Classification of food and its properties.

Class		Properties	
Physical and Physico-chemical properties	To be defined by the food itself and determined objectively. Important to define process conditions.	Mechanical Thermal Thermodynamic Mass transfer Electromagnetic Physico-chemical constants	
Nutritive/health properties	To be defined by the food itself and determined objectively. Important for food quality.	Positive health properties Negative health properties	Nutritional composition Medical properties Functional properties Toxic at any concentration Toxic after critical concentration level Excessive of imbalance uptake
Sensory properties	To be defined by the food and people. Determined subjectively and important for food quality.	Tactile Textural Color and appearance Taste Odor Sound	
Kinetic properties	Rate of changes in food.	Quality kinetic constants Microbial growth, decline, and death kinetic constants	

processing (Food, 2022). Food processing can be done in a variety of ways. A few foods, such as apples that may be plucked from the trees and consumed straight away, require almost no preparation. Others are unappealing unless they are prepared according to specified recipes. There are three types of food processing techniques: primary, secondary, and tertiary processes.

8.2.1 Primary food processing

Primary food preparation is generally the processing of raw material for food which consists of changing unprocessed agricultural materials into quick-to-eat foods. When primary processing is complete, the meal is sometimes ready to be served. Jerky made from smoked meat is a good example of this.

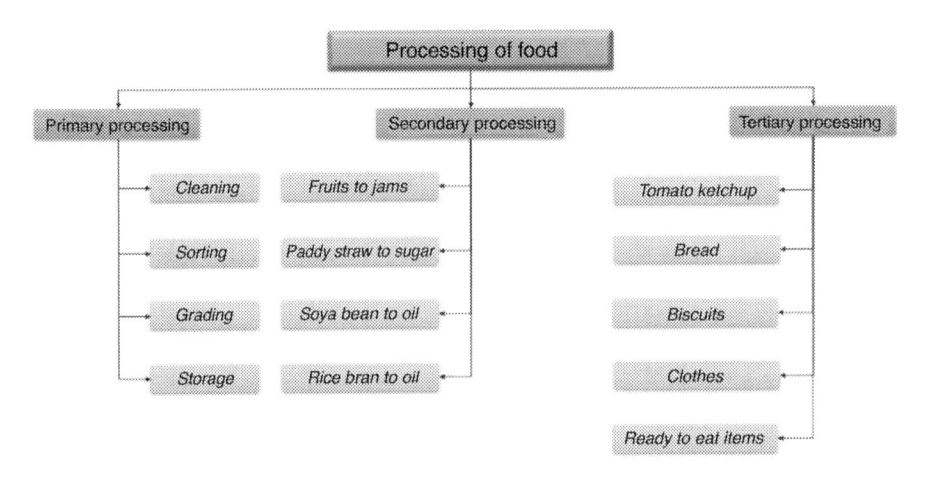

FIGURE 8.1

Various kinds of food processing processes with examples.

8.2.2 **Secondary food processing**

The method for making ready-to-eat nourishment is by utilizing fixings given through primary food preparation. Using flour to produce the batter and prepare the combination to bake bread is an example of this. In different models, grape juice is aged with wine yeast to generate wine, while ground meat is used to make hotdogs.

8.2.3 **Tertiary food processing**

This process includes a wide range of ready-to-eat foods such as frozen pizzas and packaged snacks. To get to their final, consumable structures, a few foods require distinct types of processing. When done on a large scale and in impenetrable bags, this cycle comprises tertiary food processing procedures after the food has gone through primary and secondary preparation (Food Research Lab, 2021).

Depending on the source ingredients, the processing of the food materials differs very much. Fig. 8.1 depicts all kinds of food processing with some examples.

All these processes are a series of physical processes that can be understood via various simple operations. Some of the common operations utilized in food processing are: (1) material handling; (2) cleaning; (3) separating; (4) size reduction; (5) fluid flow; (6) mixing; (7) heat transfer; (8) concentration; (9) drying; (10) forming; (11) packaging; and (12) controlling. These unit operations vary with the variety of foodstuff to be processed at each processing step (Fig. 8.2). In all these processing steps, there are various factors that influence the processing mechanism which further leads to a decrease in the quality, shelf life and nutritional value of these food materials.

Categories of food	Primary processing	Secondary processing	Tertiary processing
Fruits and vegetables	Cleaning, sorting, and cutting	Slices, pulps, and paste	Ketchups, jam, juices, and pickles
Milk	Grading and refrigerating	Cottage cheese, cream, simmered, and dried milk	Processed milk, spreadable fat, and yoghurt
Meat and poultry	Sorting and refrigerating	Cut, fried, frozen, and chilled	Ready-to-eat meals
Marine products	Chilling and freezing	Cut, fried, frozen, and chilled	Ready-to-eat meals
Grains and seeds	Seeding and grading	Flour, malt, and milling	Biscuits, noodles, flakes, cakes, and savory
Beverages	Sorting, bleaching, and grading	Leaf, dust, and powder	Tea bags, flavored coffee, soft drinks, and alcoholic beverages

FIGURE 8.2

Operation performed for different variety of food stuffs at every stage of food processing.

8.3 Factors affecting food processing

There is a vast diversity in agronomic produce with respect to general morphology and composition such as leaves, flowers, stems, fruits, and so on. Therefore, the maximum shelf life in addition to certain requirements and specifications alters among various food commodities. These factors affect the processing of the food materials at different levels starting from the preharvest factors to the consumer's uptake including the factors influencing during the collection of raw materials, during industrial processing, packaging, and transportation (Fig. 8.3). All these factors are described below briefly.

8.3.1 During pre- and postharvesting

These factors are not directly included in the food processing steps. However, these operations affect the food processing system as these conditions are directly responsible for the quality, yield, and quantity of the food to be processed. The preharvesting factors include the factors that basically affect the growth of plants. These factors involve the intrinsic and extrinsic factors. Intrinsic factors consist of genetic factors and extrinsic factors consist of environmental factors such as climatic, edaphic, biotic, physiographic, socioeconomic, and others. The genotype of the plant controls the plant growth by choosing the variety of the crop. The external factors also affect a lot of food production. The external factors such as topography, soil, and other edaphic factors are responsible for the nutritional growth of the plants and the yield of the crops. The control of climatic factors such as humidity, aeration,

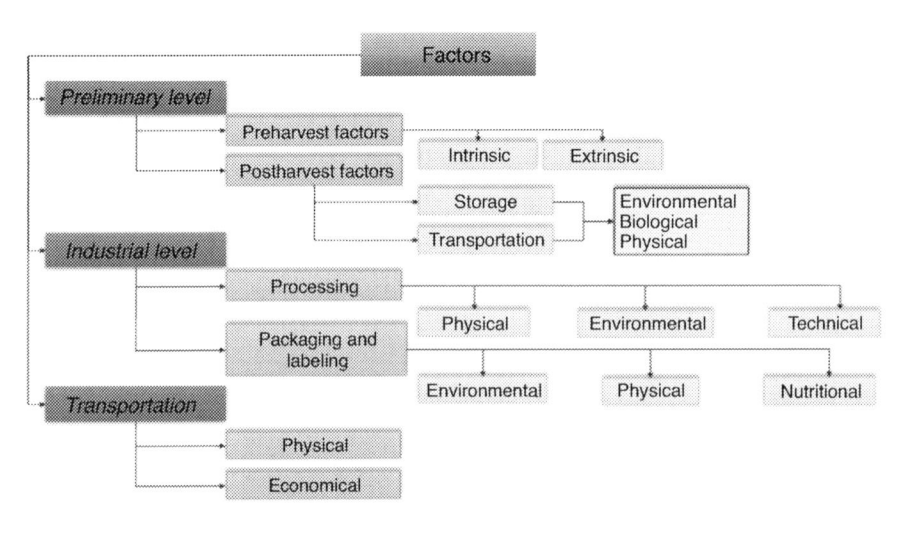

FIGURE 8.3

Factors affecting the food processing.

light, temperature, and moisture effectively helps to propagate the plant and grow it into a healthy plant (Broome, 2015). Generally, these preharvest factors affect the postharvest conditions which basically influences the fresh produce either qualitative or quantitative (Ministry of Health and Family Welfare, 2006). Despite the fact that crop research is increasingly focusing on quality rather than quantity, there has been little change in the quality of produce (Oko-Ibom and Asiegbu, 2007). A decrease in the quality of food may have a severe influence on numerous factors such as consumer acceptance and nutrient status of fruits, as well as financial gain for the growers (Arah et al., 2015). Hence, in this current chapter, we will discuss the environmental factors affecting food processing and the preharvest factors that affect the postharvest conditions. The postharvest factors to be discussed basically include: (1) living organisms such as fungi, insects, and bacteria that may contaminate the food or may feed the food; (2) biochemical activities leading to the enzymatic browning of the products, quality reduction of food, or rancidity development in fatty products; and (3) physical damages due to the inappropriate handling of the raw materials. These factors may harm the raw material during the collection, storage, and transportation before going to the food industry and also during industrial processing. These factors may vary depending on the type and nature of the food materials. For example, fruits and vegetables act as living tissues until further processed so it has become necessary to maintain the respiratory process (Grandison, 2011). The main postharvest factors utilized for further industrial processing are basically vary from the selection of raw material utilized to the factors that are associated with the quality of raw materials during storage and transportation such as temperature, humidity, and some other considerations which are described below briefly.

8.3.1.1 Consideration of raw material

Basically, the selection of the appropriate material for the processed commodities has become important attention for the food processing industries. As the raw material's quality can be upgraded rarely during processing, it has become a vital requirement to choose materials with specified properties that can meet all the processing requirements. Hence, when it comes to processed meals, it is important to remember that the quality of the finished product is crucial, not the raw ingredient. Keeping this into consideration, there are various factors such as geometric characteristics, cultivars in plant foods, breeds in animal food, harvesting time, and postharvesting handling which affect the quality of the food materials at this stage. Along with these factors, the functionality of the raw materials plays an important role in determining the quality and effectiveness of the process. For example, wheat having a protein content of 11.5%–14.0% is preferred for making white bread, and on the other hand requirement of wheat containing higher protein content (14%–16%) is utilized for whole wheat breads (Chung and Pomeranz, 2000).

8.3.1.2 Temperature

Temperature is the most important parameter of the postharvest environment which basically affects the storage life of the raw material used for industrial processing. Thorne and Alvarez (1982) reported that the environmental temperature directly deteriorates the raw material after harvesting. Holt et al. (1983) reported that the spoilage of fruits and vegetables at freezing temperature or higher temperature is due to the effect of various factors such as pathological, physiological, and physical factors. In the case of horticultural crops, postharvest losses are reported to be 25%–50% due to poor temperature management in subtropical and tropical regions or in regions where chilling amenities are not implemented (Salunkhe and Deshpande, 2012; Harvey, 1978). Inappropriate temperature management may lead to reduce the quality of the raw materials, and also affects several other characteristics such as appearance and texture, nutritional value, metabolism, and shelf life which are briefly described below:

First, color is an important parameter for most of fruits, vegetables, and crops and is the primary factor indicates the ripening of these commodities. Due to the temperature effect, some undesirable changes were observed in these commodities. For example, yellowing of broccoli has been reported due to the effect of inapt storage temperature (Tian et al., 1994). Toivonen (1997), reported that this yellowing directly affects the quality of broccoli.

Second, it is quite an obvious fact that fruits, vegetables, meat, and dairy products are the richest source of nutrients for humans. The nutritional value of these stuffs decreases with an increase in the storage temperature and time. For instance, vitamin C degrades at a much faster rate after harvesting as storage (Fennema, 1977).

Third, various metabolic activities including respiration, tissue texture, and the pigments formation or destruction of the harvested crops, fruits, and vegetables are affected due to temperature variation. Temperature variation mainly affects the respiration rate of the harvested crops which is generally influenced by lowering the temperature. In 1994, a review has been done by Singh et al. describing the various kinds of models for monitoring the quality of the stored food using time temperature-based indicators (Singh, 1994).

8.3.1.3 Humidity

Humidity conditions of the environment affect the quality and shelf life of the fresh produce in both conditions, i.e., in a lower humidity atmosphere and also in a higher humidity atmosphere. Dehydration, shrinking, diminished luster, decreased market worth, and increased vulnerability to infections may occur when produce is dealt at RH values lower than its relative water content. Furthermore, for goods sold by mass, water loss can result in a loss of salable weight and a reduction in profitability. Although a 2% weight reduction may have little effect on the product's visual quality, the expense of that lost weight could be considerable. For a pallet weighing 500 kg, a loss of 2% equates to a loss of 10 kg per pallet. If the commodity is priced at $10 per kg, the weight loss translates into a loss of $100 per pallet at retail prices. Water vapor will settle on the surfaces of the products when the relative humidity of the atmosphere is limited to saturation (100%). The link between temperature, relative humidity, and dew point (at which water condenses) is frequently determined using psychometric charts. Undesirable condensation on items can hasten the onset of sickness or physiological issues, resulting in product deterioration and loss. In order to prevent moisture loss, most fresh fruits and vegetables should be transported at high relative humidity levels (85%–95%). On the other hand, dried produce such as onions and nuts are recommended to be stored and handled in an environment with humidity of 65%–70% (Sensitech, n.d.).

8.3.1.4 Biological factors

Microorganisms that affect the raw materials' quality come in a wide variety of shapes and sizes. Bacteria cause some of the most visible and quick deterioration of proteinaceous foods such as fish, poultry, shellfish, meat, milk, and various dairy products (Petruzzi et al., 2017). Yeasts and molds grow at a slower rate than bacteria, but they are more resistant to antibiotics. They are strong spoiling agents due to their wide range of substrates and endurance under more harsh circumstances than (vegetative) bacteria (Blackburn et al., 2006). Many types of microbes that can live on food have evolved biochemical systems to break down dietary components, supplying energy toward their own expansion (Sperber, 2009). The accessible chemicals are transformed into a variety of end products that alter the physical, chemical, and sensory aspects of food (Caballero et al., 2016). The extrinsic characteristics of handling and storage, the microorganism's genetic capability, and the intrinsic characteristics of the product decide the variety of potential compounds formed which can drastically alter biochemical pathways and the conditions to meet the further processing requirements (Benner, 2014).

8.3.1.5 Other factors

There are several other factors such as odor, taint, light, and other contaminant that affects the raw materials' quality. Odors and taints can be problematic, particularly in fatty foods like meat and dairy products, as well as less visible commodities like citrus fruits with oil in their skins. Fuels and adhesives, printing materials, and other foods, such as spiced or smoked meats can produce odors and taints. Light can cause oxidation in some fatty raw materials, such as dairy products.

Furthermore, light contributes to the formation of solanine and green pigmentation in potatoes. As a result, dark storage and transportation are required.

8.3.2 **During processing and transportation**

After harvesting and collection of the raw material, the commodities are transported for further processing at the industrial level. In industries, there are various factors that influence the High water activity being prevalent in the majority of fresh food products, hence are prone to get attacked by microscopic organisms such as bacteria, fungi, and pathogens (Kader, 2013). Thus, biological deterioration results in a change of color, texture, and nutritive value of food components which majorly depends on several factors including temperature, relative humidity, atmospheric composition, and sanitation practices (Brackett, 1999). The commodities can also be subjected to mechanical injury which results in a distorted appearance of fresh produce (Kader, 2013).

Several factors affecting food quality during processing are discussed hereafter. Fresh produce goes through some processing in contrast to thermally processed or frozen foods which receive harsh treatments for the purpose of bacteriological safety. The various processing aids include human involvement, water contact, cutting, slicing and so on which have a great potential for the growth of several contaminants such as pathogens and bacteria. Despite the fact that food commodity is susceptible to getting degraded easily by several factors, the food industry somehow manages to supervise the actions of the in-plant workers and maintains the hygiene and sanitation practices successfully. Moreover, the layout of processing plants and facilities provided becomes a crucial consideration with respect to the safety of fresh produce from microscopic organisms. Fresh food commodity is considered to be as a potential source of contamination, as raw food materials inherit most of the contaminants which can possibly result in cross-contamination and becomes a threat to other food commodities. These potential hazards can preferably be minimized with proper layout and design with adequate maintenance of the industrial plants (Brackett, 1999).

In contrast to the microbiological safety of fresh produce and horticultural crops due to processing, there can be some detrimental impacts of processing too, which can lead to a decrease in the nutritional quality of foods. The processing of food component influences the vitamins and minerals stability which depends on several factors such as the duration of processing, environmental impact, and the resulting end product formed which are delivered and consumed. Depending on the chemical properties of vitamins, the degradation of fat-soluble and water-soluble vitamins varies with different sets of chemical processes. The fat-soluble vitamins get degraded by autocatalytic processes which are enhanced with the exposure to light, air, and transition metal elements such as iron and copper.

Further, if the equipment used for the handling of food materials is contaminated or possesses any metal ions during processing and does not limit contact with light and air (refers to packaging and storage) are all the factors and conditions that would greatly impact the stability of vitamins. Thermal processing of food commodities results in the loss of fat-soluble vitamins such as vitamin A and carotenoids with the exception of vitamin K. Certain water-soluble vitamins such as thiamine and cobalamin are prone to oxidation. In addition, the leaching of vitamins occurs when exposed to the aqueous medium. The loss of vitamins is also initiated during processing and storage via electromagnetic radiation and ultraviolet (UV) rays. Temperature is also one of the factors that initiates vitamin loss excluding niacin, biotin, and riboflavin which are heat stable whereas others include ascorbic acid, pantothenic acid, folate, and cobalamin. In consideration of minerals, they do not get affected by heat, light, oxidizing agents, or pH. These are generally degraded by several processes which include milling, soaking, cooking, germination, fermentation, and heat processing.

Similar to processing, transportation, and distribution of horticultural crops have profound effects on microbiological safety which includes loading workers, truck drivers, and shopkeepers as well. They play a crucial role in maintaining optimum temperature conditions while preventing cross-contamination. The other practice that initiates contamination during transportation is loading the fresh raw commodity of fresh fruits and vegetables with raw uncooked meat, live animals, or other sources of pathogenic bacteria. Inappropriate loading patterns of fresh produce can also lead to uncertain conditions of food products (Reddy and Love, 1999).

Lack of necessary engagement and scrutiny in the collection of suitable types of packaging while transporting a food commodity, infrastructure of storage space that is presence of ramps or slipways on the routes as well as type and condition of transport may reduce the quality or delay in delivery practices. Therefore, an overall analysis was conducted with respect to dairy products by Lipińska et al. (2019) using the Ishikawa diagram. This model revolves around the 5M and E diagram model which elucidates that man, management, machines, materials, methods, and the environment are responsible for the contamination during transportation. Further, it explains that sometimes it might be hard for the employees working there to make decisions regarding various issues for example to reduce or eliminate food loss for which human plays a vital role. The simultaneous practice of continuous enhancement and a planned course of action would be a great lead to overcoming the critical situations. Proper refrigeration conditions and optimum temperature are required for perishable food items such as milk and its products to be transported from farm to fork. Thus, there is a need to maintain suitable temperature conditions till the time of consumption of food items as well as minimize the travel distance because the undesired rise in temperature at the time of loading and unloading may lead to a decline in the durability of the food product especially perishable food items.

Management also comes into role as it carries equal importance. Organization of processes including loading, unloading, route adjustment, and so on should be practiced regularly. Mechanical injury can be caused by vibrations during transit together with imprudent placement of commodities. Proper regulatory conditions and arrangement of safeguard practices for food items must be implemented especially when the product to be delivered has a long travel time to avoid spoilage of food material. However, the material used for the packaging of food products must not be ignored. Inadequate packaging practices can also be the reason for the loss at this stage. Therefore, the parameters such as durability, impermeability, elasticity, and brittleness of the packaging material should also be taken into consideration. The other aspect to consider is machines. The machinery such as handling equipment, and monitoring of vehicles as there might be a problem in some parts of the machinery due to age, poor maintenance, and limited inspection protocols which can pose various accidents resulting in degradation in the quality of food products (Lipińska et al., 2019).

Contamination can also be caused by the exhaust of petroleum from the vehicle which becomes a prominent source of cross-contamination. Food commodities can also be contaminated through chemicals that are used for the purpose of disinfection while the food product is being transported at longer distance (Zunic and Peter, 2018).

Further, to elaborate Arah et al. (2015) discussed the overall postharvest factors affecting the quality attributes and shelf life stability specifically of tomatoes which are summarized hereafter. The first factor discussed was temperature, which is supposed to be very efficient in the maintenance of the quality of food products. Therefore, the metabolic activities must be ceased which initiates ripening by maintaining the low-temperature conditions of about 20°C. Moreover, the production of ethylene gets triggered due to a rise in temperature in addition to other factors which also cause of ripening of fruits. Low-temperature conditions help to maintain quality attributes such as texture, aroma, flavor, and nutrition. Below 10°C, chill injury occurs in fruits which show effects including premature softening, capricious color development, seed browning, spoilage, and off-flavor development.

The other factor involves relative humidity which is indicated as the loss of water from the horticultural produce due to the presence of moisture in the air. With the increase in relative humidity, agronomic produce maintains the overall quality attributes while a decrease in the same shows a wilting effect, softening, and so on, but on adding some moisture, shriveling of fruit can be prevented. In addition, over-saturation or 100% relative humidity encourages fungal and mold growth. Further, the consolidation of different gasses enhances the storage environment and extends the shelf life of fruits. Consequently, the physical handling of fruits after harvesting is also a factor that affects the quality of the fruits. Therefore, each and every aspect of the factors discussed above should be taken into consideration for the better quality of food commodities. Heap discusses the importance of food transportation as a link in the food chain. On a local and global scale, raw materials, food ingredients, fresh produce, and processed products are all carried by land, sea, and air. Long-distance distribution of various foods has become usual in the modern world when consumers expect year-round supplies and nonlocal products, and perishable materials may require air shipment (Heap et al., 1998).

8.3.3 During packaging

Packaging is a crucial aspect of the food production and distribution process. Microorganisms, insects, and rodents are among the threats that packaging must withstand. If oxygen and water vapor are allowed to readily penetrate the packaging, they will ruin the food. Food packaging's main functions are to protect food from outside influences and damage, to keep food safe, and to give consumers ingredient and nutritional information (Coles et al., 2003). Traceability, ease of use, and tamper detection are secondary functions that are becoming increasingly important. With constant innovation in design and printing, packaging has become a key influencer in the market and helps attract the right attention amongst a hoard of other products.

First, the purpose of packaging is to keep food contained in a cost-effective and industry-friendly manner, preserve food safety, meet customer needs and desires, and possess low environmental impact. The package acts as a silent marketer link between the food producer and the customer. Food packaging has advanced dramatically in recent decades, owing to increasing demand for high-quality, safe food and increased concern about environmental challenges. The function of food packaging in overall food production sustainability is debatable: the public assumption that packaging has substantial environmental implications clashes with scientific evidence of packing benefits in terms of possible food waste reduction. Overall, the good environmental effect of food packaging is well-known among insiders, but it needs to be highlighted by the general public, whose attitude is still rather unfavorable (Licciardello, 2017). Indeed, it is undeniable that packaging has an environmental impact during its life cycle (Huang and Ma, 2004; Ingrao et al., 2015), particularly during the raw material manufacture, processing, and end-of-life phase, which includes recycling, incineration, and landfill disposal. According to Peelman et al. (2013), the sustainability of food packaging can be accomplished at three stages: (1) at the raw materials level, where recycled materials and renewable resources can be used in order to reduce the CO_2 emissions and some fossil resources can also be used; (2) at the manufacturing level, where the use of more energy-effective protocols can lead to sustainability of food packaging; and (3) at the waste management level, where the main considerations of food packaging are their recycling, reuse, and biodegradation.

Second, the packaging of a product is a crucial type of marketing communication. Consumers rely on their buying decisions on external product qualities and appearance while shopping for everyday foods and beverages (Fenko et al., 2010). Consumers perceive intrinsic product cues such as flavor, scent, and texture differently than extrinsic product cues such as packaging material, information, and brand name (e.g., packaging material, information, brand name, and price) (Ng et al., 2013). Extrinsic product signals are processed through cognitive and psychological mechanisms, whereas intrinsic product cues are related to sensory and perceptual systems

(Cardello, 2007). Aspects of choice that occur outside of a person's conscious knowledge have a significant impact on their purchasing decisions (Fitzsimons et al., 2017). Extrinsic product signals like packaging and branding impact the way consumers perceive food products, according to a previous study (Deliza and Macfie, 1996). Furthermore, consumers' expectations are influenced by extrinsic visual signals like packaging, labeling, pricing, and nutritional information (Gunaratne et al., 2019). This is especially critical for buyers who are unfamiliar with the product. Shoppers are more likely to remember designs that are unusual and stand out.

Third, there are various chemicals that are utilized in manufacturing the packaging material that may leach into the food materials and hence lead to the reduction of the quality. The transfer of chemicals from packaging to food, on the other hand, may have a negative impact on food quality and safety. The most essential and definitely the most visible example of a material or product meant to come into contact with food is packaging. There are numerous other instances in which materials come into touch with food during its manufacture, transportation, storage, preparation, and consumption. Materials used to make storage tanks, conveyor belts, tubing, and cooking and eating utensils are among them. Food and beverages can be aggressive products that have a strong interaction with the materials they come into contact with. Food acids, for example, can corrode metals, fats, and oils can expand and leach polymers, and beverages can break down unprotected paper and carton boards. In fact, no food contact material is fully inert, and chemical elements from these materials can migrate into packaged food. When packaging materials come into contact with certain types of foods, they can all emit trace amounts of their chemical contents (Barnes et al., 2006).

8.4 Conclusion

Environmental concerns are unquestionably becoming universal. Some of these worries about agriculture and food, which were first addressed in the 1960s, are still relevant today. The majority of the chapter focuses on solid agricultural items such as fruits, vegetables, cereals, and legumes, while many of the same concerns can be applied to animal-based products such as meat, eggs, and milk as well. Food processors would like a constant supply of raw materials in an ideal world whose composition, cost, and quality remain consistent. The quality control of food materials begins on the farm and continues until the product reaches the consumer. The quality of the commodities is influenced in part by several preharvest procedures used during production. After harvest, no postharvest treatment procedure or handling practices can improve the quality of any plant material; it can only be maintained. To produce high-quality fresh produce at harvest, it is critical to understand and manage the role of preharvest factors that influence the quality of plant raw material. Furthermore, there are a variety of elements at each stage of processing that affect the food in different ways. Taking these considerations into account, the factors influencing food processing were compiled at every stage of processing. All these factors may endanger human health; therefore, the food must not change its composition; and the organoleptic features of the food must not change.

References

Arah, I.K., Amaglo, H., Kumah, E.K., Ofori, H., 2015. Preharvest and postharvest factors affecting the quality and shelf life of harvested tomatoes: a mini review. Int. J. Agron. 2015, 1–6. https://doi.org/10.1155/2015/478041.

Barnes, K., Sinclair, R., Watson, D., 2006. Chemical Migration and Food Contact Materials. Woodhead Publishing. https://doi.org/10.1533/9781845692094.

Benner, R.A., 2014. Organisms of concern but not foodborne or confirmed foodborne: spoilage microorganisms. In: Motarjemi, Y., Moy, J., Todd, E. (Eds.), Encyclopedia of Food Safety. Elsevier, Amsterdam, The Netherlands, pp. 245–250. https://doi.org/10.1016/B978-0-12-378612-8.00169-4.

Blackburn, E.H., Greider, C.W., Szostak, J.W., 2006. Telomeres and telomerase: the path from maize, *Tetrahymena* and yeast to human cancer and aging. Nat. Med. 12, 1133–1138. https://doi.org/10.1038/nm1006-1133.

Brackett, R.E., 1999. Incidence, contributing factors, and control of bacterial pathogens in produce. Postharvest Biol. Technol. 15, 305–311. https://doi.org/10.1016/S0925-5214(98)00096-9.

Broome, S.W., 2015. Factors Affecting Plant Growth. North Carolina State University.

Caballero, B., Finglas, P.M., Toldrá, F., 2016. Spoilage: yeast spoilage of food and beverages. In: Encyclopedia of Food and Health. Oxford University Press, pp. 113–117.

Cardello, A., 2007. Measuring consumer expectations to improve food product development. In: Consumer-Led Food Product Development. CABI Digital Library, pp. 223–261. https://doi.org/10.1201/9781439823903.ch10.

Chung, O., Pomeranz, Y., 2000. Cereal processing. In: Food Proteins: Processing Applications. Wiley-VWC, pp. 243–307. https://doi.org/10.3362/9781780443966.

Coles, R., McDowell, D., Kirwan, M., 2003. Food Packaging Technology. CRC Press. https://doi.org/10.1111/j.1471-0307.2005.00157.x.

Deliza, R., Macfie, H.J.H., 1996. The generation of sensory expectation by external cues and its effect on sensory perception and hedonic ratings: a review. J. Sens. Stud. 11, 103–128. https://doi.org/10.1111/j.1745-459X.1996.tb00036.x.

Fenko, A., Schifferstein, H.N.J., Hekkert, P., 2010. Shifts in sensory dominance between various stages of user-product interactions. Appl. Ergon. 41, 34–40. https://doi.org/10.1016/j.apergo.2009.03.007.

Fennema, O., 1977. Loss of vitamins in fresh and frozen foods. J. Food Technol. 31, 32–38.

Fitzsimons, G.J., Hutchinson, J.W., Williams, P., Alba, J.W., Chartrand, T.L., Huber, J., Kardes, F.R., Menon, G., Raghubir, P., Fitzsimons, G.J., Hutchinson, J.W., Williams, P., 2017. Non-conscious influences on consumer choice. Mark. Lett. 13, 269–279.

Food, 2022. Human Ecology and Family Sciences Part-1. NCERT.

Food Research Lab, 2021. Various types and methods involved in food processing. Guires Food Research Lab. https://www.foodresearchlab.com/blog/industries/various-types-and-methods-involved-in-food-Process.

Grandison, A.S., 2011. Postharvest handling and preparation of foods for processing. In: Food Processing Handbook 1. Wiley, pp. 1–30. https://doi.org/10.1002/9783527634361.ch1.

Gunaratne, N.M., Fuentes, S., Gunaratne, T.M., Torrico, D.D., Francis, C., Ashman, H., Gonzalez Viejo, C., Dunshea, F.R., 2019. Effects of packaging design on sensory liking and willingness to purchase: a study using novel chocolate packaging. Heliyon 5, e01696. https://doi.org/10.1016/j.heliyon.2019.e01696.

Harvey, J.M., 1978. Reduction of losses in fresh market fruits and vegetables. Annu. Rev. Phytopathol. 16, 321–341. https://doi.org/10.1146/annurev.py.16.090178.001541.

Heap, R., Kierstan, M., Ford, G., 1998. Food Transportation. Boom Koninklijke Uitgevers.

Holt, J.E., Schoorl, D., Muirhead, I.F., 1983. Post-harvest quality control strategies for fruit and vegetables. Agric. Syst. 10, 21–37. https://doi.org/10.1016/0308-521X(83)90014-8.

Huang, C.C., Ma, H.W., 2004. A multidimensional environmental evaluation of packaging materials. Sci. Total Environ. 324, 161–172. https://doi.org/10.1016/j.scitotenv.2003.10.039.

Ingrao, C., Lo Giudice, A., Bacenetti, J., Khaneghah, A.M., Sant'Ana, A.S., Rana, R., Siracusa, V., 2015. Foamy polystyrene trays for fresh-meat packaging: life-cycle inventory data collection and environmental impact assessment. Food Res. Int. 76, 418–426. https://doi.org/10.1016/j.foodres.2015.07.028.

Kader, A.A., 2013. Postharvest technology of horticultural crops: an overview from farm to fork. Ethiop. J. Appl. Sci. Technol. 1, 1–8.

Licciardello, F., 2017. Packaging, blessing in disguise. review on its diverse contribution to food sustainability. Trends Food Sci. Technol. 65, 32–39. https://doi.org/10.1016/j.tifs.2017.05.003.

Lipińska, M., Tomaszewska, M., Kołożyn-Krajewska, D., 2019. Identifying factors associated with food losses during transportation: potentials for social purposes. Sustainability 11 (7), 2046. https://doi.org/10.3390/su11072046.

Ministry of Health and Family Welfare, 2006. Vol I—Introduction to Food and Food Processing 2006 Across the Country. Ministry of Health and Family Welfare.

Ng, M., Chaya, C., Hort, J., 2013. The influence of sensory and packaging cues on both liking and emotional, abstract and functional conceptualisations. Food Qual. Prefer. 29, 146–156. https://doi.org/10.1016/j.foodqual.2013.03.006.

Oko-Ibom, G.O., Asiegbu, J.E., 2007. Aspects of tomato fruit quality as influenced by cultivar and scheme of fertilizer appication. Agro. Sci. 6, 71–81.

Paulus, K., 2000. Nutritional and sensory properties of processed foods. In: Food Properties and Computer-Aided Engineering of Food Processing Systems. Springer Link, pp. 177–200.

Peelman, N., Ragaert, P., De Meulenaer, B., Adons, D., Peeters, R., Cardon, L., Van Impe, F., Devlieghere, F., 2013. Application of bioplastics for food packaging. Trends Food Sci. Technol. 32, 128–141. https://doi.org/10.1016/j.tifs.2013.06.003.

Petruzzi, L., Corbo, M.R., Sinigaglia, M., Bevilacqua, A., 2017. Microbial spoilage of foods: fundamentals. In: The Microbiological Quality of Food: Foodborne Spoilers. Elsevier Ltd., pp. 1–21. https://doi.org/10.1016/B978-0-08-100502-6.00002-9.

Reddy, M.B., Love, M., 1999. The impact of food processing on the nutritional quality of vitamins and minerals. Adv. Exp. Med. Biol. 459, 99–106. https://doi.org/10.1007/978-1-4615-4853-9_7.

Salunkhe, D.K., Deshpande, S.S., 2012. Foods of Plant Origin: Production, Technology, and Human Nutrition. Springer Science & Business Media.

Sensitech, n.d. Effects of low RH on fresh produce. Effects of high RH on fresh produce takeaway. https://www.sensitech.com/en/blog/blog-articles/blog_relativehumidity.html.

Singh, R.P., 1994. Scientific principles of shelf life evaluation. In: Shelf Life Evaluation of Foods. Springer, Boston, pp. 3–26.

Sperber, W.H., 2009. Introduction to the microbiological spoilage of foods and beverages. In: Compendium of the Microbiological Spoilage of Foods and Beverages. Springer, pp. 1–18. https://doi.org/10.1007/978-1-4419-0826-1.

Thorne, S., Alvarez, J.S., 1982. The effect of irregular storage temperatures on firmness and surface colour in tomatoes. J. Sci. Food Agric. 33 (7), 671–676.

Tian, M.S., Downs, C.G., Lill, R.E., King, G.A., 1994. A role for ethylene in the yellowing of broccoli after harvest. J. Am. Soc. Hortic. Sci. 119, 276–281. https://doi.org/10.21273/jashs.119.2.276.

Toivonen, P.M.A., 1997. The effects of storage temperature, storage duration, hydro-cooling, and micro-perforated wrap on shelf life of broccoli (*Brassica oleracea* L., Italica Group). Postharvest Biol. Technol. 10, 59–65. https://doi.org/10.1016/S0925-5214(97)87275-4.

Zunic, B., Peter, S., 2018. World's Largest Science, Technology & Medicine Open Access Book Publisher. Intech, pp. 267–322.

Nanobiosensor potentialities for food toxin detection

Kamana Singh[a], Prabha Arya[a] and Ram Sunil Kumar L[b]

[a] *Department of Biochemistry, Deshbandhu College, University of Delhi, New Delhi, India,*
[b] *Department of Chemistry, Kirori Mal College, University of Delhi, New Delhi, India*

9.1 Introduction

As we all are dependent on food for our daily energy needs as well as for micro and macronutrient requirements, we must understand the significance of food safety and surveillance. Quality control of food is an important part of the food industry. It is required for the assurance of the safety of the consumers since the toxic content in food can be problematic. Food that reaches the consumers has to go through many processes, not only that the processed food has to pass through so many steps to increase its shelf life. Due to these processes and steps and also due to the food packaging materials, the food may acquire some toxic content. When crops are grown, chemical fertilizers and pesticides are used, and thus, these chemicals enter the food. Apart from these synthetic chemicals, some natural toxins are also produced by some plants or by some weeds which act as adulterants (Srivastava et al., 2018).

Genetic susceptibility of some people to the content of the food like specific protein can cause allergy at the same time there can be community-based food contamination caused by spoilage of food. Even the microorganisms can cause changes in the texture of the food, flavor, and taste. These microorganisms can be aerobic or anaerobic. The former grows on the surface of the food and makes a layer, that is thick enough for their respiration, and in the case of the latter, they can get access to the core of the food pieces. Since various categories of microorganisms from different genera can create havoc, this becomes a serious global concern. To prevent the growth of microorganisms, preservatives are added in a certain amount which is again not healthy beyond permissible limits. These microorganisms also produce toxins that can have different metabolic or biochemical mechanisms to cause infection (Siripongpreda et al., 2020). These toxins have molecular mechanisms to create different types of health conditions like diarrhea, vomiting, fever, and many other symptoms of food poisoning which may lead to death in severe conditions. The serious risk-inducing components are the heavy metals which are incorporated through many routes and their assessment is important for safe food consumption. For the detection

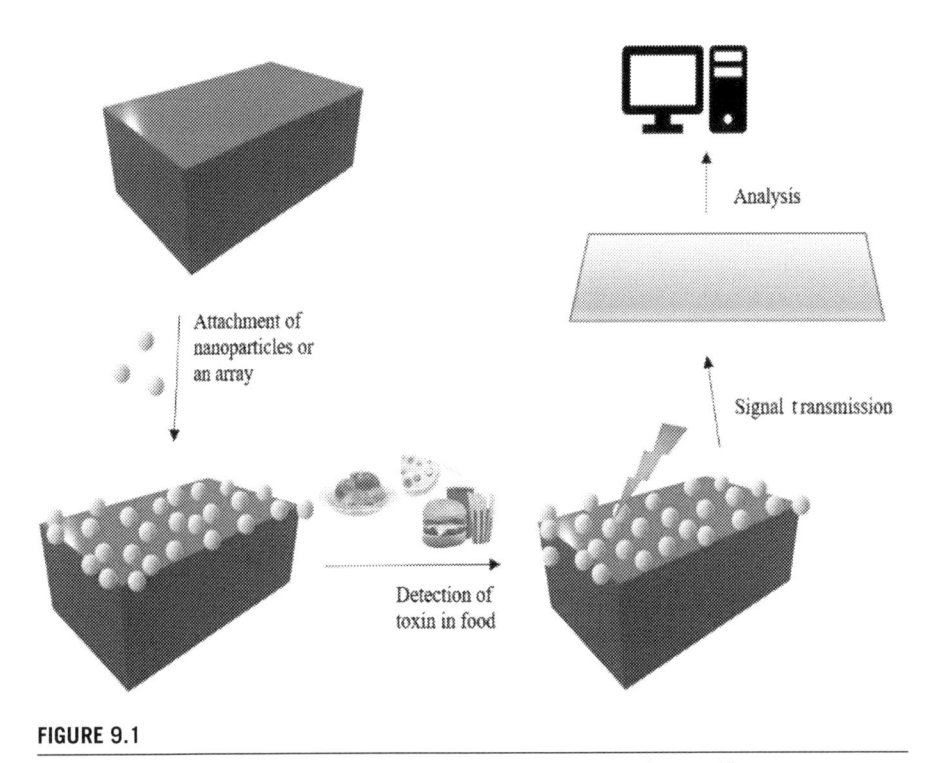

FIGURE 9.1

A typical nanobiosensors made by fabricating the nanoparticle of a specific type on an array and analyzed by different techniques.

of heavy metals such as As, Pb, Hg, and Cd, nanobiosensors are utilized that are made up of nanoparticles like silver, gold, or the nonmetallic nanoparticles of carbon (Salek Maghsoudi et al., 2021). For analysis of these toxins, traditional analytical techniques such as high performance liquid chromatography (HPLC) and gas chromatography (GC) have been used and have been reliable. These techniques require expensive instrumentation, and skilled operators and are time-consuming. But biosensors can provide an invaluable method for food diagnostics including nutrients and toxins, since they are user-friendly, portable, and need not have skilled operators, biosensors are used to detect the toxins in the food by the food industry. Nanobiosensors can be used to determine the degree of adulterants, allergens, pesticides, antibiotics, trans fats, and many other unwanted components found in foods through enzymatic and immunogenic reactions. Food toxins and nutrients can be analyzed by various types of instruments such as single-use disposable sensors, bench-top portable instruments, and large multianalyzers. Large multianalyzers which are usually situated in specialized laboratories, benchtop portable instruments, and other techniques can be used to assess food toxins at a small scale on the other hand single-use disposable sensors (Fig. 9.1) can be useful for rapid analysis and in several samples by any individual (Salek Maghsoudi et al., 2021).

9.2 **History of biosensors**

Biosensors have been in use for decades. An oxygen electrode that is modified with glucose oxidase and a Teflon membrane was applied in the first biosensor by Clarke and Lyons in 1962 for detecting glucose (Clark and Lyons, 1962). These types of biosensors were bulky. This earliest biosensor utilized the enzyme glucose oxidase to convert glucose into gluconolactone and H_2O_2. H_2O_2 is further oxidized by an enzyme coupled to this reaction to convert H_2O_2 to H_2O and hydrolyzed the color-producing agent, which is being monitored. In recent years, an attempt has been made to find out portable biosensors for the detection of different samples whether beneficial or toxic for rapid and accurate measurements.

There are different kinds of analytical techniques that can be used for the assessment of contamination. One method utilizes the identification of microorganisms in which selective media is provided to allow the expected microorganism to grow, but this technique is time-consuming. Another method utilizes the immunological approach or polymerase chain reaction (PCR)-based methods which may be error-prone. For the determination of toxins, test tube-based diagnostics were carried out and the spectroscopic method helped in earlier days. In recent years, miniaturization has taken place and small kits are made which contain all the reagents, required for detection. These kits can easily be used for qualitative analysis but in quantitative analysis, one will have to move to solution-based methods, which pose cost constraints as the materials for diagnostics are costlier. After the evolution of nanotechnology, different types of nanomaterials have been utilized to provide a scaffold that is beneficial for the assessment of toxins with minimal use of reagents and thus making diagnosis cost-effective.

9.3 **Nanotechnology as biosensor**

Biosensors are devices that are based on some specific reaction or change in pH and are mostly based on a recognition element which is biomolecules such as enzymes, antibodies, oligonucleotides, or receptors. In the earliest sensing technologies, radiolabeling was utilized but it required skills as well as was not cost-effective. Then enzymatic reactions were taken into consideration followed by antibodies and then florescence-based markers. Recently, advancements in nanotechnology provided an opportunity to imprint polymers or other receptors made artificially. In the beginning, nanoparticles were used as a scaffold for different types of reactions. Then these nanoparticles' different properties like electrical, optical, magnetic, and piezoelectric were applied to detect the toxins. In this respect, the definition of biosensors has broadened over the years to include sensors that detect biological systems that are limited to those containing a biological component.

Most of the time, enzymes are utilized for the detection of toxins which change them into detectable molecular forms like hydrogen peroxide or chromogenic substrate. Hydrogen peroxide or other product of the first reaction is then further coupled to some other signal-producing second analyte. Other most utilized sensors are based on antibodies. For the detection of specific toxins, antibodies are raised and tagged with a detectable substrate, which can be converted into a colored product, and the color produced is analyzed for the presence of the toxin. For the detection of specific antigens which can be some toxin or a surface feature of the growing microorganism itself, nucleic acid hybridization or receptors are used and analyzed by biophysical techniques like surface plasmon resonance or quartz crystal microbalance.

There is development in the detection of specific smell as well as taste changes which can come from spoilage of food which can be detected by e-nose and e-tongue, respectively. There are four components of e-nose a sampling headspace system, a sensor array, an electronic data acquisition control system, and an analysis software. The volatile analyte can be detected by a chemical sensor that senses the changes in conductance and the signal is detected by detectors. There is a wide variety of sensors used in e-nose like metal oxide, conductive polymer sensors, surface acoustic wave sensors, quartz crystal microbalance sensors, optical sensors, gas-sensitive field-effect transistors, etc. (Srivastava et al., 2018). By this method, the smell of plastic bags in which food is contained is detected (Torri and Piochi, 2016). Taste stimulants like different chemicals which stimulate the senses in the human tongue can be artificially utilized in diagnosis by e-tongue. Taste buds mimicking sensors can detect various inorganic and organic compounds (Dias et al., 2017).

The advantage of nanoparticles usage for the detection of food toxins is that these nanoparticles offer rapidity, selectivity, and cost-effectiveness with minimum user input, as the nanoparticles offer promising unique physical, optical, electrical, magnetic, and catalytic properties with the advantage that their sizes can be varied according to the requirement. As the size of the nanoparticle varies, optical properties also change, which helps in determining the preparation quality as well as in the detection signal due to the quantum effect which happens due to densely packed electrons in nanoparticles which imparts specific color bands in the spectrum. As these nanoparticles have a greater surface-to-volume ratio, they offer more area for immobilization.

In the following sections, the properties of nanoparticles, whether a single type or combination of two or more than two nanoparticles, and their applications have been discussed which have been useful in nano biosensors design.

9.4 **Types of food toxins**

Toxicity in food can be caused by one or more types of toxins which can be of natural origin, adulteration chemicals, pesticides, or heavy metals as well as metabolites produced by the microorganisms. There is a natural defense mechanism in plants that produce lectins of a specific type to deter the animals. These proteins are not toxic to plants themselves but they can be harmful to the animals or humans who are feeding on them. Similarly, some fishes like shellfish are producing toxins that in the first place are not toxic to other fishes or other animals that are eating them but these can be fatal to the humans who are eating these fishes which are not properly treated before cooking. Food items that are stored for a longer time or food items that are not properly preserved by adding preservatives along with wrong storage conditions can invite the growth of fungi species like *Fusarium, Aspergillus*, and *Penicillium*. These fungi produce secondary metabolites which can be toxins for humans. There are around 300 mycotoxins known to date, and out of these, there are few known toxins about which standards have been set for their permissible limits, like what is the range of their quantities which can be deadly or can be accepted. Some of the important mycotoxins are ochratoxins, citrinin, aflatoxins, feminizing, citreoviridin, patulin, and zearalenone (Malhotra et al., 2014). Antibiotics and drugs can enter the food through the consumption of the products from animals, who are fed these agents containing fodder (Girigoswami et al., 2021).

Even some nanoparticles may also lead to toxicity. Due to more and more products and applications in various fields, exposure to nanoparticles is increasing which can be accessed via different ways like inhalation, food intake, skin contact, and intravenous injection (Sengul and Asmatulu, 2020). Some types of metallic nanoparticles like zinc oxide and titanium dioxide can also work as toxins. These nanoparticles are used in different fields like medicine, agriculture, food, photocatalysis, and the cosmetic industry (Baranowska-Wójcik et al., 2020). Zinc oxide which is used extensively in the cosmetic industry also possesses the risk of toxicity (Sruthi et al., 2018). Silicon oxide, iron (III) oxide (Sohal et al., 2020), and cerium oxide (García et al., 2011) which are used in different industries, photocatalytic processes, and medicine, impose the risk of phytotoxicity by inhibiting seed germination. These nanoparticles have been detected by X-ray powder diffraction (XRD), transmission electron microscopy (TEM), atomic force microscopy (AFM), and confocal microscopy.

The use of plastic has been comfortable to use and through in terms of cost, handling, and disposal and also has replaced packaging material which is made up of glass and metals. Bisphenol A (BPA) is a component of many packaging materials that are found to be xenoestrogen as it can cause disturbances in the reproductive system like early puberty, a developmental disorder in the growing fetus as it causes epigenetic changes, and has been reported as neurotoxic. BPA is also considered a uremic toxin as there are reports of kidney-related problems like kidney failure in hemodialysis patients (Vandenberg et al., 2007; Bosch-Panadero et al., 2017; Kundakovic and Champagne, 2011; Ezoji et al., 2020) (Fig. 9.2).

To enhance the appearance of the food and make it more appealing, it has been an old practice to add food colors. Some spices that are good for health also add color to the food to make it more delightful but on the other hand, there are some synthetic dyes which are harmful chemicals, are also added to the food for the same purpose. Broadly this colorant can be categorized into azo compounds, triarylmethane group, the chinophthalon derivative of quinolone yellow, xanthene as erythrosine, and the indo colorant (Ai et al., 2018). Although these colorants have acceptable limits too much consumption of food containing them can be an issue. The most significant agent of food deterioration is lipid oxidation during food processing (Mpountoukas et al., 2010).

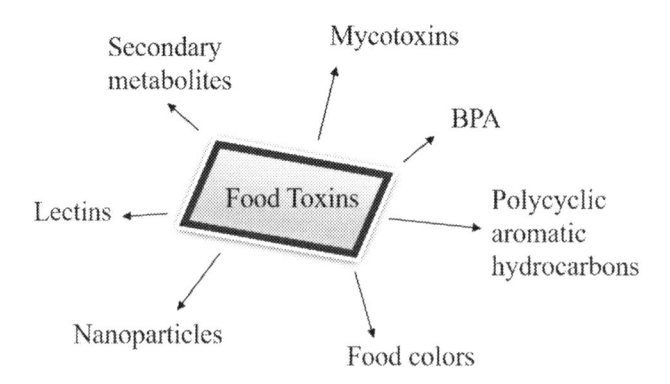

FIGURE 9.2

Different types of toxins found in food products.

In food generated toxins apart from microorganisms are trans-fat which develop as a result of the processing of food. These trans-fats are the culprit for lifestyle disorders like atherosclerosis resulting in cardiovascular diseases (Ganguly and Pierce, 2015). Similarly, polycyclic aromatic hydrocarbons generated during grilling of the food generally seep into the food items that have been found carcinogenic, teratogenic, and mutagenic (Zhang et al., 2016; Kim et al., 2013). These toxins are analyzed by either spectrophotometrically or nanobiosensors made up of different nanomaterials and composites and detected by electrochemical properties (Ronkainen et al., 2010), Raman spectra, surface-enhanced Raman spectroscopy (SERS), etc. (Zhang et al., 2016).

9.5 Carbon-based nanomaterial used in food toxin detections

For detection purposes, we rely on some techniques that can either produce some signal themselves or can be conjugated to other signal-producing entities. For example, fluorophores that produce fluorescence can be detected by appropriate detection systems and can be correlated with the amount of toxin. The disadvantage of fluorescence detection is that they are gone as soon as the exciting light source is removed (Shen et al., 2012). Since carbon-based nanomaterials have shown an advantage in detection owing to their optical properties, the above limitation is removed by their usage. Thus, there are different types of nanomaterial made up of carbon and in combination with metals which have been in practice to detect the toxin in various food items. Carbon nanotubes, graphene, fullerenes, and nanodiamonds are such carbon nanomaterials (Fig. 9.3).

9.5.1 Carbon nanotubes

Carbon nanotubes (CNTs) of single-walled or multiwalled types offer a great absorptive surface for the toxin to be analyzed. They have been used in sensing toxins as

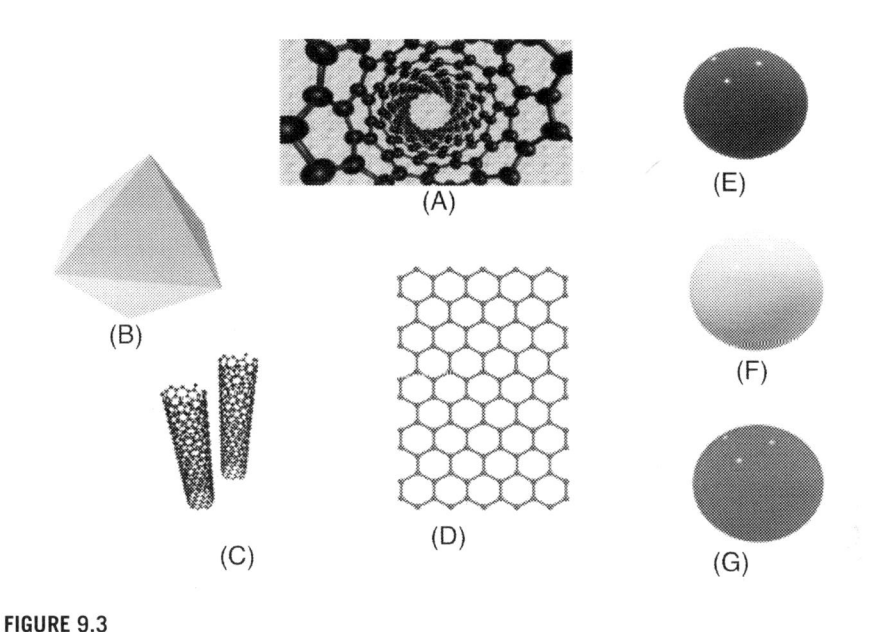

FIGURE 9.3

Various forms of carbon and metal nanoparticles; (A) buckyball, (B) nanodiamond, (C) carbon nanotubes, (D) graphite sheet, (E, F, and G) carbon, silver, and gold nanoparticles, respectively.

it has good electrical, mechanical, and chemical properties and have a structure that can be used for encasing the required chemical (Malhotra et al., 2015). CNTs have applications for the analysis of different types of toxins and allergens. For example, it has been used for the analysis of gliadin which is involved in causative agents of autoimmune disorders such as chronic diarrhea and other symptoms like fatigue, weight loss as well and anemia in people who suffer from celiac disease. Zein is a low-cost by-product from corn and is produced during ethanol production. It can form a biofilm that is biodegradable and nontoxic and is used as an anchor. Along with other nanomaterials like carbon nanotubes, graphene oxide, and laponite, zein is anchored to gliadin protein to the electrode which replaces the chemical anchors that can be reactive. To this nonreactive biofilm, antigliadin monoclonal antibodies are adhered. Gliadin from wheat flour is extracted and used for the determination of the limit of detection by binding to the antibody-coated zein-carbon nanotubes (Rouf et al., 2020).

Bisphenol A which can be found in some plastic containers used for packaging as well as in the plastic apparatuses used for food processing can be toxic. It is detected by a combination of titanium oxide nanoparticles and carbon nanotubes with chitosan as a crosslinker. The surface morphology of this sensor is checked by techniques like X-ray diffraction (XRD), field emission scanning electron microscopy (FESEM), and energy-dispersive X-ray spectroscopy (EDX) (Ezoji et al., 2020).

9.5.2 **Graphene**

Graphene has been utilized for the formation of biosensors in various structural forms. One form of graphene oxide is spun into nanofibers which are used as adsorbents for the antibiotic samples. Different types of tetracyclines are analyzed by the SPE-HPLC-FLD method in chicken tissue samples and have good prospects for the detection of antibiotics in food samples (Weng et al., 2019).

Semiconductor-based quantum dots have proved themselves useful. Specific types of quantum dots which are made up of carbon offer inertness, low toxicity, and high biocompatibility (Shen et al., 2012). For example, graphene quantum dots (GQDs) are used as potential electrochemical biosensors. These GQDs are deposited on an indium tin oxide-coated glass substrate and monoclonal antibodies of aflatoxin 1 are immobilized on the GQDs with the help of a crosslinker. The presence of Aflatoxin is detected in food by using this assembly and analyzed by ultraviolet (UV)-visible spectroscopy, photoluminescence spectroscopy, SEM, TEM, Raman spectroscopy, etc. (Bhardwaj et al., 2018).

Food color and other toxic substances like heavy metal ions, antibiotic residues, and pesticides have been studied to be detected by graphene and graphene-like 2D graphitic carbon nitride by electrosensory techniques (Magesa et al., 2019). Ochratoxin is a mycotoxin that poses major health risks (Kőszegi and Poór, 2016). Graphene surface is also utilized for the fabrication of ochratoxin sensing aptamers which sense the toxin and detection is carried out by sensing the change in voltage. This device has the benefit of rapid detection and results are reproducible (Nekrasov et al., 2019).

One interesting application of quantum dots along with platinum nanoparticles is the formation of self-propelled microsensors. These graphene quantum dots and platinum nanoparticles are prepared by standard procedure and then mixed in the presence of an analyte. For functionalization of graphene quantum dots, a solution containing graphene quantum dots was mixed with 3-aminophenylboronic acid monohydrate and N-(3-dimethylaminopropyl)-N-ethylcarbodiimide in PBS buffer and mixed in a vortex. With both the nanoparticles a Janus Micromotor is synthesized in sodium dodecyl sulphate under vigorous stirring conditions. Lipopolysaccharide from Salmonella is detected by fluorometry (Pacheco et al., 2018).

The detection of BPA is done by graphene bimetallic nanoparticles composite involving gold and silver nanoparticles and it is tested by TEM, Raman spectroscopy, and XRD. It is found that composite nanoparticles have a better detection limit than pristine metallic nanoparticles (Pogacean et al., 2016).

9.5.3 **Fullerenes**

Another form of graphitic sheets apart from the 2D dimensional structure of graphene, graphite sheets rolled up in a tube form or ball form. These structures called fullerene have the advantage of having a more surface-to-volume ratio and also provide the area for the deposition of different agents to be analyzed or part of the analytic procedure (Nakanishi et al., 2014).

Chemical warfare can lead to the incorporation of chemicals into the environment which can affect the presence of harmful chemicals in food. One example is the use of nerve agent A-234 which inhibits the enzyme acetylcholinesterase which is used to maintain the concentration of acetylcholine at synapses in neuron and neuromuscular junctions. In a specific computational study, a ball of 20 carbons and C19 balls of 19 carbon plus one carbon atom replaced with silicon molecule ($C_{19}Si$), or with transition atom scandium ($C_{19}Sc$) and aminated fullerene ($C_{20}HNH_2$) were studied and their effectiveness was assessed against detection of A-234 in the sense of changes in absorption and emission energy also noise produced and it was found that aminated fullerenes are having low emission energy and low noise for excellent detection (Motlagh et al., 2020).

9.5.4 Nanodiamonds

Nanodiamonds are used in different fields because of their tunable surface structures, excellent mechanical properties, and high surface areas. Nanodiamonds are inert, which makes them well-suited to the detection of chemicals. In a study, for the extraction of ziram which is a fungicide, carboxylated magnetic nanodiamonds are utilized as absorbents. These nanoparticles that have absorbed ziram are analyzed further by XRD, SEM, and FT-IR (Yılmaz and Soylak, 2016).

Ziram is one of the dithiocarbamates used as a broad-spectrum fungicide used in agricultural products, it is reported to delay puberty by affecting enzymes in Leydig cells in the rat (Guo et al., 2017) apart from allergic reactions reported by direct contact with the fungicide. This can be detected by nanodiamonds which have excellent mechanical properties, high surface area, more surface-to-volume ratio, and changeable surface structures along with a nontoxic nature. These nanodiamonds are made magnetic in properties by attaching carboxylic groups by treating them with concentrated acids followed by attachment of ferrous ferric oxide (Fe_3O_4) by adsorption or electrostatic attraction. These magnetized nanodiamonds are used for the detection of ziram from different samples like foodstuffs and are detected by flame atomic absorption spectrometry (Yılmaz and Soylak, 2016) (Fig. 9.4).

Phenolic compounds like monophenols (phenol and cresol) and bisphenols (hydroquinone and catechol) are pollutants that may be present in rivers and green tea. In one study these pollutants were tested by using nanodiamonds by making an electrochemically resistant electrode and a single-use nanocarbon electrode (Jiang et al., 2018).

Catechol a phenolic compound produced by phenol hydroxylation using H_2O_2 which is used in the rubber industry is toxic to human health. It can be detected by using nanodiamonds which are linked with potato starch after depositing them on glassy carbon electrodes and are electroanalyzed by sensing the voltage differences with the immobilization of the tyrosinase enzyme (Camargo et al., 2018). In another study, it has been found that nanodiamonds can change the fluorescence of luminous bacteria which is changing its bioluminescence after treatment of aflatoxin B1 which

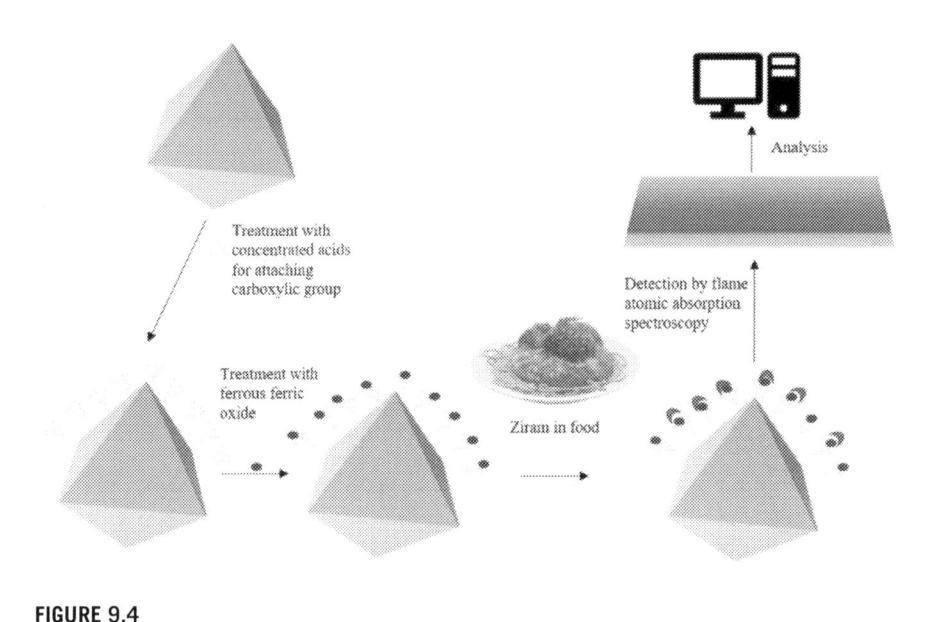

FIGURE 9.4

Strategy for the action of nanodiamonds which are conjugated with ferrous oxide via carboxylic group and analyzed for toxin detection by atomic absorption spectroscopy.

makes it possible to use nanodiamonds for the detection of aflatoxin B1 in food samples by measuring the bioluminescence (Mogilnaya et al., 2010).

9.6 Metallic nanoparticles

9.6.1 Gold nanoparticles

Gold nanoparticles have been widely used for the detection of different types of toxins like pathogens growing in food, toxins produced by microorganisms that get access to food, heavy metals in different types of food products like fruit juices, poultry products, seafood, chicken, beef, honey, etc. due to their unique optical properties (Chen et al., 2018). Gold nanoparticles have been used in different morphological forms (nanorods, nanoclusters, nanostars, nanoflowers, nanobipyramids, nanowire vesicles, nanocages, etc.) with the advantage of each of them. These nanoparticles have been utilized in different forms like attaching the analyte by H-bond or aptamer or even antibodies related to specific antigens. Enzymatic reactions that produce a detectable signal or are coupled to another detectable analyte are also used with the help of gold nanoparticles. They have been detected by color changes that can be observed without any visual aid or some specific instrument that can be used with specific physical properties like dynamic light scattering, Raman spectroscopy, photocurrent change, extinction spectra, fluorescence spectra, voltammetry, colorimetry,

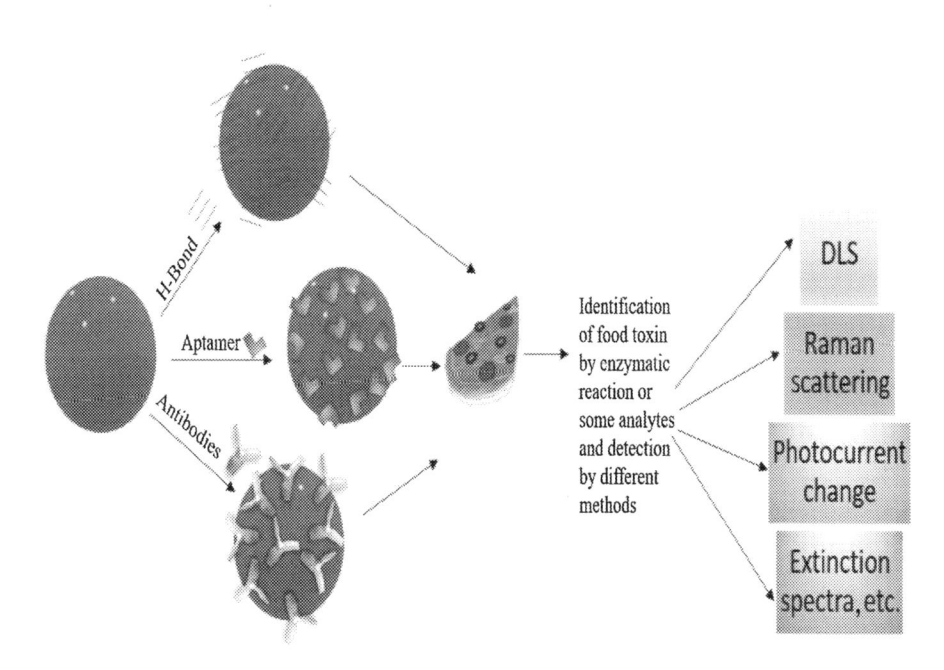

FIGURE 9.5

Gold nanoparticles provide a scaffold to H-bond, aptamers, or antibodies and are used for detection by enzymatic methods and are analyzed by various techniques.

etc. (Fig. 9.5). These methods are cumbersome and require costly instruments and skills to operate but there are some chip/strip-based sensors made with the help of gold nanoparticles that are cost-effective and user friendly (Hua et al., 2021).

In one study Aflatoxin from *Aspergillus flavus* is detected in corn and peanuts by a gold nanoparticle which is molecularly imprinted and analyzed by SPR (Akgönüllü et al., 2020).

In an interesting application, Gold nanoparticles have also been utilized in combination with carbon-based nanoparticles to detect two food dyes metanil yellow and fast green (Chen et al., 2018).

9.6.2 Silver nanoparticles

Silver nanoparticles are utilized for the detection of small nutrients and toxin development as well and adulteration of milk is also detected by using them. Melamine is an adulterant which is a type of plastic material, added to the milk to increase the false protein content but it is toxic and can increase the hospitalization of children with some reported deaths. Although melamine can also reach cattle feed through pesticides as well fertilizers which contain melamine as the source of nitrogen (Jalili, 2017). Silver nanoshells are synthesized on the surface of silica particles by using the Stober method and added in absolute ethanol for silver mirror reaction and characterized by TEM, HRTEM, STEM, and selected area electron diffraction (SAED) pattern. These silver nanoshells are utilized for the detection of lactose and melamine in the sample by applying the sample to be detected first and then adding the matrix before doing MS. After that limit of detection is decided for nutrients and toxins (Wu et al., 2018).

Similarly for the detection of Alternaria toxins (more than 30 types) produced by the fungus Alternaria which is produced during long-term storage of agricultural products. One type of toxin (AOH) is detected by silver nanoparticles which are coated with pyridine to improve the binding of AOH on the silver nanoparticle and this is further detected by SERS (Pan et al., 2018).

Zearalenone is a mycotoxin which is having estrogenic activity produced by *Fusarium gramineous* and is found in crops. This has reproductive toxicity, cytotoxicity, immunotoxicity, and genotoxicity in humans. This toxin is detected by attaching an antibody against Zearalenone and doing the fluorescence quenchometric lateral flow immunochromatographic assay. Ovalbumin is used as a donor signal probe and silver nanoparticles are used as acceptor probes. This method is used for qualitative analysis and is useful for the onsite determination of toxins and is good for detection in the presence of other toxins like aflatoxins (Li et al., 2018).

Silver nanoparticles that are coated with in situ generated food toxins like polycyclic aromatic hydrocarbons (PAH) like pyrene, anthracene, phenanthrene, etc. which have been related to carcinogenicity have been utilized for efficient analysis. PAHs are types of pollutants that are found widely and have serious health risks. These are related to disorders related to the reproductive system, embryological development, circulatory system, nervous system, and cancer. Therefore, the detection of PAHs is important with an excellent lower limit of detection. In a study, silver nanoparticle hybrid is used for direct detection of PAH along with graphene, and PAHs are analyzed by SERS (Wang et al., 2020).

9.6.3 Platinum nanoparticles

Platinum nanoparticles offer to sense the detection of a type of aflatoxin. These nanoparticles are prepared by coating the platinum on the surface of gold nanoparticles and offer a matrix that is utilized for hosting a range of analytical solutions. In one study Aflatoxin M1 is detected by these nanoparticles which are coated with aptamer (single-stranded DNA) in pasteurized milk and milk powder samples with an excellent detection limit (Jahangiri–Dehaghani et al., 2020).

In another application, platinum is utilized for the coating of the gold nanorods and used for assessing enterotoxin B produced by *staphylococcal* sp. This application utilizes aptamer for toxin detection which is either RNA or single-stranded DNA. Two types of nanoparticles are used, upconversion nanoparticles and platinum-coated gold nanorods, where the former acts as a donor while the latter as an acceptor. In another application magnetic nanoparticle is conjugated to upconversion nanoparticles via aptamer and used for detection of chloramphenicol analyzed by FT-IR (Wu et al., 2015).

In one study detection of bacteria like *Salmonella typhimurium* is directly carried out by using platinum nanoparticles. An immunologic chip is prepared by attaching the porous gold–platinum catalyst by labeling it with the target cells. A color-producing substrate is attached to be used for ELISA. In this way, the Bacterial amount can be quantified by measuring the absorbance (Zheng et al., 2020).

9.6.4 Quantum dots

Protein toxins like cholera toxin, enterotoxins, neurotoxins, and Shiga toxins are produced by different bacterial species and have been a big threat to food and health industries. Magnetic quantum dots which are attached with antibodies against these toxins are used for sensing and are coated on a nitrocellulose membrane and fluorescence is measured for detecting the presence of a particular toxin (Wang et al., 2019).

Wastage of food is also used for synthesizing carbon quantum dots by various techniques. There are different categories of carbon quantum dots produced from a variety of sources like peels of lemon, banana, grapefruit, coconut, onion, watermelon, pomegranate, shells of walnut, peanut, palm, coffee bean in plant category, skin, blood, and bones, egg shells, crab shells, from animal sources and much other raw material from other industries like wine industry, oil industry (waste frying oil and press cake), soy industry, sugar industry, tea industry, etc. and are useful for detection of toxins in food and drinks like carmine, tartrazine, glutathione, nitrite, heavy metal ions, etc. (Fan et al., 2020).

9.6.5 Polymer nanocomposites

Polymers are used for creating nanobiosensors which senses the change in voltage by its exposure to the gas/vapors produced by the contaminating bacteria. Polymers are fabricated in a 0.1 diameter droplet on the base and then incubated overnight at 60°C in an oven. Bacteria from food samples are tested for changes in voltage (Arshak et al., 2007).

pH-based sensors are effective analyzers for detecting the spoilage of food. The freshness of food can be assessed by the detection of changes in pH as during food spoilage pH changes occur.

Similarly, in food packaging, specific types of sensors are used which change the color when volatile amino acids are produced by bacterial growth and are helpful in determining the freshness of packaged food (Pavase et al., 2018).

Lead (Pb^+) ions impose serious health hazards as they can cause various health issues like central nervous system damage, muscle paralysis, high blood pressure, anemia, etc. and they are present in different environmental sources and can easily percolate in food. Detection of food with the help of gold nanoparticle helps is helpful in taking precautions. A polymer-coated nanoparticles on the porous gold electrode are good for the detection of Pb^+ and can detect around 1 nM concentration of Pb^+ and can be detected by fluorometric analysis (Ramachandran et al., 2019).

9.6.6 **Metal oxides**

Nanocomposites of metal oxides are used for the detection of different types of toxins. Oxides of tin, zinc, titanium, iron, etc. are made composite with polypyrrole, polyanaline, and SnO_2 by different methods like thin film or layer formation technique and are used in the detection of NO_2, NH_3, humidity, CO_2, H_2S, O_2, and other chemicals produced by the growing organism (Pavase et al., 2018).

9.7 **Conclusion**

Various types of food toxins are of different origins and cause different levels of toxicity from acute to chronic. Some are naturally found in specific food products, in plants or fishes, and others are generated during storage for example fungal toxins. Some chemical toxins can contaminate the food at any step. They can come from fertilizers, dairy products or they can come from antibiotics and hormones used. Antibiotics can also come from poultry sources. Some chemicals used for adulteration may cause toxicity. Food toxins are required to be detected to prevent different types of food poisoning. Various detection techniques are already available which are based on metabolic reactions, immune reactions, presence of heavy metals, etc. as well as analytical techniques like HPLC, SPR, GC, LC, and many more, but these processes are costly and time-consuming. Nanobiosensors offer many advantages like rapid and reliable detection with cost-effectiveness. The timely analysis is important for the food industry as samples that are tested negative at a time, may later contain toxins if the time taken for the analysis is too long. The nanobiosensors are easy to access, can be used for large quantities at a time, and can either be discarded or can be re-utilized. Different types of nanomaterials are being utilized for either giving the scaffold to the analytical material or they can act as sensors because of their optical properties. Carbon-based nanoparticles like carbon nanotubes, graphene, fullerenes, and nanodiamonds are already in use or are being researched for their applicability in the combination of other nanomaterials or biological materials. Some metallic nanoparticles themselves can be toxic like zinc oxide or titanium oxide but there are other metallic nanoparticles that have been used for sensors like gold nanoparticles, silver nanoparticles, platinum nanoparticles, molybdenum nanoparticles, etc., and their combinations and applications with other nanomaterials of biological material with different analytical techniques for timely and accurate detection.

References

Ai, Y., Liang, P., Wu, Y., Dong, Q., Li, J., Bai, Y., Xu, B.-J., Yu, Z., Ni, D., 2018. Rapid qualitative and quantitative determination of food colorants by both Raman spectra and surface-enhanced Raman scattering (SERS). Food Chem. 241, 427–433. https://doi.org/10.1016/j.foodchem.2017.09.019.

Akgönüllü, S., Yavuz, H., Denizli, A., 2020. SPR nanosensor based on molecularly imprinted polymer film with gold nanoparticles for sensitive detection of aflatoxin B1. Talanta 219, 121219. https://doi.org/10.1016/j.talanta.2020.121219.

Arshak, K., Adley, C., Moore, E., Cunniffe, C., Campion, M., Harris, J., 2007. Characterisation of polymer nanocomposite sensors for quantification of bacterial cultures. Sens. Actuators B: Chem. 126, 226–231. https://doi.org/10.1016/j.snb.2006.12.006.

Baranowska-Wójcik, E., Szwajgier, D., Oleszczuk, P., Winiarska-Mieczan, A., 2020. Effects of titanium dioxide nanoparticles exposure on human health: a review. Biol. Trace Elem. Res. 193, 118–129. https://doi.org/10.1007/s12011-019-01706-6.

Bhardwaj, H., Singh, C., Kotnala, R.K., Sumana, G., 2018. Graphene quantum dots-based nano-biointerface platform for food toxin detection. Anal. Bioanal. Chem. 410, 7313–7323. https://doi.org/10.1007/s00216-018-1341-y.

Bosch-Panadero, E., Mas Fontao, S., Ruiz Priego, A., Egido, J., González Parra, E., 2017. Bisphenol (A) uremic toxin to take into account in the renal disease in hemodialysis. Rev. Colomb. Nefrol. 4, 57–68. https://doi.org/10.22265/acnef.41.256.

Camargo, J.R., Baccarin, M., Raymundo-Pereira, P.A., Campos, A.M., Oliveira, G.G., Fatibello-Filho, O., Oliveira, O.N., Janegitz, B.C., 2018. Electrochemical biosensor made with tyrosinase immobilized in a matrix of nanodiamonds and potato starch for detecting phenolic compounds. Anal. Chim. Acta. 1034, 137–143. https://doi.org/10.1016/j.aca.2018.06.001.

Chen, H., Zhou, K., Zhao, G., 2018. Gold nanoparticles: from synthesis, properties to their potential application as colorimetric sensors in food safety screening. Trends Food Sci. Technol. 78, 83–94. https://doi.org/10.1016/j.tifs.2018.05.027.

Clark, L.C., Lyons, C., 1962. Electrode systems for continuous monitoring in cardiovascular surgery. Ann. N.Y. Acad. Sci. 102, 29–45. https://doi.org/10.1111/j.1749-6632.1962.tb13623.x.

Dias, L.G., Meirinho, S.G., Veloso, A.C.A., Rodrigues, L.R., Peres, A.M., 2017. Electronic tongues and aptasensors. In: Rodrigues, L., Mota, M. (Eds.), Bioinspired Materials for Medical Applications. Woodhead Publishing, pp. 371–402. https://doi.org/10.1016/B978-0-08-100741-9.00013-9.

Ezoji, H., Rahimnejad, M., Najafpour-Darzi, G., 2020. Advanced sensing platform for electrochemical monitoring of the environmental toxin; bisphenol A. Ecotoxicol. Environ. Saf. 190, 110088. https://doi.org/10.1016/j.ecoenv.2019.110088.

Fan, H., Zhang, M., Bhandari, B., Yang, C., 2020. Food waste as a carbon source in carbon quantum dots technology and their applications in food safety detection. Trends Food Sci. Technol. 95, 86–96. https://doi.org/10.1016/j.tifs.2019.11.008.

Ganguly, R., Pierce, G.N., 2015. The toxicity of dietary trans fats. Food Chem. Toxicol. 78, 170–176. https://doi.org/10.1016/j.fct.2015.02.004.

García, A., Espinosa, R., Delgado, L., Casals, E., González, E., Puntes, V., Barata, C., Font, X., Sánchez, A., 2011. Acute toxicity of cerium oxide, titanium oxide and iron oxide nanoparticles using standardized tests. Desalination 269, 136–141. https://doi.org/10.1016/j.desal.2010.10.052.

Girigoswami, K., Ghosh, M.M., Pallavi, P., Ramesh, S., Girigoswami, K., 2021. Nanotechnology in detection of food toxins – focus on the dairy products. Biointerface Res. Appl. Chem. 11, 14155–14172. https://doi.org/10.33263/BRIAC116.1415514172.

Guo, X., Zhou, S., Chen, Y., Chen, X., Liu, J., Ge, F., Lian, Q., Chen, X., Ge, R.-S., 2017. Ziram delays pubertal development of rat leydig cells. Toxicol. Sci. 160, 329–340. https://doi.org/10.1093/toxsci/kfx181.

Hua, Z., Yu, T., Liu, D., Xianyu, Y., 2021. Recent advances in gold nanoparticles-based biosensors for food safety detection. Biosens. Bioelectron. 179, 113076. https://doi.org/10.1016/j.bios.2021.113076.

Jahangiri–Dehaghani, F., Zare, H.R., Shekari, Z., 2020. Measurement of aflatoxin M1 in powder and pasteurized milk samples by using a label–free electrochemical aptasensor based on platinum nanoparticles loaded on Fe–based metal–organic frameworks. Food Chem. 310, 125820. https://doi.org/10.1016/j.foodchem.2019.125820.

Jalili, M., 2017. A review paper on melamine in milk and dairy products. J. Dairy Vet. Sci. 1 (4), 555566. https://doi.org/10.19080/JDVS.2017.01.555566.

Jiang, L., Santiago, I., Foord, J., 2018. Nanocarbon and nanodiamond for high performance phenolics sensing. Commun. Chem. 1, 1–9. https://doi.org/10.1038/s42004-018-0045-8.

J. Ronkainen, N., Brian Halsall, H., R. Heineman, W., 2010. Electrochemical biosensors. Chem. Soc. Rev. 39, 1747–1763. https://doi.org/10.1039/B714449K.

Kim, K.-H., Jahan, S.A., Kabir, E., Brown, R.J.C., 2013. A review of airborne polycyclic aromatic hydrocarbons (PAHs) and their human health effects. Environ. Int. 60, 71–80. https://doi.org/10.1016/j.envint.2013.07.019.

Kőszegi, T., Poór, M., 2016. Ochratoxin A: molecular interactions, mechanisms of toxicity and prevention at the molecular level. Toxins 8, 111. https://doi.org/10.3390/toxins8040111.

Kundakovic, M., Champagne, F.A., 2011. Epigenetic perspective on the developmental effects of bisphenol A. Brain Behav. Immun. 25, 1084–1093. https://doi.org/10.1016/j.bbi.2011.02.005.

Li, S., Wang, J., Sheng, W., Wen, W., Gu, Y., Wang, S., 2018. Fluorometric lateral flow immunochromatographic zearalenone assay by exploiting a quencher system composed of carbon dots and silver nanoparticles. Microchim. Acta. 185, 388. https://doi.org/10.1007/s00604-018-2916-1.

Magesa, F., Wu, Y., Tian, Y., Vianney, J.-M., Buza, J., He, Q., Tan, Y., 2019. Graphene and graphene like 2D graphitic carbon nitride: electrochemical detection of food colorants and toxic substances in environment. Trends Environ. Analyt. Chem. 23, e00064. https://doi.org/10.1016/j.teac.2019.e00064.

Malhotra, B.D., Srivastava, S., Ali, Md.A., Singh, C., 2014. Nanomaterial-based biosensors for food toxin detection. Appl. Biochem. Biotechnol. 174, 880–896. https://doi.org/10.1007/s12010-014-0993-0.

Malhotra, B.D., Srivastava, S., Augustine, S., 2015. Biosensors for food toxin detection: carbon nanotubes and graphene. MRS Online Proceed. Library (OPL) 1725. https://doi.org/10.1557/opl.2015.165.

Mogilnaya, O.A., Puzyr', A.P., Bondar', V.S., 2010. Growth and bioluminescence of luminous bacteria under the action of aflatoxin B1 before and after its treatment with nanodiamonds. Appl. Biochem. Microbiol. 46, 33–37. https://doi.org/10.1134/S0003683810010059.

Motlagh, N.M., Rouhani, M., Mirjafary, Z., 2020. Aminated C20 fullerene as a promising nanosensor for detection of A-234 nerve agent. Comput. Theor. Chem. 1186, 112907. https://doi.org/10.1016/j.comptc.2020.112907.

Mpountoukas, P., Pantazaki, A., Kostareli, E., Christodoulou, P., Kareli, D., Poliliou, S., Mourelatos, C., Lambropoulou, V., Lialiaris, T., 2010. Cytogenetic evaluation and DNA interaction studies of the food colorants amaranth, erythrosine and tartrazine. Food Chem. Toxicol. 48, 2934–2944. https://doi.org/10.1016/j.fct.2010.07.030.

Nakanishi, R., Nogimura, A., Eguchi, R., Kanai, K., 2014. Electronic structure of fullerene derivatives in organic photovoltaics. Org. Electron. 15, 2912–2921. https://doi.org/10.1016/j.orgel.2014.08.013.

Nekrasov, N., Kireev, D., Emelianov, A., Bobrinetskiy, I., 2019. Graphene-based sensing platform for on-chip ochratoxin A detection. Toxins 11, 550. https://doi.org/10.3390/toxins11100550.

Pacheco, M., Jurado-Sánchez, B., Escarpa, A., 2018. Sensitive monitoring of enterobacterial contamination of food using self-propelled Janus microsensors. Anal. Chem. 90, 2912–2917. https://doi.org/10.1021/acs.analchem.7b05209.

Pan, T., Sun, D.-W., Pu, H., Wei, Q., 2018. Simple approach for the rapid detection of alternariol in pear fruit by surface-enhanced Raman scattering with pyridine-modified silver nanoparticles. J. Agric. Food Chem. 66, 2180–2187. https://doi.org/10.1021/acs.jafc.7b05664.

Pavase, T.R., Lin, H., Shaikh, Q., Hussain, S., Li, Z., Ahmed, I., Lv, L., Sun, L., Shah, S.B.H., Kalhoro, M.T., 2018. Recent advances of conjugated polymer (CP) nanocomposite-based chemical sensors and their applications in food spoilage detection: a comprehensive review. Sens. Actuators B 273, 1113–1138. https://doi.org/10.1016/j.snb.2018.06.118.

Pogacean, F., Biris, A.R., Socaci, C., Coros, M., Magerusan, L., Rosu, M.-C., Lazar, M.D., Borodi, G., Pruneanu, S., 2016. Graphene–bimetallic nanoparticle composites with enhanced electro-catalytic detection of bisphenol A. Nanotechnology 27, 484001. https://doi.org/10.1088/0957-4484/27/48/484001.

Ramachandran, R., Chen, T.-W., Chen, S.-M., Baskar, T., Kannan, R., Elumalai, P., Raja, P., Jeyapragasam, T., Dinakaran, K., Kumar, G., peter, G., 2019. A review of the advanced developments of electrochemical sensors for the detection of toxic and bioactive molecules. Inorganic Chem. Front. 6, 3418–3439. https://doi.org/10.1039/C9QI00602H.

Rouf, T.B., Díaz-Amaya, S., Stanciu, L., Kokini, J., 2020. Application of corn zein as an anchoring molecule in a carbon nanotube enhanced electrochemical sensor for the detection of gliadin. Food Control. 117, 107350. https://doi.org/10.1016/j.foodcont.2020.107350.

Salek Maghsoudi, A., Hassani, S., Mirnia, K., Abdollahi, M., 2021. Recent advances in nanotechnology-based biosensors development for detection of arsenic, lead, mercury, and cadmium. Int. J. Nanomed. 16, 803–832. https://doi.org/10.2147/IJN.S294417.

Sengul, A.B., Asmatulu, E., 2020. Toxicity of metal and metal oxide nanoparticles: a review. Environ. Chem. Lett. 18, 1659–1683. https://doi.org/10.1007/s10311-020-01033-6.

Shen, J., Zhu, Y., Yang, X., Li, C., 2012. Graphene quantum dots: emergent nanolights for bioimaging, sensors, catalysis and photovoltaic devices. Chem. Commun. 48, 3686–3699. https://doi.org/10.1039/C2CC00110A.

Siripongpreda, T., Siralertmukul, K., Rodthongkum, N., 2020. Colorimetric sensor and LDI-MS detection of biogenic amines in food spoilage based on porous PLA and graphene oxide. Food Chem. 329, 127165. https://doi.org/10.1016/j.foodchem.2020.127165.

Sohal, I.S., DeLoid, G.M., O'Fallon, K.S., Gaines, P., Demokritou, P., Bello, D., 2020. Effects of ingested food-grade titanium dioxide, silicon dioxide, iron (III) oxide and zinc oxide nanoparticles on an in vitro model of intestinal epithelium: comparison between monoculture vs. a mucus-secreting coculture model. NanoImpact 17, 100209. https://doi.org/10.1016/j.impact.2020.100209.

Srivastava, A.K., Dev, A., Karmakar, S., 2018. Nanosensors and nanobiosensors in food and agriculture. Environ. Chem. Lett. 16, 161–182. https://doi.org/10.1007/s10311-017-0674-7.

Sruthi, S., Ashtami, J., Mohanan, P.V., 2018. Biomedical application and hidden toxicity of zinc oxide nanoparticles. Mater. Today Chem. 10, 175–186. https://doi.org/10.1016/j.mtchem.2018.09.008.

Torri, L., Piochi, M., 2016. Sensory methods and electronic nose as innovative tools for the evaluation of the aroma transfer properties of food plastic bags. Food Res. Int. 85, 235–243. https://doi.org/10.1016/j.foodres.2016.05.004.

Vandenberg, L.N., Hauser, R., Marcus, M., Olea, N., Welshons, W.V., 2007. Human exposure to bisphenol A (BPA). Reprod. Toxicol. 24, 139–177. https://doi.org/10.1016/j.reprotox.2007.07.010.

Wang, C., Xiao, R., Wang, Shu, Yang, X., Bai, Z., Li, X., Rong, Z., Shen, B., Wang, Shengqi, 2019. Magnetic quantum dot based lateral flow assay biosensor for multiplex and sensitive detection of protein toxins in food samples. Biosens. Bioelectron. 146, 111754. https://doi.org/10.1016/j.bios.2019.111754.

Wang, X., Xu, Q., Hu, X., Han, F., Zhu, C., 2020. Silver-nanoparticles/graphene hybrids for effective enrichment and sensitive SERS detection of polycyclic aromatic hydrocarbons. Spectrochim. Acta Part A 228, 117783. https://doi.org/10.1016/j.saa.2019.117783.

Weng, R., Sun, L., Jiang, L., Li, N., Ruan, G., Li, J., Du, F., 2019. Electrospun graphene oxide–doped nanofiber-based solid phase extraction followed by high-performance liquid chromatography for the determination of tetracycline antibiotic residues in food samples. Food Anal. Methods 12, 1594–1603. https://doi.org/10.1007/s12161-019-01495-7.

Wu, S., Qian, L., Huang, L., Sun, X., Su, H., Gurav, D.D., Jiang, M., Cai, W., Qian, K., 2018. A plasmonic mass spectrometry approach for detection of small nutrients and toxins. Nanomicro Lett. 10, 52. https://doi.org/10.1007/s40820-018-0204-6.

Wu, S., Zhang, H., Shi, Z., Duan, N., Fang, C., Dai, S., Wang, Z., 2015. Aptamer-based fluorescence biosensor for chloramphenicol determination using upconversion nanoparticles. Food Control 50, 597–604. https://doi.org/10.1016/j.foodcont.2014.10.003.

Yılmaz, E., Soylak, M., 2016. Preparation and characterization of magnetic carboxylated nanodiamonds for vortex-assisted magnetic solid-phase extraction of ziram in food and water samples. Talanta 158, 152–158. https://doi.org/10.1016/j.talanta.2016.05.042.

Zhang, M., Zhang, X., Shi, Y., Liu, Z., Zhan, J., 2016. Surface enhanced Raman spectroscopy hyphenated with surface microextraction for in-situ detection of polycyclic aromatic hydrocarbons on food contact materials. Talanta 158, 322–329. https://doi.org/10.1016/j.talanta.2016.05.069.

Zheng, L., Cai, G., Qi, W., Wang, S., Wang, M., Lin, J., 2020. Optical biosensor for rapid detection of salmonella typhimurium based on porous gold@platinum nanocatalysts and a 3D fluidic chip. ACS Sens. 5, 65–72. https://doi.org/10.1021/acssensors.9b01472.

Nanobiosensors for mycotoxins detection in foodstuff: Qualitative and quantitative assessments

Merve Çalışır[a], Erdoğan Özgür[a], Duygu Çimen[a], Aykut Arif Topçu[b], Muhammed Erkek[a], Nilay Bereli[a] and Adil Denizli[a]

[a]*Department of Chemistry, Biochemistry Division, Hacettepe University, Beytepe, Ankara, Turkey,*
[b]*Medical Laboratory Program, Vocational School of Health Service, Aksaray University, Aksaray, Turkey*

10.1 Introduction

Mycotoxin contamination is a severe global issue for food safety and quality associated with public health. Mycotoxins as secondary metabolites are naturally toxic compounds with small molecular weight, which are secretion of several fungi such as *Penicillium, Aspergillus*, and *Fusarium* (Jiang et al., 2018; Abdolmaleki et al., 2021). Exposure to mycotoxins by ingestion, contact, or inhalation may cause adverse effects on human and animal health (e.g., carcinogenicity, genotoxicity or nephrotoxicity, reproductive disorders, immune toxicity, teratogenicity, hepatotoxicity, and oxidative stress). The formation of mycotoxins could arise during cultivation, transportation, and/or storage of foods at a certain temperature, moisture, and ingredients of food material. Most mycotoxins have often a stable chemical structure and are resilient to heat treatments during food processing, making it difficult to degrade these extremely toxic compounds by conventional methods such as cooking, frying, baking, etc. Data currently accounts for 25% of all cereal crops were contaminated by mycotoxins which lead to significant losses in international and domestic trade (Abdolmaleki et al., 2021; Iqbal, 2021; Cimbalo et al., 2020; Bueno et al., 2015). According to many findings, the approximate number of mycotoxins can be more than 20,000. However, mycotoxins between 300 and 400 have been identified worldwide to date. Some of the mycotoxins, e.g., fumonisins (FBs), ochratoxins (Ots), and zearalenone (ZEN) have adverse effects on public health (Smith et al., 2016; Gallo et al., 2015); so, the development of effective, solid, early, and sensitive detection methods for controlling the contamination at the pre and postharvest stages are required to eliminate the adverse effects of these toxic compounds and also maintain favorable quality of food. Traditional analytical methods, for instance, enzyme-linked immunoassay (ELISA), gas chromatography (GC), thin layer chromatography (TLC), high-performance

Nanobiotechnology for Food Processing and Packaging. DOI: https://doi.org/10.1016/B978-0-323-91749-0.00004-6

liquid chromatography (HPLC), diode array detectors, and mass spectrometry (MS) detectors are some of the detection methods for mycotoxins (Grippi et al., 2008; Tkaczyk and Jedziniak, 2020; Njumbe Ediage et al., 2015; Pei et al., 2013). However, some drawbacks such as sophisticated instruments, are needed for skilled technicians, and high cost and consumption of time lead to the development of simple, accurate, and cost-effective techniques which enable on-site monitoring and simultaneous detection of the mycotoxins (Li et al., 2021). In view of the developments in nanotechnology, combining or integrating nano-sized materials with biosensors appears as a future strategy. In this context, we aim to introduce the detection of mycotoxins via nanobiosensor platforms.

10.2 Traditional detection methods for mycotoxin

10.2.1 Extraction and cleanup procedures of mycotoxins

The negative effects of mycotoxin on living beings were highlighted aforementioned; so, the need for an effective, easy, rapid, and reliable method is significant for the detection of these toxic secondary metabolites.

The extraction and the cleanup procedures are the most crucial steps before the detection of the mycotoxins with analytical methods. The mycotoxin to be analyzed is extracted from a matrix by using a suitable solvent and the cleanup method is used to remove the interference of the molecules from the extract to enhance the sensitivity of the experimental results (Singh and Mehta, 2020). In this sense, the sample pretreatment methods are of great importance to detect and quantify the mycotoxins.

10.2.2 Analytical methods of quantification and determination of mycotoxins

The different ratios of methanol-water and acetonitrile-water are generally preferred as extraction solvents for mycotoxins. However, the other solvents such as toluene, 1-octanol, dichloromethane, the mixtures of ethyl acetate-formic acids, and the green solvents are required depending on the structure of the mycotoxin and the extraction protocol (Agriopoulou et al., 2020; Somsubsin et al., 2018; Zhang et al., 2018; di Mavungu et al., 2009; Delmulle et al., 2006; He et al., 2020). Today, the various extraction methods including quick easy cheap rough and safe (QuEChERS), liquid-liquid extraction (LLE), solid-liquid extraction (SLE), microwave-assisted extraction (MAE), and vortex assisted low-density solvent-microextraction (VALS-ME) are utilized in increasing the extraction performance, reducing the extraction solvent, and the extraction time (Agriopoulou et al., 2020).

The cleanup procedure, as mentioned above, a significant step for mycotoxin analysis, is used for removing the interfering molecules from the matrix and enhancing the sensitivity of the experimental results (Turner et al., 2009). During the cleanup step, the various adsorbents such as solid phase extraction (SPE) cartridges including carbon-based materials, immunoaffinity columns, and bovine serum albumin columns are used for the cleanup process of the various mycotoxins (Zhang et al., 2018). Among these adsorbents, immunoaffinity columns (IACs) based on antigen–antibody interaction hold great potential for capturing the desired mycotoxin from the environment without the interference of the molecules; but these affinity columns are expensive and have a short life (Castegnaro et al., 2006).

Hence, molecularly imprinted polymers (MIPs), aptamers, and peptides are used as antibody analogs to develop the new adsorbents for the cleanup process and eliminate some disadvantages of the IACs before the determination of mycotoxins (Agriopoulou et al., 2020). For instance, MIPs and the aptamers were prepared and combined with analytical devices to detect aflatoxin B1 (AFB1), fumonisin B1 (FuB1) (Singh et al., 2021), ochratoxin A (OTA) (Liu et al., 2021), and alternariol (AOH) (Wang et al., 2022) mycotoxins.

10.2.3 Analytical methods of quantification and determination of mycotoxins

The chemical structure and the nature of the mycotoxin are highly important for the selection of the analytical method for mycotoxin detection and quantification (Soares et al., 2018).

Chromatographic approaches recently focused on mycotoxin detection and among them, TLC is utilized in mycotoxin detection because of its low cost and ease of identification of the samples without the use of sophisticated devices (Turner et al., 2009).

Some research studies for mycotoxin detection by using TCL were reported in the recent review article by Agriopoulou et al. (2020). However, the lack of sensitivity, the accuracy of TLC, and the need for clean-up steps depending on the structure of mycotoxin are the main restrictions of the TLC method (Turner et al., 2009; Lin et al., 1998; Yang et al., 2014).

The other chromatographic techniques, for instance, HPLC, ultra-high performance liquid chromatography (UPLC), and GC methods coupled with the specific detector are used to detect mycotoxins with high accuracy and sensitivity (Zhang et al., 2018). HPLC coupled with a FLD detector is a favorable chromatographic method for sensing some of the mycotoxins after the pre and postcolumn derivation protocol, but this protocol is complex and time-consuming as compared with UPLC.

Liquid chromatography tandem MS (LC-MS/MS) method is a useful method to detect mycotoxins at trace levels and this sensitive, versatile, and specific method paves the way for the detection of the multimycotoxin analysis (Rubert et al., 2012). In the previous works, the LC-MS/MS method was proved to be useful for detecting the multi mycotoxin analysis (Pallarés et al., 2017; Pantano et al., 2021; He et al., 2019; Tkaczyk and Jedziniak, 2021). Some of the other studies for mycotoxin determination are illustrated in Table 10.1.

Table 10.1 Several studies for mycotoxin detection.

Toxin	Detection method	References
Patulin	LC-MS/MS	Rosa da Silva et al. (2022)
Zearalenone	HPLC-MS/MS	Pałubicki et al. (2021)
Aflatoxins	HPLC-MS/MS	Jayasinghe et al. (2020)
17 different mycotoxins	UHPLC-MS/MS	Nualkaw et al. (2020)
α-zearalenol, zearalenone, aflatoxins (Bl, B2, G1, and G2)	UHPLC-qqQ-MS/MS	Hidalgo-Ruiz et al. (2019)
α-zearalenol, zearalenone, aflatoxins (Bl, B2, G1, and G2)	UHPLC-MS/MS	Arroyo-Manzanares et al. (2019)
Ochratoxin A	HPLC-MS/MS	Campone et al. (2018)
FuB 1 and fumonisin B2	HPLC-MS/MS	de Oliveira et al. (2017)
Multiple mycotoxins	HPLC-MS/MS	Zhao et al. (2017)
Multiple mycotoxins	HPLC-MS/MS	Hickert et al. (2015)

ELISA is the highly preferred immunoassay method depending on biomolecule affinity such as antigen–antibody complex (Pereira et al., 2014) and the ease of use, low-cost, high throughput, and portability of ELISA make it (Pleadin et al., 2012) an alternative detection method for some of the mycotoxins, e.g., AF B1 (Hafez et al., 2021; Tang et al., 2021; Zhan et al., 2021), patulin (PAT) (Przybylska et al., 2021), AF (Mohamed and Mohamed, 2021), AF M1 (Tarannum et al., 2020), citrinin (Singh et al., 2019), OTA (Bao et al., 2021), and OTB (Fadlalla et al., 2020).

10.3 **Nanobiosensors**

The development of molecular biology and technology has brought progress in many branches of the medical field. Innovative studies, especially in multidisciplinary fields, are the leading actors of developments in the field of biotechnology. Biosensors are one of the most outstanding examples of such studies. Biosensors are bioanalytical devices developed by combining the selectivity properties of biological molecules with modern electronic techniques, making use of science branches such as biology, physics, chemistry, biochemistry, materials and engineering (Scheller et al., 1991). Sensor mechanisms generally consist of a receptor that can detect the analyte, a transducer that reads the received signal, and an electronic system that processes the same signal (Nakamura and Karube, 2003). Biosensors, on the other hand, take this name by making the aforementioned sensing with biological receptors. Sense organs are actual natural biosensors established from specialized cells (receptors) that perform functions such as seeing, hearing, smelling, tasting, touching, and feeling. The artificial biosensors can easily go beyond the detection sensitivity of the sense organs (Wu et al., 2017). Developed with inspiration from the sense organs, biosensors have included nanotechnology and paved the way for even more effective analysis.

Nanobiosensors consist of unique nanomaterials, the gift of nanotechnology. The physical and chemical characteristics of material differ in bulk scale and nanoscale, and it is one of the most important properties that make nanomaterials remarkable. Extremely diverse and effective results can be achieved when designed nanomaterials are integrated into sensor systems by enabling highly specific and sensitive detection. Among those nanomaterials, nanotubes, nanofilms, nanowires, and nanoparticles can be given as examples (Luz et al., 2013).

There are many types of nanobiosensors according to different sensing methods. Electrochemical (potentiometric and amperometric), piezoelectric-based, and optical (colorimetric, fluorescent, SPR, Raman-SERS nanobiosensors) can be given as examples of these nanobiosensors. Each detection method has its advantages and disadvantages. All methods have proven themselves to be able to make very sensitive detections. In addition, electrochemical nanobiosensors stand out with their one-step target detection and being disposable. Optical nanobiosensors are among the preferred nanobiosensors with their ease of use. On the downside, electrochemical nanobiosensors are one step behind with their difficult storage conditions and relatively expensive costs. Piezoelectric and optical nanobiosensors, on the other hand, are still open to development systems due to their high price and rarity (Jurado-Sánchez et al., 2020).

Electrochemical nanobiosensors are among the most studied sensor systems. Electrochemical systems were originally developed for glucose analysis and have evolved into nanobiosensor technology. Studies that started with enzyme electrodes were later developed to increase the sensitivity of electrochemical biosensors with enzyme-linked immuno-electrochemical experiments (Heineman et al., 1979). Recent research on electrochemical nanobiosensors has focused on the improvement of electrode designs (e.g., miniaturization, efficient electron transfer, nanomaterials, and better fixation procedures). Electrochemical nanobiosensors can then be divided into amperometric/voltammetric, potentiometric, and conductivity/capacitance/impedance nanobiosensors (Thvenot et al., 1999). According to the common points with electrochemical systems, amperometric and voltammetric nanobiosensors can be classified. In fact, in the incoming understanding, amperometric sensors are a superset of voltammetric sensors. The common point of amperometric sensors in electrochemical systems is the constant potential applied to the cells in both systems. The desired current is obtained through reduction or oxidation reactions.

Optical methods are also among the oldest and most established techniques for the detection of biological and chemical analytes. Various optical techniques have been used in the production of nanobiosensors. A typical optical biosensor consists of a light source, a group of optical components, and transducers to generate a light beam with certain properties (Debnath and Das, 2020). Surface plasmon resonance (SPR) sensor comes forward in the field of nanobiosensors. Thin film on the SPR chip, where biological recognition takes place, is the key point of SPR systems, and the plasmons formed on the chip surface determine the amount of refraction of the light (Bakhshpour and Denizli, 2020; Özgür et al., 2020; Bereli et al., 2021; Saylan et al., 2018). Modifications on the chip surface determine the amount of plasmon formation and hence, the detection sensitivity. Not only the modifications but also the surface coating is one of the factors affecting the optical excitation.

Piezoelectric biosensors are based on theories developed in electricity, mass, and viscoelasticity and use instruments such as quartz crystal microbalance (QCM) sensors. Piezoelectric sensors are distinguished sensor systems in terms of versatile application, sensitivity, low cost, simplicity, and they are label-free (Buck et al., 2004). The electrical energy that emerges with the applied pressure on an insulating material sandwiched between two opposite poles, positively and negatively, is called piezoelectricity. With this logic obtained through quartz crystal plates, mechanical and electrical effects are combined. Given the nature of the piezoelectric effect, it enables a transducer to respond reliably and predictably to mechanical stress caused by an applied force, making precise determinations quite possible.

10.4 Nanobiosensors applications for mycotoxin detection

Patulin, a type of polyketide lactone is a known chemical compound that is toxic to living beings. Detecting PAT levels, therefore, is crucial for food safety (Çimen et al., 2021). Guo et al. (2017) designed electrochemical sensors to detect PAT in fruit juice using the molecular imprinting technique (MIT). In this study, 2-oxindole and Aminothiophenol (ρ-ATP) were selected as the template and the monomer, respectively. In addition, carbon dots and chitosan were used to enhance the sensing performance of the biosensor.

While performing the electropolymerization process and optimization with differential pulse voltammetry (DVP) and circular voltammetry measurements, characterization studies were examined with transmission electron microscopy (TEM), scanning electron microscopy (SEM), and atomic force microscope (AFM). The limit of detection (LOD) of PAT was found as 7.57×10^{-13} mol/L in fruit juice.

In another study for the detection of PAT in the literature, Fang et al. (2017) designed a sol–gel MIP film-based QCM sensor. The adsorption capacity of the MIP prepared for PAT determination was determined by Scatchard equation analysis. The LOD and selectivity coefficient of the sensing system were determined to be 3.1×10^{-3} µg/mL and 3.82, respectively. The sensor is reproducible and shows a good long-term stable property. Furthermore, the PAT imprinted QCM sensor has recoveries ranging from 76.9% to 91.3% in real foods, and these results follow the HPLC–MS method.

Ochratoxin A is one of the most widely used types of mycotoxins that can contaminate various agricultural products. Allowed OTA amounts in foods for animal and human consumption have been determined in various countries, and various analytical methods have been used to determine these contaminants in foods with permitted amounts. Akgönüllü et al. (2021) MIP film was formed on SPR to detect the amount of OTA in dried figs and after that, the MIP was characterized with various surface characterization methods. The detection limit for OTA determination was 0.028 ng/mL and the imprinting factor of the sensing platform was calculated as 2.85 towards the competitor molecules.

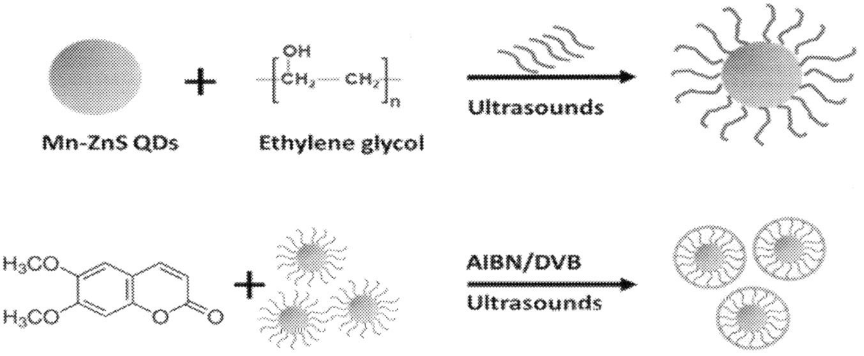

FIGURE 10.1

The preparation process of MIP-PEG-Mn-based ZnS quantum dots (Madurangika Jayasinghe et al., 2020).

In another study for OTA determination, Pereira et al. (2021) prepared a new localized SPR (LSPR) with the integration of gold nanoparticles (AuNPs). The sensitivity of the LSPR nanosensor, determined by the LSPR band λ_{max} shifts, LOD was found to be 0.001 pg/mL. It has been determined that the detection limit of the prepared LSPR nanosensor is 10 times less when compared to other methods with a very sensitive lower limit value.

Jayasinghe et al. (2020) used the selective and sensitive determination of AFs by preparing Mn-doped ZnS quantum dots (QDs) coated with MIP-based nanosensors in Fig. 10.1. For the prepared MIP nanoparticles, the total decomposition time was observed as 0.004 seconds at the intense decomposition temperature. In addition, while the phosphorescence was determined at 290 nm at room temperature, it was observed to increase from 520 nm to 720 nm after excitation. The adsorption capacity of AF imprinted QDs was higher than the unimprinted QDs and the LOD of the sensor was calculated as 56 µg/kg.

A new electrochemical method was prepared by combining the electrochemical sensor and the Fourier transform cyclic voltammetry (CV) for the determination of OTA. Norouzi et al. (2012) developed new electrochemical sensors by reducing AuNPs on graphene nanolayers oxide hybrid on the surface of a glassy carbon electrode (GCE) in an ionic liquid called 1-butyl-3-methylimidazolium tetrafluoroborate. The response of the prepared electrochemical sensor was calculated based on the charge changes below the observed peaks by integrating the current in the potential ranges determined by optimizing the experimental conditions. In this study, analysis was performed with a linear concentration range of 1–200 nm, while the lowest limit was calculated as 2.2×10^{-10} M. In addition, the most important feature of the prepared sensor is its repeatability, short response time of less than 7 seconds, stability, and high sensitivity.

Choi et al. (2011) used SPR sensors to examine both the determination of deoxynivalenol (DON) as an important mycotoxin and the feasibility of applying MIT.

In this study, the molecularly imprinted polypyrrole (MIPPy) film was formed by electropolymerization in the presence of the template molecule on a bare Au chip. The surface thickness of the MIPPy film was measured by AFM and its thickness was measured at 5 nm. LOD was found to be >1 ng/mL for standard solutions. When the selectivity studies of the molecularly imprinted SPR sensor were examined, the selectivity efficiency of DON and its acetylated analogs such as 3-ADON and 15-ADON was found to be 100%, 19%, and 44%, respectively.

Since OTA and AFT-B1 are important mycotoxins that pose significant threats to human health, so various methods have been developed to detect these mycotoxins. For instance, Kutsanedzie et al. (2020) investigated different enhancement factors (EF) by synthesizing silver nanoparticles (AgNPs) in different pH solutions. The highest EFs value for AgNP@pH-11 was found to be 1.45×10^8. Two chemometric algorithms were calculated to determine both mycotoxins using the SERS sensor in standard solutions and spiked cocoa bean samples. LOD values of OTA and AFT-B1 in the spiked cocoa bean samples were calculated as 2.63 pg/mL and 4.15 pg/mL, respectively.

Hatamluyi et al. (2020) developed a new imprinted electrochemical sensor for PAT determination. In the first step, the researchers modified the GCE with the nitrogen-containing graphene QDs (N-GQDs) and functionalized it with AuNPs Cu-metal organic framework (Au@Cu-MOF). Afterward, MIP was synthesized on the prepared electrode (Au@Cu-MOF/N-GQDs/GCE) via electropolymerization. The analysis for the molecular imprinted electrochemical sensor was performed at PAT concentrations ranging between 0.001 ng/mL and 70.0 ng/mL, and LOD was found to be 0.0007 ng/mL. In addition, it has been shown that PAT can be determined selectively, quickly, and cost-effectively when compared with traditional chromatographic methods, with a recovery rate of 97.6%–99.4% and a high sensitivity of 1.23%–4.61% in the analysis of PAT determination in apple juice samples.

Yu and Lai (2004) prepared polypyrrole films containing chloride (PPy-Cl) on the gold (Au) surface of the SPR sensor for sensing of OTA by electropolymerization in 1 mM NaCl aqueous solution at +0.76 V. By monitoring the changes in the SPR angle during the analysis, the thickness of the PPy-Cl film was between 2 nm and 5 nm. A linear relationship was observed between the bonding interaction between the prepared film and OTA, the rate of increase of the SPR angle, and the OTA concentration ranging between 0.1 µg/mL and 10 µg/mL. Analysis was performed at a binding time of <5 minutes and a sample volume of 0.2 mL.

Pacheco et al. (2015) constructed an electrochemical sensor prepared by molecular imprinted MIP and modification of a GCE with multiwalled carbon nanotubes (MWCNTs) for the detection of OTA (Fig. 10.2). MWCNTs used in the preparation of the electrochemical sensor significantly increased the sensitivity of the electrochemical sensor. OTA imprinted PPy film was prepared by CV and was synthesized by electropolymerization of pyrrole with OTA as a template molecule. In addition, the electrochemical oxidation of OTA in the prepared sensor was examined with both cyclic and DVP. The limit of quantification (LOQ) and LOD values were found 0.014 µM and 0.0041 µM, respectively. Furthermore, the

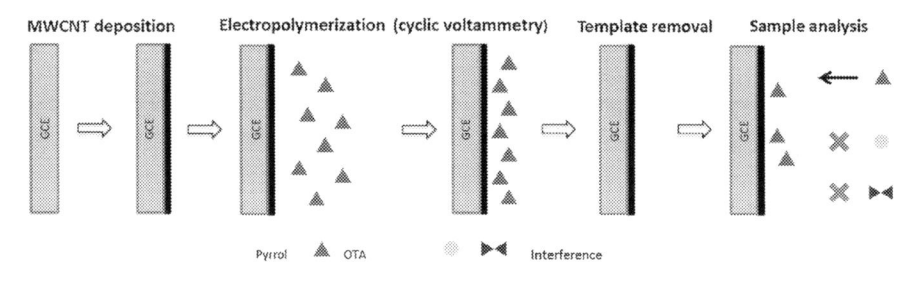

FIGURE 10.2

Illustration of preparation steps of ochratoxin A (OTA) imprinted electrochemical sensor (Pacheco et al., 2015).

amount of OTA was determined from wine and beer samples with a recovery of 84%–104%.

Chen et al. (2016) prepared the upconversion nanoparticle (UCNP)-based fluorescent probe for the dual detection of AFB1 and DON, a magnetism-induced separation and specific formation of the antibody-target complex. The developed UCNPs were optimized with various parameters and their detection range was investigated in the range of 0.001–0.1 ng/mL.

Xiang et al. (2018) developed a two-dimensional (2D) layered black phosphorene (BP) based nanosensor for voltammetric detection of OTA in grape juice and red wine samples. This sensor was used to perform voltammetric analysis of food samples using DPV.

Akgönüllü et al. (2020) developed a plasmonic detection method based on an improved SPR nanosensor for the detection of AFB1. First, the template molecule, AFB1, and functional monomer, N-methacryloyl-L-phenylalanine, are complex and the sensing surface was coated with MIP films. The sensor was characterized by a concentration of 0.0001–10.0 ng/mL of AFB1. The LOD of mycotoxin was calculated at 104 pg/mL and the imprinting factor of the sensor was found to be 5.91 as compared to the nonimprinted sensor.

In this study, Jiang et al. (2021) developed a nanosensor for the production of functional bicolor permanent luminescent nanoparticles (PLNPs) together with Fe_3O_4 magnetic nanoparticles for the measurement of AFB1 and ZEN in food samples. The sensing platform was designed with the combination of aptamer-modified PLNPs with complementary DNA-modified Fe_3O_4. According to the experimental findings, the LODs of AFB1 and ZEN in food samples were found to be 0.29 pg/mL and 0.22 pg/mL, respectively.

Rico-Yuste et al. (2021) prepared Eu (III)-doped porous MIP microspheres for selective sensing of TeA in rice extracts. The "turn-on" luminescence of Eu (III) ions at 615 nm (λex 337 nm) can be selectively detected in MIP cavities and by observation. A small-scale MIP library was created by mixing functional monomers with template and crosslinker in different proportions. For the developed sensor scans, Eu (III) and poly (DEAM-co-EDMA) were chosen as metal centers. Under optimized conditions

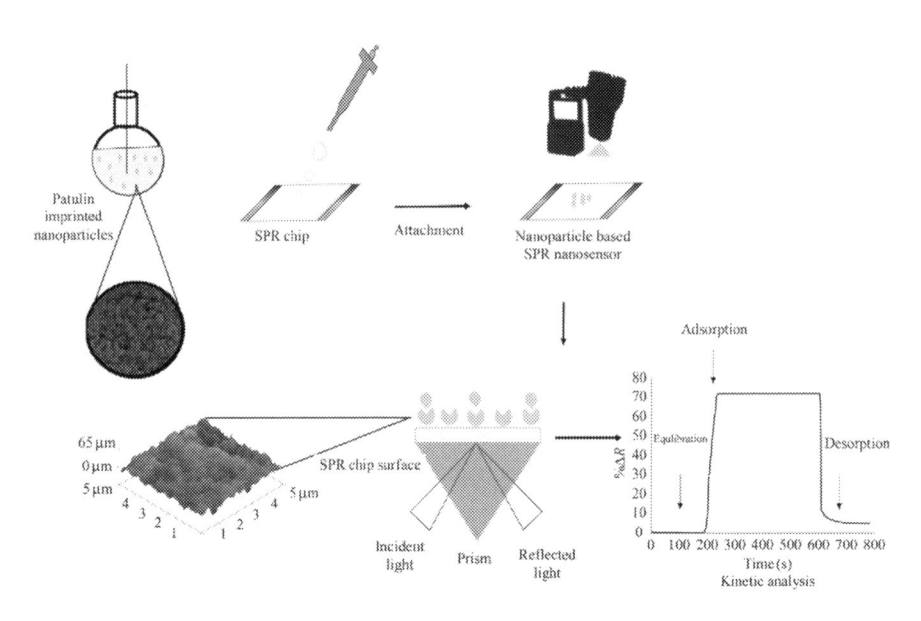

FIGURE 10.3

The design scheme of patulin (PAT) imprinted surface plasmon resonance (SPR) nanosensor (Çimen et al., 2021).

for TeA, the linear range was calculated to be 1.7–20 µg/mL, and LOD was found as 5 µg/mL.

In this study, Liu et al. (2020) developed a "turn-on" fluorescence sensor based on ZnCdSe QDs (ZnCdSe QDs) and self-assembled zinc porphyrin for sensitive and rapid detection of OTA. Using environmentally friendly dodecyl dimethyl betaine as the "soft template", zinc 5, 10, 15, 20-tetra (4-pyridyl)-21H-23H-porfin (ZNTPYP) was self-assembled onto nanorods (SA-ZNTPYP) by a green process. The LOD of the developed fluorescent sensor was 0.33 ng/mL.

In this study, Çimen et al. (2021) synthesized PAT imprinted and unprinted nanoparticles using the two-phase mini-emulsion polymerization approach, and then these nanoparticles were integrated into the sensing surface (Fig. 10.3). Kinetic studies for PAT detection were performed in the 0.5–750 nM concentration range. LOD and LOQ were calculated as 0.011 nm and 0.036 nm, respectively. In the analyzes performed during the kinetic studies, it was reported that the response time for the equilibration, adsorption, and desorption cycles were 13 minutes. SPR selectivity studies of PAT were determined against OTA and AF B1.

Hernández et al. (2020) developed a gold nanoprism aptasensor to sense the content of OTA in suspension by Raman spectroscopy (SERS). The developed aptasensor is label-free, selective, and reliable. Unlabeled SERS carries a molecular signature of the aptamer-OTA complex used. Multivariate analysis of the developed aptasensor differentiated between 10 ppb and 250 ppb (25–620 nM) of OTA solutions.

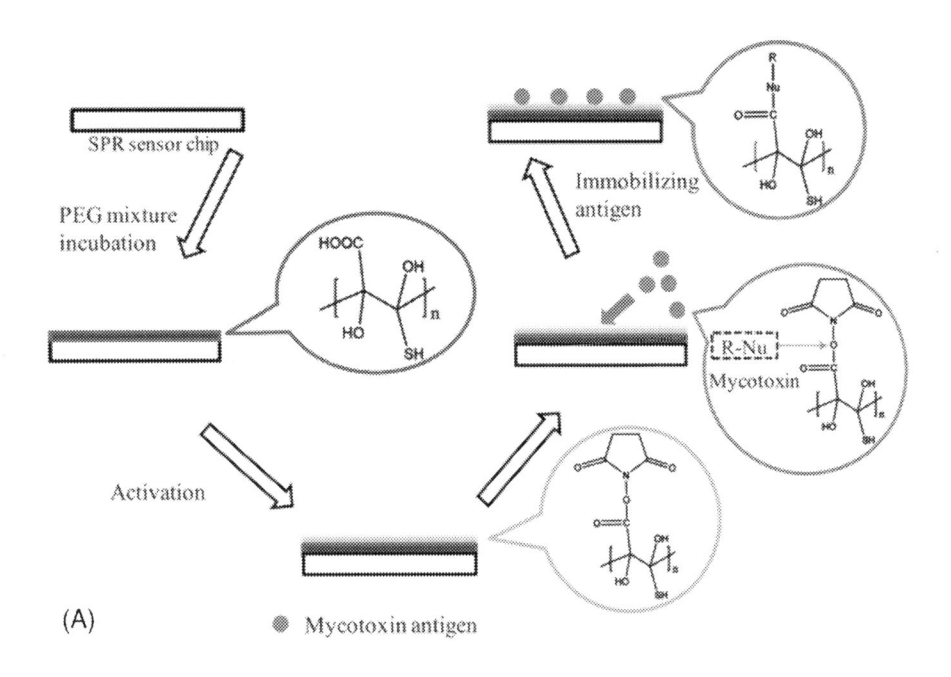

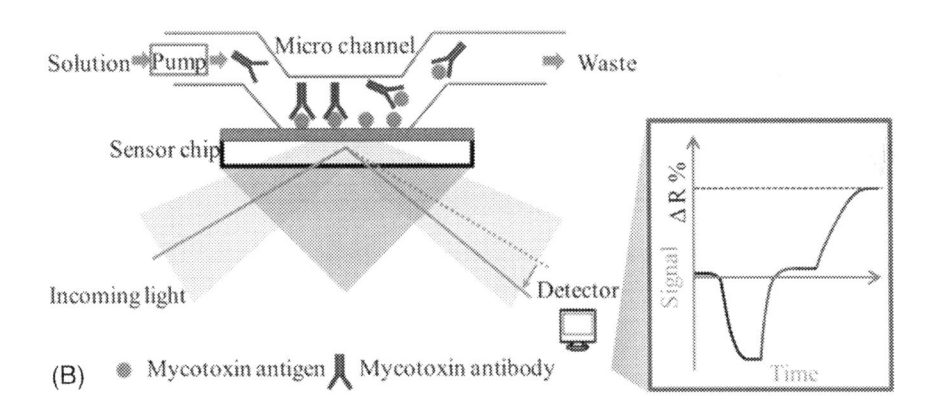

FIGURE 10.4

Schematic representation modification step (A) and the sensing process (B) (Wei et al., 2019).

Wei et al. (2019) developed the chip surface they produced using the optical-based SPR method to measure mycotoxin contaminations in wheat and maize. The chip surface was created according to the procedures described in Fig. 10.4. Detection limits for AF B1, OTA, ZEN, and deoxynivalenol were calculated as 0.59 ng/mL,

1.27 ng/mL, 7.07 ng/mL, and 3.26 ng/mL, respectively. Low cross-reactivity was obtained for all mycotoxins examined. Moreover, the sensing performance of the proposed platform was by HPLC-MS/MS.

Zhang et al. (2017) prepared a nanosensor based on Mn-doped ZnS QDs for the selective phosphorescence determination of PAT by surface molecular imprinting. When the FTIR and XRD analysis results were examined, it was observed that MIP was grafted onto it as desired. The detection limit of MIP-QDs to identify PAT was calculated as 0.32 μmol/L.

In this study, Su et al. (2022) created two OTA-specific fluonanobodies (FluoNbs) with a fused nanobody at the carboxyl (SGFP-Nb) or amino (Nb-SGFP) terminus of the super folder green fluorescent protein. When compared with two specific fluonanobodies, SGFP-Nb exhibited better performance and during the preparation of FN, SGFP-Nb and OTA-QDs acted as a donor and an acceptor, respectively, and the FN was characterized by the concentration range of 5–5000 pg/mL OTA concentration. According to the experimental findings, the sensor showed a good recognition performance towards OTA with a low LOD value.

In this study, Altunbas et al. (2020) developed a terbium (Tb^{3+}) chelated nanoparticle sensor and investigated the time-dependent fluorescence detection of OTA, which is common and toxic in foodstuffs. The design of the sensor was developed for luminescent OTA detection. In coordination of OTA with the Tb^{3+} center on the nanoparticle surface, a detection limit of 20 ppb was calculated by the researchers.

10.5 Conclusion

Mycotoxins are naturally toxic compounds produced by fungi and these metabolites show toxic effects, e.g., carcinogenic, immunogenic, and immunotoxic for humans and animals. Mycotoxins are frequently found in daily food products and their side effects can emerge in all parts of the biosystems as a chain reaction. Hence, the detection of mycotoxins plays a significant role in protecting public health.

The extraction and cleanup procedures are crucial steps before the determination of the mycotoxins from the contaminated food samples by using analytical techniques such as HPLC, HPLC-MS/MS, UHPLC-MS/MS, and ELISA. These analytical approaches are sensitive, selective toward the mycotoxin of interest, and allow for detecting the multimycotoxin analysis, but these methods need costly equipment, skilled personnel, and are laborious as well.

Biosensors are analytical devices, that competent in recognizing the target molecules with high sensitivity and provide many advantages, for instance, quick response time, ease of operation, without the need for costly types of equipment, and a skilled person. Hence, these sensing platforms have paved the way for various applications including mycotoxin detection.

In this context, we summarized the traditional methods for mycotoxin detection with the latest research articles and tried to emphasize some advantages of biosensor platforms among the other analytical approaches.

References

Abdolmaleki, K., Khedri, S., Alizadeh, L., Javanmardi, F., Oliveira, C.A.F., Mousavi Khaneghah, A., 2021. The mycotoxins in edible oils: an overview of prevalence, concentration, toxicity, detection and decontamination techniques. Trends Food Sci. Technol. 115, 500–511.

Agriopoulou, S., Stamatelopoulou, E., Varzakas, T., 2020. Advances in analysis and detection of major mycotoxins in foods. Foods 9, 518.

Akgönüllü, S., Armutcu, C., Denizli, A., 2021. Molecularly imprinted polymer film based plasmonic sensors for detection of ochratoxin A in dried fig. Polym. Bull. 79, 4049–4067.

Akgönüllü, S., Yavuz, H., Denizli, A., 2020. SPR nanosensor based on molecularly imprinted polymer film with gold nanoparticles for sensitive detection of aflatoxin B1. Talanta 219, 121219.

Altunbas, O., Ozdas, A., Yilmaz, M.D., 2020. Luminescent detection of Ochratoxin A using terbium chelated mesoporous silica nanoparticles. J. Hazard. Mater. 382, 121049.

Arroyo-Manzanares, N., Hamed, A.M., García-Campaña, A.M., Gámiz-Gracia, L., 2019. Plant-based milks: unexplored source of emerging mycotoxins. A proposal for the control of enniatins and beauvericin using UHPLC-MS/MS. Food Addit. Contam. Part B Surveill 12, 296–302.

Bakhshpour, M., Denizli, A., 2020. Highly sensitive detection of Cd(II) ions using ion-imprinted surface plasmon resonance sensors. Microchem. J. 159, 105572.

Bao, K., Liu, X., Xu, Q., Su, B., Liu, Z., Cao, H., Chen, Q., 2021. Nanobody multimerization strategy to enhance the sensitivity of competitive ELISA for detection of ochratoxin A in coffee samples. Food Control 127, 108167.

Bereli, N., Bakhshpour, M., Topçu, A.A., Denizli, A., 2021. Surface plasmon resonance-based immunosensor for igm detection with gold nanoparticles. Micromachines 12, 1092.

Buck, R.P., Lindner, E., Kutner, W., Inzelt, G., 2004. Piezoelectric chemical sensors (IUPAC technical report). Pure Appl. Chem. 76, 1139–1160.

Bueno, D., Istamboulie, G., Muñoz, R., Marty, J.L., 2015. Determination of mycotoxins in food: a review of bioanalytical to analytical methods. Appl. Spectrosc. Rev. 50, 728–774.

Campone, L., Piccinelli, A.L., Celano, R., Pagano, I., Russo, M., Rastrelli, L., 2018. Rapid and automated on-line solid phase extraction HPLC–MS/MS with peak focusing for the determination of ochratoxin A in wine samples. Food Chem. 244, 128–135.

Castegnaro, M., Tozlovanu, M., Wild, C., Molinié, A., Sylla, A., Pfohl-Leszkowicz, A., 2006. Advantages and drawbacks of immunoaffinity columns in analysis of mycotoxins in food. Mol. Nutr. Food Res. 50, 480–487.

Chen, Q., Hu, W., Sun, C., Li, H., Ouyang, Q., 2016. Synthesis of improved upconversion nanoparticles as ultrasensitive fluorescence probe for mycotoxins. Anal. Chim. Acta 938, 137–145.

Choi, S.W., Chang, H.J., Lee, N., Chun, H.S., 2011. A Surface plasmon resonance sensor for the detection of deoxynivalenol using a molecularly imprinted polymer. Sensors 11, 8654–8664.

Cimbalo, A., Alonso-Garrido, M., Font, G., Manyes, L., 2020. Toxicity of mycotoxins in vivo on vertebrate organisms: a review. Food Chem. Toxicol. 137, 111161.

Çimen, D., Bereli, N., Denizli, A., 2021. Patulin imprinted nanoparticles decorated surface plasmon resonance chips for patulin detection. Photonic Sens. 12, 117–129.

Çimen, D., Bereli, N., Yavuz, H., 2021. Sensors for the detection of food contaminants. In: Nanosensors for Environment, Food and Agriculture, vol. 1. Springer, pp. 169–182.

Debnath, N., Das, S., 2020. Nanobiosensor: current trends and applications. In: NanoBio Medicine. Springer, Singapore, pp. 389–409.

Delmulle, B., De Saeger, S., Adams, A., De Kimpe, N., Van Peteghem, C., 2006. Development of a liquid chromatography/tandem mass spectrometry method for the simultaneous determination of 16 mycotoxins on cellulose filters and in fungal cultures. Rapid Commun. Mass Spectrom. 20, 771–776.

de Oliveira, G.B., de Castro Gomes Vieira, C.M., Orlando, R.M., Faria, A.F., 2017. Simultaneous determination of fumonisins B1 and B2 in different types of maize by matrix solid phase dispersion and HPLC-MS/MS. Food Chem. 233, 11–19.

di Mavungu, J.D., Monbaliu, S., Scippo, M.L., Maghuin-Rogister, G., Schneider, Y.J., Larondelle, Y., Callebaut, A., Robbens, J., van Peteghem, C., de Saeger, S., 2009. LC-MS/MS multi-analyte method for mycotoxin determination in food supplements. Food Addit. Contam. Part A Chem. Anal. Control Expo. Risk Assess. 26, 885–895.

Fadlalla, M.H., Ling, S., Wang, R., Li, X., Yuan, J., Xiao, S., Wang, K., Tang, S., Elsir, H., Wang, S., 2020. Development of ELISA and lateral flow immunoassays for ochratoxins (OTA and OTB) detection based on monoclonal antibody. Front. Cell. Infect. Microbiol. 10, 80.

Fang, G., Yang, Y., Zhu, H., Qi, Y., Liu, J., Liu, H., Wang, S., 2017. Development and application of molecularly imprinted quartz crystal microbalance sensor for rapid detection of metolcarb in foods. Sens. Actuators B 251, 720–728.

Gallo, A., Giuberti, G., Frisvad, J.C., Bertuzzi, T., Nielsen, K.F., 2015. Review on mycotoxin issues in ruminants: occurrence in forages, effects of mycotoxin ingestion on health status and animal performance and practical strategies to counteract their negative effects. Toxins 7, 3057–3111.

Grippi, F., Crosta, L., Aiello, G., Tolomeo, M., Oliveri, F., Gebbia, N., Curione, A., 2008. Determination of stilbenes in Sicilian pistachio by high-performance liquid chromatographic diode array (HPLC-DAD/FLD) and evaluation of eventually mycotoxin contamination. Food Chem. 107, 483–488.

Guo, W., Pi, F., Zhang, H., Sun, J., Zhang, Y., Sun, X., 2017. A novel molecularly imprinted electrochemical sensor modified with carbon dots, chitosan, gold nanoparticles for the determination of patulin. Biosens. Bioelectron. 98, 299–304.

Hafez, E., Abd El-Aziz, N.M., Darwish, A.M.G., Shehata, M.G., Ibrahim, A.A., Elframawy, A.M., Badr, A.N., 2021. Validation of new ELISA technique for detection of aflatoxin B1 contamination in food products versus HPLC and VICAM. Toxins 13, 747.

Hatamluyi, B., Rezayi, M., Beheshti, H.R., Boroushaki, M.T., 2020. Ultra-sensitive molecularly imprinted electrochemical sensor for patulin detection based on a novel assembling strategy using Au@Cu-MOF/N-GQDs. Sens. Actuators B 318, 128219.

He, J., Zhang, B., Zhang, H., Hao, L.L., Ma, T.Z., Wang, J., Han, S.Y., 2019. Monitoring of 49 pesticides and 17 mycotoxins in wine by QuEChERS and UHPLC–MS/MS analysis. J. Food Sci. 84, 2688–2697.

He, T., Zhou, T., Wan, Y., Tan, T., 2020. A simple strategy based on deep eutectic solvent for determination of aflatoxins in rice samples. Food Anal. Methods 13, 542–550.

Heineman, W.R., Anderson, C.W., Halsall, H.B., 1979. Immunoassay by differential pulse polarography. Science 204, 865–866.

Hernández, Y., Lagos, L.K., Galarreta, B.C., 2020. Development of a label-free-SERS gold nanoaptasensor for the accessible determination of ochratoxin A. Sens. Bio-Sens. Res. 28, 100331.

Hickert, S., Gerding, J., Ncube, E., Hübner, F., Flett, B., Cramer, B., Humpf, H.U., 2015. A new approach using micro HPLC-MS/MS for multi-mycotoxin analysis in maize samples. Mycotoxin Res. 31, 109–115.

Hidalgo-Ruiz, J.L., Romero-González, R., Martínez Vidal, J.L., Garrido Frenich, A., 2019. A rapid method for the determination of mycotoxins in edible vegetable oils by ultra-high performance liquid chromatography-tandem mass spectrometry. Food Chem. 288, 22–28.

Iqbal, S.Z., 2021. Mycotoxins in food, recent development in food analysis and future challenges: a review. Curr. Opin. Food Sci. 42, 237–247.

Jayasinghe, G.D.T.M., Domínguez-González, R., Bermejo-Barrera, P., Moreda-Piñeiro, A., 2020. Miniaturized vortex assisted-dispersive molecularly imprinted polymer micro-solid phase extraction and HPLC-MS/MS for assessing trace aflatoxins in cultured fish. Anal. Methods 12, 4351–4362.

Jiang, K., Huang, Q., Fan, K., Wu, L., Nie, D., Guo, W., Wu, Y., Han, Z., 2018. Reduced graphene oxide and gold nanoparticle composite-based solid-phase extraction coupled with ultra-high-performance liquid chromatography-tandem mass spectrometry for the determination of 9 mycotoxins in milk. Food Chem. 264, 218–225.

Jiang, Y.Y., Zhao, X., Chen, L.J., Yang, C., Yin, X.B., Yan, X.P., 2021. A dual-colored persistent luminescence nanosensor for simultaneous and autofluorescence-free determination of aflatoxin B1 and zearalenone. Talanta 232, 122395.

Jurado-Sánchez, B., Moreno-Guzmán, M., Perales-Rondon, J.V., Escarpa, A., 2020. Nanobiosensors for food analysis. In: Handbook of Food Nanotechnology. Elsevier, pp. 415–457.

Kutsanedzie, F.Y.H., Agyekum, A.A., Annavaram, V., Chen, Q., 2020. Signal-enhanced SERS-sensors of CAR-PLS and GA-PLS coupled AgNPs for ochratoxin A and aflatoxin B1 detection. Food Chem. 315, 126231.

Li, R., Wen, Y., Wang, F., He, P., 2021. Recent advances in immunoassays and biosensors for mycotoxins detection in feedstuffs and foods. J. Anim. Sci. Biotechnol. 12, 108.

Lin, L., Zhang, J., Wang, P., Wang, Y., Chen, J., 1998. Thin-layer chromatography of mycotoxins and comparison with other chromatographic methods. J. Chromatogr. A 815, 3–20.

Liu, L., Huang, Q., Tanveer, Z.I., Jiang, K., Zhang, J., Pan, H., Luan, L., Liu, X., Han, Z., Wu, Y., 2020. "Turn off-on" fluorescent sensor based on quantum dots and self-assembled porphyrin for rapid detection of ochratoxin A. Sens. Actuators B 302, 127212.

Liu, Y., Su, Z., Wang, J., Gong, Z., Lyu, H., Xie, Z., 2021. Molecularly imprinted polymer with mixed-mode mechanism for selective extraction and on-line detection of ochratoxin A in beer sample. Microchem. J. 170, 106696.

Luz, R.A.S., Iost, R.M., Crespilho, F.N., 2013. Nanomaterials for biosensors and implantable biodevices. In: Nanobioelectrochemistry: From Implantable Biosensors to Green Power Generation. Springer-Verlag, Berlin-Heidelberg, pp. 27–48.

Madurangika Jayasinghe, G.D.T., Domínguez-González, R., Bermejo-Barrera, P., Moreda-Piñeiro, A., 2020. Room temperature phosphorescent determination of aflatoxins in fish feed based on molecularly imprinted polymer: Mn-doped ZnS quantum dots. Anal. Chim. Acta 1103, 183–191.

Mohamed, Z.T., Mohamed, R.Y., 2021. Quality and quantity determination of aflatoxin in maize by using ELISA and HPLC. Int. J. Agric. Stat. Sci. 17, 335–339.

Nakamura, H., Karube, I., 2003. Current research activity in biosensors. Anal. Bioanal. Chem. 377, 446–468.

Njumbe Ediage, E., Van Poucke, C., De Saeger, S., 2015. A multi-analyte LC-MS/MS method for the analysis of 23 mycotoxins in different sorghum varieties: the forgotten sample matrix. Food Chem. 177, 397–404.

Norouzi, P., Larijani, B., Ganjali, M.R., 2012. Ochratoxin a sensor based on nanocomposite hybrid film of ionic liquid-graphene nano-sheets using coulometric FFT cyclic voltammetry. Int. J. Electrochem. Sci. 7, 7313–7324.

Nualkaw, K., Poapolathep, S., Zhang, Z., Zhang, Q., Giorgi, M., Li, P., Logrieco, A.F., Poapolathep, A., 2020. Simultaneous determination of multiple mycotoxins in swine, poultry and dairy feeds using ultra high performance liquid chromatography-tandem mass spectrometry. Toxins 12, 253.

Özgür, E., Topçu, A.A., Yılmaz, E., Denizli, A., 2020. Surface plasmon resonance based biomimetic sensor for urinary tract infections. Talanta 212, 120778.

Pacheco, J.G., Castro, M., Machado, S., Barroso, M.F., Nouws, H.P.A., Delerue-Matos, C., 2015. Molecularly imprinted electrochemical sensor for ochratoxin A detection in food samples. Sens. Actuators B 215, 107–112.

Pallarés, N., Font, G., Mañes, J., Ferrer, E., 2017. Multimycotoxin LC-MS/MS analysis in tea beverages after dispersive liquid-liquid microextraction (DLLME). J. Agric. Food Chem. 65, 10282–10289.

Pałubicki, J., Kosicki, R., Twarużek, M., Ałtyn, I., Grajewski, J., 2021. Concentrations of zearalenone and its metabolites in female wild boars from woodlands and farmlands. Toxicon 196, 19–24.

Pantano, L., Scala, L.La, Olibrio, F., Galluzzo, F.G., Bongiorno, C., Buscemi, M.D., Macaluso, A., Vella, A., 2021. Quechers LC–MS/MS screening method for mycotoxin detection in cereal products and spices. Int. J. Environ. Res. Public Health 18, 3774.

Pei, S.C., Lee, W.J., Zhang, G.P., Hu, X.F., Eremin, S.A., Zhang, L.J., 2013. Development of anti-zearalenone monoclonal antibody and detection of zearalenone in corn products from China by ELISA. Food Control 31, 65–70.

Pereira, R.H.A., Keijok, W.J., Prado, A.R., de Oliveira, J.P., Guimarães, M.C.C., 2021. Rapid and sensitive detection of ochratoxin A using antibody-conjugated gold nanoparticles based on localized surface plasmon resonance. Toxicon 199, 139–144.

Pereira, V.L., Fernandes, J.O., Cunha, S.C., 2014. Mycotoxins in cereals and related foodstuffs: a review on occurrence and recent methods of analysis. Trends Food Sci. Technol. 36, 96–136.

Pleadin, J., Peri, N., Zadravec, M., Sokolovi, M., Vuli, A., Jaki, V., Mitak, M., 2012. Correlation of deoxynivalenol and fumonisin concentration determined in maize by ELISA methods. J. Immunoassay Immunochem. 33, 414–421.

Przybylska, A., Chrustek, A., Olszewska-Słonina, D., Koba, M., Kruszewski, S., 2021. Determination of patulin in products containing dried fruits by enzyme-linked immunosorbent assay technique patulin in dried fruits. Food Sci. Nutr. 9, 4211–4220.

Rico-Yuste, A., Abouhany, R., Urraca, J.L., Descalzo, A.B., Orellana, G., Moreno-Bondi, M.C., 2021. Eu(III)-templated molecularly imprinted polymer used as a luminescent sensor for the determination of tenuazonic acid mycotoxin in food samples. Sens. Actuators B 329, 129256.

Rosa da Silva, C., Tonial Simões, C., Kobs Vidal, J., Reghelin, M.A., Araújo de Almeida, C.A., Mallmann, C.A., 2022. Development and validation of an extraction method using liquid chromatography-tandem mass spectrometry to determine patulin in apple juice. Food Chem. 366, 130654.

Rubert, J., Soler, C., Mañes, J., 2012. Application of an HPLC-MS/MS method for mycotoxin analysis in commercial baby foods. Food Chem. 133, 176–183.

Saylan, Y., Yilmaz, F., Özgür, E., Derazshamshir, A., Bereli, N., Yavuz, H., Denizli, A., 2018. Surface plasmon resonance sensors for medical diagnosis. In: Nanotechnology Characterization Tools for Biosensing and Medical Diagnosis. Springer, Berlin-Heidelberg, pp. 425–458.

Scheller, F.W., Hintsche, R., Pfeiffer, D., Schubert, F., Riedel, K., Kindervater, R., 1991. Biosensors: fundamentals, applications and trends. Sens. Actuators B Chem. 4, 197–206.

Singh, A.K., Lakshmi, G.B.V.S., Fernandes, M., Sarkar, T., Gulati, P., Singh, R.P., Solanki, P.R., 2021. A simple detection platform based on molecularly imprinted polymer for AFB1 and FuB1 mycotoxins. Microchem. J. 171, 106730.

Singh, G., Velasquez, L., Huet, A.C., Delahaut, P., Gillard, N., Koerner, T., 2019. Development of a sensitive polyclonal antibody-based competitive indirect ELISA for determination of citrinin in grain-based foods. Food Addit. Contam. Part A Chem. Anal. Control. Expos. Risk Assess. 36, 1567–1573.

Singh, J., Mehta, A., 2020. Rapid and sensitive detection of mycotoxins by advanced and emerging analytical methods: a review. Food Sci. Nutr. 8, 2183–2204.

Smith, M.C., Madec, S., Coton, E., Hymery, N., 2016. Natural co-occurrence of mycotoxins in foods and feeds and their in vitro combined toxicological effects. Toxins 8, 94.

Soares, R.R.G., Ricelli, A., Fanelli, C., Caputo, D., De Cesare, G., Chu, V., Aires-Barros, M.R., Conde, J.P., 2018. Advances, challenges and opportunities for point-of-need screening of mycotoxins in foods and feeds. Analyst 143, 1015–1035.

Somsubsin, S., Seebunrueng, K., Boonchiangma, S., Srijaranai, S., 2018. A simple solvent based microextraction for high performance liquid chromatographic analysis of aflatoxins in rice samples. Talanta 176, 172–177.

Su, B., Zhang, Z., Sun, Z., Tang, Z., Xie, X., Chen, Q., Cao, H., Yu, X., Xu, Y., Liu, X., Hammock, B.D., 2022. Fluonanobody-based nanosensor via fluorescence resonance energy transfer for ultrasensitive detection of ochratoxin A. J. Hazard. Mater. 422, 126838.

Tang, W., Qi, Y., Li, Z., 2021. A portable, cost-effective and user-friendly instrument for colorimetric enzyme-linked immunosorbent assay and rapid detection of aflatoxin B$_1$. Foods 10, 2483.

Tarannum, N., Nipa, M.N., Das, S., Parveen, S., 2020. Aflatoxin M1 detection by ELISA in raw and processed milk in Bangladesh. Toxicol. Rep. 7, 1339–1343.

Thvenot, D.R., Toth, K., Durst, R.A., Wilson, G.S., 1999. Electrochemical biosensors: recommended definitions and classification (technical report). Pure Appl. Chem. 71, 2333–2348.

Tkaczyk, A., Jedziniak, P., 2020. Dilute-and-shoot HPLC-UV method for determination of urinary creatinine as a normalization tool in mycotoxin biomonitoring in pigs. Molecules 25, 2445.

Tkaczyk, A., Jedziniak, P., 2021. Development of a multi-mycotoxin LC-MS/MS method for the determination of biomarkers in pig urine. Mycotoxin Res. 37, 169–181.

Turner, N.W., Subrahmanyam, S., Piletsky, S.A., 2009. Analytical methods for determination of mycotoxins: a review. Anal. Chim. Acta 632, 168–180.

Wang, S., Gao, H., Wei, Z., Zhou, J., Ren, S., He, J., Luan, Y., Lou, X., 2022. Shortened and multivalent aptamers for ultrasensitive and rapid detection of alternariol in wheat using optical waveguide sensors. Biosens. Bioelectron. 196, 113702.

Wei, T., Ren, P., Huang, L., Ouyang, Z., Wang, Z., Kong, X., Li, T., Yin, Y., Wu, Y., He, Q., 2019. Simultaneous detection of aflatoxin B1, ochratoxin A, zearalenone and deoxynivalenol in corn and wheat using surface plasmon resonance. Food Chem. 300, 125176.

Wu, C., Du, Y.W., Huang, L., Galeczki, Y.B.S., Dagan-Wiener, A., Naim, M., Niv, M.Y., Wang, P., 2017. Biomimetic sensors for the senses: towards better understanding of taste and odor sensation. Sensors (Switzerland) 17, 2881.

Xiang, Y., Camarada, M.B., Wen, Y., Wu, H., Chen, J., Li, M., Liao, X., 2018. Simple voltammetric analyses of ochratoxin A in food samples using highly-stable and anti-fouling black phosphorene nanosensor. Electrochim. Acta 282, 490–498.

Yang, J., Li, J., Jiang, Y., Duan, X., Qu, H., Yang, B., Chen, F., Sivakumar, D., 2014. Natural occurrence, analysis, and prevention of mycotoxins in fruits and their processed products. Crit. Rev. Food Sci. Nutr. 54, 64–83.

Yu, J.C.C., Lai, E.P.C., 2004. Polypyrrole film on miniaturized surface plasmon resonance sensor for ochratoxin A detection. Synth. Met. 143, 253–258.

Zhan, S., Hu, J., Li, Y., Huang, X., Xiong, Y., 2021. Direct competitive ELISA enhanced by dynamic light scattering for the ultrasensitive detection of aflatoxin B1 in corn samples. Food Chem. 342, 128327.

Zhang, L., Dou, X.W., Zhang, C., Logrieco, A.F., Yang, M.H., 2018. A review of current methods for analysis of mycotoxins in herbal medicines. Toxins 10, 65.

Zhang, W., Han, Y., Chen, X., Luo, X., Wang, J., Yue, T., Li, Z., 2017. Surface molecularly imprinted polymer capped Mn-doped ZnS quantum dots as a phosphorescent nanosensor for detecting patulin in apple juice. Food Chem. 232, 145–154.

Zhao, H., Chen, X., Shen, C., Qu, B., 2017. Determination of 16 mycotoxins in vegetable oils using a QuEChERS method combined with high-performance liquid chromatography-tandem mass spectrometry. Food Addit. Contam. Part A Chem. Anal. Control Expos. Risk Assess. 34, 255–264.

Sustainability and environmental issues in food processing

11

Geetanjali[a] and Ram Singh[b]

[a]*Department of Chemistry, Kirori Mal College, University of Delhi, New Delhi, India,*
[b]*Department of Applied Chemistry, Delhi Technological University, New Delhi, India*

11.1 Introduction

Food processing involves many types of alteration to their raw state. All types of food products like meat, fruits, vegetables, coffee and cocoa, dairy, etc. are to be processed before being commercialized. The processes like cooking, preservation, and mixing with other food are included in the food processing methods. The processed food has been directly or indirectly remained a part of our diet (Floros et al., 2010; Petrus et al., 2021). In ancient Greece, processed food such as wine, bread, and olive oil were the three most important foods. Processing allowed the transformation of many nonedible raw materials into safe, nutritious, and stable food like soy, cocoa, olives, and others (Floros et al., 2010). Many raw food products are not stable toward physical, chemical, biological, and thermal changes. Hence, the processing makes them overcome these limitations.

The food processing is associated with several environmental concerns for all three components of ecology: soil, water, and air (Clark, 2010). There are many challenges associated with food processing that require considerable attention (Teixeira, 2018) (Fig. 11.1). The available food systems use a lot of land, produce large volumes of greenhouse gases, and pollute/consume water affecting ecosystems. The successful sustainable management of the environmental concern aroused due to food processing has been a subject of research.

Sustainable management is the method in which the needs of the present generation are fulfilled without compromising the requirements of future generations (Mensah and Casadevall, 2019; Fomo et al., 2020). Sustainability assessment should consider all the aspects like economic, environmental, and societal. A sustainable food chain must be both reasonable and flexible, so that it ensures a supply of food without affecting environmental safety. A number of steps are involved in the process of the food chain which includes food production (contributes 17%–32% of global greenhouse gas emissions), food processing that includes both transport and distribution (responsible for 25% of global water consumption), retail, consumption,

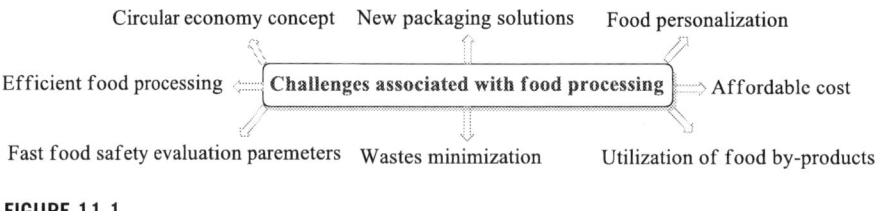

FIGURE 11.1

Challenges associated with food processing.

and end of life (My Dieu, 2009). In this chapter, we discuss sustainability and environmental issues related to food processing industries.

11.2 **Food processing**

Food processing is a process of converting fresh foods (like fruit, vegetables, milk, cereals, etc.) into food products that are suitable for the consumption of human beings using different methods (Monteiro et al., 2010) (Fig. 11.2). From the prehistoric ages, cooking (like roasting, smoking, steaming, oven baking, etc.), fermentation, sun drying, and preserving with salt are some methods used for food processing but these days one or a combination of various processes like washing, chopping, pasteurizing, ultraheat treatment, freezing, fermenting, high-pressure processing (HPP), or modified atmosphere packaging, cooking, and many more are used to process food (Floros et al., 2010). Food processing also includes adding ingredients like artificial colors, salt, sugar, and fats to food to extend its shelf life (Dwyer et al., 2012; Weaver et al., 2014) to maintain or improve its safety, freshness, taste, texture, or appearance.

The initial concept of food processing was to preserve food for a longer duration and solve the issue with a long storage period. However, soon the other side of food processing becomes more prominent like environmental problems and food-health imbalance (Knorr and Watzke, 2019). Packaged food added waste packaging materials to the environment. The packaging materials are going to the landfills where they are posing environmental pollution and affecting the people who live nearby. Nowadays, frequent debates are taking place about food processing methods and their environmental impacts and sustainability.

The sustainable food processes will boost the socio-economic condition of humankind for the long term (Romanello and Veglio, 2022). A better food processing is always linked to sustainable drivers (Murphy et al., 2014) (Fig. 11.3). Food Safety and Standards Authority of India (FSSAI) looks after all types of legislative aspects related to food in India being a statutory body. At the global level, the Food and Agriculture Organization (FAO), United Nations (UN), Group of 20 (G20), and individual countries own legislation always suggest or frame guidelines for food and food process sustainability (Candel, 2014). Exclusive law for sustainable and environmentally friendly food processing is still required.

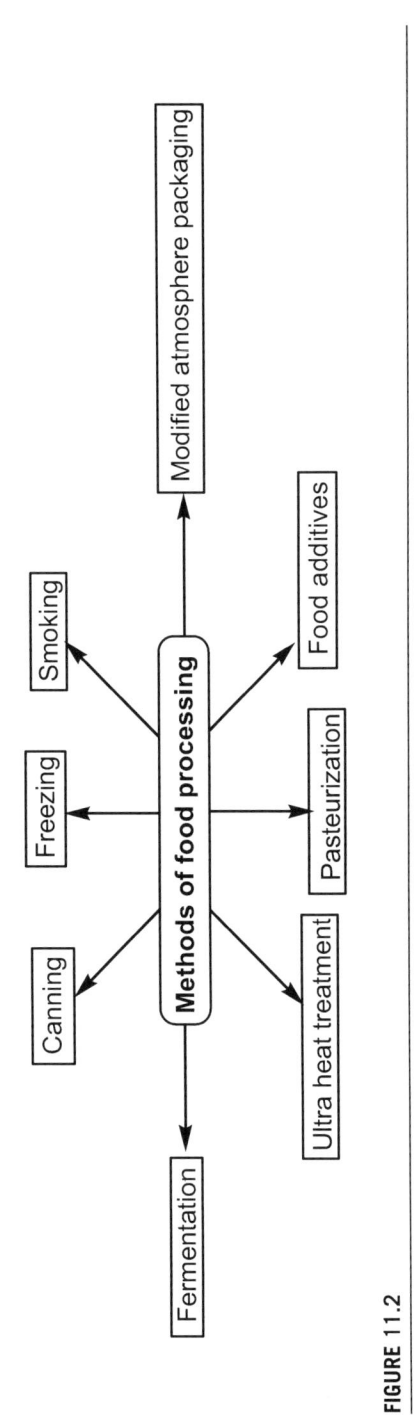

FIGURE 11.2

Food processing methods.

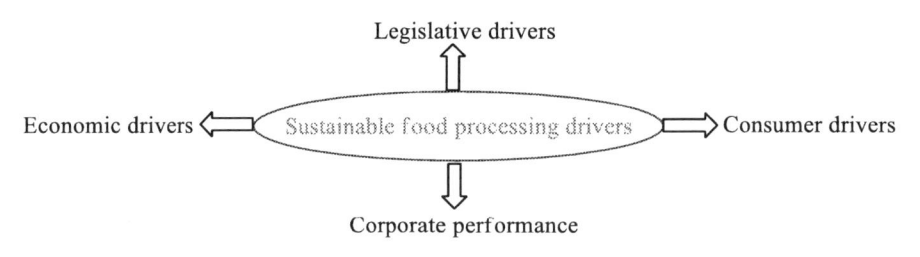

FIGURE 11.3

Sustainable food processing drivers.

The economics plays a very important part in the development of sustainable food processing. This is because, the execution of the process requires high starting investment or expensive modification in their existing plants (Darbari et al., 2018; Notarnicola et al., 2012). However, once set up, this will have a better impact in the long run. Another economic issue is the cost of sustainable products which is on the higher side than the other products (Bhaskaran et al., 2006).

Consumers can also play a positive role in the adoption of sustainable and environmentally friendly products (Salim et al., 2018). The consumer always settles for cheap products and is least bothered about the indirect effects on them due to adverse environmental issues caused due to the manufacturing of cheap products (Simpson and Jewitt, 2019). However, with proper awareness programs, the sale of sustainable and environmentally friendly products is getting higher, but this will take time to switch over completely on this. The cooperation from the corporate world, suppliers, and managers to adopt sustainability practices will add to its success (Smith, 2008).

11.3 Environmental impact of food processing

11.3.1 Energy efficiency

Energy inputs are a very important constituent of food processing industries. The food production is directly related to energy input (Woods et al., 2010). Energy efficiency improvement, waste heat recovery, design of new energy-efficient operations, and increase in the use of renewable energy are some important aspects in the food industry to increase the sustainability of food processing in the last decades (Wang, 2014). Energy conservation in the food industry can be done by replacing the conventional energy-intensive food processes with novel technologies like thermodynamic cycles such as power cogeneration cycle, absorption refrigeration cycle, heat pipe, heat pump, and heat (Sun and Wang, 2001; Ozyurt et al., 2004; Kuzgunkaya and Hepbasli, 2007), nonthermal food processes such as high-pressure processing, irradiation, and pulsed electric fields (PEFs) treatment (Toepfl et al., 2006; Brown et al., 2008). The methods such as infrared heating, ultrasonication, and microwave are also categorized under energy-efficient processes (Yang et al., 2010; Nguyen et al., 2013).

11.3.2 **Solid waste**

Solid waste is another major challenge to food processing industries which include waste food, noneatable part, packaging, etc. One of the major sinks of these solid wastes has been the landfill areas but in recent times, this has caused a serious environmental concern. The solid wastes lead to greenhouse gas emissions and water contamination (Ridoutt et al., 2010). The food processing industries like meat processing, poultry processing, fruit processing, confectionery, etc. are the major contributors to solid wastes (My Dieu, 2009). Another challenge is to anticipate the composition of waste produced, so that a particular process for their management can be done.

11.3.3 **Liquid consumption and waste**

The food production chain, starting from cultivation to end-product utilizes a large amount of water (Drastig et al., 2010). This poses a challenge to environmentalists because more than 20% of the world's population lacks clean drinking water (Milman and Short, 2008). This is further going to deteriorate due to the increasing population and more processed food demands.

The major contributor toward wastewater is the processing of cattle products as the whole process is water-intensive. The water is mixed with hair, fat, blood, etc. has been discharged, many times without proper treatment, leading to enormous water pollution.

11.4 **Sustainable processes for food processing**

The limitations of traditional and existing food processing methods led to the acquisition of green and sustainable techniques for pasteurization, processing, and extraction methods (Chemat et al., 2017). These methods take care of the environment through energy conservation, reduction in time factor, minimization of water usage, etc. (Chemat et al., 2011). The use of less energy without compromising the production output is a sustainable process. A possible solution is material recycling which is related to greenhouse gas emissions and energy consumption (Kullmann et al., 2021). The integration of different processes is another possibility to reduce energy-based inputs (Edwards, 1989). The integration is also as per the possibility of a circular economy which allows the use of by-products, waste minimization, and input recycling (Jurgilevich et al., 2016). Some of the important sustainable processes are discussed in the following sections.

11.4.1 **Nonthermal processing**

Processing of foods with the application of heat is known as thermal processing while processing without the application of heat is known as nonthermal processing. Thermal processing of foods is a slow and energy-intensive process, on the other hand, nonthermal processing is more energy efficient and comparatively faster since processing times are usually very short (Wang, 2014). Nonthermal processing techniques retain the color, taste, appearance, and nutrition content (Toepfl et al., 2006; Boeckel et al., 2010), and also extend the shelf life of the food by preventing or destroying microorganisms (Morris et al., 2007). Nonthermal processing technique save water, increases reliability, lowers emissions, and improves the quality of the product (Masanet et al., 2008). Some of the emerging techniques include HPP, ultrasound (US), PEF treatment, pulsed light, ultraviolet (UV) light, irradiation, oscillating magnetic field, etc. (Pereira and Vicente, 2010). The PEF treatment utilizes the biological weaknesses of cells and works at lower temperatures (Jäger et al., 2017; Raso et al., 2016).

Membrane separation processes for sustainable food processing are also prevalent which work in continuous mode to give better product quality (Picart-Palmade et al., 2019). This avoids stepwise cleaning procedures, the use of chemical additives, and involves working temperatures less than 80°C. Due to low working temperature, the product quality remains intact for heat-sensitive products and minimizes waste generation. Membrane separation process also helps in purification, concentration, extraction, and fractionation leading to the recovery and reuse of by-products (Hausmann et al., 2014; Sirkar et al., 2015; Macedonio and Drioli, 2017). In this process, membrane selection is important which works as a thin selective wall to allow the movement of solvent or solutes with the help of some driving force. The beverage processing industries are looking toward the use of membrane technologies (Dornier et al., 2018). Filtration methods like microfiltration (MF) and ultrafiltration (UF) have replaced the traditional filtration methods leading to low waste and better product quality. The juice concentration uses the reverse osmosis method (Picart-Palmade et al., 2019). However, membrane fouling is the main limitation of this process. The research toward overcoming this limitation has been widely studied (Abou-Ghazala et al., 2000; Teissié et al., 2002).

11.4.2 Efficient use of water in food industry

The efficient use of water or wastewater management is the primary requirement of food processing industries to minimize environmental impact and cut production costs (Nemati-Amirkolaii et al., 2019). One of the appropriate concepts of optimal water utility is based on the heat pinch analysis, known as water pinch analysis (Linnho and Flower, 1978). Many factories and industries follow this approach to minimize water and other fluid utility (Wang and Smith, 1994). The water savings of 63%–72% have been achieved in the fruit juice production industry with the help of this protocol (Almató et al., 1997). In the corn refinery industries, mathematical optimization and pinch analysis together led to the saving of 30% of water use (Bavar et al., 2018).

11.4.3 Low-temperature industrial surplus heat utilization

The utilization of surplus heat is based on the concept of using waste generated from one process becomes input for another process. The low-temperature industrial surplus heat is a waste which is going underutilized due to quality problems (Reyes-Lúa et al., 2021; Woolley et al., 2018). All the low-temperature processes in food processing industries like fish rearing, greenhouse production, and moisture removal may utilize this energy.

The limitation of this type of energy is the availability of data which is not being calculated directly but rather calculated based on industrial CO_2 emissions (Ammar et al., 2012). This has been estimated that the world surplus heat is approximately 42% with temperatures less than 100°C (Forman et al., 2016). As per another report, 60% of industrial heat corresponds to temperatures below 60°C in the United States (Johnson et al., 2008). This energy may be utilized in food processing industries. The Nordic countries have been utilizing this type of energy in fish production and greenhouse gases (Reyes-Lúa et al., 2021). The drying of seaweed is a field where low-temperature surplus heat is useful.

11.5 Assessment methods for environmental sustainability

Environmental sustainability is a complex word and requires the cooperation between different stakeholders to work in a collective way. For the food processing industries, the three highly studied assessment methods for environmental sustainability include ecological footprints, carbon footprints, and life cycle assessment (LCA) (Jungbluth et al., 2012).

11.5.1 Ecological footprint

According to the Global Footprint Network (2012), the ecological footprint is "a measure of how much area of biologically productive land and water an individual, population, or activity requires to produce all the resources it consumes and to absorb the waste it generates, using prevailing technology and resource management practices." The care is always taken that the ecological footprint must be assessed transparently and accurately. The results are intuitive and scientifically robust (Cerutti et al., 2011). However, this method does not cover all the environmental-related parameters including the geographic specificity (Samuel-Fitwi et al., 2012). Different components of ecosystems like aquatic, marine, forest, etc. behave differently making the assessment process less robust (Samuel-Fitwi et al., 2012). The food process industries related to aquaculture, wheat production, intensive agriculture, wine production, etc. are associated with ecological footprints. The presence of moisture, soil condition, temperature, etc. in any ecological system affects the chemical constituents present in the plant species (Singh and Geetanjali, 2018). This also affects the quality of food and processed products. Individual assessment method is not sufficient to provide the overall sustainability and environmental impact as they may not cover several indicators.

11.5.2 **Carbon footprint**

Due to recent climate change issues, the assessment with respect to carbon footprint has become a major parameter to study. This is related to the release of greenhouse gases during any process, herein, for food process industries. The emission data is analyzed for the complete life cycle of any process in carbon footprint analysis (Murphy et al., 2014; Weidmann and Minx, 2008). There have been drafts and guidelines for governing carbon footprints like ISO 14067. Similar to the ecological footprints, the carbon footprint is also not foolproof to calculate environmental product information. Only the calculation of carbon dioxide emission value cannot reflect the total fact about sustainability and environmental impacts (Samuel-Fitwi et al., 2012). In the food processing industries, food crop cultivation and dairy production are examples where carbon footprints play an important role in assessments (Rotz et al., 2010; Röös et al., 2010).

11.5.3 **Life cycle assessment**

Life cycle assessment covers some of the limitations of ecological and carbon footprints. The environmental impact is being assessed based on resource use, services, and by taking into account each step including the disposal process. The LCA covers from raw materials to processed products and finally till disposal (Berlin, 2003). This has been concluded that LCA gives a complete environmental aspect view for all types of industrial services including the recently included agri-food production sector.

The food processing industries also use this LCA method for their assessment of environmental sustainability. There are many articles that cover the LCA studies in food processing (de Vries and de Boer, 2010; Milani et al., 2011). The companies like Arla Food, a Danish company, and Dannon in the United States have already performed LCA studies. Some of the companies already switched to the more environmentally friendly type of packaging for processed foods that reduces more than 20% of carbon footprints (Murphy et al., 2014).

11.6 **Food preservation, shelf life, and environment suitability**

Every living organism requires food to survive and hence, this requires food availability. Food has its own shelf life, beyond which rancidity starts. Rancid food is not suitable for eating. To increase the shelf life of food, the food processing industries use different types of food preservatives and processes (Sharma, 2015; Inetianbor et al., 2015; Amit et al., 2017) (Fig. 11.4). The synthetic food preservatives have hazardous side effects if used for a longer period. On the basis of shelf life, the food has been grouped into three categories: nonperishable, semiperishable, and perishable. Food preservatives are required mainly for the latter two categories. Many people have reviewed food preservation techniques (Pardo and Zufía, 2012; Sharma, 2015;

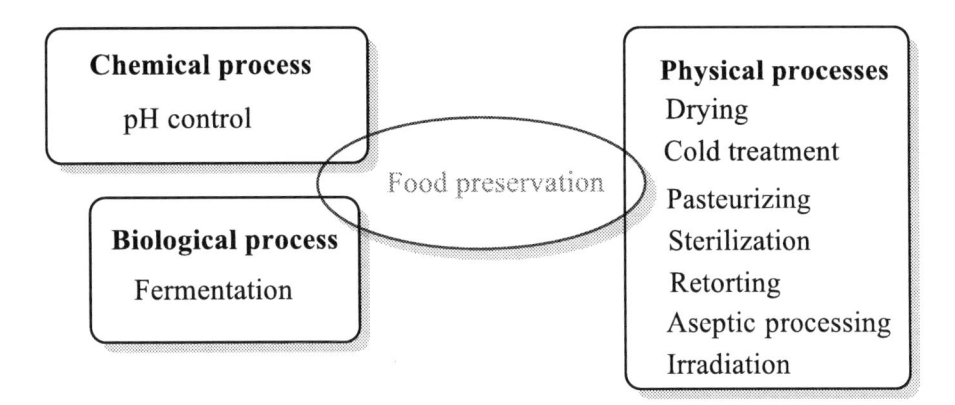

FIGURE 11.4

Methods of food preservation.

Inetianbor et al., 2015; Amit et al., 2017). Traditional methods of food preservation lack sustainability. The LCA method has been used to assess the environmental sustainability of food preservation.

Drying or dehydration removes moisture from food, so that the food contains a very low amount of moisture and gets preserved for a longer period of time (Leniger and Beverloo, 1975; Syamaladevi et al., 2016). Low water content does not allow the microorganisms to grow and spoil the food. Freezing is a cold treatment of food stuff which inhibits microbial growth by slowing down the biochemical reaction at low-temperature (George, 2008; Velez-Ruiz and Rahman, 1999). Similarly, in the chilling process, the foodstuff is kept in the temperature range of 1°C–8°C. This low-temperature slows down or stops the microbiological changes and improves the shelf life of food. Keeping food in the refrigerator is a common example of this case. The spoilage-causing microorganisms and enzymes also get destroyed by heating to a particular temperature and enhance the food shelf life. This process is also known as pasteurization (Laudan, 2009; Cavazos-Garduño et al., 2016). However, the high heat destroys important minerals and vitamins. Sterilization is also a high-heat process. In the retorting process, the food along with its container subjected to high heat treatment (Kirk-Othmer, 2007; Tucker, 2007). All the food preservative methods have their limitations, and they are not free from side effects on human health after prolonged use. To keep the food for a longer time without compromising the quality is a big challenge.

11.7 **Conclusion**

There are many foods that cannot be eaten in raw form and require processing. Other types of food require more shelf life to reach from one place to another place as commercial items. The processes like cooking, preservation, and mixing with other food are included in the food processing methods. Processed food has directly or indirectly remained a part of our diet. Many challenges are associated with food processing industries. The environmental concerns including more energy usage, waste liquid production, and excessive postconsumer packaging materials accumulations forced the stakeholders to think of more sustainable processes.

Sustainability along with quality and safety is the main priority of food processing industries. A sustainable food chain must be both reasonable and flexible, so that it ensures a secure, environmentally sustainable, and healthy supply of food. Use of energy-efficient processes, minimization of water consumption, use of biodegradable packaging materials, and application of circular economy are some of the sustainable processes which must be adopted by the food processing industries. No doubt, a lot of sustainable steps are being taken in recent times but keeping in view the increasing demand for processed food, more appropriate environmentally friendly steps are the need of the present day.

Acknowledgment

The authors are thankful to their institutions for infrastructural support.

References

Abou-Ghazala, A., Schoenbach, K., 2000. Biofouling prevention with pulsed electric fields. IEEE Trans. Plasma Sci. 28, 115–121.

Almató, M., Sanmartí, E., Espun, A., Puigjaner, L., 1997. Rationalizing the water use in the batch process industry. Comput. Chem. Eng. 21, 971–976.

Amit, S.K., Uddin, M.M., Rahman, R., Islam, S.M.R., Khan, M.S., 2017. A review on mechanisms and commercial aspects of food preservation and processing. Agric. Food Secur. 6, 51.

Ammar, Y., Joyce, S., Norman, R., Wang, Y., Roskilly, A.P., 2012. Low grade thermal energy sources and uses from the process industry in the UK. Appl. Energy 89, 3–20.

Bavar, M., Sarrafzadeh, M.H., Asgharnejad, H., Norouzi-Firouz, H,, 2018. Water management methods in food industry: corn refinery as a case study. J. Food Eng. 238, 78–84.

Berlin, J., 2003. Life cycle assessment (LCA): an introduction. In: Mattson, B, Sonesson, U (Eds.), Environmentally-Friendly Food Processing. Woodhead Publishing Limited, Cambridge.

Bhaskaran, S., Polonsky, M., Cary, J., Fernandez, S., 2006. Environmentally sustainable food production and marketing: opportunity or hype? Br. Food J. 108, 677–690.

Boekel, M.V., Fogliano, V., Pellegrini, N., Stanton, C., Scholz, G., Lallje, S, Somoza, V., Knorr, D., Jasti, P.R., Eisenbrand, G., 2010. A review on the beneficial aspects of food processing. Mol. Nutr. Food Res. 54, 1215–1247.

Brown, Z.K., Fryer, P.J., Norton, I.T., Bakalis, S., Bridson, R.H., 2008. Drying of foods using supercritical carbon dioxide-investigations with carrot. Innov. Food Emerg. Technol. 9, 280–289.

Candel, J.J.L., 2014. Food security governance: a systematic literature review. Food Secur. 6, 585–601.

Cavazos-Garduño, A., Serrano-Niño, J.C., Solís-Pacheco, J.R., Gutierrez-Padilla, J.A., González-Reynoso, O., García, H.S., Aguilar-Uscanga, B.R., 2016. Effect of pasteurization, freeze-drying and spray drying on the fat globule and lipid profile of human milk. J. Food Nutr. Res. 4, 296–302.

Cerutti, A.K., Bruun, S., Beccaro, G.L., Bounous, G., 2011. A review of studies applying environmental impact assessment methods on fruit production systems. J. Environ. Manage. 92, 2277–2286.

Chemat, F., Huma, Z., Khan, M.K., 2011. Applications of ultrasound in food technology: processing, preservation and extraction. Ultrason. Sonochem. 18, 813–835.

Chemat, F., Rombaut, N., Meullemiestre, A., Turk, M., Perino, S., Fabiano-Tixier, A.-S., Abert-Vian, M., 2017. Review of green food processing techniques: preservation, transformation, and extraction. Innov. Food Sci. Emerg. Technol. 41, 357–377.

Clark, J.H., 2010. Introduction to green chemistry. In: Proctor, A (Ed.), Alternatives to Conventional Food Processing. Royal Society of Chemistry, London, pp. 1–10.

Darbari, J.D., Agarwal, V., Sharma, R., Jha, P.C., 2018. Analysis of Impediments to Sustainability in the Food Supply Chain: An Interpretive Structural Modeling Approach. Springer, Berlin.

de Vries, M., de Boer, I.J.M., 2010. Comparing environmental impacts for livestock products: a review of life cycle assessments. Livest. Sci. 128, 1–11.

Dornier, M., Belleville, M.-P., Vaillant, F., 2018. Membrane technologies for fruit juice processing. In: Rosenthal, A., Deliza, R., Barbosa-Carvosa, G.V., Welti-Chanes, J. (Eds.), Fruit Preservation: Novel and Conventional Technologies. Springer Publishing Co, New York, pp. 211–248.

Drastig, K., Prochnow, A., Kraatz, S., Klauss, H., Plöchl, M., 2010. Water footprint analysis for the assessment of milk production in Brandenburg (Germany). Adv. Geosci. 27, 65–70.

Dwyer, J.T., Fulgoni, V.L., Clemens, R.A., David, B.S., Marjorie, R.F., 2012. Is 'Processed' a four-letter word? The role of processed foods in achieving dietary guidelines and nutrients recommendations. Adv. Nutr. 3, 536–548.

Edwards, C.A., 1989. The importance of integration in sustainable agricultural systems. Agric. Ecosyst. Environ. 27, 25–35.

Floros, J.D., Newsome, R., Fisher, W., Barbosa-Cánovas, G.V., Chen, H., Dunne, C.P., German, J.B., Hall, R.L., Heldman, D.R Karwe, M.V., Knabel, S.J., Labuza, T.P., Lund, D.B., Newell-McGloughlin, M., Robinson, J.L., Sebranek, J.G., Shewfelt, R.L, Tracy, W.F., Weaver, C.M., Ziegler, G.R., 2010. Feeding the world today and tomorrow: the importance of food science and technology. Compr. Rev. Food Sci. Food Saf. 9 (5), 572–599.

Fomo, G., Madzimbamuto, T.N., Ojumu, T.V., 2020. Applications of nonconventional green extraction technologies in process industries: challenges, limitations and perspectives. Sustainability 12, 5244.

Forman, C., Muritala, I.K., Pardemann, R., Meyer, B., 2016. Estimating the global waste heat potential. Renew. Sustain. Energ. Rev. 57, 1568–1579.

George, M., 2008. Food Biodeterioration and Preservation. In: Tucker, G.S. (Ed.). Blackwell Publisher: Singapore.

Hausmann, A., Sanciolo, P., Vasiljevic, T., Kulozik, U., Duke, M., 2014. Performance assessment of membrane distillation for skim milk and whey processing. J. Dairy Sci. 97, 56–71.

Inetianbor, J.E., Yakubu, J.M., Ezeonu, S.C., 2015. Effects of food additives and preservatives on man: a review. Asian J. Sci. Technol. 6, 1118–1135.

Jäger, H., Knorr, D., 2017. Pulsed electric fields treatment in food technology: challenges and opportunities. In: Miklavcic, D. (Ed.), Handbook of Electroporation. Springer, Cham, pp. 2657–2680.

Johnson, I., Choate, W.T., Davidson, A., 2008. Waste Heat Recovery. Technology and Opportunities in U.S. Industry. EERE Publication and Product Library, Laurel, Technical Report.

Jungbluth, N., Büsser, S., Frischknecht, R., Flury, K., Stucki, M., 2012. Feasibility of environmental product information based on life cycle thinking and recommendations for Switzerland. J. Cleaner Prod. 28, 187–197.

Jurgilevich, A., Birge, T., Kentala-Lehtonen, J., Korhonen-Kurki, K., Pietikäinen, J., Saikku, L., Schösler, H., 2016. Transition towards circular economy in the food system. Sustainability 8, 69.

Knorr, D., Watzke, H., 2019. Food processing at a crossroad. Front. Nutr. 6, 85.

Kullmann, F., Markewitz, P., Stolten, D., Robinius, M., 2021. Combining the worlds of energy systems and material flow analysis: a review. Energ. Sustain. Soc. 11, 13 Art.

Kuzgunkaya, E.H., Hepbasli, A., 2007. Exergetic performance assessment of a ground-source heat pump drying system. Int. J. Energy Res. 31, 760–777.

Laudan, R., 2009. Food and Nutrition: Lifespan, Human to Pesticides. Marshall Cavendish, New York.

Leniger, H.A., Beverloo, W.A., 1875. Food Process Engineering. Springer, Netherlands.

Linnho, B., Flower, J.R., 1978. Synthesis of heat exchanger networks: I. systematic generation of energy optimal networks. AIChE J. 24, 633–642.

Macedonio, F., Drioli, E., 2017. Membrane engineering for green process engineering. Engineering 3, 290–298.

Masanet, E., Worrell, E., Graus, W., Galitsky, C., 2008. Energy Efficiency Improvement and Cost Saving Opportunities for the Fruit and Vegetable Processing Industry, an Energy Star Guide for Energy and Plant Managers. Environmental Energy Technologies Division, Technical Report.

Mensah, J., Casadevall, S.R., 2019. Sustainable development—meaning, history, principles, pillars, and implications for human action: literature review. Cogent Soc. Sci. 5, 1.

Milani, F.X., Nutter, D., Thoma, G., 2011. Environmental impacts of dairy processing and products: a review. J. Dairy Sci. 94, 4243–4254.

Milman, A., Short, A., 2008. Incorporating resilience into sustainability indicators: an example for the urban water sector. Glob. Environ. Chang. 18, 758–767.

Monteiro, C.A., Levy, R.B., Claro, R.M., Castro, Inês Rugani Ribeiro de, Cannon, G, 2010. A new classification of foods based on the extent and purpose of their processing. Cadernos de Saúde Pública 26, 2039–2049.

Morris, C., Brody, A.L., Wicker, L., 2007. Non-thermal food processing/preservation technologies: a review with packaging implications. Packag. Technol. Sci. 20, 275–286.

Murphy, F., McDonnell, K., Fagan, C.C., 2014. Sustainability and environmental issues in food processing. In: Clark, S., Jung, S., Lamsal, B. (Eds.), Food Processing: Principles and applications. John Wiley & Sons, Ltd, pp. 207–232.

My Dieu, T.T., 2009. Food processing and food waste. In: Baldwin C. (ed.) Sustainability in the Food Industry. Ames, IA: Wiley-Blackwell, pp. 23–60.

Nemati-Amirkolaii, K., Romdhana, H., Lameloise, M.-L., 2019. Pinch methods for efficient use of water in food industry: a survey review. Sustainability 11, 4492.

Nguyen, L.T., Choi, W., Lee, S.H., June, S., 2013. Exploring the heating patterns of multiphase foods in a continuous flow, simultaneous microwave and ohmic combination heater. J. Food Eng. 116, 65–71.

Notarnicola, B., Hayashi, K., Curran, M.A., Huisingh, D., 2012. Progress in working towards a more sustainable agri-food industry. J. Cleaner Prod. 28, 1–8.

Othmer, K., 2007. Food and Feed Technology, vol. 1. Wiley-Interscience, New Jersey.

Ozyurt, O., Comakli, O., Yilmaz, M., Karsli, S., 2004. Heat pump use in milk pasteurization: an energy analysis. Int. J. Energy Res. 28, 833–846.

Pardo, G., Zufía, J., 2012. Life cycle assessment of food-preservation technologies. J. Cleaner Prod. 28, 198–207.

Pereira, R.N., Vicente, A.A., 2010. Environmental impact of novel thermal and non-thermal technologies in food processing. Food Res. Int. 43, 1936–1943.

Petrus, R.R., Sobral, P.J.A., Tadini, C.C., Gonçalves, C.B., 2021. The NOVA classification system: a critical perspective in food science. Trends Food Sci. Technol. 116, 603–608.

Picart-Palmade, L., Cunault, C., Chevalier-Lucia, D., Belleville, M.-P., Marchesseau, S., 2019. Potentialities and limits of some non-thermal technologies to improve sustainability of food processing. Front. Nutr. 5, 130.

Raso, J., Frey, W., Pataro, G., Knorr, D., Teissie, J., Miklavcic, D., 2016. Recommendations and guidelines on key information to be reported in studies of application of PEF technology in food and biotechnology processes. Innov. Food Sci. Emerg. Technol. 37, 312–321.

Reyes-Lúa, A., Straus, J., Skjervold, V.T., Durakovic, G., Nordtvedt, T.S., 2021. A novel concept for sustainable food production utilizing low temperature industrial surplus heat. Sustainability 13, 9786.

Ridoutt, B.G., Juliano, P., Sanguansri, P., Sellahewa, J., 2010. The water footprint of food waste: case study of fresh mango in Australia. J. Cleaner Prod. 18, 1714–1721.

Romanello, R., Veglio, V., 2022. Industry 4.0 in food processing: drivers, challenges and outcomes. Br. Food J. 124, 375–390.

Röös, E., Sundberg, C., Hansson, P.A., 2010. Uncertainties in the carbon footprint of food products: a case study on table potatoes. Int. J. Life Cycle Assess. 15, 478–488.

Rotz, C.A., Montes, F., Chianese, D.S., 2010. The carbon footprint of dairy production systems through partial life cycle assessment. J. Dairy Sci. 93, 1266–1282.

Salim, H.K., Padfield, R., Lee, C.T., Syayuti, K., Papargyropoulou, E., Tham, M.H., 2018. An investigation of the drivers, barriers, and incentives for environmental management systems in the Malaysian food and beverage industry. Clean Technol. Environ. Policy 20, 529–538.

Samuel-Fitwi, B., Wuertz, S., Schroeder, J.P., Schulz, C., 2012. Sustainability assessment tools to support aquaculture development. J. Cleaner Prod. 32, 183–192.

Sharma, S., 2015. Food preservatives and their harmful effects. Int. J. Sci. Res. Publ. 5, 1–2.

Simpson, G.B., Jewitt, G.P.W., 2019. The development of the water-energy-food nexus as a framework for achieving resource security: a review. Front. Environ. Sci. 7, 8.

Singh, R., Geetanjali, 2018. Chemotaxonomy of medicinal plants: possibilities and limitations. In: Mandal, S.C., Mandal, V., Konishi, T. (Eds.), Natural Products and Drug Discovery – An Integrated Approach. Elsevier, pp. 119–136.

Sirkar, K.K., Fane, A.G., Wang, R., Wickramasinghe, S.R., 2015. Process intensification with selected membrane processes. Chem. Eng. Process.: Process Intensif. 87, 16–25.

Smith, B.G., 2008. Developing sustainable food supply chains. Philos. Trans. Royal Soc. B Biol. Sci. 363, 849–861.

Sun, D.W., Wang, L.J., 2001. Novel refrigeration cycles. In: Sun, D.W. (Ed.), Advances in Food Refrigeration. Leatherhead Publishing, pp. 1–69.

Syamaladevi, R.M., Tang, J., Villa-Rojas, R., Sablani, S., Carter, B., Campbell, G., 2016. Influence of water activity on thermal resistance of microorganisms in low-moisture foods: a review. Compr. Rev. Food Sci. Food Saf. 15, 353–370.

Teissié, J., Eynard, N., Vernhes, M., Bénichou, A., Ganeva, V., Galutzov, B., Cabanes, P.A., 2002. Recent biotechnological developments of electropulsation: a prospective review. Bioelectrochemistry 55, 107–112.

Teixeira, J.A., 2018. Grand challenges in sustainable food processing. Front. Sustain. Food Syst. 2, 19.

Toepfl, S., Mathys, A., Heinz, V., Knorr, D., 2006. Review: potential of high hydrostatic pressure and pulsed electric fields for energy efficient and environmentally friendly food processing. Food Rev. Int. 22, 405–423.

Tucker, G.S., 2007. Food Biodeterioration and Preservation, first ed. Wiley-Blackwell, New Jersy.

Velez-Ruiz, J.F., Rahman, M.S., 1999. Food preservation by freezing. In: Rahman, M.S. (Ed.), Handbook of Food Preservation. CRC Press, New York.

Wang, L., 2014. Energy efficiency technologies for sustainable food processing. Energ. Effic. 7, 791–810.

Wang, Y.P., Smith, R., 1994. Wastewater minimisation. Chem. Eng. Sci. 49, 981–1006.

Weaver, C.M., Dwyer, J., Fulgoni, V.L., King, J.C., Leveille, G.A., MacDonald, R.S., Ordovas, J., Schnakenberg, D., 2014. Processed food: contributions to nutrition. Am. J. Clin. Nutr. 99, 1525–1542.

Weidmann, T., Minx, J., 2008. A definition of 'carbon footprint'. In: Pertsova, C.C. (Ed.), Ecological Economics Research Trends. Nova Science Publishers, New York, pp. 1–11.

Woods, J., Williams, A., Hughes, J.K., Black, M., Murphy, R., 2010. Energy and the food system. Philos. Trans. Royal Soc. B Biol. Sci. 365, 2991–3006.

Woolley, E., Luo, Y., Simeone, A., 2018. Industrial waste heat recovery: a systematic approach. Sustain. Energ. Technol. Assess. 29, 50–59.

Yang, J., Bingol, G., Pan, Z., Brandl, M.T., McHugh, T.H., Wang, H., 2010. Infrared heating for dry-roasting and pasteurization of almonds. J. Food Eng. 101, 273–280.

Common techniques in food processing technologies 12

Abel Inobeme [a], John Tsado Mathew [b], Alexander Ajai (Ikechuku) [c], Charles Oluwaseun Adetunji [d], Jonathan Inobeme [e], Munirat Maliki [a], Mathew Adefusika Adekoya [f], Elija Shaba (Yanda) [c], Olori Eric [a], Sadiq Akhor (Oshoke) [a] and Chinenye Eziukwu [a]

[a] *Department of Chemistry, Edo State University Uzairue, Edo State, Nigeria,* [b] *Department of Chemistry, Ibrahim Badamasi Babangida University Lapai, Niger State, Nigeria,* [c] *Department of Chemistry, Federal University of Technology, Minna, Nigeria,* [d] *Applied Microbiology, Biotechnology and Nanotechnology Laboratory, Department of Microbiology, Edo State University Uzairue, Iyamho, Edo State, Nigeria,* [e] *Department of Geography, Ahmadu Bello University, Zaria, Nigeria,* [f] *Department of Physics, Edo State University Uzairue, Edo State, Nigeria*

12.1 Introduction

Food processing is one of the fastest-growing areas of food technology. Throughout the history of human development, man has continually been challenged with the search for food and the need to either eat them in their raw form or convert them into some other forms. Food technology is a broad area that encompasses the act of producing and processing food. It is a technique aimed at converting raw food materials into properly cooked and effectively preserved edible foods for humans and other animals. The food sector has become increasingly dynamic and competitive and strives to introduce freshly prepared food with remarkable qualities. In other to meet this objective, various food producers have at their disposal several methods and technologies available for processing their foods. It is any conscious change in food material prior to its availability for consumption (Knorr and Watzke, 2019). It can, therefore, be as simple as drying or freezing or as complex as the formulation of a frozen diet using the right amount of component nutrients. Food processing involves the transformation of food materials into forms that can be utilized. It cuts across the act of processing essential raw materials into food through different chemical and physical processes (Aguilera, 2018). Several innovations in the area of food processing have given rise to new and more attractive products such as freeze/dry coffee, concentrated fruit juice, and instant foods. There are various activities involved in the process which include liquefaction, canning, cooking, mincing, pickling, emulsification, and macerating. Clean harvested plant materials or slaughtered or butchered products from animals are required for the production of food products that

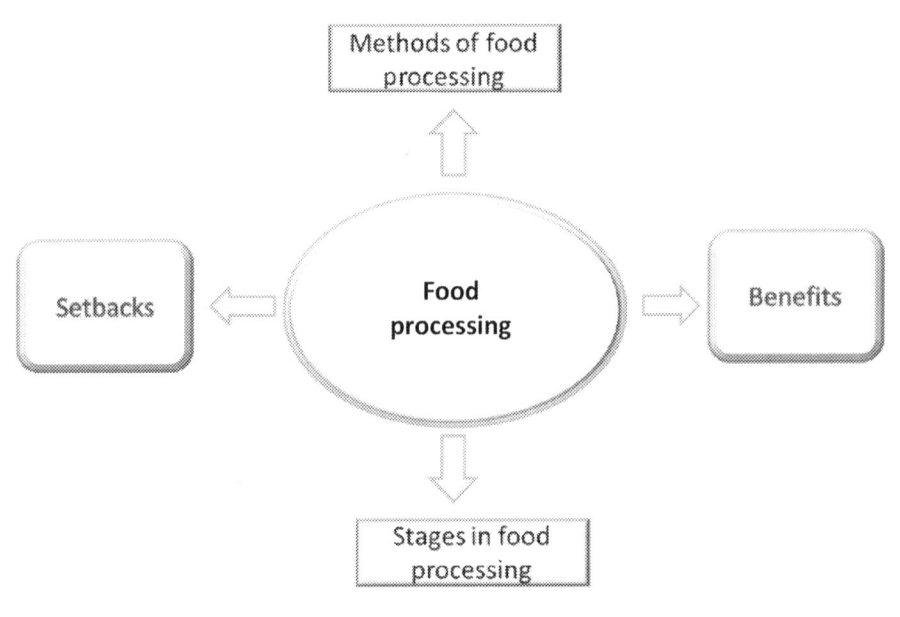

FIGURE 12.1

Schematic representation of chapter.

are of required market qualities (Watzke, 2018). In some cases, the processing of food materials could also result in a reduction in the nutritional qualities of the food and also introduce various organic and inorganic contaminants of health concern into the food. Food processing has been identified to have certain primary objectives which include improvement in the shelf life of food materials for efficient preservation, prevention of microbial contamination of food materials, increase in the overall qualities of food by adding value to their attractiveness and marketability, ensuring efficient storage and easy transportation of food materials over long distance and for long duration, and the provision of employment opportunity for a larger portion of the population (ISO, 2017). The preservation of food is vital in food processing and is carried out in order to deter the growth and activities of various microorganisms, bacteria, and fungi. It involves the reduction of the rate of fat oxidation which could bring about the occurrence of rancidity (Jermann et al., 2015). This chapter, therefore, highlights the concept of food processing (Fig. 12.1). It discusses the various stages and methods involved in the transformation of food into higher market value with greater attractiveness. Some of the limitations associated with the processing of food are also presented.

12.2 Stages in food processing

Food processing can be classified into primary, secondary, and tertiary stages. In primary food processing, the target is to make the food edible for human consumption.

In primary food processing, agricultural products such as livestock and raw materials like wheat kernels are converted into edible forms. This basically includes traditional processes like drying, milling, winnowing, threshing, shelling of nuts, and butchering of meat from animals. Other activities in primary processing include milk pasteurizing, cutting and deboning of meat, smoking and freezing, candling of eggs, and homogenizing (Awad et al., 2012).

In secondary food processing, the food is turned into common foods such as bread. There are many criticisms for the tertiary stage because it involves the addition of enough salt, sugar, and reduction of fiber, reducing the quality of the food from a health point of view. Secondary food processing consists of the daily processes during which food is made from various ingredients which are ready for usage. For example, baking bread, making wine from grains, and fermentation of fish amongst others. Secondary processed meat includes sausage that is made from grinding meat that has already passed through the primary processing stage. Most of the known secondary methods of food processing are described commonly as cooking methods.

Tertiary process of food processing is the large-scale or commercial production of food materials described as processed food (Raso et al., 2016; Jager and Knorr, 2017).

12.3 **Traditional techniques of food processing**

12.3.1 **Homogenization**

Homogenization is a vital technique commonly employed and a vital operational step in the manufacturing and processing of various dairy food materials. The use of homogenization aids in the performing of numerous roles which include dispersion, emulsification, reduction in size of particles, mixing, dissolution, and encapsulation. Homogenization is achieved through the use of various instruments. In ultrasonic homogenization, a mechanical wedge or transducer is employed for the generation of vibration at a very high frequency and the resulting output is not very audible. This results in the production of shock waves and cavitations within the mixture. As the frequency decreases, the overall efficiency falls. The transducer technique has been employed on a small scale during the production of emulsion for disruption of cells, while the wedge technique is employed in the emulsification of food products during processing (Hu et al., 2017).

12.3.2 **Pasteurization**

Pasteurization is one of the most prominent approaches to food processing and involves the use of heat treatment in the mild state of the food material or product, to kill pathogens and various microorganisms present, thereby, increasing the overall shelf life of the food. This technique of food processing and preservation has been used for several centuries. It is more commonly utilized for the preservation of milk. The use of this technique in the preservation of milk has helped in the reduction of various diseases that are easily contagious through the consumption of contaminated milk and related products. This technique for food preservation is also suitable for the processing of other food materials such as almonds, canned food, cider, juices, beer, cheeses, eggs, and butter (Kumar et al., 2019).

12.3.3 Canning

Canning is a technique of preservation that involves the placement of food in similar containers or jars after which they are heated to a temperature at which the various microbes present have the ability to spoil the food and food products are killed. The process involves the driving out of air at the temperature of operation, resulting in the formation of a vacuum seal. The formation of the vacuum seal aids in the prevention of air from going back into the food product as well as the microorganisms present in the air. After the initial processes such as cleaning and other preliminary treatments, the can filling is done automatically using machines. The cans are filled with solid materials to be preserved followed by a liquid component usually syrup or brine which helps in replacing the quantity of air in the can (Hu et al., 2017).

12.3.4 Drying

Drying aids in the removal of moisture from food and food products, so that various microorganisms such as mold, yeast, and bacteria are prevented from growing and spoiling the food products. The drying of food also aids in slowing down the metabolic activities of various enzymes though does not subject them to total deactivation. After the food is ready for consumption, the water is then added back to help in the returning of food to its original shape. Drying has been used for a long as a suitable substitute for freezing and canning and also used in complementing the previously mentioned methods. The drying of food is simple, easy, and highly economical (Kumar et al., 2019).

12.3.5 Smoking

Smoking is a technique that is related to drying but unique in that is, it also aids in positive and desirable impacts on the flavor and taste of the food. It has been widely and successfully used for various fish and meat products in this regard. Smoking helps in keeping microbes away from the food material during the process of drying them. The primary target of smoking is to remove moisture from food and food products. The removal of moisture helps in the prevention of fungal growth which has the potential of ruining the food. There are different methods of smoking food (Inobeme et al., 2019a, b).

12.4 Benefits of food processing

There are several benefits that are associated with the processing of food. Some of these benefits are discussed. Generally, the processing of food helps in increasing

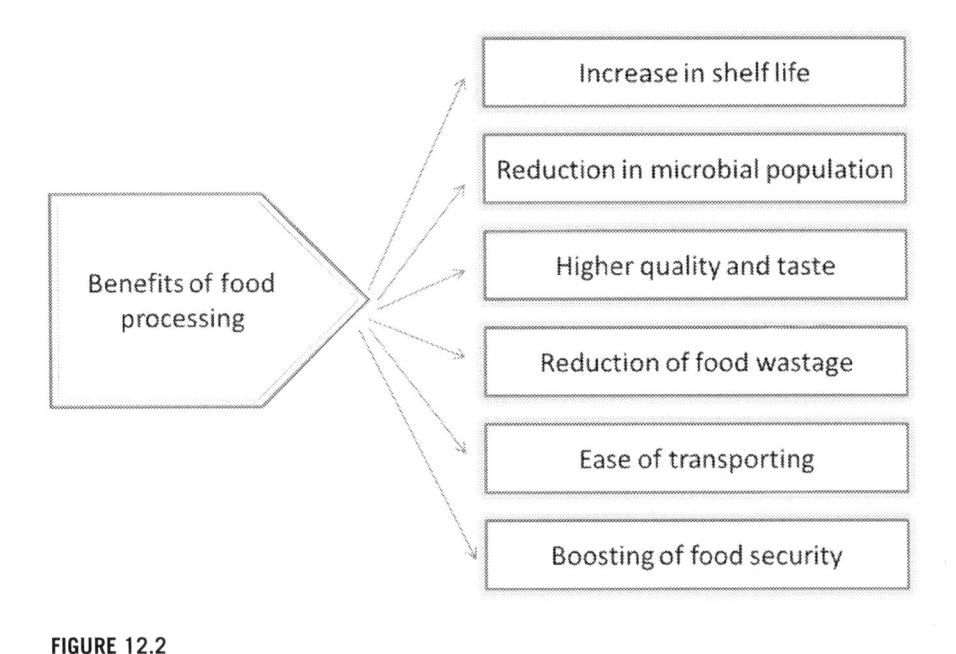

FIGURE 12.2

The benefits of food processing.

their shelf life, so that they can be stored for a long period of time and easily transported over long distances. The processing of food materials also brings about the reduction of the population of microbes such as bacteria that are responsible for various diseases and food poisoning. This is because in most of the food processing mechanisms, there is dehydration of the food material as well as alteration of the pH of the medium, thus, preventing the growth and activities of various microorganisms. Processing also brings about a significant reduction in health inequality and related health concerns (Gibney et al., 2017). Some of the processes of food processing are also vital in the reduction of food wastage, thereby ensuring a reduction in the overall environmental impact of agriculture and aiding food security. Processing of food helps in the reduction of the incidence of diseases that are foodborne. Fresh food materials in their raw forms tend to harbor germs and various microbes and, hence, can cause disease outbreaks when compared to processed ones. The act of food processing also enhances higher quality of the food taste. Processed food also tends to free people from the long time needed for cooking and preparing food from its raw form before consumption. Food processing, therefore, makes food safer for consumption through the destruction of toxins and the inhibition of various pathogens. Heating processes like pasteurization, also help in the destruction of various microbes (Monteiro, 2015). The various benefits of food processing are shown in Fig. 12.2.

There are certain performance parameters for food processing. Such parameters must be taken into consideration during the design of processes for the food industry.

These include hygiene, energy efficiency, waste minimization, and required labor. The hygiene parameter is a measure of the number of microorganisms for each mL of the finished product. The energy efficiency expresses the energy, for instance, a ton of steam connected with each ton of sugar that is produced (Frauster et al., 2018).

12.5 Importance of food processing and preservation

Almost all food materials are processed in one way or the other prior to consumption. Thus, virtually all foods undergo various kinds of processing with the primary objectives of improving their overall quality, keeping them free of microbes, and thereby ensuring human safety. With the growing human population and the quest for novel approaches to food storage and preservation, food processing, therefore, becomes paramount. There are various groups of microorganisms present in the environment. The majority of these disease-causing microorganisms are conveyed through the consumption of contaminated food materials such as milk, dairy products, meat, fruits, and vegetables amongst others. Processing of food, therefore, aids in ensuring food free from pathogens which helps in curtailing the spread of various diseases. Processing of food ensures the safety of humans from food poisoning.

12.6 Methods of food processing

Various methods are available for the processing of food. Some of these methods basically include removal of outer layers from the raw food materials, slicing or mincing, fermentation of food materials, liquefaction, emulsification, mixing, cooking, and introduction of gases as in drinks, spray drying, packaging, pasteurization, and proofing.

Some of the methods commonly employed for the preservation and processing of food include are detailed below.

12.6.1 Drying

This is one of the oldest approaches to food processing. It involves the exposure of the food materials to sun energy, so as to reduce their moisture, thereby drying them. The moisture in the food materials is evaporated, thereby making it difficult for microorganisms to thrive. Moisture removal can also be achieved through the blowing of hot air. Drying has been effectively employed in the processing of various grains such as rice, wheat, barley rye, grams, and oats (Raso et al., 2016).

12.6.2 Cooling

This also aids the slowing down of microbial and enzymatic activities which are responsible for the decomposition and spoilage of food. Food products such as fish, dairy products, and meats are commonly stored in refrigerators, so as to increase their shelf life.

12.6.3 Sugaring

This is mainly used for the processing and preservation of fruits. It can be effectively used for plums, apples, and peaches. During this process, the food products are cooked and sugar is continually added until the formation of crystals after which it is stored in a dry atmosphere. At present, sugar is used in combination with alcohol for the production of some brands of spirits and alcohols (Knorr and Khoo, 2014).

12.6.4 Vacuum pack

This involves the packing of food products in bottles and air bags that are tight within a vacuum space. Oxygen is paramount for the activities of microorganisms which are responsible for food spoilage. Thus, in vacuum pack method, the vacuum environment does not contain oxygen, hence, microorganisms cannot thrive. This method has been used successfully for the preservation of nuts.

12.6.5 Smoking

Smoking is a traditional approach to food preservation that has been employed for a long. It involves the introduction of smoke from various sources such as wood, and charcoal amongst others. The process of smoking brings about a reduction in the moisture of the food material and also introduces a characteristic flavor which is the reason why a larger population demands some smoked food products. Smoking is a simple process that also enhances the taste of the food products. It is commonly used for the processing and preservation of meat, fish, and related products.

12.6.6 Freezing

This approach has been widely employed in the preservation of food and food products. It is a suitable method for broad varieties of food materials such as meat, fish, and related products. It is suitable for products containing moisture, under freezing conditions there is total solidification which stops the activities of microbes and enzymes associated with the decomposition of food. Freezing, however, should be used with caution for some food materials as rapid freezing could adversely affect the texture.

12.6.7 Pickling

This involves the use of edible and antimicrobial liquid for the preservation of food. There are two major types of pickling, which are the thermal and fermentation pickling. In fermentation pickling, there is the production of organic compounds through the action of the microorganisms present in the liquid. The food can also be stored in a liquid which does not permit the growth of microorganisms. The food material is cooked inside chemicals that can attack microbes. This involves a careful selection of chemical substances which do not support the growth of microorganisms but are edible by humans. Commonly used chemicals for this purpose include ethanol, brine, vegetable oil, and vinegar. Pickling is used for the preservation of vegetables like pepper, tomatoes, and cabbage. It can also be used for eggs and beef.

12.6.8 Salting

Salting involves the addition of salt to the food material, thereby aiding the dehydration of the food and creating an environment that is not conducive for microbes. It involves osmotic processes during which water is drained out of the food material. This is mostly employed for the processing of fish and meat products.

12.6.9 Poaching

This is a gradual and gentle method of processing food in which the food is immersed inside a hot liquid at a moderate temperature. The low-temperature functions best for food items that are delicate and there is preservation of flavor and moisture without the need for oil and fat. This method is suitable for the processing of eggs, fish, fruits, and poultry. Pouching is very vital in the preservation of fish such as tilapia, sole, and cod. In most cases, court bouillon, a unique broth is added for the purpose of adding flavor. In the processing of fruits, a sweetened liquid is employed for apples and pears for a special dessert (Priyadarshini et al., 2019).

12.6.10 Simmering

This is a relatively gentle approach to food processing which is carried out at a relatively higher temperature when compared to poaching. The temperature employed lies lower than the boiling point and gives rise to tiny bubbles. In order to achieve a simmer, water is first heated to its boiling point and then the temperature is lowered. Simmering can be carried out for vegetables, grains, rice, and meat.

12.7 Setbacks in food processing

There are certain drawbacks that are connected with the processing of food which have become issues of concern. The processing of food materials using some methods involves the addition of various chemical substances which are artificial supplements and ingredients. Most of these substances tend to be toxic to tissues on accumulation. Common ingredients for preservation include various compounds of sodium and nitrogen which are of serious health concern, especially for a majority of the human population with certain degenerative diseases such as high blood pressure, and diabetes amongst others. The increasing use of sugar endangers the diabetics. Unhealthy diets containing high amounts of fats, salt, and added sugar as in those in processed food can bring about an increase in cancer risk and type 2 diabetes. There are also a large number of resources that are used up in order to make the food products to be pleasant for consumption and this usually brings about overconsumption of some ingredients (Bari et al., 2017). Most processed food also contains a large amount of white sugar, and this is considered unsafe for a larger portion of the human population. The processing of food results in a decrease in their nutritional quality as well as the introduction of trace amounts of some contaminants. The quantity of nutrients lost during the processing of food depends on the method adopted and the nature of the food. For example, the vitamin C content of raw fruits is significantly higher than that of the processed canned drink. This is due to the thermal degradation of vitamin C during heat treatment (Chemat et al., 2017).

12.8 Nonthermal technologies for food processing

Nonthermal techniques in food processing refer to the various technologies of food processing that do not involve the use of heat. They are numerous with the most prominent being the use of pulsed light, pulsed electric field, oscillating magnetic field (OMF), cold plasma, irradiation, ultrasound, and high-pressure processing. Each approach has its own merits as well as inherent limitations. The utilization of nonthermal approaches in the preservation of food and food products is targeted at retaining some of the unique properties of the food such as taste, color, nutritional constituents, and appearance (Ahmad et al., 2021).

12.8.1 Pulse light technique

This is a nonthermal technique in food processing and preservation which involves the decontamination of food materials such as meat, meat products, fruit juices, and vegetables through the use of light of very high intensity within a short period of time. It involves the use of pulsed light with wavelengths lying between 200 nm and 1100 nm. Some of the major benefits of this technique are that it brings about a remarkable reduction in the microbes within the shortest time, it is an environmentally friendly approach, and highly flexible in its application.

12.8.2 Cold plasma

This is a highly recent and novel approach to nonthermal food processing. It makes use of gases that are highly energetic and reactive which immediately deactivate the pathogens present in meat, fruits, poultry, and dairy and vegetable products. This method is highly sanitizing and flexible since it makes use of various carrier gases such as nitrogen, air, oxygen, helium, and electricity. Being a new technique, there are limited studies on the impact of this approach on the general properties of the food material (Jadhav et al., 2021).

12.8.3 Ultrasound

Ultrasound is a nonthermal process which can be employed for the inactivation of microbes as well as various enzymes present in food in order to enhance their shelf life while retaining their quality. The quick formation and collapsing of the bubbles produced by the waves from the ultrasound are responsible for the antimicrobial potential of the process (Jadhav et al., 2021).

12.8.4 Irradiation

This involves the use of ionizing radiation on food products. It is a technology that is aimed at extending the shelf life as well as the safety of food products through reducing various pathogens and insects present in the food products. Irradiation, therefore, helps in making food safer for human consumption.

12.8.5 Magnetic field processing

Magnetic fields are currently being applied for the preservation and processing of food. The primary essence is the reduction of microbial activities. There are various kinds of equipment that are used for achieving this, such as OMF, static magnetic field (SMF) and this depends on the type of equipment that is used.

12.8.6 High pressure techniques

This is a novel approach to food processing that does not involve the use of heat. In this process, the food products are subjected to very high pressure for a very short duration, ranging from seconds to minutes. The foods are pressurized through indirect and direct methods through the use of media that transmit pressure such as water. The process brings about the inactivation of microbes and enzymatic activities (Jadhav et al., 2021).

12.9 Conclusion

The present chapter has highlighted the various methods of food processing as well as the advantages and limitations of processing foods. Some processes of food processing such as heat treatment could bring about the introduction of various organic and inorganic compounds such as polycyclic aromatic hydrocarbons, polycyclic amines, and other chlorinated compounds. It is, therefore, paramount to continually monitor the impact of the different methods of food processing on the introduction of various contaminants into the food. There is also a need to continually evaluate the nutritional values of processed food for effective deduction of the impact of the process on the nutritional quality of the food. There is a further need for studies focusing on the optimization of various methods of processing to promote the beneficial impacts and to eliminate undesirable outcomes.

References

Aguilera, J.M., 2018. Relating food engineering to cooking and gastronomy. Compr. Rev. Food Sci. Food Saf. 2018, 12361. https://doi.org/10.1111/1541-4337.12361.

Ahmad, J., Ali, M.Q., Arif, M., Iftikhar, S., Robina, H., 2021. Traditional and modern techniques for food preservation. Int. J. Mod. Agric. 10 (3), 2305–7246.

Awad, T.S., Moharram, H.A., Shaltout, O.E., Asker, D., Youssef, M.M., 2012. Applications of ultrasound in analysis, processing and quality control of food: a review. Food Res. Int. 48, 410–427. https://doi.org/10.1016/j.foodres.2012.05.004.

Bari, L., Grumezescu, A., Ukuku, D., Dey, G., Miyaji, T., 2017. New food processing technologies and food safety. J. Food Qual. 2017, 3535917. https://doi.org/10.1155/2017/3535917.

Chemat, F., Rombaut, N., Meullemiestre, A., Turk, M., Perino, S., Fabiano-Tixier, A.-S., Abert-Vian, M., 2017. Review of green food processing techniques. Preservation, transformation, and extraction. Innov. Food Sci. Emerg. Technol. 41, 357–377.

Fauster, T., Schlossnikl, D., Rath, F., Ostermeier, R., Teufel, F., Toepfl, S., et al., 2018. Impact of pulsed electric field (PEF) pretreatment on process performance of industrial French fries production. J. Food Eng. 235, 16–22. doi:10.1016/j.jfoodeng.2018.04.023. https://doi.org/10.1016/j.ifset.2017.04.016.

Gibney, M.J., Forde, C.G., Mullally, D., Gibney, E.R., 2017. Ultra-processed foods in human health: a critical appraisal. Am. J. Clin. Nutr. 106 (3), 717–724. doi:10.3945/ajcn.117.160440. Epub 2017 Aug 9. Erratum in: Am. J. Clin. Nutr. 2018 Mar 1; 107 (3), 482–483.

Hu, Y., Ting, Y., Yu, J., Hsieh, A., 2017. Techniques and methods to study functional characteristics of emulsion systems. J. Food Drug Anal. 25 (1), 16–26.

Inobeme, A., Ajai, A.I., Mann, A., Iyaka, Y.A., 2019a. Effect of smoking on heavy metal content of fish. J. Faculty Food Eng. 4 (2), 395–403.

Inobeme, A., Ajai, A.I., Mann, A., Iyaka, Y.A., 2019b. Determination of polycyclic aromatic hydrocarbons and heavy metal contents of barbecue beef, fish and chicken. J. Faculty Food Eng. 4 (2), 395–403.

International Standard Organization, 2017. Definition and Technical Criteria for Food Ingredients to be Considered as Natural. International Standard Organization.

Jadhav, H.B., Annapure, U.S., Deshmukh, R., 2021. Non-thermal technologies for food processing. Front. Nutr. Sec. Food Chem. 8, 1–14. https://doi.org/10.3389/fnut.2021.657090.

Jäger, H., Knorr, D., 2017. Pulsed electric fields treatment in food technology: challenges and opportunities. In: Miklavcic, D. (Ed.), Handbook of Electroporation. Springer, Cham, pp. 2657–2680.

Jermann, C., Koutchman, T., Margas, E., Leadley, C., Ros-Polski, V., 2015. Mapping trends in novel and emerging food processing technologies around the world. Innov. Food Sci. Emerg. Technol. 31, 14–27.

Khoo, C.S., Knorr, D., 2014. Grand challenges in nutrition and food science technology. Front. Nutr. 1, 4. https://doi.org/10.3389/fnut.2014.00004.

Knorr, D., Watzke, H., 2019. Food processing at a crossroad. Front. Nutr. 6. https://doi.org/10.3389/fnut.2019.00085.

Kumar., A., 2019. Food preservation: traditional and modern techniques. Acta Sci. Nutr. Health 3 (12), 45–49.

Monteiro, C.A., 2009. Nutrition and health. The issue is not food, nor nutrients, so much as processing. Public Health Nutr. 12, 729–731. https://doi.org/10.1017/S1368980009005291.

Priyadarshini, A., Rajauria, G., O'Donnell, C.P., Tiwari, B.K., 2019. Emerging food processing technologies and factors impacting their industrial adoption. Crit. Rev. Food Sci. Nutr. 59 (19), 3082–3101. doi:10.1080/10408398.2018.1483890, Epub 2018 Nov 21. PMID: 29863891.

Raso, J., Frey, W., Pataro, G., Knorr, D., Teissie, J., Miclavcic, D., 2016. Recommendations and guidelines on key information to be reported in studies of application of PEF technology in food and biotechnology processes. Innov. Food Sci. Emerg. Technol. 37, 312–321. https://doi.org/10.1016/j.ifset.2016.08.003.

Watzke, H., 2018. New maps for healthy dietary trajectories and food product innovations. Sight Life Magaz. 32, 78–83. https://sightandlife.org/wp.

Microbial assessment and other considerations in food processing

13

Arezoo Ebrahimi and Anna Abdolshahi

Food Safety Research Center (Salt), Semnan University of Medical Sciences, Semnan, Iran

13.1 Introduction

Microbial food safety has been noticed as a world concern. In response to the rise of foodborne disease and microbial food spoilage, efforts have focused on improving food safety. In this case, food industries were forced by regulations in the agri-food chain to design food processing according to food safety targets. The control actions for keeping food products at acceptable limits should be paid attention to every production stage by food processors (Sevindik and Uysal, 2021). Knowing the optimum conditions that influence microbial growth and their changes in food products is necessary for ideal performance. A microbial assessment scheme can be considered to evaluate all food processing conditions and check the critical control point. Microbial contamination of food could occur and be distributed by different routes. Therefore, the origins of contamination must be assessed in every food production chain. A food safety management system (FSMS) could conduct an overall assessment (Caffrey et al., 2019; Sharif et al., 2018).

Biochemical mechanisms have evolved by various microbial agents that can grow on food to digest food components and provide energy for their growth. Nevertheless, in a particular food, typically, just one or a few types of spoiling agents will grow enough to become the predominant microorganism (Mossel and Ingram, 1955). The nutritious nature of foods promotes microbial growth, including human bacterial pathogens, which contribute to food-borne illnesses (Le Guyader and Atmar, 2008).

13.2 Microbial growth condition

Microbial growth in foods can be affected by the most critical factors, including intrinsic, extrinsic, and interactions among these items (Fig. 13.1).

13.2.1 Intrinsic factors

Parameters, which are associated with foods, are characterized as intrinsic factors. Some noteworthy examples of these factors include moisture content or water activity,

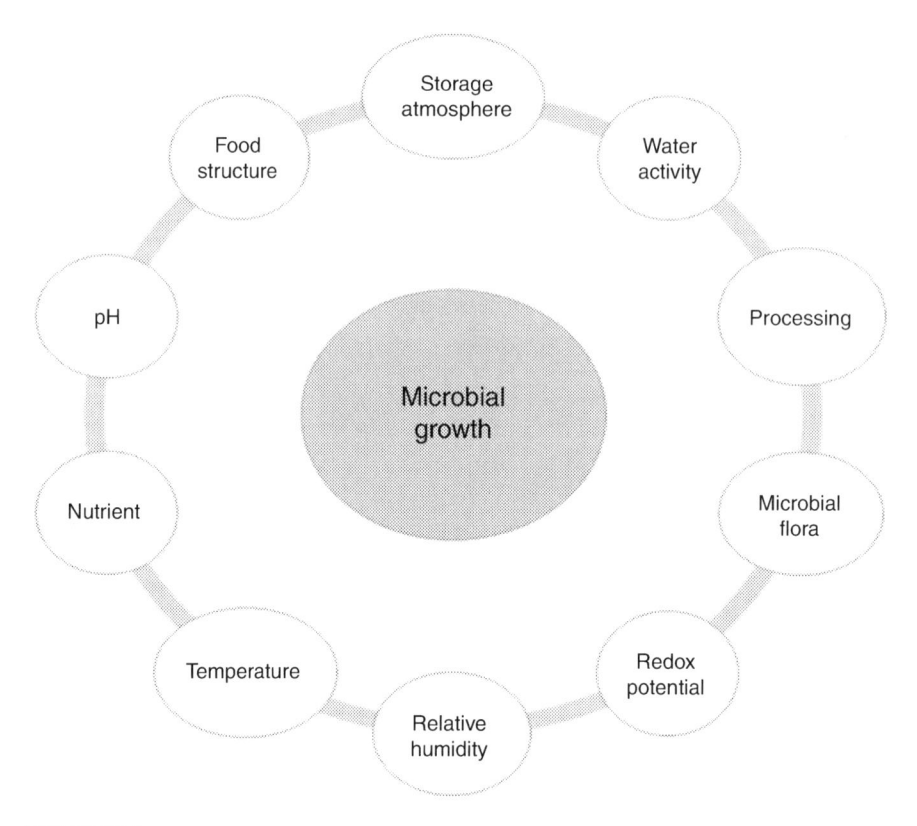

FIGURE 13.1

The main factors influence on microbial growth.

hydrogen ion concentration or pH levels, food composition, antimicrobial components, biological structures, and oxidation-reduction (redox) potential. Each one of these is detailed below.

13.2.1.1 Moisture content or water activity (a_w)

The water requirements of microbial components in food products are usually labeled the water activity (a_w) of the foods. It is determined as the water vapor pressure of food divided by the vapor pressure of pure water at the same temperature (Jay, 2000). The optimum a_w for the growth of the most spoilage bacteria is above 0.91, while for spoilage molds, it is even 0.80. Food-borne bacteria, halophiles (salt-tolerant), have been recorded to grow at a_w values of 0.75, whereas osmophilic yeasts and xerophilic molds can grow at a_w values of 0.61 and 0.65.

13.2.1.2 Hydrogen ion concentration (pH)

In food systems, the growth and preservation of the microorganisms can be critically affected by the pH levels. Most microorganisms have optimum growth around neutral pH levels, with a few being able to grow in acidic conditions (Jay et al., 2008). In general, the widest range of growth belongs to molds, followed by yeasts, whereas bacteria have a confined pH range, especially pathogenic food-borne bacteria (Jay, 2000). Moreover, Gram-positive bacteria are observed to be more resistant to extreme pH levels than Gram-negative ones (Ray and Bhunia, 2008).

13.2.1.3 Composition of foods

The growth and performance of microorganisms depend on specific nutritional components. The main key parameters, that have to be considered, are water content, energy substances in food (carbohydrates, lipids, and proteins), nitrogen substances (amino acids and nucleotides), vitamins, minerals, and growth parameters (Jay et al., 2008). In general, almost all organisms will utilize simple components such as amino acids before any breakdown of the more complex nutrients. All microbial agents require water as a conveying medium for metabolic reactions, though water is not fundamentally considered a nutrient. Gram-positive bacteria (i.e., *Listeria* and *Staphylococcus*) have the most outstanding nutritional requirements, followed by yeasts; however, the lowest dietary needs are relevant to Gram-negative bacteria and molds.

13.2.1.4 Antimicrobial substances

Many antimicrobial components are naturally present in foods, whereas others are formed during food processing or artificially added to the food. These certain products delay or inhibit the growth of food-borne microbes that cause disease and spoilage. Essential oils are an example of these defense strategies identified in garlic, cloves, cinnamon, sage, and thyme (Wareing, 2010). Some examples of these compounds in animal-based foods include lysozyme in eggs, the Lactoperoxidase system, lactoferrin, and conglutin in cow's milk (Mossel and Ingram, 1955).

13.2.1.5 Biological structures

Natural biological structures are commonly present in most raw foods derived from plants or animals, providing an excellent barrier against the passage and side effects of spoiling microorganisms (Jay, 2000; Margesin and Schinner, 2001). Examples of such structures include the outer coverings on fruits, nuts, testa of seeds, and the coating layers of eggs.

13.2.1.6 Redox potential (Eh)

The oxidation-reduction potential of a substance determines the ease of the substrate to lose or gain electrons. A potential difference between the donor compound and the acceptor will be generated by transferring electrons between these two compounds, representing the redox potential of these two factors. The type of microbial agents and chemical reactions produced in the food will be influenced by the reducing and oxidizing power of the food. In food systems, the Eh is contributed by oxygen concentration, type of microorganisms associated, chemical composition, and processing treatments (Jay et al., 2008).

13.2.2 Extrinsic factors

Factors regulated through external conditions are referred to as extrinsic factors. Examples of such agents include gaseous environments, relative humidity, temperature, processing operations, and the presence of other microorganisms. Generally, a series of both internal and external parameters is applied to keep food safety and quality. The main extrinsic parameters are described below.

13.2.2.1 Impact of storage atmosphere of the food

Like oxygen (O_2), carbon dioxide (CO_2) is one of the most critical atmospheric gases utilized to control the growth of microbial substances in foods. Gaseous environments with specific levels of O_2, CO_2, and N_2 can be adjusted according to the product and indexed microbe. Modified atmosphere packaging (MAP) controls and preserves the development of disease-causing and spoiling microorganisms and extends the shelf life of packaged food, without using the chemical preservatives (Gould, 2000). Other technologies used include the addition of carbon dioxide (DAC), controlled-atmosphere packaging (CAP), controlled-atmosphere storage (CAS), and hypobaric storage. The other atmospheric gas is ozone (O_3) which has been observed to kill food-related microorganisms.

13.2.2.2 Relative humidity

Relative humidity (RH) of the storage condition is able to affect the quality of food because it leads to changes in the a_w levels. Eventually, there will be a moisture equilibrium between the food and the surroundings; therefore, moisture may evaporate from or condense on the surface of the food. So, when the a_w of food is vital for its shelf life or safety, it should be stored in an environment that does not significantly alter this property.

13.2.2.3 Impact of storage temperature

In food systems, enzymatic reactions and microbial proliferation are influenced by temperature. Food-borne pathogens and spoiling microorganisms can generally grow at developed limits temperatures. The minimum temperature for microbial growth has been recorded as $-34°C$; the highest is above $100°C$. Some spore-producing bacteria for instance *Bacillus stearothermophilus* and *Clostridium tetani* can grow at temperatures above $100°C$. On the basis of the growth temperature range of microorganisms, the following classification has been developed: psychrophiles (psychrotrophic) ($0°C$ to $20°C–30°C$), mesophiles ($20°C–45°C$), and thermophiles ($55°C–65°C$) (Jay et al., 2008).

13.2.2.4 Other microbial flora

Multiple microbial agents in a food system can compete for the same nutrients and produce toxic by-products that can inhibit the growth of contending species. These secondary metabolites include bacteriocins, antibiotics, organic acids, and harmful metabolites toxic to many bacteria. The chemical properties of the food product can also change, leading to the prevention of some microbial agents (Kornacki, 2010).

13.2.2.5 Processing steps

Treatments like heating, cooling, and drying are various processes affecting food composition and types and number of spoiling and pathogenic microorganisms that remain in the food after treatment. By storing contaminated raw food containing heat-resistant and heat-sensitive microbes at room temperature, it can be spoiled by heat-sensitive microorganisms, as these conditions cause fast growth of mesophiles strains.

13.2.3 Interactions of factors

The microbial growth in foods can also be influenced by the interaction of the multiple parameters explained above, and these effects may be synergistic or additive. This phenomenon has been referred to the food-preservation strategy as the "hurdle concept." The essential parameters employed in this technology are the intrinsic, extrinsic, and processing factors. Predictive food microbiology has constantly demonstrated that many inhibitory factors (hurdles) that may not inhibit microbial growth are considered individually adequate if combined. The hurdle technology is used in foods such as jams and jellies. During the processing, heating, low a_w, low pH, and anaerobic packaging are applied to prevent the growth of different organisms even if it is stored at room temperature.

13.3 Food-borne pathogen microorganisms

A robust connection between food consumption and human illnesses was recognized so early. Hippocrates (460 B.C.) described an association between food consumed and human disease (Hutt and Hutt, 1984). Food-borne pathogens, such as bacteria, viruses, and parasites, are biological agents that can cause food-borne illness events when ingested.

Food-borne illness outbreaks are described as the incidence of some cases of close diseases associated with the ingestion of a common food (Centers for Disease Control and Prevention. Summary of Notifiable Diseases — United States, 2012). Food-borne illnesses occur by consumption of pathogenic agents in food or beverages. These causatives establish themselves and usually multiply in the human host, or toxigenic agents establish themselves in food and produce toxins, and then will be ingested by the human host. Therefore, food-borne diseases are generally categorized into: (1) food-borne infections and (2) food-borne intoxications. In food-borne intoxications, the time from ingestion until symptoms occur is much faster than in food-borne infections (Fig. 13.2).

Most food-borne diseases can be inhibited by food processing or cooking the food to destroy pathogens. Every day we face new challenges in the fight against foodborne

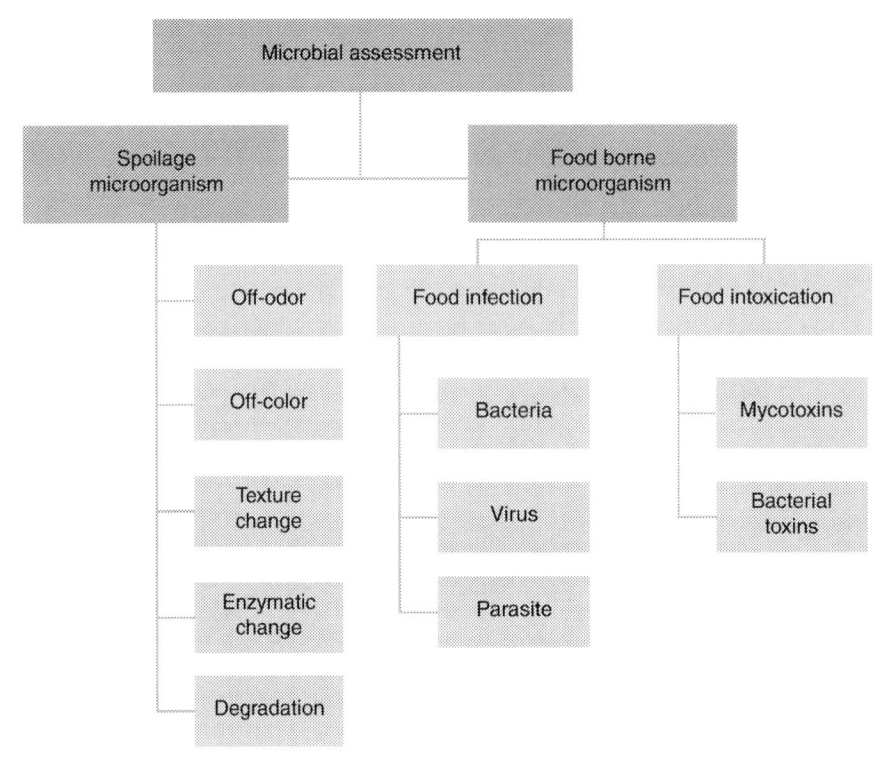

FIGURE 13.2

Microbial assessment in food product.

illnesses due to the globalization of the food market, the variation of climate, and people's food consumption patterns as consumers these days prefer fresh or minimally processed foods (Schelin et al., 2011).

Food can be contaminated by many different disease-causing microorganisms, so they are responsible for more than 200 various food-borne diseases that have been identified (Mead et al., 1999).

Although viruses cause more than 50% of all food-borne diseases, hospitalizations, and deaths are usually due to bacterial agents. According to the results of a report in the European Union (EU) for 2015, most of the outbreaks were caused by bacterial agents, particularly *Salmonella* spp. and *Campylobacter* spp. (21.8% and 8.9% of all outbreaks, respectively). Bacterial toxins were the second reason for food disease outbreaks (19.5% of the total outbreaks), whereas 9.2% of actual outbreaks accounted for viruses. Parasites and other causative agents, mainly histamine, were ranked less than 3% of the reported outbreaks. Furthermore, the causative agent in a third of the total attacks (34%) remained unknown (European Food Safety Authority and European Centre for Disease Prevention and Control, 2016). The EU summary report on trends and sources of zoonoses.

In 2008, it was estimated that gastrointestinal diseases were responsible for 2.0 million deaths occurred worldwide (Fleury et al., 2008). However, the exact rate of mortality from food-borne diseases is difficult to determine (Helms et al., 2003).

13.3.1 Characteristics of bacterial agents, viruses, and parasites

13.3.1.1 Bacterial agents

As the most common sources of food-borne illnesses, bacterial agents come in a great variety of shapes, types, and properties. Spore-forming bacteria survive harsh conditions like heat and antibiotic treatment (e.g., *Bacillus cereus, Bacillus subtilis, Clostridium botulinum*, and *Clostridium perfringens*). Others such as *C. botulinum* and *Staphylococcus aureus* can produce heat-resistant toxins (Bacon and Sofos, 2003; Bintsis, 2017).

Salmonella was described in the 20th century as the leading documented cause of food-borne illnesses followed by *Campylobacter*, entero-pathogenic *Escherichia coli*, and *Listeria* (Nyenje et al., 2012). In the United States, the bacteria that frequently cause food-borne toxicities include *B. cereus, S. aureus*, and *C. botulinum*. In the last decade of the 20th century; however, the most often source of bacterial food-borne diseases in France was recorded by *Salmonella, Campylobacter*, and *Listeria*, respectively (Vaillant et al., 2005).

Mesophilic bacteria with the optimal growth temperature 20°C–45°C constitute the majority of pathogens. Nevertheless, some food pathogens (characterized psychrotrophs) like *Yersinia enterocolitica* and *Listeria monocytogenes* are able to grow under refrigerated environments or even less than 10°C (Bacon and Sofos, 2003). For example, *L. monocytogenes* will grow at temperatures just above the freezing point, albeit slowly. Some strains of *C. botulinum, Salmonella* spp., *B. cereus, S. aureus*, and *E. coli* O157:H7 may also be capable of growth slowly under refrigerated temperatures.

13.3.1.2 Viruses

Viruses are particulate and need other living cells to grow and multiply. So, they are not able to survive outside the host for long periods. Even though more than 100 types of enteric viruses have been implicated in food-borne diseases, noroviruses, Norwalk viruses, and hepatitis A are the most frequent food-borne virus pathogens.

The mentioned viruses are frequently transmitted through food, especially seafood. For example, in the shellfish-growing waters, which are progressively subject to sewage discharges and human fecal contamination, the shellfish collect these viruses in their digestive tract. Thus, it is more difficult to remove the viruses during thermal inactivation and processes proposed to cleanse them (DiGirolamo et al., 1970; Grohmann et al., 1981). Furthermore, unlike other seafood, they are usually eaten raw and have their digestive tracts in place.

13.3.1.3 Parasitic protozoa

They are single-celled creatures without a firm cell wall but a separate nucleus. These unicellular organisms are more significant than bacteria and only multiply in their hosts and the virulent form of them is named *cyst. Toxoplasma gondii, Cyclospora cayatenensis*, and *Cryptosporidium parvum* are protozoa that have been responsible for food-borne illnesses.

13.3.1.4 Multicellular parasites

These organisms live at the expense of the host. We termed them in the animal kingdom macro parasites because they are large enough to be seen without a microscope. They may be observed in foods in their immature forms, such as eggs and larvae. Trichinosis is now known as a major pathogen associated with undercooked pork (Mattiucci et al., 2013). Cestodes or tapeworms are other parasites of concern (usually associated with pork, beef, or fish), flatworms or nematodes (associated with fish), or flukes (Schmidt et al., 2003).

13.4 Food spoilage microorganisms

A complex process that causes foods to become unacceptable or undesirable for the consumers, due to changes in sensory characteristics, is called food spoilage. Changes in appearance, taste, smell, or texture in spoiled foods cause them to be rejected, while they may still be safe to eat because there are no pathogens or toxins present. Some ecological studies have proposed that these noxious chemicals produced by microorganisms repel large animals thus the food remains for themselves (Burkepile et al., 2006). In addition, some of the spoiled foods are discarded, resulting in environmental and resource costs (Rawat, 2015).

Investigation of food spoiling microorganisms is helpful in two ways. The first way is attention to biochemical properties and the laboratory tests employed to characterize and differentiate microorganisms broadly. Description of the groups of similar microbes involved in spoiling is the second way (de W Blackburn, 2006).

It should be noted that water, soil, or the intestinal tracts of animals are often mentioned as typical habitats for spoilage agents, and they can be dispersed through the water and air by the activity of tiny animals, primarily insects (Rawat, 2015). Different microorganisms that use food as a source of energy and carbon, mediate chemical reactions that lead to unpleasant sensory changes in foods, with fungi representing the most serious group of spoilage strains. The almost countless biological agents involved in spoiling foods are as follows.

13.4.1 Microorganisms involved in spoilage

Spoiling microorganisms include prokaryotes and eukaryotes. Bacteria, as the most prominent type of prokaryotes, are one-celled without rigid nuclei and other organelles, while eukaryotes, including yeasts (unicellular) and molds (multicellular), have certain nuclei and other organelles. Some of these agents are frequently detected in various spoiled foods; however, others are more selective in choosing the foods they utilize. Some spoiling microbes, like molds and lactic acid bacteria (LAB), secrete chemicals that prevent competitors (Gram et al., 2002).

13.4.1.1 Molds

One of the basic requirements for metabolic metabolism of molds is oxygen. They are well adapted for growth across a broad range of temperatures. Some can grow at deficient a_w levels (0.7–0.8) and a pH range of 3–8. A wide range of foods at every food chain step from field to fork can also be attacked by them. Remarkably, even the spoilage of bottled mineral water has been reported by them (Criado et al., 2005). In the following lines, representative genera of food spoilage molds are briefly explained.

Zygomycetes generally cause rots in bread and a variety of stored fruit and vegetables. The most common spoilage species are *Rhizopus* and *Mucor*. Despite the beneficial effects of *Penicillium* in producing antibiotics and blue cheese, various species are essential spoiling agents, causing spoilage in different fruits and vegetables, and cereals. Some can produce mycotoxins, including ochratoxin, patulin, penitent, and citreoviridin. Furthermore, *Byssochlamys* a high heat resistance fungus, is the most crucial spoilage mold of pasteurized juices. *Aspergillus* is more resistant to low a_w and high temperatures and grows faster than *Penicillium* spp. *Aspergilli* spoil a wide range of foods, for example, grains, peanuts, dried beans, and some spices. Many of them generally produce mycotoxins such as aflatoxins, cyclopiazonic acid, and ochratoxin. Other molds have been isolated from spoiled food; however, they are not significant causes of spoilage. For example, *Fusarium* spp. Produce important mycotoxins but not important spoilage molds.

13.4.1.2 Yeasts

Yeasts can be defined in two wide-ranging categories: oxidative and fermentative. They grow best at water activities above 0.9 and low pHs. As the most widely recognized spoilage yeasts, facultative anaerobic fermentative organisms produce off-odors, carbon dioxide, and ethanol from simple sugars. Some of them (i.e., *Zygosaccharomyces*) are the most halotolerant, and osmotolerant species are able to grow at water activities as low as 0.60 (Martorell et al., 2005). *Debaryomyces* can tolerate salt concentrations up to 24% and are frequently isolated from salt brines. This species is also the most important spoiling agent in salad dressings (Mandrell et al., 2006). *Saccharomyces* spp. Are involved in the spoilage of alcoholic beverages by producing off-flavors, gassiness, and turbidity associated with acetic acid and hydrogen sulfide. Representative genera of aerobic film yeasts include *Candida*, *Mycoderma*, and *Debaryomyces* are capable of growth on fermented foods and metabolize alcohols and organic acids, taking the metabolic characteristics of molds and the morphological characteristics of yeasts. *Candida* can spoil dairy products and fruit and vegetables (Casey and Dobson, 2003). *Dekkera/Brettanomyces* generally are involved in the spoilage of fermented foods such as alcoholic beverages and dairy products by producing volatile phenolic compounds responsible for off-flavors (Couto et al., 2005).

13.4.1.3 Bacteria

Three major genera of spore-forming bacteria are important in food spoilage, including *Bacillus*, *Clostridium*, and *Alicyclobacillus*. They are the predominant spoilage microbes in heat-treated foods like pasteurized foods and improperly sterilized foods due to the survival of their spores against high processing temperatures. Flat sour spoilage of low or high pH canned foods can be caused by *Bacillus* spp. With little or no gas production (Boor and Fromm, 2006). The principal spoilage species of *Clostridium*, including *C. butyricum, C. sporogenes,* and *C. thermosaccharolyticum* are typically involved in the early blowing of cheeses, and spoilage of canned or vacuum-packaged products. Medicinal flavors in canned low-acid foods are usually associated with *Alicyclobacillus* (Chang and Kang, 2004).

Lactic acid bacteria species such as *Leuconostoc, Lactobacillus, Pediococcus,* and *Oenococcus* are helpful in producing fermented foods like salami and yogurt under low pH, oxygen, and temperature become the primary spoiling agents on a variety of foods. These bacteria's growth causes undesirable changes, including bloater damage in pickles, gas formation in cheeses, and greening of meat.

The predominant food spoilage *Pseudomonas* and related genera are *Xanthomonas campestris, Shewanella putrefaciens,* and four species of *Pseudomonas* (*P. fragi, P. fluorescens, P. viridiflava,* and *P. lundensis*). *X. campestris, P. viridiflava,* and *P. fluorescens. Pseudomonads* are the primary spoiling bacteria in eggs, fresh meat, poultry, and seafood.

Enterobacteriaceae include many spoilage organisms such as *Escherichia, Serratia, Erwinia, Enterobacter, Citrobacter,* and *Proteus.* Enteric bacteria are usually associated with eggs, fresh vegetables, meat, poultry, and fish spoilage.

Other bacteria can spoil chilled, high-protein foods, including dairy products and meat. *Psychrobacter and Acinetobacter* are some examples of dominant spoiling bacteria on a variety of spoiled mat and fish. *Acinetobacter* has also been isolated in spoiled soft drinks. These two bacteria are considered to have a low potential concerning health risks and spoilage association.

13.5 **Food safety management**

The food safety management system is developed based on the process control program, procedures, monitoring, and documentation. The results of control of all hazards in the food production chain should be considered in this system. In any food production process from raw material receiving to product distribution, the handlers must be trained to perform operations safely. The FSMS is defined as a series of actions including procedures, training, and monitoring that conducted by food manufacturers to control risk factors (de W Blackburn and McClure, 2009). In this area, the hazard analysis critical control point (HACCP) plan and prerequisite control program would improve food safety management skills. To establish an FSMS, all control points should be monitored. In case of any noncontrolled issues, corrective actions must ensure food products' safety. Effective safety management involves knowing the type of defects, including minor, major, and critical defects. The realization of how a defect is critical for health is crucial in this management system. However, the microbial aspect of food safety has been a world concern and numerous food-borne outbreaks have been recorded annually. Some microbial intoxication and

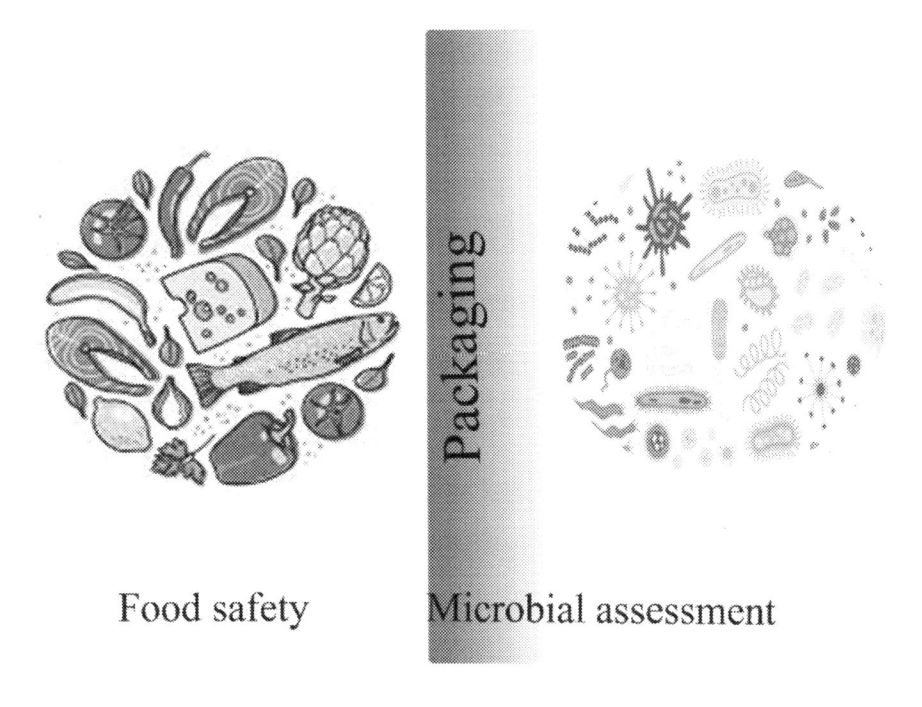

Food safety   Packaging   Microbial assessment

FIGURE 13.3

The impact of packaging on food safety.

infection, referred to as food consumption, make a force for such FSMS in food production lines. FEMS system controls all activities within defined acceptable limits to ensure product and process safety. In this regard, quality assurance (QA) standards like ISO standard series are available. Also, international standards are set explicitly for food processing and food industries. One of the paramount considerations in food processing is accordance with standards. For this reason, good manufacturing practices (GMP) and good hygienic practices (GHP) should be considered in all food processes (Jacxsens et al., 2009; Mensah and Julien, 2011; Sofos, 2002).

Packaging is one of the critical stages in food industries that play a key role in QA. Food packaging generally protects food against influential factors such as oxygen, temperature, light, microorganisms, humidity, physical damage, biological, and chemical contaminants. The quality and safety of a food product are associated with its preservation (Fig. 13.3). Besides, food's shelf-life could be extended by inhibiting food losses and spoilage. Since oxidation, microorganism growth, respiration, and metabolism are leading causes of food deterioration, controlling these events in packaged food directly affects food quality (Bari and Ukuku, 2015; Bomba and Susol, 2020; Stevens and Hood, 2019).

On the other hand, these factors may influence consumer acceptance and economics. By increasing consumer knowledge, the demand for safe products is on the rise. The efforts have resulted in considerable development in packaging materials and packaging methods (Han et al., 2018). Up to now, innovations in packaging industries have provided quality, safety, and shelf-life of food.

13.6 Conclusion

Microbial assessment by evaluating the foodborne and spoilage microorganism's growth in food serves as an applied tool in food safety issues. Regarding a high number of reported foodborne diseases annually, the implementation and development of predictive and detective strategies should assist in food processing. In this case, the sampling of food products, the microbial analysis, the identification of microbiological parameters, and the data analysis are the main actions of the microbial assessment program. The other considerations contribute to the food production line and each process. Packaging is one of the critical process that directly affect the microbial condition of a final food product. Any neglect in the packaging methods and properties of packaging materials also the pre- and postpreservation of packaged food could result in food deterioration. In conclusion, the microbial assessment is a scientific basis for ensuring health and safety.

References

Bacon, R.T., Sofos, J.N., 2003. Characteristics of biological hazards in foods. In: Food Safety Handbook, vol. 10. Wiley, pp. 157–195.

Bari, M.L., Ukuku, D.O., 2015. Foodborne Pathogens and Food Safety. CRC Press.

Bintsis, T., 2017. Foodborne pathogens. AIMS Microbiol. 3 (3), 529.

Bomba, M.Y., Susol, N.Y., 2020. Main requirements for food safety management systems under international standards: BRC, IFS, FSSC 22000, ISO 22000, Global GAP, SQF. Sci. Messen. LNU Vet. Med. Biotechnol. Ser. Food Technol. 22 (93), 18–25.

Boor, K., Fromm, H., 2006. Food Spoilage Microorganisms. CRC Press, Boca Raton.

Burkepile, D.E., Parker, J.D., Woodson, C.B., Mills, H.J., Kubanek, J., Sobecky, P.A., Hay, M.E., 2006. Chemically mediated competition between microbes and animals: microbes as consumers in food webs. Ecology 87 (11), 2821–2831.

Caffrey, N., Invik, J., Waldner, C., Ramsay, D., Checkley, S., 2019. Risk assessments evaluating foodborne antimicrobial resistance in humans: a scoping review. Microb. Risk Anal. 11, 31–46.

Casey, G., Dobson, A., 2003. Molecular detection of *Candida krusei* contamination in fruit juice using the citrate synthase gene cs1 and a potential role for this gene in the adaptive response to acetic acid. J. Appl. Microbiol. 95 (1), 13–22.

Centers for Disease Control and Prevention. Summary of Notifiable Diseases — United States. http://www.cdc.gov.

Chang, S.-S., Kang, D.-H., 2004. *Alicyclobacillus spp.* in the fruit juice industry: history, characteristics, and current isolation/detection procedures. Crit. Rev. Microbiol. 30 (2), 55–74.

Couto, J.A., Neves, F., Campos, F., Hogg, T., 2005. Thermal inactivation of the wine spoilage yeasts *Dekkera/Brettanomyces*. Int. J. Food Microbiol. 104 (3), 337–344.

Criado, M.V., Pinto, V.E.F., Badessari, A., Cabral, D., 2005. Conditions that regulate the growth of moulds inoculated into bottled mineral water. Int. J. Food Microbiol. 99 (3), 343–349.

de W Blackburn, C., 2006. Food Spoilage Microorganisms. Woodhead Publishing.

de W Blackburn, C., McClure, P.J., 2009. Foodborne Pathogens: Hazards, Risk Analysis and Control. Elsevier.

DiGirolamo, R., Liston, J., Matches, J., 1970. Survival of virus in chilled, frozen, and processed oysters. Appl. Microbiol. 20 (1), 58–63.

European Food Safety Authority and European Centre for Disease Prevention and Control, 2016. The European Union summary report on trends and sources of zoonoses, z. a. EFSA J. 14, 4634–4865.

Fleury, M.D., Stratton, J., Tinga, C., Charron, D.F., Aramini, J., 2008. A descriptive analysis of hospitalization due to acute gastrointestinal illness in Canada, 1995–2004. Can. J. Public Health 99 (6), 489–493.

Gould, G.W., 2000. Preservation: past, present and future. Br. Med. Bull. 56 (1), 84–96.

Gram, L., Ravn, L., Rasch, M., Bruhn, J.B., Christensen, A.B., Givskov, M., 2002. Food spoilage—interactions between food spoilage bacteria. Int. J. Food Microbiol. 78 (1–2), 79–97.

Grohmann, G., Murphy, A., Christopher, P., Auty, E., Greenberg, H., 1981. Norwalk virus gastroenteritis in volunteers consuming depurated oysters. Aust. J. Exp. Biol. Med. Sci. 59 (2), 219–228.

Han, J.W., Ruiz-Garcia, L., Qian, J.P., Yang, X.T., 2018. Food packaging: a comprehensive review and future trends. Compr. Rev. Food Sci. Food Saf. 17 (4), 860–877.

Helms, M., Evans, S., Vastrup, P., Gerner-Smidt, P., 2003. Short and long term mortality associated with foodborne bacterial gastrointestinal infections: registry based study commentary: matched cohorts can be useful. BMJ 326 (7385), 357.

Hutt, P.B., Hutt, P.B.I., 1984. A history of government regulation of adulteration and misbranding of food. Food Drug Cosm. Law J. 39, 2.

Jacxsens, L., Kussaga, J., Luning, P., Van der Spiegel, M., Devlieghere, F., Uyttendaele, M., 2009. A microbial assessment scheme to measure microbial performance of food safety management systems. Int. J. Food Microbiol. 134 (1-2), 113–125.

Jay, J.M., 2000. Modern Food Microbiology. Aspen Publishers, Inc., Maryland.

Jay, J.M., Loessner, M.J., Golden, D.A., 2008. Modern Food Microbiology. Springer Science & Business Media.

Kornacki, J.L., 2010. What factors are required for microbes to grow, survive, and die? In: Principles of Microbiological Troubleshooting in the Industrial Food Processing Environment. Springer, pp. 103–115.

Le Guyader, F.S., Atmar, R.L., 2008. Binding and inactivation of viruses on and in food, with a focus on the role of the matrix. In: Food-Borne Viruses: Progress and Challenges, pp. 189–208.

Mandrell, R.E., Gorski, L., Brandl, M.T., 2006. Microbiology of Fresh Fruits and Vegetables. Taylor and Francis Group, New York.

Margesin, R., Schinner, F., 2001. Potential of halotolerant and halophilic microorganisms for biotechnology. Extremophiles 5 (2), 73–83.

Martorell, P., Fernández-Espinar, M.T., Querol, A., 2005. Molecular monitoring of spoilage yeasts during the production of candied fruit nougats to determine food contamination sources. Int. J. Food Microbiol. 101 (3), 293–302.

Mattiucci, S., Fazii, P., De Rosa, A., Paoletti, M., Megna, A.S., Glielmo, A., De Angelis, M., Costa, A., Meucci, C., Calvaruso, V., 2013. Anisakiasis and gastroallergic reactions associated with *Anisakis pegreffii* infection, Italy. Emerg. Infect. Dis. 19 (3), 496.

Mead, P.S., Slutsker, L., Dietz, V., McCaig, L.F., Bresee, J.S., Shapiro, C., Griffin, P.M., Tauxe, R.V., 1999. Food-related illness and death in the United States. Emerg. Infect. Dis. 5 (5), 607.

Mensah, L.D., Julien, D., 2011. Implementation of food safety management systems in the UK. Food Control 22 (8), 1216–1225.

Mossel, D., Ingram, M., 1955. The physiology of the microbial spoilage of foods. J. Appl. Bacteriol. 18 (2), 232–268.

Nyenje, M.E., Odjadjare, C.E., Tanih, N.F., Green, E., Ndip, R.N., 2012. Foodborne pathogens recovered from ready-to-eat foods from roadside cafeterias and retail outlets in Alice, Eastern Cape Province, South Africa: public health implications. Int. J. Environ. Res. Public Health 9 (8), 2608–2619.

Rawat, S., 2015. Food spoilage: microorganisms and their prevention. Asian J. Plant Sci. Res. 5 (4), 47–56.

Ray, B., Bhunia, A., 2008. Fundamental Food Microbiology, fourth ed. CRC. Taylor and Francis, Boca Raton.

Schelin, J., Wallin-Carlquist, N., Thorup Cohn, M., Lindqvist, R., Barker, G.C., 2011. The formation of *Staphylococcus aureus* enterotoxin in food environments and advances in risk assessment. Virulence 2 (6), 580–592.

Schmidt, R.H., Goodrich, R.M., Archer, D.L., Schneider, K.R., 2003. General overview of the causative agents of foodborne illness. EDIS 2003 (6), 1–5. https://doi.org/10.32473/edis-fs099-2003.

Sevindik, M., Uysal, I., 2021. Food spoilage and microorganisms. Turk. J. Agricult. Food Sci. Technol. 9 (10), 1921–1924.

Sharif, M.K., Javed, K., Nasir, A., 2018. Foodborne illness: threats and control. In: Foodborne Diseases. Elsevier, pp. 501–523.

Sofos, J.N., 2002. Microbial control in foods: needs and concerns. Control of Foodborne Microorganisms. CRC press, pp. 1–11.

Stevens, K., Hood, S., 2019. food safety management systems. In: Food Microbiology: Fundamentals and Frontiers, pp. 1007–1020.

Vaillant, V., De Valk, H., Baron, E., Ancelle, T., Colin, P., Delmas, M., Dufour, B., Pouillot, R., Le Strat, Y., Weinbreck, P., 2005. Foodborne pathogens and disease. Fall 2, 221–232.

Wareing, P., 2010. Factors affecting the growth of micro-organisms in foods. In: Micro-Facts. Royal Society of Chemistry, pp. 1–7.

Sustainable novel food packaging in current scenario: An analysis to environmental perspectives

14

Roopa Rani[a]**, Jaya Tuteja**[b] **and Arpit Sand**[a]

[a]*Department of Sciences, School of Science, Manav Rachna University, Faridabad, Haryana, India,* [b]*School of Basic Sciences, Galgotias University, Greater Noida, Gautam Buddh Nagar, Uttar Pradesh, India*

14.1 Introduction

Food, the most important aspect for the survival of life, must be hygienic, nutritionally balanced, and also easily digestible. Thus, its prosperous availability, distribution, consumption, and assimilation are important for good health. Food, generally consumed by humans, consists of many components, possessing several nutrients essentially required for the growth of the body and also building the true health of a person. However, in the current scenario of modernization, urbanization, and industrialization, the life of people has become very tedious making them less available for their daily routine work and engaging them toward the hectic schedule, to earn their livelihood (Rani et al., 2012). However, to be in a state of good health, every person wants to provide one's family with a balanced and nourishing diet. But, in the modern era when people are more focused on their work, especially their career opportunities, they have hardly got time to cook at home, due to which eating habits are also shifting from "cook at home" to "ready to eat food" approach very rapidly. In search of harmonization, many probabilities of packed food have now flooded the market, maybe it is vegetarian or nonvegetarian, but almost everything, either in the form of cooked or semi-cooked state, can now be accessible at the market. Sometimes, in such cases people are getting food with poor quality ingredients or even when its life has been lapsed. Although many times, people are also not aware of the facts of adulteration in food items or their diminished nutritive power, etc. which in turn affect their lives. For such practices, the packaging of food items becomes an important perspective that should aim at proper, rapid, and reliable food distribution that can help reduce malnutrition in our society.

Based on the above facts, packaging has become one of the most important procedures in the food manufacturing industry. The main purposes of food packaging are—to secure it from leakage or breakage until it is finally consumed; to protect it from the damage created by dust, microorganisms, and other contaminants; to make handling more convenient throughout the processes of storage, distribution, transportation, etc.; subsidizing the circulation of food item efficiently; packaging also provide the information of the quality of food item and the condition in which it was stored from the date of manufacture through the labeling or temper indication (Marsh et al., 2007). The packaging system generally clarifies the shelf life (duration of keeping a food item before it becomes unacceptable for consumption) of the food product. The packaging materials used for any particular food item must not only be protective but also accessible with appropriate charge in a specific area.

14.2 Food packaging

The UK Institute of Food Packaging defines the process of packaging in three different ways—first "a coordinated system of preparing goods for transport, distribution, storage, retailing, and end-use," second "a means of ensuring safe delivery to the ultimate consume in sound condition at minimum cost," and third "a techno-economic function aimed at minimizing cost of delivery while maximizing sales." A packaging system can be an enclosure of a food product either in the form of wrapping or in any one of these forms like a box, carry bag, can, tray, tub, tube, pouches, bottles, containers, etc. (Valentas et al., 1997).

The need for the packing system of food items can be related to its nontoxicity, protection against harmful ultraviolet (UV)-infrared radiations (IR), must provide confrontation to physical damage, protection against microorganisms, environmental toxicants, high temperatures, high humidity, can be easy to exposed, easily dispensable, economic, good appearance, disposable, etc.

The functions, shapes, and materials of different packing supplies used for various food products may vary depending on the need and the nature of the food products. It becomes a crucial factor to maintain the balance between the function, quality, and shape of packaging materials. In order to maintain hygienic, healthy, freshness, and quality of food products, efficient packing becomes an important factor will maintain the food quality (Sarkar and Kuna, 2020). The basic functions of the packaging systems can be:

- Protection: It helps in protecting food items from external sources of contamination like rodents, insects, birds, animals, microorganisms, moisture, heat, etc.
- Distribution: It helps in the easy distribution of food materials to different places through an efficient packaging system.
- Unitization: To compile several small packings into a large packing so that the transportation and distribution can be easily made.
- Containment: To pack the food items in a fixed packing of standard weight with proper labeling having "date of packaging," "use before" date, "MRP," and other product details.
- Preservation: The packing system also preserves the food product from the environment like high temperature, moisture, microorganisms, enzymes, etc.

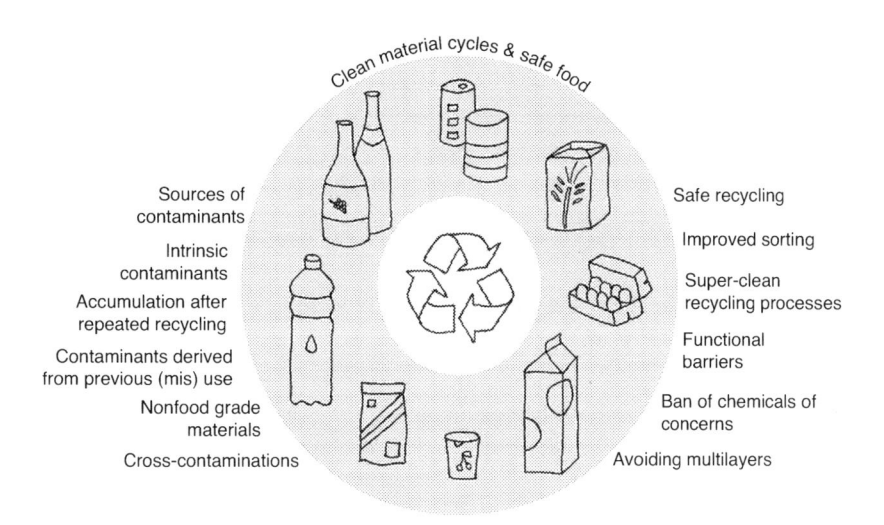

FIGURE 14.1

Different packaging materials and challenges associated with them (Geueke et al., 2018). https://ars.els-cdn.com/content/image/1-s2.0-S0959652618313325-fx1_lrg.jpg.

- Communication: The basic information printed over the packing reveals the legal requirements, objectives for marketing, and eco-labeling of all the food products. Sometimes, it provides brand communication too.
- Presentation: The packing materials also provide information about the shape, size, food materials type, color, merchandising display units, etc. to enhance the product authenticity.
- Promotion: It helps in promoting the food products too depending upon the offers that can be displayed over the packing systems.
- Economy: The packing system must be economical and should provide value to the food items packed inside it.
- Standard: It provides information about the products like their weight, standard quality; and requirements by the Bureau of Indian Standards (BIS), Food Safety and Standards Authority of India (FSSAI), or any other government authorized bodies.

14.3 Types of food packaging

There are many types of packaging systems used to pack food items for the preservation, communication, distribution, transportation, etc. for the global availability of food products and to eliminate the situation of malnutrition. The packing system can be of different materials like paper/paperboard (cellulose), glass containers, metal containers, plastic wares, edible films, pouches, etc. (Krochta, 2006; Robertson, 2013) as shown in Fig. 14.1.

14.3.1 **Plastic packaging**

Plastics are generally polymeric forms of organic macromolecules obtained by polymeric reactions like poly-addition, poly-condensation, polymerization, etc. between different monomeric units. The monomeric units can be propylene, ethane, ethylene terephthalate, ethylene vinyl acetate, vinyl chloride, amide, olefins, styrene, etc. Their polymeric units can form different types of plastics like thermosetting plastic and thermoplastic polymeric plastics. Thermosetting plastics have high melting points and low tensile strength and their shape and size do not change easily like on applying heat, current, and low mechanical stress, etc. while thermoplastics have low tensile strength and high melting points and hence cannot tolerate mechanical stress, heat or current effects, etc. Different forms of plastics have differences in their transition temperatures, electrical properties, mechanical properties, environmental protection, etc.

Plastic is used as a packing material due to many advantages:

- Impedance to micro-organisms
- Not much reactive to the chemical composition of food especially inorganic components (acids and bases)
- These containers are flexible or rigid depending upon the necessity and can be easily made in different shapes
- Easily dispensing the food products too

Types of plastic packaging

- Polyvinyl chloride (PVC): Vinyl chloride polymer is resistant to fats and oils, light weighted, transparent, economical, and nontoxic, and are generally used to make fruit juice and vegetable oil or mineral water bottles, cling films for food, tray for cooked food packing, etc. It can soften at a temperature of 80°C–90°C subject to its composition. The flexibility can be altered by adding a different number of plasticizers which can help in reducing surface friction. There can be added some colors or pigments to produce a variety of containers for numerous food items. These containers can have high barriers to oxygen and
- Polystyrene (PS): It can have good thermal insulation and protection from external agents like temperature and moisture.
- Polyamide (PA): It extends the shelf life of the food product by protecting it from harmful UV radiation, humidity, and oxygen content and also maintains the taste of the food product.
- Polyethylene (PE): It is widely used as a food container since it keeps food safe for a longer time. It can be used to prepare pouches, plastic wrappings, and lids, and can also be used to promote brands. It can be of different types based on their density like low-density polyethylene, ultra-low-density polyethylene, medium-density polyethylene, and high-density polyethylene (Stewart, 1996).
    - Low-density polyethylene is elastic, lenient, and sometimes glossy and can be used to prepare pouches or squeezable bottles or tubes for food products.
    - Ultra-low-density polyethylene is a lenient thermoplastic material that can be used to store milk products like cheese, coffee packing, etc.

FIGURE 14.2

Various types of glass packaging jars and bottles.

- ◦ Medium-density polyethylene can be used to make dispensing bottles, plastic carry bags, packing bakery products, etc.
- ◦ High-density polyethylene can be used to store juice, milk, butter tubs, prepare grocery bags, store water, etc.
- Polypropylene (PP): It is chemically inert to commonly used chemicals including organic and inorganic, aromatic or aliphatic, etc. It can act as a barrier to water vapors, oils, and fat.
- Polyethylene terephthalate (PET): PET has higher melting points than PP and can be easily used for keeping carbonated soft drinks, juices, edible oils, mineral water, etc.

14.3.2 Glass packaging

Glass is defined as "an inorganic product of fusion which has cooled to a rigid state without crystallizing" as per the American Society for Testing Materials (2010). Glass can be described as neither a solid nor liquid but can occur in a glassy state. The molecular arrangement in glass is not ordered to a long range, but cannot be categorized under purely amorphous materials. However, it has many properties similar to amorphous solids having significant cohesion forces to possess mechanical rigidity.

The chemical composition of glass is silica (SiO_2) whose network can frame the structure of glass materials. It has a high melting point and high viscosity and thus it is hard to reshape it easily. The designing of glass materials into different packaging forms can be made easy when we dope silica with small amounts of soda (Na_2O) which can result in dismantling a few Si-O-Si bonds resulting in a reduction of the viscosity and lowering the melting temperature too. Finally, lime (CaO) can be added to stabilize the network and provide stability to the ultimate products formed. Different packing materials made up of glass (Fig. 14.2) with variant colors can have colorants added to the system (Boyd, 1994).

Types of glass packaging

- Bottled packaging: Glass materials can be used in the form of bottles for packing different types of food materials as well as many other products. The bottles can be of different shapes, sizes as well as colors depending upon the nature of the food materials to be kept in these bottles. The glass bottles can be transparent which can be used to keep soft drinks, sauces, juices, etc. However, other colors of the glass bottles can be helpful in keeping various other food materials with specific purposes. The transparency of the glass can be built by adding silica with lime, soda, alumina (Al_2O_3), magnesia (MgO), and potash (K_2O). Green, blue, brown, and dark blue colors can also be produced by adding some amounts of iron oxide (Fe_2O_3) or chromium oxide (Cr_2O_3) or their mixtures.
- Containers: Glass containers can be used to keep other food materials like pickles, sauces, vinegar, jams, some pharmaceutical liquid syrups, etc. They can be of various shapes and sizes and many times they are reusable too.

The main aspects of glass containers are that they can be reused, reshaped, and remolded for other uses (Sharma, 2012). The glass containers have advantages in that they can be helpful in maintaining the quality of the products, can predict the transparency of the food product, their surface textures, and colors, these containers are susceptible to decorative aspects, impermeability to microbes, maintaining chemical reliability of the food product, etc. They can be heat resistant, microwave friendly, easily dispensable, tamper apparent, protection from UV, and can provide strength to the container of food materials. Hence glass packaging can also play a versatile role in food packaging systems.

The sustainability aspect plays a vital role in deciding the food packaging using glass materials. Since the glass materials are not biodegradable but can only be reused or recycled for different purposes. Therefore, the glass containers can be recollected at the site of manufacturing and reprocessed for various purposes. Since glass materials can remain as such for thousands of years in the environment, therefore, its reusable concept plays a vital role. However, nowadays, the use of glass has also reduced due to some of its disadvantages easily breakable, massive, not easily portable, etc., and thus it has been replaced with many convenient packing materials wherever feasible like paper packing, cloth packing, etc.

14.3.3 Metal food packaging materials

It has been estimated that the total market for metal food packaging in the world is around 410 billion units per annum; among which drink cans contribute majorly which is around 320 billion units and processed food cans contribute around 75 billion units (Page et al., 2011). Since long back around 200 years back tin cans have been and are still used in major amounts even today.

Considering the purpose of food packaging there are four metals which have been/are being used majorly are tin, steel, aluminum, and sometimes chromium along with tin/steel. It is the most commonly used of all packaging forms available owing to excellent physical and chemical protection, its ability to make it more creative and decorative, its acceptability to various customers, and its recyclability. Considering the above advantages, the metal packaging industry possesses various negative effects on health and the environment. If we consider both the pros and cons of metal packaging the benefits come with the impact on cost, or expensive. The disadvantages to the environment are to release of carbon dioxide from metal manufacturing plants discharge of harmful chemicals as by-products. Generally metal cans are manufactured with aluminum metals; apart from aluminum, other metals like steel, tinplate, and tin-free steel can be used. The advantage of using metal cans includes availability in various shapes and sizes. It has been seen that 90% of the aluminum cans that are being used are round in shape; this is possibly due to the enhanced demand for beverages in the form of healthy drinks, energy drinks, soft drinks carbonated and noncarbonated, fruits/vegetable juices, etc. In this chapter, we will focus on the requirements or expectations of metal cans and how we can achieve those requirements. Here we will also list a few materials that can be used for food packaging in detail.

The basic functionalities we expect from a metal can are: (1) it should be capable of preserving and protecting the food without alterations; (2) the metal should be stable enough to bear environmental conditions, (3) the can opening should be easily removable and wide enough to transfer the product in exact manner, and (4) the can material should be made from recycling materials. Along with the existence of all these qualities; it is of utmost importance that these qualities persist till the end of the tasks or the said shelf life is covered.

For instance, many of the food and beverages are considered to be held at ambient temperature conditions but they might be subjected to some amount of heat owing to the prolonged shelf life of the product, for this reason, the container must have the capability to hold on the same. The major thing to consider over here is that after heating and coming back to the normal temperature again will create a negative pressure or vacuum in the can. Similarly, the carbonated drinks increase the pressure after can closing the resulting pressure provides superior physical support to the container.

14.3.3.1 *Metal container design requirements*

For designing any metal container, its specifications, shape, required performance, capacity, and durability are required. It has been estimated that the cost of metal packaging is entirely dependent on the cost of metal as it contributes around 60%–70% of the total cost of the container. Thus, the cost of metal and its amount in particular containers is the deciding factor of packing cost, the cost is also predicted on the basis of metal thickness, surface area, and temper, where thickness is chosen on the basis of required physical parameters such as handling, storage, and processing of food. You must have observed that generally all types of food and drinks can be in circular shape and its surface area is basically determined by the volume it needs to contain.

14.3.3.2 Raw material for the production of cans

Owing to nontoxicity, pocket-friendly, high-strength behavior steel and aluminum are the two common metals that are generally used for making cans for food and beverages packaging.

14.3.3.3 Steel

Stainless steel with a low content of carbon is used as a black plate on which further coating is done via the electrolytically coating method. Thin or thick thickness can be chosen on the basis of internal food requirements and external environmental conditions. Tin coating over steel not only makes it creative and decorative but also prevents the can from corrosion in this way they tin coated steel very well useful for food packaging (Cvetkovski, 2012). As a suitable alternative to tin-coated steel electrolytically chromium or chromium oxide-coated steel (ECCS) is used which has also been approved by product standard EN 10202:2001 (European Committee for Standardization, 2001). The major difference among both is that the ECCS has the same coatings on both sides of the surface whereas tin-coated steel can be differently coated on each side of the surface coil.

The chromium coating on steel provides excellent strength against atmospheric oxidation and it also helps in lacquer adhesion. It is always recommended to use an additional organic coating or can coating with ECCS (Oldering, 2007).

14.3.3.4 Aluminum

Aluminum is a metal that is widely used in the food packaging industry, aluminum alloy may contain the presence of zinc, magnesium, copper, iron, etc. Aluminum is also found in the earth's crust as the highest metallic constituent of the earth's crust which is 8.8%. Aluminum is of keen interest in today's world as it resists the probability of corrosion and when aluminum is exposed to air it is converted into Al_2O_3 which is a colorless and tough film and generally not soluble with various chemicals and suitable to use for food packaging.

Everything comes with its advantages and disadvantages, here the disadvantage of using aluminum is its cost, and the average cost of using aluminum is more than any of the steel of coated steel packaging. Second aluminum is its inability to undergo welding that can only be used for seamless containers. Aluminum is considered one of the lightest food packaging materials that can be utilized for packing seafood, soft-drink cans, etc. in its pure form only. Only aluminum metal does not provide much strength to enhance the strength. We can add magnesium to make aluminum alloy (Ahvenainen, 2003). Aluminum metal is used in foil form to pack home food like chapati, dosa such flatbread items on a daily basis, it can also be used for laminated cans and can be added with paper and plastic for various wafer's packaging. It is also used for making foil, cans, laminated and metalized packaging material in combination with paper and plastics. Aluminum adds beauty and strength to paper and plastic packaging. The utmost important advantage of using aluminum is its recyclability; aluminum can be easily converted to form new products.

14.3.3.5 Aluminum foil

Aluminum foil exists in 99% purity and which can be prepared in the desired thickness and length of aluminum foil. Generally, the thickness of aluminum foil varies from 4 microns to 150 microns based on the purpose. Commercially aluminum foil was introduced in the year of 1913 for wrapping confectionary items like candies, chewing gums, chocolates, etc. Aluminum foil production is done by casting the rectangular blocks of aluminum metal followed by its scalping to remove oxides. The oxides removed blocks are further subjected to various hot or cold-pressed rollers for preparing the desired thickness. The final step is softening or annealing after which the foil is ready to use (Morris, 2011; Bayus, 2016; Sun, 2008).

14.3.3.6 Tin

Tin individually does not have a major application in food packaging but it's being used majorly as a coating for steel. The advantage of using tin-coated steel is: (1) tin acts as an oxygen scavenger by blocking the interaction of oxygen with iron present in steel and (2) tin provides efficient corrosion resistance. It has been seen that tin is heavier and more pocket-friendly than aluminum and possesses a magnetic property which helps easy segregation upon recycling of tin. Tin can also sustain at high temperatures which makes them suitable for use in sterilized precast, heating purposes, and hot beverages packing. In this respect, Catala et al. (2005) mentioned that for sterilization purposes it is better to use coating along with a tin plate for better results (Ertl, 2018).

14.3.3.7 Tin coated steel

It is one of the most important coated steels known so far for food packaging applications. This generally contains a steel base and a tin plate on the sides. For the preparation of tin-coated steel material: the dipping method was used historically on the other hand modernization leads to the importance of the commercial electroplating method for making tin-coated steel. Electroplating comes with the advantage of placing different coating thicknesses on both sides of the can.

14.3.3.7 Coated steel

Steels are iron alloys with carbon content in the range of 0.2%–2%, which helps to bind iron in rigid spaces to enhance the mechanical properties of steel. These are also known as carbon steel and possess high tensile strength but for specially food packaging it is recommended to use a carbon content of not more than 1%. For this purpose, various other options have taken its place which are known as coated steel such as tin-free steel, tin plate, and polymer-coated steel.

14.3.3.8 Polymer-coated steel

We all are aware of the fact that steel is an alloy of iron which is susceptible to corrosion very easily thus various efforts have been made to make it corrosion-resistant by applying various coatings to it (Catala et al., 2005). One such coating that is now of major interest to various researchers is polymer coating which includes polymers like polythiopen, polyaniline, and polypyrrole. Enhancement of performance has been seen through the use of bilayered polyaniline (Catala et al., 2005). Leivo et al. (2004) performed numerous thermally sprayed coatings of synthetic polymers to steel and results were quite impressive and could resist corrosion against liquid emersions and salt spray tests in 2004. Synthetic polymers they use for this purpose are listed as perfluoroalkoxy alkane (PFA), fluorinated perfluoroethylenepropylene (FEP), ethylene chlorotrifluoroethylene (ECTFE), and polyvinylidene fluoride (PVDF) (Boelen et al., 2004). Steel coated with polymer shows high corrosion-resistant properties with a beautiful appearance and strong barrier against moisture.

14.3.3.9 Use of coatings for metal packaging

From the above discussion, we have understood the role of coatings. But this coating is again of two types segregated: internal coating which is in contact with food and external coating which is exposed to the atmosphere.

Internal or in-food contact coating is applied to protect the food from metal, for example, dark-colored fruits decolorize while coming in contact with metal on a similar note beer reacts with the metal if not coated. Outside coating is applied for the purpose to make it attractive and to save it from environmental degradations. Another purpose of outside enamel ink or varnish painting is to put labeling, consumer information, and product names.

14.3.3.10 Types of metal packaging

Three-piece cans: These three-piece cans include two lids bottom end, an up lid, and a central flat metal sheet roll. These three pieces can be manufactured easily in any desired dimensions of diameter or length. The body or wall of the can is formed by a metal sheet rolled in the form of a cylinder and joined by welding or seaming. Out-of-welding and seaming welding is considered a better option as it consumes less metal. Seaming also possesses lead contamination to food, in earlier times seaming was generally used to close the cylindrical walls but after 1980 seaming has been replaced by welding. The cylindrical wall is thus painted with a protective lacquer and decorated followed by the attachment of the bottom end. After filling the items in the can the lid is attached to finally pack the can. These types of cans are usually used for noncarbonated beverages, fruits dipped in sugar syrup, and fresh juices.

Two-piece cans: It contributes as a great initiative in the field of cans making. It consists of one lid and a flat metal sheet rolled in such a way it makes a cup shape without any significant joint done by welding or seaming. These two pieces are way easier to make, hygienic, and contain more space for writing brand names and make them more decorative. These cans are much lighter than three-piece cans and pocket-friendly with better integrity. These cans can be further divided into different categories based on their manufacturing (1) drawn and ironed wall and (2) drawn and redrawn wall cans:

1. Drawn ironed wall: DWI cans are generally made up of steel and aluminum and used for packaging carbonated beverages, beer, and cold drinks. The fizz and carbonated generated pressure provide strength to thin-walled cans and resistance against external atmospheric pressures. To make this DWI cans a drawn cup is stretched to make a cylinder with a decreased thickness generally achieved by wall ironing of the metal while stretching. The difference between a normal three-piece can and a DWI can is the neck diameter which is being reduced in DWI cans at the top.
2. Drawn and redrawn walls: DRD is applied when larger cans are required; their multiple operations of drawing and redrawn are carried out for lengthening the can; the rest of the treatment is similar to DWI cans discussed before.

14.3.3.11 Metal food packaging

From the above discussions we have understood the role of metal in food packaging, how they can be used, what are the expectations from the metals and how can we meet those. Now we are going to discuss the sustainability aspects over here.

Use of natural resources: Considering the metal used for food packaging it has been noticed that these elements (aluminum and iron ore) are naturally available in sufficient quantities. In short, we are using natural resources in limited quantities without hitting the future generation requirements as the metal food packages can be recycled a number of times and reused multiple times. The more we recycle cans and metals, the more environmentally friendly packaging we can make. Recycling metal also impacts reducing CO_2 emission thus it can be concluded that more recycling will ensure the limited use of natural resources and also partially prohibits the emission of CO_2 as the steel and aluminum manufacturing release CO_2 in the environment.

Light-weight and portable packaging: Metal cans are considered a relatively light packaging methodology considering their strength and the variety of food they can handle. Additionally, the continuous researches in this area has made it possible to prepare the lightest metal packaging material or metal cans. The lightening of metal can lead to less use of natural resources which ultimately impacts cost as well. These days' research has been shifted to make the existing procedure more and more environmentally friendly considering the scenario of global warming and lack of resources. Thus, minimizing waste material is also a major concern these days, thus cans are designed in such a way that they can be reused and recycled with almost the same amount of the original thus designing metal cans is also an important concern. These are designed in such a way as to minimize the waste from cut-offs, finished scrap products, and web scrap. DWI cans being the lightest are of new interest to various food packaging agencies. If we look through the statistics the demand for metal cans for drinks and food has increased to a significant extent but on a similar note the recycling of these cans has also increased which ultimately reduces the demand for virgin metal resources and decreases CO_2 emission. Thus, it is necessary to recycle more and more metal cans to keep our environment clean for all.

Greenhouse gas emission: It is not hard to believe that food packaging is also one of the major contributors to greenhouse gas emissions. Metal cans are the ones used for food packaging to supply fresh food, and wet food to places where it is not possible to grow in desired quantities. Metal cans represent the best alternative to supply food instead of the freezing method, as freezing emits chlorofluorocarbon gas emission which is a greenhouse gas. These metal cans exhibit excellent barrier properties against light, air water, and atmospheric heat. These can be used at a certain temperature for food processing and then can again be stored at ambient temperature. No such food additives/preservatives are required if food is being packed in metal cans, moreover, the food can be stored for years without any degradation in these metal cans. Metal cans are a perfect solution for underdeveloped countries where food and other agriculture items are being forced to waste owing to limited sources, there These heat-processed cans can serve as a perfect route to preserve food without putting in additional energy. In addition to that the food can be easily transported to farther places in its fresh form.

14.3.4 **Paper packaging**

Paper-based food packaging was started in the 17th century and its extended usage was further noticed in the year 19th century (Coles et al., 2003). The main component of paper packaging is cellulose fibers, which come from the lignocellulosic biomass or simply from the logs of trees. The cellulose fibers network gives strength and moisture-absorbing capability to paper and paper boards (cardboards). The cellulose fibers are obtained from wood by reaction with sulfate and sulfite. The resulting pulpy material is further bleached and treated with other chemicals to strengthen the paper product. The application is of paper packaging mainly inclusive of milk packaging in cartons, pastry or cakes packaging in paper boxes, these Amazons and Flipkart used packaging in cardboard for groceries and other packed food items, as wrapping paper in replacement of aluminum foil, paper cups for serving hot and cold beverages, tissue papers, disposable plates for having food at various places, etc (Poças et al., 2010).

It is clearly evident that food packaging paper and normal paper are two different things, you can give it a try of using a normal A4 sheet or newspaper for food packaging what you will observe is: that it will tear off quickly in coming contact with small moisture, its durability is very low, keep aside hot, warm food cannot be wrapped in plain paper. Paper when used for the purpose of packaging is always treated, laminated, or coated with other ingredients such as waxes, polymer, resins, lacquers, etc. to improve its durability and efficiency.

Now, that we have understood the application of paper-based packaging the next thing that comes is how safe and environment friendly this packaging is. When we pack our food in paper there are chances of transporting food migrants from packaging to food which ultimately can cause health hazards (Ohtsuka, 2012). Numerous additives that are used to provide strength to paper may migrate to food products when they come in direct contact with the food and can cause severe health hazards (Leivo et al., 2004). The migrant level or amount and their respective toxicity are also important to calculate the health hazards and are also a bit complex to study (Grob et al., 2006). Another important factor to take into account is the ink used in paper, the ink components are the major constituent among food migrants which can cause damage to the liver, kidneys lungs. Thus, research in this area is of keen interest to develop an ink that cannot migrate to the food component.

Coming to the environment and sustainability impact using paper for packaging looks environment friendly but it is actually not. The first reason is segregation, in India still people are not segregating their garbage, which ultimately leads to the spoiling of those papers; next is the waste generated in the manufacturing of paper which means the paper-making industry in general induces tons of waste at various levels of production. Looking at the scenario, landfilling and waste disposal is the only output of paper packaging rather than recycling. Recycling can be done for paper packaging but that again demands proper separation of paper from other garbage, the energy involved in recycling the paper such as pyrolysis, paper production, and others moreover there is a 12% Goods and Services Tax (GST) applied on the waste paper. The additives used for the strengthening of paper also create much interference in the recycling of paper. But on the same side recycling is equally necessary to decrease the burden on fresh timber and to reduce global warming.

Types of paper used in food packaging are:

1. Greaseproof paper: It is mainly used to pack greasy items as suggested by name like fried snacks, cookies, candies, and energy bars and we can say that it is a biodegradable alternative to plastics. The thing to be taken care of in its manufacturing is to hydrate more or hydrate for a longer duration than normal in the manufacturing process is called beating. This results in the breaking of fibers and turning into gelatinous and this is further packed densely to provide resistance to oils and other greases.
2. Kraft paper: It is not greasing proof so basically found application in the packaging of dry food items like flour, lentils, sugar, dry fruits, etc. These are of various types like natural brown or bleached white, heavy or light weighted, etc. The kraft paper is also known as sulfate paper as it is being prepared by sulfate treatment.
3. Sulfite paper: As discussed above sulfite treatment is done and known as sulfite paper, its basic application is in packing confectionary items by glazing or laminating them with plastic to make the paper attractive and printable.

Paper board: The thicker and heavier version of paper is known as paper board which is inclusive of multiple layers, mainly used for shipping items from one place to another, they can be inform of cardboard, boxes, trays, etc. Here the food does not come in contact with the paper board as packed foods are used to further pack in the boxes for transportation. They again can be manufactured in different types based on the requirement.

1. Fiberboard: The fiberboard is strong and consists of two layers, the inner layer is made up of a bleached whiteboard and the outer is made up of kraft paper. It provides a good barrier against compression and other atmospheric conditions. The strength can be further improved by lamination with plastic or aluminum; these boxes are used to pack dry food items like cornflakes, coffee, etc.

2. Whiteboard: It is made of multilayers of bleached chemical pulp and is the inner part of cartons; here whiteboard is generally laminated with wax of some polymers to provide some resistance.
3. Paper laminates: They can be coated or uncoated papers derived from sulfite or kraft paper and are generally used to pack the products of premixes like soups, herbs, and spices.

14.4 Recycling

Recycling is the most important nowadays when the world is struggling for resources and dumping in an open environment is deteriorating environmental health. It is also considered green as it generates valuable things from waste items. For recycling proper sorting followed by chemical processing and manufacturing are the important parameters, we also need to take care of costing as well. The recycling process cost should not be much higher than the actual cost of production from virgin material.

It has been seen that virgin paper can be recycled up to 6–7 times further for usage (Biedermann-Brem et al., 2016; Muncke, 2011; Ervasti et al., 2016). For example, freshly manufactured paper can be used for official documentation; first recycled paper can be used for personal uses (reading, taking printouts) as the quality slightly degrades (Augusto and Wenzel, 2007). Going further with a similar pattern recycled paper can be used for newspapers. If we keep on recycling the paper for further uses the saved virgin resources of wood, pulp, and trees can be further utilized to fulfill our fuel demands by generating bioethanol (Wang et al., 2013).

14.5 Conclusion

Food packaging is primarily carried out to maintain its authenticity, quality, hygiene, safety, etc. Every material used for food packaging whether it is glass, metal, paper or plastic can meet the criteria of keeping food material safe, secure, and preserved but the major concern is the sustainability associated with the food packaging system being used. The ultimate disposal of food packaging systems, their eco-friendly nature, impact on the environment on decomposing, etc. plays a key role in deciding the type of packaging systems which we must adopt. Nowadays, the Indian Government has completely banned the use of single plastic which is mainly being used in packaging and transporting food and beverages. These single-use plastics have been replaced by cloth bags, cardboard boxes, and fiber bags which are definitely greener and cleaner. Conclusively, the waste generated by food packaging materials can be reduced by selectively categorizing the greener materials and reviewing the amount and impact it produces when disposed of in the environment. Sensitizing common people will also produce a revolutionary impact in this area, thus the government's initiative to promote eco-friendly packaging is significant. The cumulative efforts by the government, industries, and most prominently "the consumers" will surely impart continuous improvement and reduce the burden on the environment.

References

Ahvenainen, R. (Ed.), 2003. Novel Food Packaging Techniques. Elsevier.

American Society for Testing and Materials. 2010. Standard terminology of glass and glass products, 2010. American Society for Testing and Materials, C162-05

Bayus, J., Ge, C., Thorn, B., 2016. A preliminary environmental assessment of foil and metalized film centered laminates. Resour. Conserv. Recycl. 115, 31–41.

Biedermann-Brem, S., Biedermann, M., Grob, K., 2016. Required barrier efficiency of internal bags against the migration from recycled paperboard packaging into food: a benchmark. Food Addit. Contam. Part A 33 (4), 725–740.

Boelen, B., den Hartog, H., van der Weijde, H., 2004. Product performance of polymer coated packaging steel, study of the mechanism of defect growth in cans. Prog. Org. Coat. 50 (1), 40–46.

Boyd, D.C., Danielson, P.S., Thompson, D.A., 1994. Glass. In: Kroschwitz, J. (Ed.), Kirk-Othmer Encyclopedia of Chemical Technology, vol. 12. John Wiley & Sons, New York, pp. 555–628.

Catala, R., Alonso, M., Gavara, R., Almeida, E., Bastidas, J.M., Puente, J.M., De Cristaforo, N., 2005. Titanium-passivated tinplate for canning foods. Food Sci. Technol. Int. 11 (3), 223–227.

Coles, R., McDowell, D., Kirwan, M.J., 2003. Food Packaging Technology, vol. 5. CRC Press.

Cvetkovski, S., 2012. Stainless steel in contact with food and bevarage. Metall. Mater. Eng. 18 (4), 283–294.

Ertl, K., Goessler, W., 2018. Aluminium in foodstuff and the influence of aluminium foil used for food preparation or short time storage. Food Addit. Contam. Part B 11 (2), 153–159.

Ervasti, I., Miranda, R., Kauranen, I., 2016. A global, comprehensive review of literature related to paper recycling: a pressing need for a uniform system of terms and definitions. Waste Manage. 48, 64–71.

European Committee for Standardization, 2001. European Committee for Standardization (ECS Report), Cold reduced tinmill products – Electrolytic tinplate and electrolytic chromium/chromium oxide coated steel, European Standard, EN10202, 1–48. https:// standards.iteh.ai/catalog/standards/cen/eb353920-e059-4b31-a83c-2b0deb0069d0/en-10202-2001. (Accessed 5 June 2023).

Geueke, B., Groh, K., Muncke, J., 2018. Food packaging in the circular economy: overview of chemical safety aspects for commonly used materials. J. Clean. Prod. 193, 491–505.

Grob, K., Biedermann, M., Scherbaum, E., Roth, M., Rieger, K., 2006. Food contamination with organic materials in perspective: packaging materials as the largest and least controlled source? A view focusing on the European situation. Crit. Rev. Food Sci. Nutr. 46 (7), 529–535.

Krochta, J.M., 2006. Food packaging. In: Heldman, D., Lund, D., Sabliov, C. (Eds.), Handbook of Food Engineering. CRC Press, Boca Raton, FL, USA, pp. 859–940.

Lee, D.S., Yam, K.L., Piergiovanni, L., 2008. Food Packaging Science and Technology. CRC Press.

Leivo, E., Wilenius, T., Kinos, T., Vuoristo, P., Mäntylä, T., 2004. Properties of thermally sprayed fluoropolymer PVDF, ECTFE, PFA and FEP coatings. Prog. Org. Coat. 49 (1), 69–73.

Marsh, K., Bugusu, B., 2007. Food packaging—roles, materials, and environmental issues. J. Food Sci. 72 (3), R39–R55.

Morris, S.A., 2011. Food and Package Engineering. John Wiley & Sons.

Muncke, J., 2011. Endocrine disrupting chemicals and other substances of concern in food contact materials: an updated review of exposure, effect and risk assessment. J. Steroid Biochem. Mol. Biol. 127 (1-2), 118–127.

Ohtsuka, T., 2012. Corrosion protection of steels by conducting polymer coating. Int. J. Corros. 2012, 1–7.

Oldring, P.K., Nehring, U., 2007. Packaging Materials: Metal Packaging for Foodstuffs. ILSI Europe.

Page, B., Edwards, M., May, N., 2011. Metal Packaging: Food and Beverage Packaging Technology, 2nd ed. John Wiley & Sons, pp. 107–135.

Poças, M.F., Oliveira, J.C., Pereira, J.R., Hogg, T., 2010. Consumer exposure to phthalates from paper packaging: an integrated approach. Food Addit. Contam. 27 (10), 1451–1459.

Rani, R., Medhe, S., Raj, K.R., Srivastava, M.M., 2012. High performance thin layer chromatography for routine monitoring of adulterants in milk. Natl. Acad. Sci. Lett. 35 (4), 309–313.

Robertson, G.L., 2013. Food Packaging Principles and Practice. CRC press, Taylor and Francis Group, Boca Raton, London, New York.

Sarkar, S., Kuna, A., 2020. Food packaging and storage. In: Singh, P., Kumar, S. (Eds.), Research Trends in Home Science and Extension, vol. 3. Akinik, pp. 27–51.

Sharma, H., 2012. Food Packing Technology. Anand Agriculture University, Gujrat. https://www.hzu.edu.in/agriculture/Food-Packaging-Technology.pdf.

Stewart, B., 1996. Packaging as an Effective Marketing Tool. CRC Press.

Valentas, K.J., Rotstein, E., Singh, R.P., 1997. Handbook of Food Engineering Practice. CRC Press.

Villanueva, A., Wenzel, H., 2007. Paper waste–recycling, incineration or landfilling? A review of existing life cycle assessments. Waste Manage. 27 (8), S29–S46.

Wang, L., Sharifzadeh, M., Templer, R., Murphy, R.J., 2013. Bioethanol production from various waste papers: economic feasibility and sensitivity analysis. Appl. Energy 111, 1172–1182.

Nanobiotechnology for sustainable food waste management

15

Srishti Sharma [a,b] **and Namrata Singh** [c,d]

[a] *School of Studies in Chemistry, Pt. Ravishankar Shukla University, Raipur, Chhattisgarh, India,*
[b] *Department of Chemistry, Dr. Ghanshyam Singh P.G. College, Varanasi, Uttar Pradesh, India,*
[c] *Department of Engineering Sciences, Ramrao Adik Institute of Technology, DY Patil Deemed-to-be University, Navi Mumbai, Maharashtra, India,* [d] *Department of Chemistry, Faculty of Science, University of Hradec Kralove, Hradec Kralove, Czech Republic*

15.1 Introduction

Biodegradable waste generated due to discharges (both precooked and left over) from the food industry, domestic uses, and hospitality sectors is known as "food waste." Food waste is waste and residue in the food value chain from primary production, processing, and distribution till the final consumption. Agro-farm products, for example, husks and cereal straws are primary productions. According to the United Nations Environment Programme (UNEP) Food Waste Index report 2021, about 80% of food waste is responsible for global greenhouse gas emissions. It has been concluded from various reports that food waste can escalate in the next 25 years due to commercial and population progress, mainly in Asian countries. National strategy and execution of food waste prevention at a multifaceted level is the need of the hour. The global increase in population would have never been a problem if the advancements of necessities had been developed parallelly. However, it was surprising to note that the world suffers from food shortage not because of less supply or low productivity but due to the wastage of food (Manikandababu et al., 2022; Sharma et al., 2022). From the initial stage of food production to processing and ultimately consumption of food, the entire sector engenders an excessive amount of waste (Fan et al., 2022). As the name suggests, food waste is not the left out or the contaminated food but the food which is eventually consumable yet wasted (Parfitt et al., 2010). This food waste occurs at different stages of the food chain starting right from production in agricultural fields till it reaches consumers in the packed form (Krishnan et al., 2020). Fig. 15.1 shows various traditions responsible for generating food waste. This food waste needs to be reduced and reused to control this food loss. Researchers are now making use of nanobiotechnology to limit this food waste (Jafarizadeh-Malmiri et al., 2019) to achieve triple goals of sustainability, food safety, and food security. Nanobiotechnology is a hybrid technology formed by

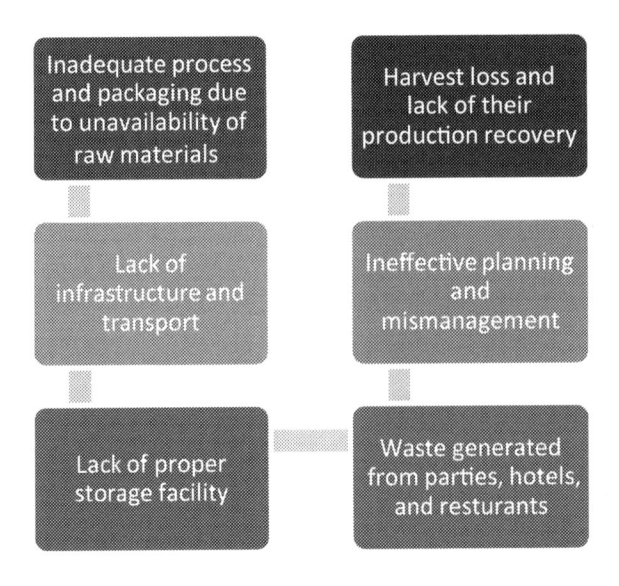

FIGURE 15.1

A representation of various ways responsible for generating food waste.

integrating nanotechnology with biotechnology (No and Park, 2010). Once the renowned physicist Richard Feynman said, "There's Plenty of Room at the Bottom", we never knew it would be a beacon in the field of science (Toumey, 2009). Nanotechnology has reverberated its presence in the farfetched regime of optics and electronics, the food sector and agriculture, pharmaceuticals and molecular biology, medicine and drug delivery, computational technology, etc. (Solgaard, 2009; Sozer and Kokini, 2009; Silva, 2004). The significant reasons for preferring nanotechnology in the food sector are their minimal chemical inputs and enhanced productivity (Pushparaj et al., 2022). Proper management and reduction of risk in the food sector can be achieved by applying a nanotechnology-based approach. Nanotechnology has the potential to regulate food properties at the nano-scale level, although there is little research reported in this domain. Due to this, there is a lack of awareness in the food sector regarding nanomaterial techniques. The combination of the two very crucial disciplines (engineering and molecular biology) led to the evolution of a very versatile and highly applicable branch of study that is "nanobiotechnology." This domain mainly deals with systems that can perform physical and chemical analysis at the molecular level. Hence, nanobiotechnology has become a perfect ground for researchers to find solutions for many unfolded theories related to all the sectors of various subjects. The whole food sustainability sector is facing top five setbacks and these are (1) challenges regarding food safety and security, (2) quality of food due to diversity, (3) fiscal challenges in food chain systems, (4) ecological challenges due to food waste management and processing, and (5) novelty in food production and engineering challenges.

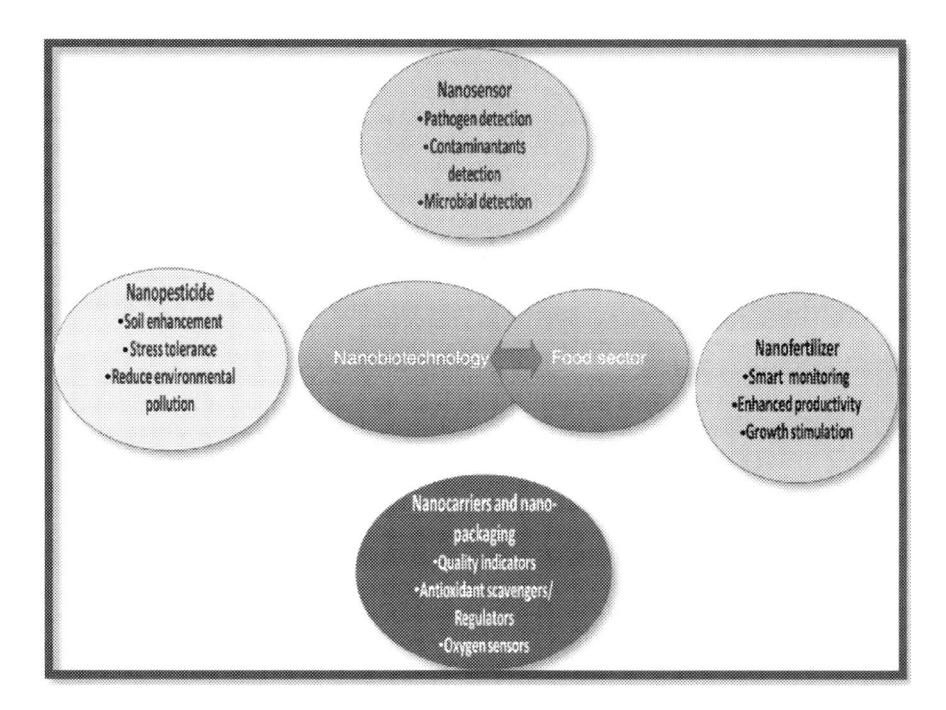

FIGURE 15.2

Illustration of diverse techniques of nanobiotechnology employed in the food sector.

In this chapter, we have tried to unfold the role of this domain in the food sector and its advancements with special emphasis on food waste management. This book's chapter provides insightful viewpoints on nanobiotechnology for reducing food waste and increasing the shelf life of food products. Special emphasis is given to different nanosystems, their understanding, and applications. A systematic analysis of safety concerns and the impact of extensive usage of NPs on human health and the environment is the key highlight of this chapter.

15.2 Role of nanobiotechnology in sustainable food waste management

In the food sector, nanobiotechnology has proved its potential by introducing a number of techniques that have laid breakthrough advancements (Fig. 15.2). Techniques such as nanosensors, nanopesticides, nanofertilizers, nanocarriers have a deep impact at the different stages of food production. Nanobiotechnology has dramatically reduced environmental pollution and has enhanced the shelf life of food by improved methods of storage and preservation. Nowadays, globally these techniques are employed in the food sector. A detailed description of these techniques has been provided in this chapter. Vital nanomaterials in food management are dendrimers, micelles, liposomes, nanocapsules, nanofilms, nanospheres, nanoconjugates, (Lugani et al., 2021) etc. Organic nanoparticles (NPs) are considered to be safer and more effective in the human body from a digestion perspective (McClements and Xiao, 2017). Metallic NPs commonly used in food and nutrition industries are silica, zinc, silver, etc. (Martirosyan and Schneider, 2014). Bonilla et al. (2016) have reported polymer-functionalized fluorescent quantum dots as food pathogen detectors. The major application of NPs in food management lies in their ability to control and deliver bioactive materials effectively. Additives, enzymes, nutrients, etc. can be easily administered through biotechnology (Lugani et al., 2021). The technique of storage, processing, and marketing of food products in the presence of nanomaterial has become quite efficient (Singh et al., 2017). Scavenging ethylene gas (Pradhan et al., 2015) for ripening of fruits and vegetables has already proven to be an effective measure. Quick absorption contaminants from food that may result in foul odor and improper ripening of fruits can be controlled by some eco-friendly nanocomposites (Vasile, 2018). Food waste management has been possible because of various effective systems of nanobiotechnology such as nanocapsules, nanosensors, nanocomposites, nanoparticles, nanofibers, and nanotubes (Duncan, 2011).

15.2.1 **Role of nanosensors**

Nanosensors help in the detection of those signals which are the output of some physical quantities (Yonzon, et al., 2005). Nanosensors contribute to food waste management by detecting pathogens, food contaminants, adulterants, dyes, fertilizers, and pesticides (Omanović-Mikličanina and Maksimović, 2016; Srivastava et al., 2018). They also help in monitoring the taste and odor of the food (Jafarizadeh-Malmiri et al., 2019). Such nano-based platforms are cost-effective strategies for limiting food waste. There are two considerable prospects for sustainable food waste management in the purview of nanotechnology (Fuertes et al., 2016).

1. Chemicals monitoring using sensors.
2. Pathogen detection and control using sensor-based technology.

These nano-enabled probes and sensors have immense potential to enhance agricultural productivity (Xin et al., 2020). Nanosensors and probes have supremacy over traditional sensors as they can identify extant of any chemical or microbes due to highly sensitive electrical and optical properties (Huang et al., 2011). Other accountable roles of nanosensors are: (1) the identification of colorants, (2) adulterants, (3) organic molecules in food, and (4) they assist in the quantification of hydrocarbons, flavor compounds, antioxidants, contaminants, vitamins, etc. (Purkayastha and Manhar 2016; Kuswandi et al., 2017).

Other techniques such as noninvasive gas sensing technique (food spoiling gases such as oxygen, volatile organics, gaseous amines, carbon dioxide, etc.) have emerged with the help of nanoprobes to help in exposing foodborne pathogens. Biosynthesis has come up with various solutions for various designs and advancements in NPs. This is because biological synthesis can prevent the use of toxic chemicals and hazardous substances since most of the reactions are performed at neutral pH and ambient temperatures. The most attractive part is that biosynthesized NPs are nonharmful and biocompatible in nature. The drawback of chemical synthesis of nanomaterial lies in extreme pH, pressure, and temperature conditions. Many NPs are used as natural flavor enhancers in the food industry, e.g., titanium dioxide. NPs have also been widely used in food preservation and packing. Food treatment and storage have been made easy with the concept of nanotechnology.

Food waste management being the biggest challenge of the food industry can be easily overcome by nano techniques and innovative approaches. Due to some attractive properties like cost effectiveness, mechanical and thermal properties, "nanoclay" is popular for food packing. Edible covering with nanomaterials has proved to be beneficial in food safety management and security. Nanomaterials being extremely sensitive are used in food safety and risk assessment, e.g., toxins, pathogens, and pesticides detection (Arduini et al., 2016). An interesting application of ZnO quantum dots has been reported by Sahoo et al. (2018) for the detection of pesticides involving glyphosate, atrazine, aldrin, and tetradifon since strong binding has been observed between pesticides leaving groups and QDs. Scattering of lights if detected by certain techniques can effectively assure the presence of *Escherichia coli* in food. The mechanism of the sensor is binding to a specific protein which in turn can bind to *E. coli* which is present in food (Bhattacharya et al, 2007). Detection of bacteria and other pathogens in food is a crucial step in food waste management. Surface-enhanced Raman spectroscopy (SERS) (Naja et al., 2007) technology with silver nanoclloids can be an effective technique to detect bacteria since silver nanocolloids can augment Raman signals. Also, silver nanocolloids, silver NPs (Siddiqui and Alrumman, 2021), carbon nanotubes (Zuo et al., 2013), and magnetic beads (Holzinger et al., 2014) are widely used in food pathogen detection.

15.2.2 Role of nanopesticides

Nanopesticides are responsible for restoring efficacy and suppressing the ill effects of pesticides by reducing their residual activity (Agathokleous et al., 2020). A few examples of nanopesticides are polymer-stabilized bifenthrin nanoparticles, chitosan nanogel, etc. (Shah et al., 2016; Kah et al., 2013). There are diverse benefits of nanopesticides like enhanced targeted and controlled delivery, pesticide stability and bioavailability, increased solubility of active ingredients, etc. Among their several advantages, they also help in cost cutting, reduce environmental pollution by conserving water and energy, and lower the chances of denature degradation of pesticides. Research on nanopesticides is going at a full pace to introduce their potential for their targeted and on-time release due to excellent penetration properties. The residual activity of pesticides can be reduced in nanomediums and they can enhance biological activity (Selyutina et al., 2020; Prasad et al., 2019). Easy penetration and augmented solubility in plant cuticles are possible in the presence of saccharides-NPs.

These nanopesticide delivery systems have more affinity with plasma membrane and other targets which is the rationale for their promising applicability (Alfei, 2020). Nanosystems using glycyrrhizin and arabinogalactan nanocomposites containing pesticides tebuconazole, imidacloprid, imazalil, and prochloraz have been synthesized by Selyutina et al. (2020). The authors have reported augmented pesticide penetration and solubility in corn and rape seeds.

15.2.3 Role of nanofertilizers

Nanofertilizers are formed by attaching fertilizers on the surface of nanoparticles for enhancing the cultivation, productivity, and nutritional value (Solanki et al., 2015; Chhipa, 2017). Nanofertilizers are composed of nutrients, thus, when they bind to the roots, it helps in the increase in the nutritious value (Singh, 2017). The nutrients in nanofertilizers are kept in the encapsulated way which attributes them to higher surface tension and are cost-effective (Mahanta et al., 2019). They deliver ions in the soil in a controlled-release manner, hold the nutrients firmly, and assist in decreasing their leaching (Preetha and Balakrishnan, 2017).

The advantages related to the usage of nanobiofertilizers (NFs) are numerous as they can enrich soil microbial composition, increase nutrient stability, and supply as well as reduce their leaching that ultimately promotes growth, productivity, and quality (Zulfiqar et al., 2019; Kalia and Kaur, 2019). However, the issues related to the interaction of nanobiofertilizers and crop restricts their frequent employment in fields (Batista and Singh, 2021). Quick physiological maturity and yield in wheat plants have been achieved with nanobiofertilizers along with an increase in spike number and length (Mardalipour et al., 2014). Still, there is a need to explore the interactions among NPs/NFs and plants in order to have smooth and better growth with high grain yield (Babu et al., 2021).

15.2.4 Role of nanocarriers and nano-packaging

With the development of plastics, the purpose of easy packaging and handing was fulfilled but their disposal in landfills laid an adverse impact on the environment and reduced soil fertility. Thus, the researchers introduced bioplastics formed out of starch (Peighambardoust et al., 2019; Marvizadeh et al., 2017). Their tensile properties, easy degradation, low thermal stability, and high flexibility of bioplastics influenced their application as an alternative to conventional plastics. Nano-based food packaging attributes safety to foodstuffs for a longer period (Kuswandi, 2017; Wesley et al., 2014). This technique has made possible physical–chemical stability for a longer duration. Physical–chemical permanence and its microbial adulteration during storage is a huge problem in the food industry. Environmental factors like light, moisture, and pH can compel a lot of changes in food products, still, we can overcome this by using edible thin film packing. Highly applicable examples of such products are carrageenan, chitosan, gelatin, polylactic acid, polyglycolic acid, alginate, etc. (Lagaron and Lopez-Rubio, 2011). Gases like oxygen and ethylene can damage food products and affect the shelf life and appearance of food content, this can be controlled by using edible thin films and coatings in food products (Venancio et al., 2005). Falguera et al. (2011) have reviewed edible films and coatings in the food industry, their trends, and applications. Biologically synthesis of silver and gold nanoparticles from *Fusarium* sp., *Pseudomonas struzeri, Pseudomonas aeruginosa*, and *Penicillium* sp. have been reported and their use in antimicrobial packaging has been discussed (Sadowski et al., 2008; Shateri-Khalilabad and Yazdanshenas, 2013). To remove chemicals and pathogens from food, oxygen and moisture leakage need to be reduced because this keeps food fresh for a long duration. For this purpose, silicate, titanium oxide, and zinc oxide metallic nanoparticles can be effectively used (Dwivedi and Jaykus, 2011). Zinc oxide nanoparticles (Espitia et al., 2012; Carbone et al., 2016), nanoceramics (Wang and Webster, 2015), and polymeric nanoparticles (Andriotis et al., 2021) have been used in food packaging and food waste management. The application of essential oils in food packing is an important area of research in food waste management projects (Ribeiro-Santos et al., 2017). Silver nanoparticles functionalized with chitosan manifested enhanced abilities for food preservation (Priyadarshi and Rhim, 2020). There is plenty of evidence for the application of nanoclays in food industries, packing, and food waste management (de Abreu et al., 2007). Carbon nanotubes and nanovascular systems like nanocapsules have recently emerged as effective nano-tools in food products.

15.3 Applications of nanobiotechnology in food and agriculture

Nanobiotechnology has been a boon in transforming the best out of waste. Nanobiotechnology-derived materials have promising opportunities and could figure out various challenges occurring in the food sector. This technique has helped us to produce value-added products from the nutritious extracts of food waste. It has been found that food waste is an excellent source of dietary fiber, minerals, organic acids, etc. (Torres-León et al., 2018; Ravindran and Jaiswal, 2016). Nanobiotechnology has roles in taste enhancement of food, elongates shelf life of food products, the nanoformulations can improve crop production, etc. In Table 15.1, various nanoparticles synthesized from agrowaste have been tabulated. Nanobiotechnology and its derived products have positively affected the crop industry. Such techniques have led to enhanced breeding, germination, and fruiting, moreover, it has reduced the plant life cycle and elevated stress tolerance (Servin et al., 2015; Jasrotia et al., 2018). Along with these properties, they have successfully minimized the postharvest decaying wastage proving their vitality in the agricultural sector (Yadollahi et al., 2009).

15.4 Effective strategies

Food processing is the channeling of food waste from various streams into usable energy. In the urgent need for effective food waste technology which includes proteins,

Table 15.1 Antimicrobial activities of nanoparticles synthesized from agrowaste.

S. no.	Nanoparticle	Source	Positive effects or use	References
1.	Ag Nps	Peel of pomegranate	*Staphylococcus aureus, Pseudomonas aeruginosa*, and *Escherichia coli* antibacterial activity	Shanmugavadivu et al. (2014)
2.	ZnO Nps	Green tea leaves	Antibacterial and antifungal activity against *E. coli, Staphylococcus aureus*, and *Aspergillus niger*	Irshad et al. (2018)
3.	Fe Nps	–	Improve child health by enhancing bioavailability and reactivity with this nutritional drink	Miller and Senjen (2019)
4.	Nanosilicates	–	Preservation and packing of food acting as a gas reduces food deterioration and rancidity	Jayas (2007)
5.	Nanobarcodes	–	Testing of food quality	Coles and Frewer (2013)
6.	Sodium and potassium ions-doped CaP NPs	Wheat	Increased yield while utilizing a considerably less amount of nitrogen	Ramírez-Rodríguez (2020)
7.	TiO_2 and ZnO NPs	Tomato	Enhancement in growth nanoparticulate fertilizer	Ramírez-Rodríguez (2020)
8.	CS–urea NPs (1000 mg/L) and plant mycorrhiza	Brassica oleracea spraying of plants cut	Nanobiofertilizers *Brassica oleracea* plants eliminate chemical nitrogen fertilizer inputs	Shams (2019)
9.	Micronutrient NFs Cu NPs	Alfalfa	Microorganisms necessary for elemental absorption have grown faster	Cota-Ruiz et al. (2020)
10.	Micronutrient NFs Biosynthesized Zn NPs Pearl millets	–	Crop production has increased. Primary growth indices and plant metabolic activities have increased significantly	Tarafdar et al. (2014)

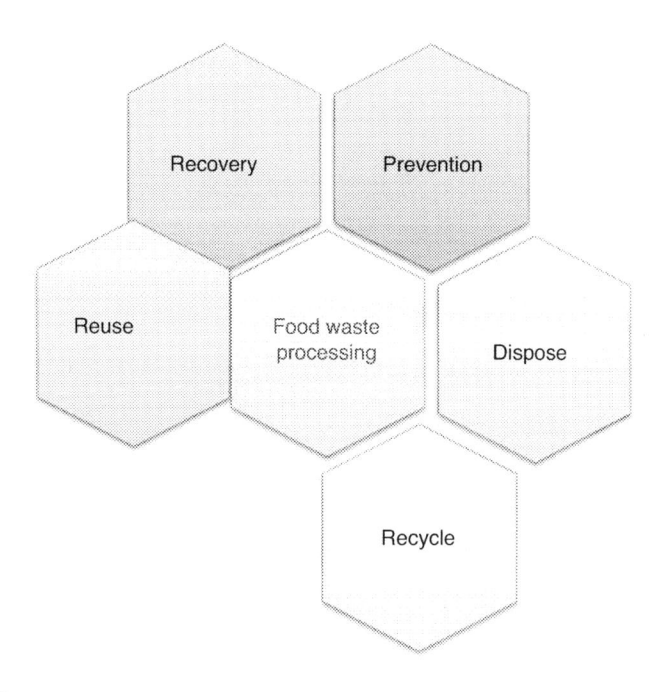

FIGURE 15.3

Effective strategy for management of food waste.

carbohydrates, nutraceuticals, and lipids, a sustainable approach is important. In this context, biodegradable plastics, bioactive compounds, biofuels, nanomaterials, and enzymes have been successful so far to a great extent. Food waste is not only limited to solids but also in water which further imposes serious health and environmental concerns if not directed properly. Water sensing through various technologies should be the major tool to overcome water-waste problems. Microbial enzymes play a significant role in the food industry for waste management as they are the greenest technique to process and recycle food waste. Biocatalysis of food waste should be mild reactions with lower generation of waste. This can be effectively achieved by microbial enzymes since they can aid in faster reactions with high biodegradability and specificity. Biosensing and enzyme action together have the capabilities of complete food waste recycling. There are many naturally available and synthesized biocatalysts (Ushani et al., 2020; Ng et al., 2020) which can be used in food waste management strategies. Five major strategies to achieve triple goals can be explained on the basis of five plans (1) source reduction, i.e., reduce food losses and food wastes, (2) redistribute the loss or wasted food to those who are hungry and needy, (3) recycle the food waste by feeding animal compost, (4) recover the energy nutrients, and (5) incineration. Fig. 15.3 depicts various strategies for the management of food waste. Food waste and security must be in alignment with food safety to ultimately achieve sustainability. Sustainability in food waste management overall means food availability, sufficient food production, food access, food nutrition, quality, safety, and stability. Food fraud is an alarming issue that may need immediate attention (Visciano and Schirone, 2021). There are several scandals in the food supply chain that are challenging national food security and safety. Landfills are posing hazardous effects on flora and fauna environment and health aspects because they release very offensive odor and polluted air (methane). Upon the availability of rainwater, these landfills pollute water and soil.

15.5 Conversion of food waste to biofuel

Food waste has a sufficient amount of lipids, carbohydrates, and proteins. Food waste can account for 30% of average lipid content and 50% of total carbohydrate content. This food waste can be hydrolyzed enzymatically for further breakdown and finally produce biodiesel. Household food waste is also a very good source for bioethanol production. Mainly liquefaction and saccharification are the two most common methods followed by fermentation for food waste conversion to biodiesel. Technologies like anaerobic digestion, esterification, hydrolysis and fermentation, hydrogenation and thermal gasification are being exploited for food waste conversion to biodiesel (Fig. 15.4). This strategy will solve two purposes at the same time because this can be used to overcome commercial shortage of fuel as well. From environmental concerns, this biological recycling of food waste is preferable. Ethanol production from food waste using saccharomyces cerevisiae H058 has been documented by Yan et al. (2013). Biocatalysts and several microorganisms can aid in large-scale hydrolysis of food waste in an environmentally friendly and cost-effective manner. Food waste can also be digested and decomposed biologically to produce biogas, a very effective and eco-friendly fuel to meet global energy demands. There have been a lot of research reports on the use of nanomaterials and nanocatalysts for the manufacturing of biodiesel (Rodríguez-Couto, 2019). Nanocatalysts can be reused several times during production with almost no loss of efficiency. Important microorganisms in biodiesel production are classified into four groups including yeasts, fungi, microalgae, and bacteria. Various nanoparticles like metal oxide nanoparticles, nanofibers, carbon nanotubes, and silica have been used for biofuel production (biogas, bioethanol, and biofuel) (Kamla et al., 2021).

15.6 Agrowaste management

Agrowaste management is also considered to be very significant in food waste management at the very initial stage. Agrowastes can be used to develop a completely green and sustainable route for the synthesis of NPs with low or no toxic effects on the environment. It has been reported that agrowaste acts as a reducing and stabilizing agent in NP synthesis (Periakaruppan et al., 2021). Due to irresponsible disposal

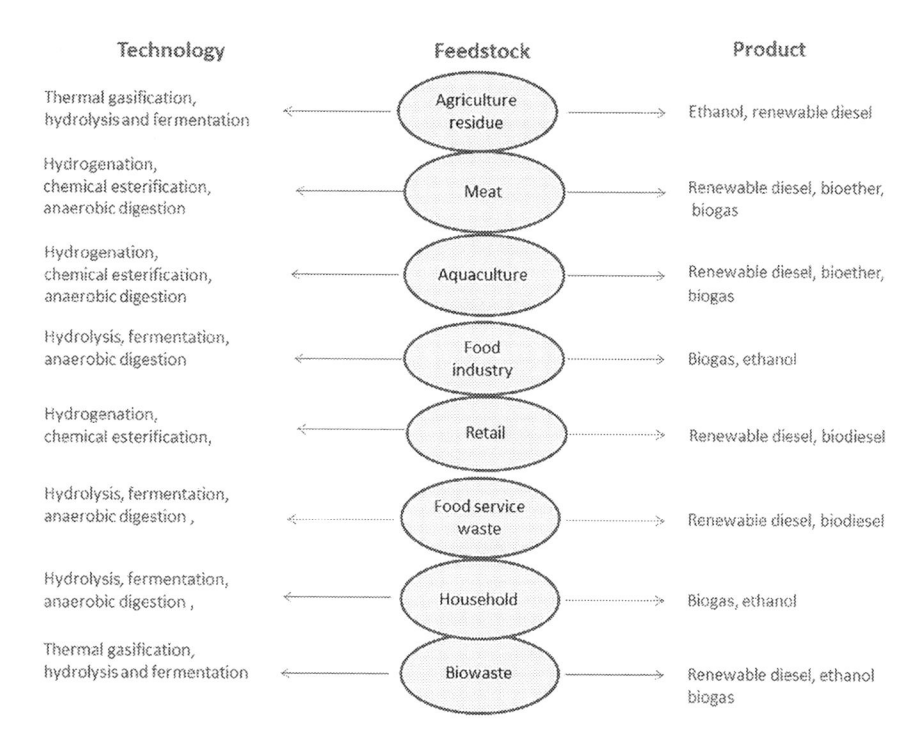

FIGURE 15.4

Routes for usable fuel energy production from food waste.

or burning of otherwise useful agromass in fields postharvest, creates pollution like smog and hence, poses serious health impacts. Hence, the concept of agrofood waste management needs to be advertised, so that these can be recycled properly (Kumar et al., 2021).

15.7 **Safety issues**

Apart from the advantages, profits, and benefits of nanobiotechnology, their use has been restricted and suffers trust issues, and they are not allowed to be used at the wide level (Fig. 15.5). Nanomaterials have unique properties which are still under investigation, thus, if using them in food leads to any kind of chemical reaction, it may be hazardous to human health. It has been found that nanomaterials involved in food packaging have caused toxic effects on the skin and lungs. The high absorption of nanomaterials causes bioaccumulation within the living system. Reports have announced increased reactive oxygen species (ROS), damaged DNA, and decreased ATP content. The high deposition of nanomaterials in the environment can adversely affect the flora and fauna, marine ecosystem. There are innumerable obstacles in

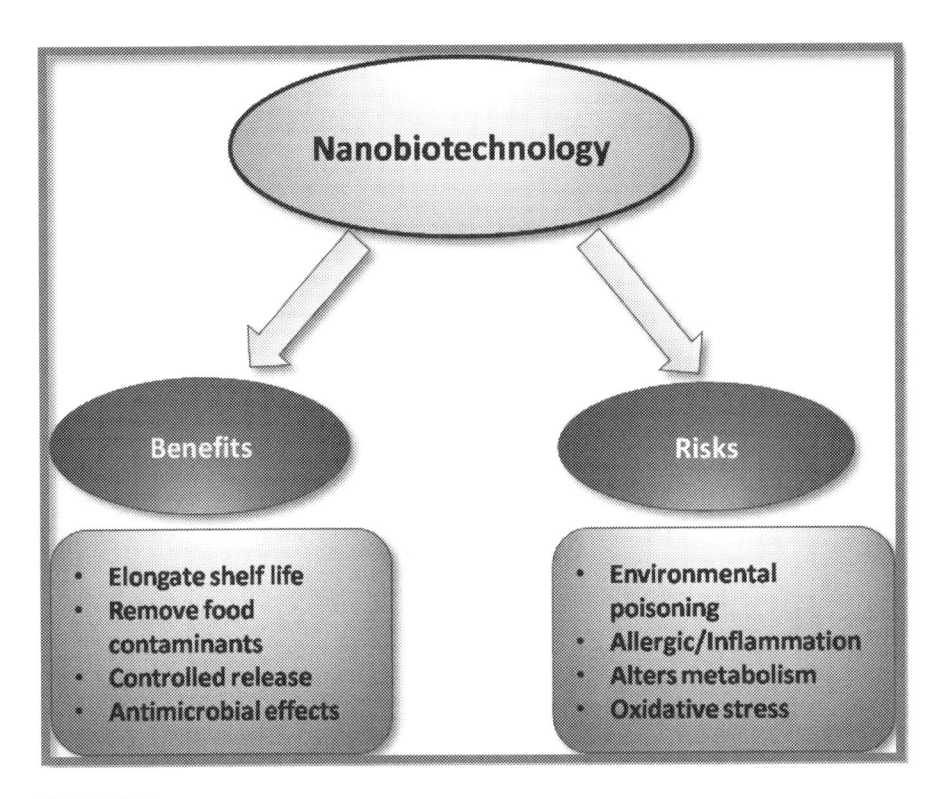

FIGURE 15.5

Benefits and risks associated with nanobiotechnology with respect to the food sector.

the execution of nano-based food safety solutions and antimicrobial agents. Gaining the trust of the public is extremely vital which could be achieved via efficacious communication and tools of management which nano-engineered organisms usually lack. Over usage of nanomaterials to enhance food productivity or as an additive to increase food shelf life has led to serious consequences. The physicochemical properties of nanomaterials decide the safety of their use since their ecotoxicity has often been reported (Kleandrova et al., 2014) and extensively discussed (Handy et al., 2012). Cytotoxicity and genotoxicity of NPs in the human body have been major areas of concern, despite the phenomenal applications of NPs (Valdiglesias et al., 2013; Medici et al., 2021).

Due to their unique chemical and physical properties, NPs can cause enormous toxic effects on the human body and also on the environment. Due to their exploitation in the food industry as preservatives, color and flavor enhancers, private and government sectors are concerned about the associated safety risks. These NPs can cause bioaccumulation and pose a threat to ecology and human health directly. Proper management of these safety issues becomes the utmost responsibility of people in the food industry. Flora and fauna are now being directly impacted by the toxic effects of NPs (Exbrayat et al., 2015; Remédios et al., 2012).

Food safety and security are crucial to enhance shelf life, provided pathogens and contaminants can be a major threat to health. The advancement in this field lies in the application of different nanomaterials, nanotubes, and quantum dots-based biosensors for the detection of pathogens. Nanobiosensors can act as magical bio tools to analyze pathogens and toxins present in food. Fluorescent dye and magnetic NPs-based biosensors have been studied for detecting campylobacter (Luo and Stutzenberger, 2008; Shan et al., 2015). Proper food waste management can be done by intelligent application of nanobiotechnology. Wisely designed biosensors can aid in the detection of food pathogens like Listeria monocytogenesin and mycotoxins (Välimaa et al., 2015). It is well known that ROS and oxidative stress can be responsible for cell death and dysfunctions, hence, NPs risk issue in food applications needs to be understood well. Kim et al. (2013) made an effort to study the negative effects of silver NPs in human lungs. Toxic effects of carbon nanotubes used in packing have also been reported in human lungs (Mills and Hazafy, 2009). Metal NPs are another threat to the food sector since it has been observed that they trigger cell toxicity. In spite of these evidence against the safety of NPs, a careful and systematic research is required to completely understand the direct toxic effects.

15.8 **Conclusion and future recommendations**

Food waste is adding 3.3 billion tons of CO_2 into the atmosphere every year along with food and land resource wastage. Anaerobic digestion can be an alluring option to strengthen the world's energy security by employing food waste to generate biogas while addressing waste management and nutrient recycling. There are myriad opportunities for NPs application in the food sector with several specialities and hence, proper management and risk assessment becomes extremely crucial. NPs have supreme potential of enhancing the nutritional value of food, maintain food consistency and upgrade performance, increase the shelf of food products, lower the fat band sugar content, detection of pathogens and check contaminants, and aid in the effective and safe packing of food. Innovative ideas and their ground implementations must be analyzed effectively to find a solution for food waste management using biotechnology. The current book's chapter precisely illustrates the utilization of nanobiotechnology for handling one of the prominent disguised challenges of the food sector, i.e., how to manage food waste in a sustainable manner. We can limit food waste conventionally by improvising the food inventory control, providing the remaining food to the either malnourished or the starving community. The food waste could also be fed to animals. The other ways of supervising the waste are their commercial composting and their management on an industrial scale. Better quality, longer shelf lives, desirable nutritional value, food safety, etc. have been raised by the proper application of nanotechnology and biotechnology in food waste management. On the other hand, food packing has also aided in enhancing food quality, flavor, nutrition, taste, etc. Nanofood concept is now not a dream but soon be a reality for consumers. Food management and safety includes a huge role of food companies as they must precisely pursue the guidelines issued by legislative bodies, for example, World Health Organization (WHO) and Food and Drug Administration (FDA) to assess the safety, security, packing, storage, and usage of supplements in food. These bodies should check for any toxic impurities or hazardous impacts of food. These authorized bodies are responsible for checking safety, security, packing, storage, etc. It is also important to frame strict guidelines and regulations to be followed for NPs application in the food sector. Synthesis of NPs with greener and safer techniques should be mandatory and promoted. Such kinds of efforts can aid in minimizing the risk factors associated with the food sector and food safety. In the long run, we must achieve triple goals of sustainability, food safety, and food security to reduce food losses and carbon footprints.

Acknowledgment

Srishti Sharma sincerely acknowledges the financial assistance of the Department of Science and Technology (DST), New Delhi, India for the research grant under the scheme DST INSPIRE fellowship (Vide Order No. DST/INSPIRE/03/2016/000619). Namrata Singh is grateful to the Faculty of Science, University of Hradec Kralove (VT2019-2021) and the excellence project UHK for their support. Namrata Singh is also thankful to the Department of Engineering Sciences, Ramrao Adik Institute of Technology, D.Y. Patil Deemed to be University for providing research facilities and support.

References

Agathokleous, E., Feng, Z., Iavicoli, I., Calabrese, E.J., 2020. Nano-pesticides: a great challenge for biodiversity? The need for a broader perspective. Nano Today 30, 100808.

Alfei, S., 2020. Nanotechnology applications to improve solubility of bioactive constituents of foods for health-promoting purposes. In: Nano-Food Engineering. Springer, Cham, pp. 189–257.

Andriotis, E.G., Papi, R.M., Paraskevopoulou, A., Achilias, D.S., 2021. Synthesis of D-limonene loaded polymeric nanoparticles with enhanced antimicrobial properties for potential application in food packaging. Nanomater 11, 191.

Arduini, F., Cinti, S., Scognamiglio, V., Moscone, D., 2016. Nanomaterials in electrochemical biosensors for pesticide detection: advances and challenges in food analysis. Microchim. Acta 183, 2063–2083.

Babu, S., Singh, R., Yadav, D., Rathore, S.S., Raj, R., Avasthe, R., Yadav, S.K., Das, A., Yadav, V., Yadav, B., Shekhawat, K., 2021. Nanofertilizers for agricultural and environmental sustainability. Chemosphere 292, 133451.

Batista, B.D., Singh, B.K., 2021. Realities and hopes in the application of microbial tools in agriculture. Microb. Biotechnol. 14, 1258–1268.

Bhattacharya, S., Jang, J., Yang, L., Akin, D., Bashir, R., 2007. BioMEMS and nanotechnology-based approaches for rapid detection of biological entities. J. Rapid Methods Autom. Micribiol. 15, 1–32.

Bonilla, J.C., Bozkurt, F., Ansari, S., Sozer, N., Kokini, J.L., 2016. Applications of quantum dots in food science and biology. Trends Food Sci. Technol. 53, 75–89.

Carbone, M., Donia, D.T., Sabbatella, G., Antiochia, R., 2016. Silver nanoparticles in polymeric matrices for fresh food packaging. J. King Saud Univ. Sci. 28, 273–279.

Chhipa, H., 2017. Nanofertilizers and nanopesticides for agriculture. Environ Chem. Lett. 15, 15–22.

Coles, D., Frewer, L.J., 2013. Nanotechnology applied to European food production: a review of ethical and regulatory issues. Trends Food Sci. Technol. 34, 32–43.

Cota-Ruiz, K., Ye, Y., Valdes, C., Deng, C., Wang, Y., Hernández-Viezcas, J.A., Duarte-Gardea, M., Gardea-Torresdey, J.L., 2020. Copper nanowires as nanofertilizers for alfalfa plants: understanding nano-bio systems interactions from microbial genomics, plant molecular responses and spectroscopic studies. Sci. Total Environ. 742, 140572.

de Abreu, D.P., Losada, P.P., Angulo, I., Cruz, J.M., 2007. Development of new polyolefin films with nanoclays for application in food packaging. Eur. Polym. J. 43, 2229–2243.

Duncan, T.V., 2011. Applications of nanotechnology in food packaging and food safety: barrier materials, antimicrobials and sensors. J. Colloid Interface Sci. 363, 1–24.

Dwivedi, H.P., Jaykus, L.A., 2011. Detection of pathogens in foods: the current state-of-the-art and future directions. Crit. Rev. Microbiol. 37, 40–63.

Espitia, P.J.P., Soares, N.D.F.F., dos Reis Coimbra, J.S., de Andrade, N.J., Cruz, R.S., Medeiros, E.A.A., 2012. Zinc oxide nanoparticles: synthesis, antimicrobial activity and food packaging applications. Food Bioproc. Technol. 5, 1447–1464.

Exbrayat, J.M., Moudilou, E.N., Lapied, E., 2015. Harmful effects of nanoparticles on animals. J. Nanotechnol. 2015, 1–10.

Falguera, V., Quintero, J.P., Jiménez, A., Muñoz, J.A., Ibarz, A., 2011. Edible films and coatings: structures, active functions and trends in their use. Trends Food Sci. Technol. 22, 292–303.

Fan, L., Ellison, B., Wilson, N.L., 2022. What food waste solutions do people support? J. Clean. Prod. 330, 129907.

Fuertes, G., Soto, I., Carrasco, R., Vargas, M., Sabattin, J., Lagos, C., 2016. Intelligent packaging systems: sensors and nanosensors to monitor food quality and safety. J. Sens. 2016, 1–8.

Handy, R.D., Cornelis, G., Fernandes, T., Tsyusko, O., Decho, A., Sabo-Attwood, T., Metcalfe, C., Steevens, J.A., Klaine, S.J., Koelmans, A.A., Horne, N., 2012. Ecotoxicity test methods for engineered nanomaterials: practical experiences and recommendations from the bench. Environ. Toxicol. Chem. 31, 15–31.

Holzinger, M., Le Goff, A., Cosnier, S., 2014. Nanomaterials for biosensing applications: a review. Front. Chem. 2, 63.

Huang, Y., Dong, X., Liu, Y., Li, L.J., Chen, P., 2011. Graphene-based biosensors for detection of bacteria and their metabolic activities. J. Mater. Chem. 21, 12358–12362.

Irshad, S., Salamat, A., Anjum, A.A., Sana, S., Saleem, R.S., Naheed, A., Iqbal, A., 2018. Green tea leaves mediated ZnO nanoparticles and its antimicrobial activity. Cogent Chem. 4, 1469207.

Jafarizadeh-Malmiri, H., Sayyar, Z., Anarjan, N., Berenjian, A., 2019. In: Nano-Sensors in food nanobiotechnology. In: Nanobiotechnology in Food: Concepts, Applications and Perspectives. Springer, Cham, pp. 81–94.

Jasrotia, P., Kashyap, P.L., Bhardwaj, A.K., Kumar, S., Singh, G.P., 2018. Scope and applications of nanotechnology for wheat production: a review of recent advances. Wheat Barley Res. 10, 1–14.

Jayas, D.S., 2007. Sensors for grain storage. In: 2007 ASAE Annual Meeting. American Society of Agricultural and Biological Engineers 1.

Kah, M., Beulke, S., Tiede, K., Hofmann, T., 2013. Nanopesticides: state of knowledge, environmental fate, and exposure modeling. Crit. Rev. Environ. Sci. Technol. 43, 1823–1867.

Kalia, A., Kaur, H., 2019. Nano-biofertilizers: harnessing dual benefits of nano-nutrient and bio-fertilizers for enhanced nutrient use efficiency and sustainable productivity. In: Nanoscience for Sustainable Agriculture. Springer, Cham, pp. 51–73.

Kamla, M., Sushil, A., Karmal, M., 2021. Nanotechnology: a sustainable solution for bioenergy and biofuel production. J. Nanosci. Nanotechnol. 21, 3481–3494.

Kim, J.S., Peters, T.M., O'Shaughnessy, P.T., Adamcakova-Dodd, A., Thorne, P.S., 2013. Validation of an in vitro exposure system for toxicity assessment of air-delivered nanomaterials. Toxicol. In Vitro 27, 164–173.

Kleandrova, V.V., Luan, F., González-Díaz, H., Ruso, J.M., Melo, A., Speck-Planche, A., Cordeiro, M.N.D., 2014. Computational ecotoxicology: simultaneous prediction of ecotoxic effects of nanoparticles under different experimental conditions. Environ. Int. 73, 288–294.

Krishnan, R., Agarwal, R., Bajada, C., Arshinder, K., 2020. Redesigning a food supply chain for environmental sustainability: an analysis of resource use and recovery. J. Clean. Prod. 242, 118374.

Kumar, S., Tripathi, G., Mishra, G.V., 2021. A comparative study on agrowaste conversion into biofertilizer employing two earthworm species. Appl. Ecol. Environ. Res. 9, 280–285.

Kuswandi, B., 2017. Environmental friendly food nano-packaging. Environ. Chem. Lett. 15, 205–221.

Kuswandi, B., Futra, D., Heng, L.Y., 2017. Nanosensors for the detection of food contaminants. In: Nanotechnology Applications in Food. Academic Press, pp. 307–333.

Lagaron, J.M., Lopez-Rubio, A., 2011. Nanotechnology for bioplastics: opportunities, challenges and strategies. Trends Food Sci. Technol. 22, 611–617.

Lugani, Y., Sooch, B.S., Singh, P., Kumar, S., 2021. Nanobiotechnology applications in food sector and future innovations. In: Microbial Biotechnology in Food and Health. Academic Press, pp. 197–225.

Luo, P.G., Stutzenberger, F.J., 2008. Nanotechnology in the detection and control of microorganisms. Adv. Appl. Microbiol. 63, 145–181.

Mahanta, N., Ashok, D., Montrishna, R., 2019. Nutrient use efficiency through nano fertilizers. Int. J. Chem. Stud. 7, 2839–2842.

Manikandababu, C.S., Jagadeeswari, M., Priyanka, R., Preethi, S., Rithika, V., Kumar, J.R., 2022. Webpage portal for crowd sourcing on food waste management. In: Sentimental Analysis and Deep Learning. Springer, Singapore, pp. 813–827.

Mardalipour, M., Zahedi, H., Sharghi, Y., 2014. Evaluation of nano biofertilizer efficiency on agronomic traits of spring wheat at different sowing date. Biol. Forum 6, 349.

Martirosyan, A., Schneider, Y.-J., 2014. Engineered nanomaterials in food: implications for food safety and consumer health. Int. J. Environ. Res. Public Health 11, 5720–5750.

Marvizadeh, M.M., Oladzadabbasabadi, N., Nafchi, A.M., Jokar, M., 2017. Preparation and characterization of bionanocomposite film based on tapioca starch/bovine gelatin/nanorod zinc oxide. Int. J. Biol. Macromol. 99, 1–7.

McClements, D.J., Xiao, H., 2017. Is nano safe in foods? Establishing the factors impacting the gastrointestinal fate and toxicity of organic and inorganic food-grade nanoparticles. NPJ Sci. Food 1, 1–13.

Medici, S., Peana, M., Pelucelli, A., Zoroddu, M.A., 2021. An updated overview on metal nanoparticles toxicity. In: Seminars in Cancer Biology, 76. Academic Press, pp. 17–26.

Miller, G., Senjen, R., 2019. Nanotechnology in Food and Agriculture. Jenny Stanford Publishing, pp. 417–444.

Mills, A., Hazafy, D., 2009. Nanocrystalline SnO_2-based, UVB-activated, colourimetric oxygen indicator. Sens. Actuators B Chem. 136, 344–349.

Naja, G., Bouvrette, P., Hrapovic, S., Luong, J.H., 2007. Raman-based detection of bacteria using silver nanoparticles conjugated with antibodies. Analyst 132, 679–686.

Ng, H.S., Kee, P.E., Yim, H.S., Chen, P.T., Wei, Y.H., Lan, J.C.W., 2020. Recent advances on the sustainable approaches for conversion and reutilization of food wastes to valuable bioproducts. Bioresour. Technol. 302, 122889.

No, H.J., Park, Y., 2010. Trajectory patterns of technology fusion: trend analysis and taxonomical grouping in nanobiotechnology. Technol. Forecast. Soc. Change 77, 63–75.

Omanović-Mikličanina, E., Maksimović, M., 2016. Nanosensors applications in agriculture and food industry. Bull. Chem. Technol. Bosnia Herzegovina 47, 59–70.

Parfitt, J., Barthel, M., Macnaughton, S., 2010. Food waste within food supply chains: quantification and potential for change to 2050. Philos. Trans. Royal Soc. B Biol. Sci. 365, 3065–3081.

Peighambardoust, S.J., Peighambardoust, S.H., Pournasir, N., Pakdel, P.M., 2019. Properties of active starch-based films incorporating a combination of Ag, ZnO and CuO nanoparticles for potential use in food packaging applications. Food Packag. Shelf Life 22, 100420.

Periakaruppan, R., Li, J., Mei, H., Yu, Y., Hu, S., Chen, X., Li, X., Guo, G., 2021. AGRO-WASTE mediated biopolymer for production of biogenic nano iron oxide with superparamagnetic power and antioxidant strength. J. Clean Prod. 311, 127–512.

Pradhan, N., Singh, S., Ojha, N., Shrivastava, A., Barla, A., Rai, V., Bose, S., 2015. Facets of nanotechnology as seen in food processing, packaging, and preservation industry. Biomed. Res. Int. 2015, 1–17.

Prasad, R., Kumar, V., Kumar, M., Choudhary, D.K. (Eds.), 2019. Nanobiotechnology in Bioformulations. Springer International Publishing, Cham.

Preetha, P.S., Balakrishnan, N., 2017. A review of nano fertilizers and their use and functions in soil. Int. J. Curr. Microbiol. App. Sci 6, 3117–3133.

Priyadarshi, R., Rhim, J.W., 2020. Chitosan-based biodegradable functional films for food packaging applications. Innov. Food Sci. Emerg. Technol. 62, 102346.

Purkayastha, M.D., Manhar, A.K., 2016. Nanotechnological applications in food packaging, sensors and bioactive delivery systems. Nanosci. Food Agric. 2, 59–128.

Pushparaj, K., Liu, W.C., Meyyazhagan, A., Orlacchio, A., Pappusamy, M., Vadivalagan, C., Robert, A.A., Arumugam, V.A., Kamyab, H., Klemeš, J.J., Khademi, T., 2022. Nano- from nature to nurture: a comprehensive review on facets, trends, perspectives and sustainability of nanotechnology in the food sector. Energy 240, 122732.

Ramírez-Rodríguez, G.B., Dal Sasso, G., Carmona, F.J., Miguel-Rojas, C., Pérez-de-Luque, A., Masciocchi, N., Guagliardi, A., Delgado-López, J.M., 2020. Engineering biomimetic calcium phosphate nanoparticles: a green synthesis of slow-release multinutrient (NPK) nanofertilizers. ACS Appl. Bio. Mater. 3, 1344–1353.

Ravindran, R., Jaiswal, A.K., 2016. Exploitation of food industry waste for high-value products. Trends Biotechnol. 34, 58–69.

Remédios, C., Rosário, F., Bastos, V., 2012. Environmental nanoparticles interactions with plants: morphological, physiological, and genotoxic aspects. J. Bot. 2012, 751686.

Ribeiro-Santos, R., Andrade, M., Sanches-Silva, A., 2017. Application of encapsulated essential oils as antimicrobial agents in food packaging. Curr. Opin. Food Sci. 14, 78–84.

Rodríguez-Couto, S., 2019. Green nanotechnology for biofuel production. In: Sustainable Approaches for Biofuels Production Technologies. Springer, Cham, pp. 73–82.

Sadowski, Z., Maliszewska, I.H., Grochowalska, B., Polowczyk, I., Kozlecki, T., 2008. Synthesis of silver nanoparticles using microorganisms. Mater. Sci.-Pol. 26, 419–424.

Sahoo, D., Mandal, A., Mitra, T., Chakraborty, K., Bardhan, M., Dasgupta, A.K., 2018. Nanosensing of pesticides by zinc oxide quantum dot: an optical and electrochemical approach for the detection of pesticides in water. J. Agri. Food Chem. 66, 414–423.

Selyutina, O.Y., Khalikov, S.S., Polyakov, N.E., 2020. Arabinogalactan and glycyrrhizin based nanopesticides as novel delivery systems for plant protection. Environ. Sci. Pollut. Res. 27, 5864–5872.

Servin, A., Elmer, W., Mukherjee, A., De la Torre-Roche, R., Hamdi, H., White, J.C., Bindraban, P., Dimkpa, C., 2015. A review of the use of engineered nanomaterials to suppress plant disease and enhance crop yield. J. Nanoparticle Res. 17, 1–21.

Shah, M.A., Wani, S.H., Khan, A.A., 2016. Nanotechnology and insecticidal formulations. J. Food Bioeng. Nanoproc. 1, 285–310.

Shams, A.S., 2019. Foliar applications of nano chitosan-urea and inoculation with mycorrhiza on kohlrabi (*Brassica oleracea* Var. *Gongylodes L.*). J. Plant Prod. 10, 799–805.

Shan, S., Lai, W., Xiong, Y., Wei, H., Xu, H., 2015. Novel strategies to enhance lateral flow immunoassay sensitivity for detecting foodborne pathogens. J. Agric. Food Chem. 63, 745–775.

Shanmugavadivu, M., Kuppusamy, S., Ranjithkumar, R., 2014. Synthesis of pomegranate peel extract mediated silver nanoparticles and its antibacterial activity. Am. J. Adv. Drug Deliv. 2, 174–182.

Sharma, M.K., Dhaka, V.S., Shekhawat, R.S., 2022. Intelligent agro-food chain supply. In: Internet of Things and Analytics for Agriculture, 3. Springer, Singapore, pp. 65–91.

Shateri-Khalilabad, M., Yazdanshenas, M.E., 2013. Bifunctionalization of cotton textiles by ZnO nanostructures: antimicrobial activity and ultraviolet protection. Text. Res. J. 83, 993–1004.

Siddiqui, S., Alrumman, S.A., 2021. Influence of nanoparticles on food: an analytical assessment. J. King Saud Univ. Sci. 33, 101530.

Silva, G.A., 2004. Introduction to nanotechnology and its applications to medicine. Surg. Neurol. 61, 216–220.

Singh, M.D., 2017. Nano-fertilizers is a new way to increase nutrients use efficiency in crop production. Int. J. Agric. Sci. 7, 0975–3710.

Singh, T., Shukla, S., Kumar, P., Wahla, V., Bajpai, V.K., Rather, I.A., 2017. Application of nanotechnology in food science: perception and overview. Front. Microbiol. 8, 1501.

Solanki, P., Bhargava, A., Chhipa, H., Jain, N., Panwar, J., 2015. Nano-fertilizers and their smart delivery system. In: Nanotechnologies in Food and Agriculture. Springer, Cham, pp. 81–101.

Solgaard, O., 2009. Photonic Microsystems: Micro and Nanotechnology Applied to Optical Devices and Systems. Springer Science & Business Media.

Sozer, N., Kokini, J.L., 2009. Nanotechnology and its applications in the food sector. Trends Biotechnol. 27, 82–89.

Srivastava, A.K., Dev, A., Karmakar, S., 2018. Nanosensors and nanobiosensors in food and agriculture. Environ. Chem. Lett. 16, 161–182.

Tarafdar, J.C., Raliya, R., Mahawar, H., Rathore, I., 2014. Development of zinc nanofertilizer to enhance crop production in pearl millet (*Pennisetum americanum*). Agric. Res. 3, 257–262.

Torres-León, C., Ramírez-Guzman, N., Londoño-Hernandez, L., Martinez-Medina, G.A., Díaz-Herrera, R., Navarro-Macias, V., Alvarez-Pérez, O.B., Picazo, B., Villarreal-Vázquez, M., Ascacio-Valdes, J., Aguilar, C.N., 2018. Food waste and byproducts: an opportunity to minimize malnutrition and hunger in developing countries. Front. Sustain. Food Syst. 2, 52.

Toumey, C., 2009. Plenty of room, plenty of history. Nat. Nanotechnol. 4, 783–784.

Ushani, U., Sumayya, A.R., Archana, G., Banu, J.R., Dai, J., 2020. Enzymes/biocatalysts and bioreactors for valorization of food wastes. In: Food Waste to Valuable Resources. Academic Press, pp. 211–233.

Valdiglesias, V., Costa, C., Kiliç, G., Costa, S., Pásaro, E., Laffon, B., Teixeira, J.P., 2013. Neuronal cytotoxicity and genotoxicity induced by zinc oxide nanoparticles. Environ. Int. 55, 92–100.

Välimaa, A.L., Tilsala-Timisjärvi, A., Virtanen, E., 2015. Rapid detection and identification methods for *Listeria* monocytogenes in the food chain: a review. Food Control 55, 103–114.

Vasile, C., 2018. Polymeric nanocomposites and nanocoatings for food packaging: a review. Material 11, 1834.

Venancio, E.C., Consolin Filho, N., Constantino, C.J., Martin-Neto, L., Mattoso, L.H., 2005. Studies on the interaction between humic substances and conducting polymers for sensor application. J. Braz. Chem. Soc. 16, 24–30.

Visciano, P., Schirone, M., 2021. Food frauds: global incidents and misleading situations. Trends Food Sci. Technol. 114, 424–442.

Wang, Q., Webster, T.J., 2015. Nanoceramics processing: revolutionizing medicine. Adv. Process. Manuf. Technol. Nanostruc. Multifunc. Mater. 594, 213.

Wesley, S.J., Raja, P., Raj, A.A., Tiroutchelvamae, D., 2014. Review on-nanotechnology applications in food packaging and safety. Int. J. Eng. Res. Technol. 3, 645–651.

Xin, X., Judy, J.D., Sumerlin, B.B., He, Z., 2020. Nano-enabled agriculture: from nanoparticles to smart nanodelivery systems. Environ. Chem. 17, 413–425.

Yadollahi, A., Arzani, K., Khoshghalb, H., 2009. The role of nanotechnology in horticultural crops postharvest management. In: Southeast Asia Symposium on Quality and Safety of Fresh and Fresh-Cut Produce, pp. 49–56.

Yan, S., Chen, X., Wu, J., Wang, P., 2013. Pilot-scale production of fuel ethanol from concentrated food waste hydrolysates using saccharomyces cerevisiae H058. Bioprocess. Biosyst. Eng. 36, 937–946.

Yonzon, C.R., Stuart, D.A., Zhang, X., McFarland, A.D., Haynes, C.L., Van Duyne, R.P., 2005. Towards advanced chemical and biological nanosensors: an overview. Talanta 67, 438–448.

Zulfiqar, F., Navarro, M., Ashraf, M., Akram, N.A., Munné-Bosch, S., 2019. Nanofertilizer use for sustainable agriculture: advantages and limitations. Plant Sci. 289, 110270.

Zuo, P., Li, X., Dominguez, D.C., Ye, B.C., 2013. A PDMS/paper/glass hybrid microfluidic biochip integrated with aptamer-functionalized graphene oxide nano-biosensors for one-step multiplexed pathogen detection. Lab Chip 13, 3921–3928.

Food packaging

Potential significance of nanobiotechnology in food packaging

16

Narender Ranga[a], Rohit Ranga[b], Krishan Kumar[b] and Ekta Poonia[b]

[a]*Department of Physics, D.C.R. University of Science and Technology, Sonipat, Haryana, India,*
[b]*Department of Chemistry, D.C.R. University of Science and Technology, Sonipat, Haryana, India*

16.1 Introduction

In order to address key issues, maintain integrity, and actively stop food spoiling, nanobiotechnology is anticipated to play a significant role in the development of future food (Nile et al., 2020). Food packaging contributes significantly to food security by lowering losses, as well as to food safety and trade, which is essential for the growth of many economies. Animal leaves and skins were once used to transport, retain, and preserve food before the use of food packaging became widespread. The use of various materials in the production of food packaging and the development of contemporary packaging types, such as intelligent packaging or smart packaging (gas, radio frequency identification, microwave cooking indicators, temperature indicator, time, and others) as well as active packaging, are results of this industry's advancements (antimicrobials, moisture absorbers, and oxygen sensors). These days, simple packaging materials consisting of paper or plastic are employed. In industrialized nations, these materials make up the majority of the waste, and the expense of recycling them is rising along with the harm done to the marine environment. The attention to potential solutions is centered on "smart packaging." Smartness packaging refers to a variety of practical techniques that can be customized for different items, such as different kinds of food, drinks, pharmaceuticals, household goods, etc. (Nicoletti and Del Serrone, 2017).

The majority of materials used in the packaging industry today are petroleum-based polymeric polymers that cannot be degraded (Keskin et al., 2017). Because of this, nondegradable food packaging materials pose a significant environmental threat on a global scale. By extending shelf life and lowering packaging waste, which in turn improves food quality, the adoption of bio-based packaging materials, such as biodegradable and edible films from renewable resources, could therefore, at least in part, solve the waste problem. In this regard, it is feasible to maintain the product's quality and freshness for the duration of its commercialization and consumption by using the right materials and packaging techniques. By developing tiny ingestible capsules or nanoparticles that release their contents on demand at

Nanobiotechnology for Food Processing and Packaging. DOI: https://doi.org/10.1016/B978-0-323-91749-0.00017-4

specific locations in the body, many food researchers are attempting to optimize the delivery of medications or delicate micronutrients in common foods (Sumit, 2012). The importance of nanobiotechnology in food packaging is highlighted in this chapter, along with a discussion of bio-based packaging.

16.2 History of packaging

Natural materials like leaves have been used in packaging. Products like woven fabrics and ceramics were later produced in bulk (Shaikh et al., 2022). Glass and wood packaging have reportedly been used for commerce for almost 5000 years (Kale et al., 2007). A packaging innovation frequently requires the convergence of several technological breakthroughs before it is implemented. These have included advancements in postharvest technologies, new retail formats, and home appliances like refrigerators, freezers, and microwaves. They have also included transportation and infrastructural innovations. For instance, the invention of the microwave oven sparked the creation of convenient food packing for a variety of cuisines. For an innovation to be successful, the sociocultural, demographic, consumer lifestyle, and economic environment must also create enough market demand.

Primary, secondary, and tertiary packaging are the three degrees of packaging according to a particular function (Lindh et al., 2016). Because it is suggested for various packing scenarios, it is crucial to differentiate between the three levels.

1. Primary packaging: The protection, preservation, containment, and education of the consumers are the basic goals of primary packaging. The principal packaging of a liquid product is frequently illustrated by a bottle.
2. Secondary packaging: The majority of this packaging consists of cardboard boxes, though they can also be made of a plastic material. A typical example is the packaging of milk, where each carton would be primary packaging and the cardboard box holding the pack of cartons would be secondary packaging.
3. Tertiary packaging: In terms of the brand's standpoint, tertiary packaging can also be very important. Tertiary packaging is used to preserve the goods during storage and transportation; it is not frequently seen on store shelves.

16.3 Food packaging

One of the most important stages in ensuring food safety is food packaging (Stringer and Hall, 2007). The main goals of food packaging are to stop contamination and deterioration, increase sensitivity by allowing enzyme activity, and stop weight loss (Primožič et al., 2021). The primary use of nanotechnology in the food business is reportedly in food packaging. It has been shown that adding nanoparticles to shaped materials and films improves their qualities, particularly their toughness, recycling properties, optical properties, barrier properties, flame resistance, and temperature resistance. Antimicrobial nanomaterials are a significant portion of the modern packaging idea intended to introduce potent nanoparticles that can be incorporated into a food package (Thiruvengadam et al., 2018). Nanotechnology has significant potential to provide benefits not just within food products but also surrounding them. It may be used to develop stronger flavors and color quality as well as to detect germs in packaging and improve food safety. Nanotechnology does, in fact, quickly open new opportunities for innovation in the food sector, but uncertainties and health concerns are also beginning to surface.

16.4 Nanobiotechnology in food packaging

16.4.1 Advanced nanobiocomposite packaging

Nanobiocomposite materials, which are suitable for food packaging, are the foundation of advanced packaging (Sadjadi et al., 2019). The biodegradable films made of nanocomposite packaging materials could be used for food packaging to regulate the exchange of gases or moisture, extending shelf life, ensuring food safety, and preserving nutritious contents and sensory quality (Qu et al., 2022). By separating food from its surroundings and preventing food from deteriorating due to microbes, gas conditions, and relative ambient humidity, bio-based packaging materials are more environmentally friendly than plastic packaging (Kuswandi and Moradi, 2019).

Recent research has shown that a novel class of bio-nanocomposite materials can be a reliable replacement for existing materials by enhancing their mechanical and barrier properties (Kumar et al., 2017). A biopolymer matrix that has been fortified by nanoparticles with at least one dimension in the nanometer range makes up the bio-nanocomposites (1–100 nm). Due to interest in sustainable development, biopolymers are being examined as viable replacements for traditional plastic packaging materials. The majority of studies on the characteristics and creation of biopolymer films concentrate on using them as edible films (Mellinas et al., 2016). The various nanocomposites found in nature may serve as inspiration for research. For instance, aragonite, a carbonate mineral, is a natural nanocomposite found in seashells that has excellent mechanical strength and toughness. Another example of nanocomposites in nature is bone tissue. Its structure might serve as a guide for creating biomimetic materials (Pina et al., 2015). The fibril of mineralized collagen serves as its basic unit. Nanoscale hydroxyapatite ($Ca_5(PO_4)_3OH$) plates dispersed in a collagen matrix give bone its ordered layered structure. Due to interest in sustainable development, biopolymers in Fig. 16.1 are being examined as viable replacements for traditional plastic packaging materials.

16.4.2 Active packages

One of the most innovative food preservation solutions is active packaging. Its functionality depends on the inherent qualities of the polymer or the unique qualities of the additives used in the packaging systems. Chitosan is a type of biopolymer

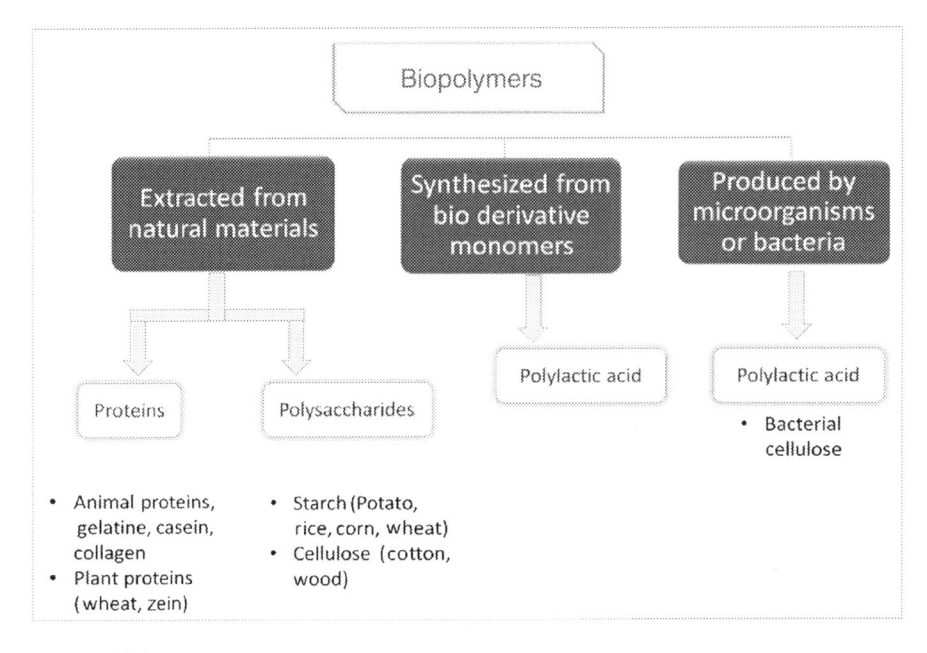

FIGURE 16.1

Schematic overview of biopolymers.

with antibacterial and antifungal properties. These properties are linked to changes in the cellular permeability of microorganisms brought on by electrostatic interactions between the amine groups in chitosan and the electronegative charges on cellular surfaces (Xing et al., 2015). In order to achieve the desired result with lower concentrations, active agents can be incorporated into the packaging material, coated on its surface, or also inside certain package-related elements, such as bottle caps, pads, labels, and bags. This limits the number of unfavorable odors and flavors that are transferred to foods. If necessary, these active ingredients are added to alter the composition or organoleptic properties of food while also releasing or absorbing chemicals into or out of the packaged food or its environment. The use of natural additives is currently preferred over synthetic ones, which are occasionally linked to specific health hazards (Carocho et al., 2015). Even some agro-industrial waste or by-products, such as those from the wine, beer, dairy, and meat sectors as well as those produced from the processing of fruits and vegetables, offer practical and affordable sources of powerful active chemicals. To prevent their loss or destruction during the production of the material and to ensure their activity in the packaging, it is important to take into account the thermal resistance and action mechanism of these additives. The moisture-regulating components, oxygen scavengers, CO_2 scavengers and emitters, and antimicrobials make up the active packaging systems. Adaptive packaging solutions are created based on the storage's intended use.

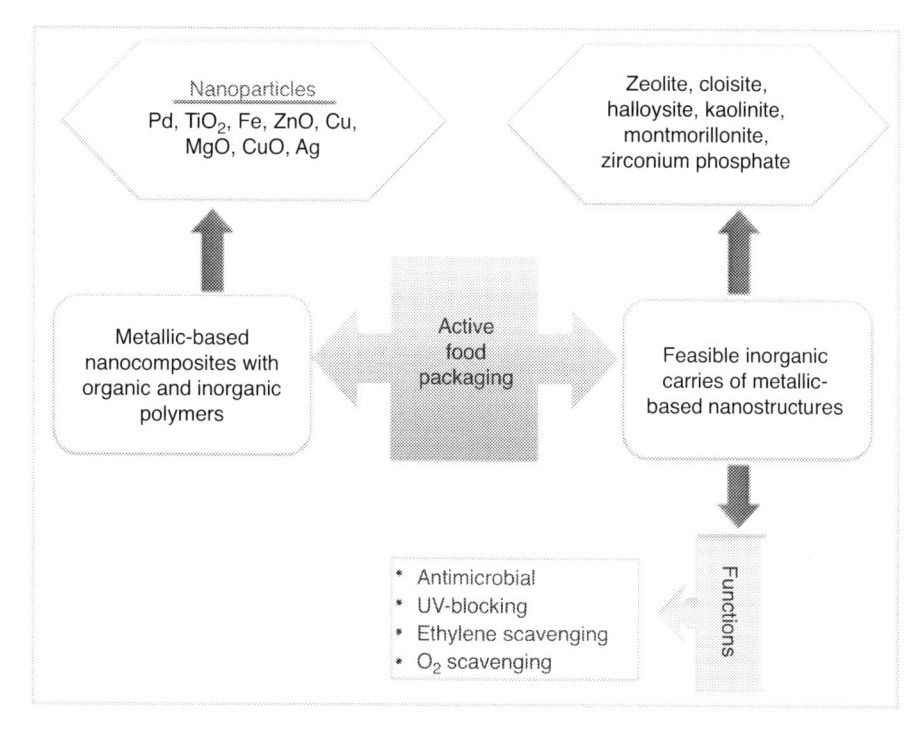

FIGURE 16.2

Active food nano packaging.

One of the crucial systems used to distribute, store, and maintain meat products in a cold environment is the modified atmospheric packaging (MAP) (Kerry et al., 2006). The nano packaging for active foods is depicted below as explained in Fig. 16.2.

16.4.3 Antimicrobial nanocomposite packaging

Nanoparticles can be integrated into polymeric matrices to create nanocomposites, which can give materials specialized properties like high barrier, antioxidant capacity, antibacterial activity, or better mechanical performance (Li et al., 2019). Due to their potent antibacterial and/or antifungal properties, various nanoparticles have been studied, including nanoclays, silver, titanium dioxide, and zinc oxide (Espitia et al., 2012). Improving food safety, preserving product color, flavor, and odor, preventing spoiling, and prolonging shelf life are the driving forces behind antimicrobial packaging. The food manufacturer may have more possibilities with antimicrobial packaging. During handling, shipping, and storage, refrigeration and the management of temperature excursions are also crucial. Antimicrobials can be used in one of three ways to prevent bacterial growth inside the package. One option is to include the antimicrobial agent as an ingredient in the food product itself. Packaging rules do not

Table 16.1 Different nanoparticles and nanomaterials used in antimicrobial nanocomposites.

Nanoparticles	Main features
Silver nanoparticles	High thermal stability, diverse and prolonged antimicrobial effect. Ag^+ interacts with proteins and enzymes, cell wall and plasma membrane constituents, and microbial DNA.
Metal oxide nanoparticles	Nanoparticles of ZnO, MgO, CaO, CuO, Al_2O_3, AgO, and CeO_2 have been studied as constituents of antimicrobial composites. The ZnO showed good antibacterial behavior in both biodegradable and nonbiodegradable polymers. TiO_2 has also been used, although its antimicrobial activity is dependent on ultraviolet (UV) and UV visible effects.
Copper nanoparticles	Copper nanoparticles can be obtained by physical or chemical methods using heat and UV radiation or reducing agents, respectively. Antifungal and antibacterial activity has been confirmed by the addition of biodegradable and nonbiodegradable polymers.
Nanoemulsions	Techniques such as high-pressure homogenization and sonication may be considered to obtain nanoemulsions since very low-size droplets have been related to higher antimicrobial activity.

apply to this food additive method. The agent can also be added to the headspace. This is how sachets and MAP (define MAP) function. The initial environment provided by MAP can be used to restrict microbial development, however, if the headspace atmosphere varies over time, the rate of microbial growth may increase. Additionally, as MAP often has low oxygen levels, it is essential to inhibit the growth of anaerobic microbes. The agent might be included in the packaging as a third tactic. Antimicrobial additives must typically come in touch with food to be effective, hence, they must either be of food-grade material or securely attached to the package surface (Lopez-Rubio et al., 2004). The majority of microbial development happens on top of solid and semisolid food products. The antimicrobial agent must transfer from the package to the food's surface through a certain process.

Researchers used blends made of starch and gelatin to investigate the antibacterial effects of neem extract. They noticed a decline in the drug's efficiency when used to treat *Pseudomonas, Escherichia coli, Staphylococcus aureus*, and *Salmonella typhimorium*. Metallic nanoparticles are used in biodegradable bio-packaging to increase barrier and mechanical qualities as well as the shelf life of the product (Asgher et al., 2020). It is well known that the antimicrobial properties of azadirachtin play a major role in the antibacterial activity of neem products. The investigation of the metabolomic and biological activity of neem products enables us to comprehend how crucial it is to manage the entire phytocomplex. The various nanoparticles and nanomaterials employed in antimicrobial nanocomposites are listed in Table 16.1.

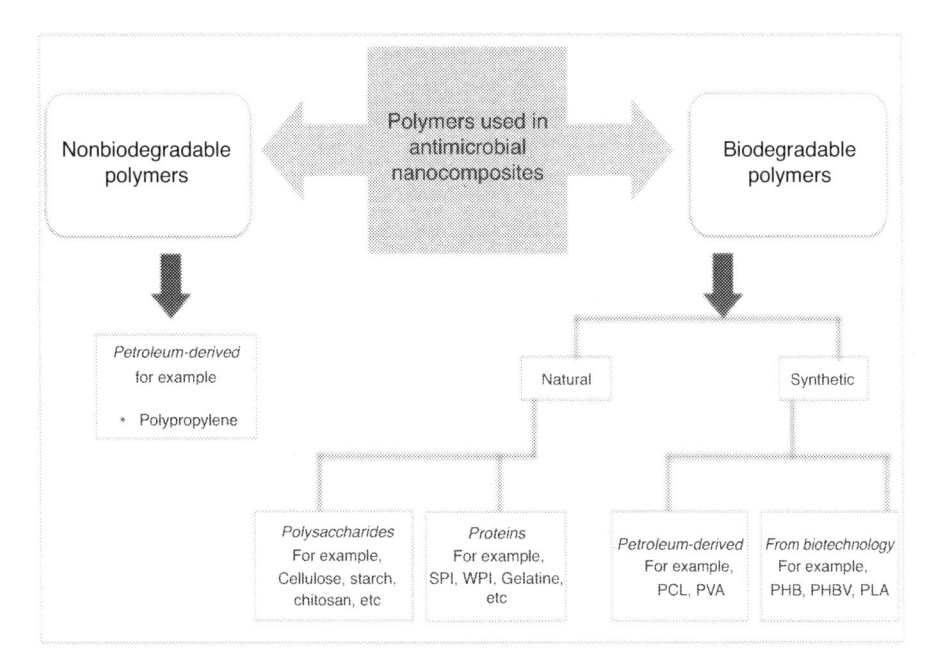

FIGURE 16.3

Polymers used in antimicrobial nanocomposites.

In recent years, a large number of polymers have been evaluated in order to create antimicrobial nanocomposites (Tamayo et al., 2016). In reality, nanoparticles and other additives like antioxidants and plasticizers are employed to enhance various polymer properties. This is because, when compared, these materials provide various advantages but also several disadvantages. For instance, polyolefins have strong mechanical performance and serve as excellent moisture transfer barriers. In contrast, they perform less well than polysaccharide-based films at preventing oxygen transfer (Nesic and Seslija, 2017). Different types of classifications may apply to the polymers employed in antibacterial nanocomposites. As a result, they can be divided into groups based on their intended application (e.g., in surface treatment, tissue engineering, food packaging, etc.), their source (for example, natural, petroleum-derived, or created by microbes), or their biodegradability. Because biodegradable polymers are used in the majority of current publications, polymers are categorized in Fig. 16.3 according to their biodegradability.

16.4.4 Bio-based food packaging materials

"Bio-based" refers to polymers whose starting elements are entirely or largely obtained from the biomass of plants, animals, or microorganisms. They might or might not be biodegradable (Fig. 16.4). The ability of a material to break down under the influence of enzymes produced by microorganisms is referred to as biodegradability.

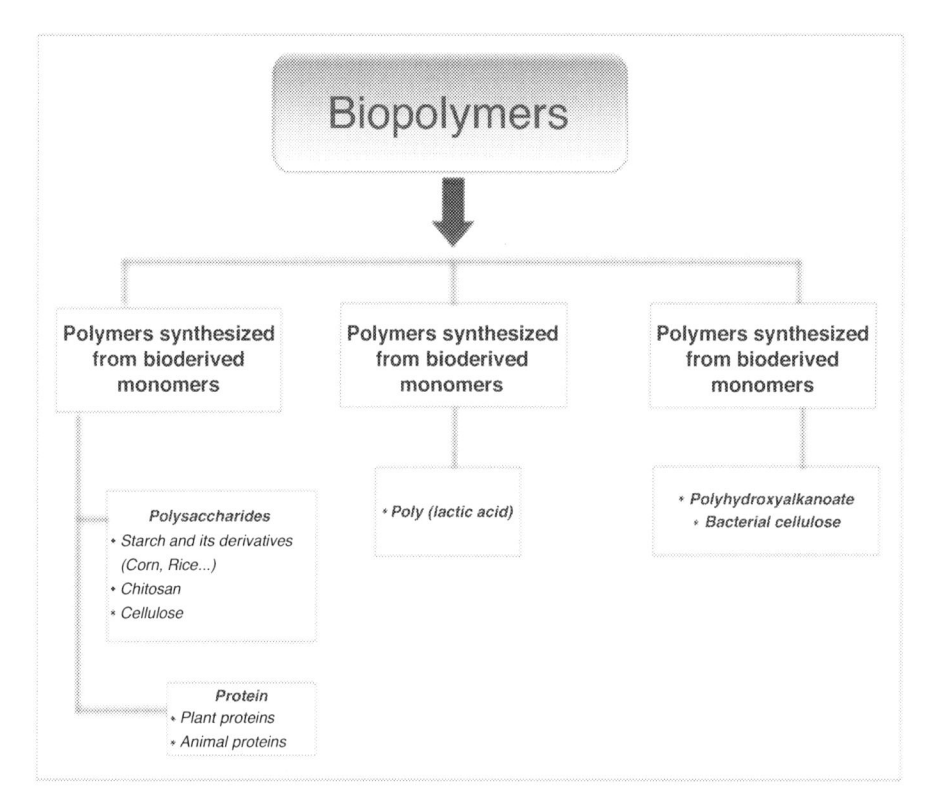

FIGURE 16.4

Categories of bio-based materials.

Bio-based and biodegradable materials are divided into three primary groups based on their point of origin: polymers that are directly taken from natural materials, polymers that are traditionally synthesized from bioderived monomers, and polymers made by microorganisms as shown in Fig. 16.4 (Kolybaba et al., 2006).

16.4.5 Nanoencapsulation

Nanoformulations used to encapsulate bioactive nutraceutical compounds improved their bioavailability and biodistribution. The method of nanoencapsulation involves packing substances into extremely small structures, either through nano-structuring, nano-emulsification, or nanocomposites that enable the core's-controlled release (Nile et al., 2020). Different types of nanoencapsulations, including nanoparticle micelles, nanospheres, and liposomes, have been used depending on the need. They can be utilized as dietary supplements, to mask off-putting flavors, to increase bioavailability, and to effectively disperse insoluble supplements without the need for emulsifiers or surfactants.

The nanoencapsulated curcumin and quercetin in turmeric extract were stabilized using polylactide (PLA)-based nanoparticles (Singh et al., 2017). As antidiabetic nutraceutical agents, nontoxic natural sweets and sativoside nanoparticles were employed. By encasing the polyphenols (epicatechin and catechin) in tea in bovine serum albumin (BSA) nanoparticles, their stability and bioavailability were improved (Tyagi et al., 2017). The gradual release of bioactive components due to their nanoformulation preserves their antioxidant capacity and increases the potency of the bioactive compounds (Croitoru et al., 2019). The majority of natural phytochemicals are environment-sensitive.

Bioactive substances (carbohydrates, proteins, phytochemicals, vitamins, and antioxidants) are entrapped within nanoparticles, which not only provide protection but also enhance the activity and stability of the bioactive substances. The nanocapsules can be engineered to incorporate nanoadditives, antimicrobials, and detoxifying in the animal feeds, whilst the nanocoating can be employed as transporters for functional components during nanoencapsulation. Some of the substances are bifunctional molecules that can be employed as emulsifiers or surfactants and can be used to encapsulate bioactive substances, medications, or antibiotics. Archaeosomes, nanoliposomes, and nanocochleates are frequently used in lipid-based nanoencapsulation systems. Nutrients, enzymes, food antimicrobials, and food additives may be carried in nanoliposomes. Probiotic nanoencapsulation has also been documented in the past. Probiotics are the live bacterial species combinations that are added to the diet. Cheese, fruit-flavored beverages, yogurts, and puddings with a yogurt-like fermentation process are examples of common probiotic foods.

16.4.6 **Nanocomposite**

Nanocomposite materials are made up of two or more phases, one of which functions as a continuous phase and the other as a dispersed phase (Sharma and Dhanjal, 2016). Typically, a continuous phase of polymer is employed, and a scattered phase of reinforcing material. Nanocomposites are often defined as materials with a reinforcing material smaller than 100 nm and at least one dispersed organic or inorganic phase of a polymer.

Polylactide, chitosan, cellulose acetate, thermoplastic starch, and separated whey protein are examples of biodegradable polymers that are frequently used to create nanocomposites since they are derived from renewable sources (Grujić et al., 2017). Different nanoparticles including SiO_2, clay, TiO_2, $KMnO_4$, nanocellulose, nanofibrillated cellulose, and carbon nanotubes, can be assimilated to alter the mechanical and barrier properties of the packaging material. According to reports, adding nanomaterials to the matrices of polymers can improve the materials' mechanical properties, including ductile strength, shear strength, barrier potential, thermal stability, and optical qualities.

Gelatin, silver nanoparticles, and organoclay were used to make antimicrobial nanocomposite films, which demonstrated strong antibacterial action. The nanocomposite layer even has substantial antibacterial activity against the foodborne pathogen.

The usage of gelatin-based nanocomposite films was found to increase food quality and shelf life by battling and eliminating bacterial invaders. Silver nanoparticles added to methylcellulose improved the material's mechanical characteristics by directing the strengthening and toughening effects of the metal into the polymer. The antibacterial activity that methylcellulose-silver nanocomposite sheets displayed against a variety of microorganisms (Thomas et al., 2007).

The amount of monomeric acid, silver nanoparticles, and size of the nanocomposites all have an impact on the antimicrobial activity of hydrogel-silver nanocomposites. Even while nanoparticles can control bacteria, there is still much that must be learned about how well metal nanoparticles work against harmful bacteria and how they are regulated.

16.5 Benefits of nanomaterials in food packaging

Foods can be packaged with bioactive materials to help with oxidation and prevent the formation of bad flavors and undesirable textures (Thiruvengadam et al., 2018). The food and packaging industries continue to encounter difficulties such as nonsustainable manufacturing, lack of recyclability, and inadequate mechanical and barrier qualities. Plastics are still widely employed in food packaging despite the fact that glass and metal are good barrier materials that can be used to prevent unwanted mass transport because of their lightweight, formability, affordability, and adaptability (Mihindukulasuriya and Lim, 2014).

Given that the packaging sector uses more than 40% of all plastic, and that food packaging makes up 50% of this 40%, it is clear that antimicrobial films have been launched to extend the shelf life of food and dairy goods as well as some intriguing revolutionary nanotechnology items as explained in Fig. 16.5. Additionally, materials for food preservation and packaging have become crucial in the food sector.

Resources for bioactive packaging must be ready to keep bioactive chemicals, such as probiotics, prebiotics, bioavailable flavonoids, and encapsulated vitamins in top condition until they are released into the food product in a regulated manner (Thiruvengadam et al., 2018).

16.6 Future prospective

Although nanoparticles are produced all over the world, very few nations have the requisite regulatory framework in place to allow for the use of nanotechnology in food items. Finding any conclusions about the efficacy of nanosystems is challenging due to a lack of scientific investigation. When used as a food ingredient, nanoparticles are more dangerous than when they are used in food packaging. During their production and use, nanomaterials pose a constant risk of entering the food chain through the air, water, and soil. This could result in DNA damage, disruption of cell membranes,

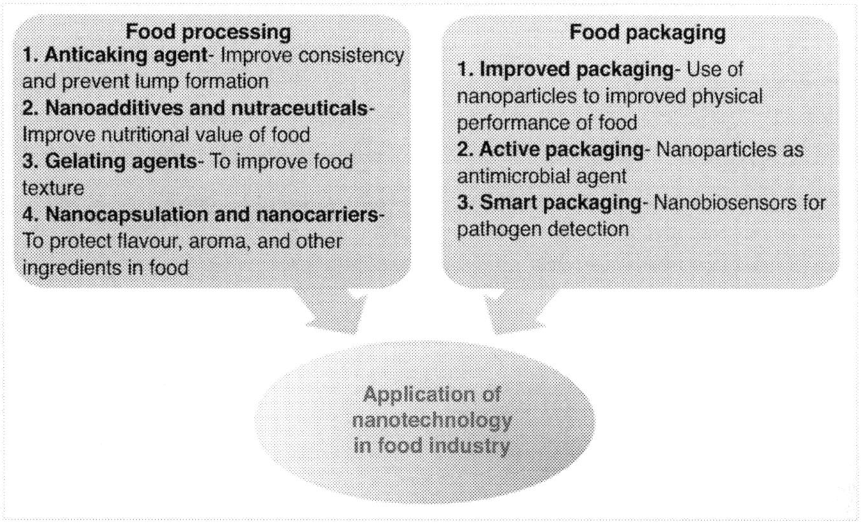

Food processing
1. **Anticaking agent-** Improve consistency and prevent lump formation
2. **Nanoadditives and nutraceuticals-** Improve nutritional value of food
3. **Gelating agents-** To improve food texture
4. **Nanocapsulation and nanocarriers-** To protect flavour, aroma, and other ingredients in food

Food packaging
1. **Improved packaging-** Use of nanoparticles to improved physical performance of food
2. **Active packaging-** Nanoparticles as antimicrobial agent
3. **Smart packaging-** Nanobiosensors for pathogen detection

Application of nanotechnology in food industry

FIGURE 16.5

Schematic diagram showing the role of nanotechnology in different aspects of food sectors.

and cell death. There has not been much in-vivo research done yet on how nanofoods affect both human and animal health.

In order to promote customer acceptance, proper labeling and restrictions should be recommended for the marketing of nanofoods. Therefore, if handled and regulated properly, the use of these nanotechnologies can significantly contribute to enhancing food processing and product quality, which will enhance human health and well-being. As a result of the increased demand for food preservation and transportation brought on by globalization, bio-based materials for food packaging will be needed to perform better in future research. When discussing the actual application of these bio-based materials, a number of issues need to be taken into account.

For the purpose of packaging specific meals, it is important to select bio-based materials that are suitable for the job and further, tailor their qualities via physical or chemical processes.

Another emerging trend will be to learn more about the sustainability and biodegradability of these materials in order to reduce their long-term environmental impact. More work is needed to create environmentally friendly procedures that use fewer or no harmful organic solvents. For food packaging, genuine bio-based materials that are eco-friendly, long-lasting, and biodegradable are needed. In conclusion, nanobiotechnology will serve as a crucial strategy for overcoming the current challenges associated with active food packaging solutions.

16.7 Conclusion

In conclusion, nanobiotechnology has grown significantly within the food sector. One of the industries where nanotechnology is expected to have a significant future impact is the food industry. Nanotechnology is a new and emerging innovation with incredibly extraordinary properties in the food supply chain (precision farming methods, smart feed, bioavailability/nutrient values, improvement of food texture and quality, crop production, packaging, labeling, and use of agrochemicals such as nanoherbicides, nanofertilizers, and nanopesticides) across the global agricultural sector. Nanofood packaging materials have the potential to extend shelf life, improve food safety, warn consumers when food is contaminated or destroyed, mend packaging tears, and uniformly release additional ingredients to extend the shelf life of the food inside the container.

New technologies that aim to personalize and customize items are what the future holds. The first step will be to increase food quality and safety. Finally, nanotechnology makes it possible to modify current food systems and processing in order to ensure product safety, foster a culture of healthy eating, and improve food quality nutritionally.

In order to avoid using metals, such as gold or silver, and to limit the use of environmentally harmful plastics in the inclusion of the texture, new promising methods for the utilization of organic biodegradable innovative materials, such as algae, fern, chitosan, and others, can also be taken into consideration. Through the use of edible or biodegradable materials, plant extracts, and nanobiomaterials, the environmental effect of packaging for food might significantly decrease with more intense development of sustainable or green food packaging. When developing new food packaging materials, carbon, energy, water, and land footprints must also be taken into account in order to prevent regrettable substitutes to previously existing ones.

References

Asgher, M., et al., 2020. Bio-based active food packaging materials: sustainable alternative to conventional petrochemical-based packaging materials. Food Res. Int. 137, 109625.

Carocho, M., Morales, P., Ferreira, I.C.F.R., 2015. Natural food additives: *Quo Vadis?* Trends Food Sci. Technol. 45 (2), 284–295.

Croitoru, A., et al., 2019. Evaluation and exploitation of bioactive compounds of walnut, *Juglans regia*. Curr. Pharm. Des. 25 (2), 119–131.

Espitia, P.J.P., et al., 2012. Zinc oxide nanoparticles: synthesis, antimicrobial activity and food packaging applications. Food Bioproc. Technol. 5 (5), 1447–1464.

Grujić, R., Vujadinović, D., Savanović, D., 2017. Biopolymers as food packaging materials. In: Advances in Applications of Industrial Biomaterials, pp. 139–160.

Kale, G., et al., 2007. Compostability of bioplastic packaging materials: an overview. Macromol. Biosci. 7 (3), 255–277.

Kerry, J.P., O'grady, M.N., Hogan, S.A., 2006. Past, current and potential utilisation of active and intelligent packaging systems for meat and muscle-based products: a review. Meat Sci. 74 (1), 113–130.

Keskin, G., et al., 2017. Potential of polyhydroxyalkanoate (PHA) polymers family as substitutes of petroleum based polymers for packaging applications and solutions brought by their composites to form barrier materials. Pure Appl. Chem. 89 (12), 1841–1848.

Kolybaba, M., et al., 2006. Biodegradable polymers: past, present, and future. In: ASABE/CSBE North Central Intersectional Meeting. American Society of Agricultural and Biological Engineers.

Kumar, N., Kaur, P., Bhatia, S., 2017. Advances in bio-nanocomposite materials for food packaging: a review. Nutr. Food Sci. 47 (4), 591–606.

Kuswandi, B., Moradi, M., 2019. Improvement of food packaging based on functional nanomaterial. In: Nanotechnology: Applications in Energy, Drug and Food. Springer, Cham, pp. 309–344.

Li, K., et al., 2019. Bioinspired interface engineering of gelatin/cellulose nanofibrils nanocomposites with high mechanical performance and antibacterial properties for active packaging. Compos. B Eng. 171, 222–234.

Lindh, H., et al., 2016. Elucidating the indirect contributions of packaging to sustainable development: a terminology of packaging functions and features. Packag. Technol. Sci. 29 (4–5), 225–246.

Lopez-Rubio, A., et al., 2004. Overview of active polymer-based packaging technologies for food applications. Food Rev. Int. 20 (4), 357–387.

Mellinas, C., et al., 2016. Active edible films: current state and future trends. J. Appl. Polym. Sci. 133 (2), 42631.

Mihindukulasuriya, S.D.F., Lim, L.-T., 2014. Nanotechnology development in food packaging: a review. Trends Food Sci. Technol. 40 (2), 149–167.

Nesic, A.R., Seslija, S.I., 2017. The influence of nanofillers on physical–chemical properties of polysaccharide-based film intended for food packaging. In: Food Packaging. Academic Press, pp. 637–697.

Nicoletti, M., Del Serrone, P., 2017. Intelligent and smart packaging. In: Future Foods. IntechOpen, p. 68773.

Nile, S.H., et al., 2020. Nanotechnologies in food science: applications, recent trends, and future perspectives. Nano-Micro Lett. 12 (1), 1–34.

Pina, S., Oliveira, J.M., Reis, R.L., 2015. Natural-based nanocomposites for bone tissue engineering and regenerative medicine: a review. Adv. Mater. 27 (7), 1143–1169.

Primožič, M., Knez, Ž., Leitgeb, M., 2021. (Bio) Nanotechnology in food science: food packaging. Nanomaterials 11 (2), 292.

Qu, P., et al., 2022. Microporous modified atmosphere packaging to extend shelf life of fresh foods: a review. Crit. Rev. Food Sci. Nutr. 62 (1), 51–65.

Sadjadi, M.A.S., et al., 2019. State of nano technology in novel food packaging and new application opportunities. Int. J. Bio-Inorg. Hybrid Nanomater. 8 (4), 143–152.

Shaikh, A.-U.H., Sadaqat, Raza, A., 2022. Financial statement and ratio analysis of Force Motors and Yamaha Motors: an empirical study. J. Develop. Soc. Sci. 3 (4), 406–416.

Sharma, D., Dhanjal, D.S., 2016. Bio-nanotechnology for active food packaging. J. Appl. Pharm. Sci. 6 (9), 220–226.

Singh, R., Kumari, P., Kumar, S., 2017. Nanotechnology for enhanced bioactivity of bioactive phytomolecules. In: Nutrient Delivery. Academic Press, pp. 413–456.

Stringer, M.F., Hall, M.N., 2007. A generic model of the integrated food supply chain to aid the investigation of food safety breakdowns. Food Control 18 (7), 755–765.

Sumit, G., 2012. Nanotechnology in food packaging a critical review. Russ. J. Agric. Soc.-Econ. Sci. 10 (10), 14–24.

Tamayo, L., et al., 2016. Copper-polymer nanocomposites: an excellent and cost-effective biocide for use on antibacterial surfaces. Mater. Sci. Eng. C 69, 1391–1409.

Thiruvengadam, M., Rajakumar, G., Chung, I.-M., 2018. Nanotechnology: current uses and future applications in the food industry. 3 Biotech 8 (1), 1–13.

Thomas, V., et al., 2007. A versatile strategy to fabricate hydrogel–silver nanocomposites and investigation of their antimicrobial activity. J. Colloid Interface Sci. 315 (1), 389–395.

Tyagi, N., et al., 2017. Cancer therapeutics with epigallocatechin-3-gallate encapsulated in biopolymeric nanoparticles. Int. J. Pharm. 518 (1–2), 220–227.

Xing, K., et al., 2015. Chitosan antimicrobial and eliciting properties for pest control in agriculture: a review. Agron. Sustain. Dev. 35 (2), 569–588.

Progress and challenges of nanobiotechnology in food packaging

17

Kantrol Kumar Sahu[a], Monika Kaurav[b], Ramakant Joshi[c], Rakesh Raj[d], Pooja Mongia[e] and Sunita Minz[f]

[a] *Institute of Pharmaceutical Research, GLA University, Mathura, Uttar Pradesh, India,* [b] *KIET School of Pharmacy, KIET Group of Institutions, Ghaziabad, Uttar Pradesh, India,* [c] *Department of Pharmaceutics, ShriRam College of Pharmacy, Morena, Madhya Pradesh, India,* [d] *DSEU Meerabai Maharani Bagh Campus, Delhi Skill and Entrepreneurship University, New Delhi, India,* [e] *Department of Pharmaceutics, Delhi Institute of Pharmaceutical Sciences and Research University, New Delhi, India,* [f] *Department of Pharmacy, Indira Gandhi National Tribal University, Amarkantak, Madhya Pradesh, India*

17.1 Introduction

Packaging is the skill of arranging a material for sale, storage, and handy transportation of any product. It is the science of designing new techniques for the protection of products from handling hazards as well as environmental damage with easy manageable and quick identification (Paine, 1991; Robertson, 1993). If packaging is related to food products, it becomes more crucial to make it a quality product where product loss and safe storage are two important considerations with the aid of transportation-friendly and attractive advertisements (Steenis et al., 2017). Food packaging has huge possibilities to grow with the help of nanotechnology where unique properties are created for innovative materials and products of great importance (Parisi et al., 2015; Neethirajan and Jayas, 2011).

Polymer plays a significant part in food packaging nanomaterials. However, its quality can be enhanced by the incorporation of some special compounds in the polymer matrix. Okada et al. (1988) were the first to show the presence of high mechanical strength and temperature stability in polymeric nanomaterial by incorporating silicate in it (Okada et al., 1988).

Biodegradable materials can also be used in nanocomposites for packaging applications. Biological molecules such as polysaccharides, proteins, waxes, etc. have tremendous capability as they have well-ordered structures and are easy to procure. They have the capability to self-assemble into nanometric size range and thus can be applicable for packaging as nanomaterial (Rhim et al., 2013). They are nontoxic, biodegradable, and compatible with almost every product. These qualities knock the possibilities for its wide range of usage in this field. Starch and chitosan are some of

Nanobiotechnology for Food Processing and Packaging. DOI: https://doi.org/10.1016/B978-0-323-91749-0.00022-8

these materials that have been used successfully in various types of packaging (Moura et al., 2009; Corre et al., 2010). Cellulose-based nanostructures such as nanofibrillated cellulose (NFC) and cellulose nanocrystals (CNC) are also having high mechanical and thermal stability. It is employed to strengthen packaging materials with enhanced barrier capacity and multifunctional capabilities (Abdollahi et al., 2013; Bilbao-Sainz et al., 2011). One online survey detailed that the use of nanotechnology has a great impact on food packaging and it is a real concern for consumers (Roosen et al., 2015). One other study showed that consumers are eager to believe novel packaging materials in the form of food nanotechnology (Giles et al., 2015).

In 1993, AlliedSignal (United States) is a company that presented the application of nanomaterials in food packaging. They provided an explanation of how melting polymer with an exfoliated coating material obtained from reactive organo-silanes produced extruded films and film laminates from a polymer nanocomposite (Maxfield and Christiani, 1993). Nanotechnology has a huge potential to solve and support all aspects of food packaging. It includes rapid detection for any bacterial growth or environmental factor monitoring which ensures product quality. It also has nanodevices for monitoring the bioavailability of chemical constituent inside the packaging (Neethirajanand, 2011).

17.2 Traditional versus intelligent packaging

17.2.1 Traditional packaging system

Traditional food packaging is simple and compatible. Environmental protection and physical support are two main criteria of a food packaging system. It ensures product safety against external atmosphere and stimuli which ease its distribution, transport, and storage. The shelf life of the product is enhanced by a proper packaging system where that acts as a barrier to microbial entry and also ensures physiochemical changes. Apart from these features, an ideal packaging should be ease of use also. It should be conveniently stored and opened quickly by customers. Its design and construction should be in such a way that it is easily approachable and has a suitable brand identity.

Traditional food packaging being simplest and passive provides a rigid barrier and is designed to protect foodstuffs against harsh environmental conditions. These include protection against environmental oxygen, microorganisms, bad odors, and even light. Preserving food quality for a long period as well as safe handling are another advantage of traditional packaging. Moreover, the packaging material here shows the minimum interaction with the food products and assures that they are inert enough for any compatibility (Dainelli et al., 2008).

17.2.2 Intelligent packaging

This type of packaging is an upgraded and advanced form of traditional packaging. According to their functions, it is also classified as clever, smart, active, or interactive covering. IP is generally related to monitoring and ensuring the safety of the environment inside the packaging or in the surroundings of the packaging. Active packaging provides increased food protection while smart packaging is the combination of beneficial concepts behind active and intelligent technology. External information provided by IP is through smart indication (Aday and Yener, 2015). IP also provides detailed information about "hazard analysis and critical control points" (HACCP) and "quality analysis and critical control points" (QACCP) systems (Vanderroost et al., 2014).

17.3 **Active packaging of food and beverage**

Activated packaging refers to green technology that preserves or increases food stability while keeping them authentic, hygienic, and pure. A packaging system that intentionally integrates a characteristic that could discharge or absorb compounds keen on or from the merchandise or affect the food is termed active packaging, according to European Regulation No 450/2009 (European Commission, 2009). These methods come in two different varieties: the active capture system (absorber) and the active release system (emitter). However, the latter adds molecules such as antimicrobials, carbon dioxide, and polyphenols to the packed food or its surroundings while removing undesirable contaminants that are present in the food or its surroundings such as O_2, CO_2, and moisture or ethylene or odor. Table 17.1 shows the main active packaging methods used in food applications and their related advantages.

Understanding the deterioration of foods is important before using active packaging systems. This is because food's internal and external characteristics, like its pH, water, and nutrients, the number of antimicrobial compounds that are found in it, its redox potential, respiration rate, and its biological structure all influence how long it lasts. Indirectly affecting food degradation are these elements, such as chemical, biological, physical, and microbiological ones. In many cases, to address these principles, alternative active packaging solutions can be adopted to enhance the superiority and storage period.

17.3.1 **Oxygen scavengers**

Oxygen can harm various foods. It has been shown to accelerate the process of fusting in baked goods, deteriorate the flavor of certain fish such as salmon and trout add to the coloration of cooked meats and herbs, as well as assisting in the breakdown of vitamins. Cooked meat, nuts, fried foods, cheeses, and lipids are sensitive to oxidative acidity, causing flavors, and odors. Food composition, temperature, O_2 concentration, and the existence of antioxidants like metal ions and light all have a role. The elimination of O_2 can aid in the quality preservation of foods (Potter et al., 2008).

The efficiency of oxygen capture in packs depends largely on food pH balance and oxygen diffusion through packaging materials. The ability of the oxygen purifier depends on its activation mode, purification capacity, and oxygen discharge rate.

Table 17.1 Applications of active packaging.

S. no.	Active packaging types	Food types	Advantages
Systems for active scavenging (absorber)			
A.	Oxygen scavenger	Fruit and vegetable juices, seeds and nuts, meat products (cooked), oils and fats-containing food products	Prevention from color loss, Retard mold growth, maintain the concentration of vitamin C, protection from rancidity and browning
B.	Moisture scavenger	Fish and meat, mushrooms, maize, grains, seeds, tomatoes, strawberries	Prolonging the shelf life by preventing moisture loss, prevent from browning or color loss thus improves appearance
C.	Ethylene absorber	Climacteric fruits and vegetables	Slow down the process of ripening then extend stability
Active releasing system (emitter)			
A.	Antioxidant releaser	Suitable for meat which is fresh and fish (fatty), proper fit for storage of oil and fat-containing instant powders, fried products, seeds, and nuts	Prevent from oxidative damage thus improving food quality
B.	Carbon dioxide emitter	Suitable for nonvegetables such as fresh meat and fish	Retard microbial growth and enhance shelf
C.	Packaging systems for antimicrobials	Meat (fresh, processed meat, and ready-to-eat), smoked fish, fruits and vegetables, cereals and grain, sea foods and milk products	Inhibit microbial growth and prolong shelf life

When selecting oxygen scavengers, keep in mind first about the food target stability, second the quantity of O_2 in the container, and the materials used for packaging (Potter et al., 2008).

The best O_2 purifier is a small bag of iron-based powder and catalyst. In food packages, this type of chemical generally interacts with water to generate a reactive hydrogen reduction agent. This agent is used for allowing oxygen from the food package to be scavenged and permanently converted into oxygen dioxide. If you put iron powder in a can and you warn yourself "not to eat," you can keep it away from food. The typical residual oxygen level that can be achieved with this type of oxygen scavenger is 0.03%–3.3%. However, this type of oxygen scavenger can leave behind less than 0.01%. Iron-based oxygen scavengers were initially introduced in 1976 in Japan under the brand name Ageless by Mitsubishi Gas Chemical Co. Ltd.

(Japan). Fine iron powder is inside the sachets, which are included in sea salt and natural zeolites that have been soaked in NaCl solution. The scavenger, according to Mitsubishi, may lower O_2 concentration to less than 0.01%. Nonmetallic O_2 disposal devices have also been created to limit the risk of metal contamination in foodstuffs. Intriguingly, new metal detectors may now phase out the scavenger signal while still keeping great sensitivity to ferrous and nonferrous metallic contaminants, which also answers the problem of online metal detectors being accidentally activated (Anon, 1995).

17.3.1.1 CO_2 scavenger and releasers

CO_2 can be obtained or created using a wide variety of commercially available sachets and labels. Carbon dioxide disposal products include coffee, sour goods, cheese, fresh meat, and poultry. Fresh roasting or crushing coffee is too moist and contains too much oxygen, so it cannot be packed due to a loss of volatile aroma and taste. Due to the process of roasting, fresh coffee can produce large amounts of CO_2; most of that goes to waste during the process of grinding, while a few remains intent which is released gradually after packaging. Due to the carbon dioxide generated by the roasting of coffee, if the bag is re-packed directly after roasting, the bag breaks down (Subramaniam, 1998).

17.3.1.2 Ethylene scavengers

Among fruits, vegetables, and flowers, ethylene (C_2H_4) speeds breathing rates and ages them. Ethylene is responsible for citrus fruit, bananas, and tomatoes gaining color, and tiny carrots developing their roots more quickly and producing harsh flavors in bulk cucumbers, in addition to forming pineapple blooms. A variety of physiological diseases and softening of products are common effects of ethylene in horticulture, so suppressing its negative effects is generally the goal. As a result, numerous experiments have been conducted to add ethylene absorbers to clean products. Potassium permanganate ($KMnO_4$) can be employed efficiently because of the large surface area supplied by a solid substrate like alumina or silica gel. Purple $KMnO_4$ turns brown during ethanol oxidation, which indicates that it does still have some capability for scavenging ethylene. The ethyl scavenger powder pumice stones can be placed in films and caught when they come out of fruit. Zeolites also absorb ethylene, water vapor, and the scent of food and beverage packaging technology. EverFresh Bags (United States) is treated with natural minerals that absorb ethyl gas released by fresh foods. Another method by which ethylene can be easily separated is by scraping carbon-based scavengers using different metal catalysts. Ethylene is scavenged by carbon and decomposed. Several ways are employed, including incorporating them into sachet-like packaging for inclusion in produce packs or incorporating them into either paper bags or corrugated cardboard boxes for produce storage. In laboratory testing, charcoal has been shown to reduce the softening of a variety of fruits, including bananas and kiwis (Waite, 2003). Longer Stayfresh bags reduce the natural aging of food and prevent humidity and germs. When fruits are exposed to gases like ethylene, ammonia, and hydrogen sulfide, they develop faster and are removed from Biofresh (Grofit Plastic, Israel).

17.3.1.3 Ethanol emitters

Ethanol is widely recognized as a germicide. In addition to the prevention of mold, it can also inhibit bacterial and fungal growth. Direct application of ethanol to food goods, immediately before packing, has been shown to increase shelf life. Researchers reported that the spraying of 95% ethanol on baked goods significantly prolongs the product's storage period by increasing concentrations from 0.5 to 0.5% (w/w). However, ethanol production is more practical and safer. Ethanol-emitting films and sachets can be used to make ethanol (Rooney, 1995). Instead of being sprayed on top of the goods, the ethanol might be gently delivered through an air sachet inside the bag. Alcohol has antifungal and antibacterial properties and is used for both fish and cheese treatment, although its primary use is baking. Ethanol inhibits the breakdown of molds like *Aspergillus, Penicillium*, etc., and the growth of bacteria like *Staphlococcus* and *Salmonella*, according to numerous studies (Potter et al., 2008). There is contradictory evidence regarding the ability of ethanol emitters to suppress yeast growth. Many ethanol-emitted films and fillers are patented mainly by Japanese companies. Freund Industrial Co.'s Ethicap, Antimould 102, and Negamold, Nippon Kayaku Co. Ltd.'s Oitech, Ueno Seiyaku Co. Ltd.'s ET Pack, and Ageless type SE are among them (Mitsubishi Gas Chemical Co. Ltd.). All of these films and packaging have a carrier substance that allows the controlled release of vapor produced from ethanol. The sachets with ethanol and water contain silicon powder absorbed by ethanol and water. When product humidity makes active the emitter, the ethanol is evolved into the air and packaging.

17.3.1.4. Moisture absorbers

Excessive humidity is the main cause of food destruction. By inhibiting microbial development and moisture-related texture and taste degradation, various absorbers or desiccants safeguard quality and extend the life period of food. As well as bags and pads, you can find moisture absorbers in sheets, blankets, and sheets. Food packaging contains enough condensation to cause fruit and vegetable spoilage, as well as a significant amount of droplets of food, poultry, and fish. Fresh air is produced, and water is discharged into packaging materials. The packing material absorbs this water. Degradation occurs more quickly if moisture is present inside the packaging material and its permeability is higher. It is also possible for baked and dry goods to become deteriorate and lower in quality if moisture is allowed to penetrate the package. Excessive moisture from the drop of meat and fish is undesirable, as it degrades the product's quality and causes deterioration. Reduce the package's moisture content by using a moisture absorber. Dehumidizer, like silica gel and CaO, can be used as absorption materials, as well as activated clays and minerals. Moisture-absorbing properties of silica gel allow it to be used in humid and hot conditions. The level of saturation is estimated with a silica gel that changes the color. Silicate-based molecular sieves have a higher absorption rate than silicon gels and clays. The smell can also be absorbed by a molecular sieve. Clay is a natural material that absorbs moisture. When exposed to moisture, it does not break or decompose. Activated carbon can also be used to remove the odor and moisture from packages (Yildirim et al., 2018).

17.3.1.5 Flavor/odor absorbers

It has long been recognized that packaging can affect food flavor and aroma, especially with the inclusion of undesirable flavor components. For example, after 2 weeks in an aseptic orange juice container, considerable amounts of limonene have been shown to be cut (Rooney, 1995). These absorption agents can be used to eliminate unpleasant odors and tastes caused by oxidation and biochemical reactions occurring when products get older. These substances are emitted and identified by the consumer when the packaging is opened. This type of scavenger has a relatively small commercial potential and is now banned in the European Union (EU). Consumers use smells as the main detector tool to determine whether products are protected from intake (Vermeiren et al., 2003). There is a matter to focus on the removal of the scent will lead customers to ingest foods that are not safe to eat. However, because pasteurized citrus juices can be used, this product has potential. After pressing oranges and pasteurizing them, tetraterpene limonin is released into the juice, causing a bitter taste. Some types of oranges, like navel, are more likely to have these bitter flavors. Such liquids are passed through cellulose triacetate and nylon bead columns as a method of obtaining them (Rooney, 1995). Flavanol glycosides are naringins that cause the bitterness of grape and lemon juice. Adsorbents like cellulose triacetates or paper (acetylated) can be used in the orange juice packing material to remove these chemicals. Naringinase enzymes, which are made up of rhamnosidase and glucosidase, are found in the acetate layer and hydrolyze naringin to cleave it. To minimize limonene, low-density polyethylene (LDPE) and cellulose acetate can be utilized (Waite, 2003).

17.3.1.6 Lactose and cholesterol removers

In the present world, where health consciousness is on the rise, it is now necessary to eliminate particular food components, such as lactose and cholesterol. The container surface can contain lactase to hydrolyze lactase in milk and dairy products. The cholesterol reduction enzyme changes cholesterol to pregnenolone, which is not absorbed by the intestine and hence remains unabsorbed in the body, lowering the quantity of cholesterol absorbed (Majid et al., 2018).

17.3.1.7 Antioxidant release

Oxygen can harm meals by imparting unwanted flavors, smells, and color changes. Products high in fat and oil, such as nuts, crackers, and processed meat, are particularly prone to oxidation. Antioxidants can be added to packaging to help reduce oxidation (Nerin et al., 2006). Butylated hydroxytoluene (BHT) and butylated hydroxyanisol (BHA) are integrated into and bound to packing materials, allowing them to be released from storage via diffusion. As an antioxidant, alpha-tocopherol can also be isolated from high-density polyethylene (HDPE). It produces less than BHT. Many antioxidants are lost during the releasing process, while the remainder are absorbed by food and provide protection. The foods in the package have an effect on the release rate, according to the results of the tests. The release rate is affected by fat, alcohol, and acid concentration (Nerín et al., 2008; Ganiari et al., 2017; Gómez-Estaca et al., 2014).

17.4 Nanoparticles in active packaging

Silver nanoparticles (AgNPs), nanoclay, nano-zinc oxide (nano-ZnO) (Hirvikorpi et al., 2010), nano-titanium dioxide (nano-TiO_2), titanium nitride nanoparticles (nano-TiN), and other synthetic nanomaterials are now used as functional additives for the packaging of food (Mohanty et al., 2009). Because of differences in features and chemical structure, each nanomaterial has different qualities for the host material, resulting in a wide range of packaging applications (Rubilar et al., 2014). Nanomaterials, including zinc oxide as well as titanium dioxide, are commonly used as photocatalysts to break down organic molecules as well as bacteria, while antimicrobial agents include AgNPs, nano-clays, and layered silicates (Majeed et al., 2013). The photocatalytic interaction of nano-ZnO and nano-TiO_2 produces reactive oxygen species (ROS), which causes bacterial cell cytoplasm oxidation and cell death (Bodaghi et al., 2013). Since olden times, silver (Ag) has been used as an antibacterial agent. The nanodimension of its antimicrobial agent increases, and various studies have been conducted using Ag nanoparticles included in food packaging materials for its action of antimicrobial activity (Panea et al., 2014; Azlin-Hasim et al., 2016; Li et al., 2017). According to recent research, Ag nanoparticles incorporating into various polymeric matrices, along with additional additives, can significantly extend the shelf life of many foods. According to Li and colleagues (2017), rice which is stored in LDPE without Ag/TiO_2 had a bad mildew condition even after 1 month also, with total plate counts (TPC) increasing from 4.84 log cfu/g to 7.15 log cfu/g, whereas rice if stored in a nanocomposite based on LDPE with significant amount of Ag/TiO_2 had a low TPC of 5.48 log cfu/g. According to Mihaly Cozmuta and colleagues (2015), the microbiological safety of bread stored in Ag/TiO_2-based packaging prevented the multiplication of yeast/molds, such as *Bacillus cereus* and *Bacillus subtilis*. Bread's shelf life was extended when evaluated along with openly kept bread or in frequent usage of packaging containing plastic by slowing the rate of deterioration of the main nutritious components. Azlin-Hasim and colleagues (2016) created nanocomposites from PVC and silver nanoparticles that significantly prolonged the shelf life of chicken breast fillets while also resulting in lower lipid oxidation, whereas Panea and colleagues (2014) reported a reduction in MO but enhanced lipid oxidation.

Emamifar and colleagues (2010) looked into the antibacterial properties of nano-silver as well as zinc oxide-treated LDPE for orange juice packaging (ZnO). This method of extending the shelf life of packed orange juice was extremely effective (up to 28 days). The initial bacterial counts of *Staphylococcus aureus* and *Salmonella typhimurium* were reduced by 2 log after 24 hours of incubation at 81°C, according to Akbar and Anal (2014), who used ZnO as an antibacterial agent applied to proper active packaging sheets for packaging of fresh poultry meat. After 6 days of incubation, there were no viable *S. aureus* cells and no viable *S. typhimurium* cells were observed.

As an antimicrobial nanoparticle in LDPE, titanium dioxide (TiO_2) reduced mesophilic bacteria from 3.14 log CFU/g to 2 log CFU/g during the 17-day storage period, whereas LDPE cell loads rose from 3.19 log CFU/g to 4.02 log CFU/g. Additionally, the yeast population decreased from 2.45 log CFU/g to 2 log CFU/g, whereas the control sample grew from 2.1 log CFU/g to 3.37 log CFU/g (Bodaghiet al., 2013). *Pseudomonas* spp. were also killed when copper (Cu) was incorporated into PLA and used for the packaging of fiordilatte cheese (isolated from spoilt fiordilatte cheese). Bacterial growth was inhibited when the active films were used (Conte et al., 2013).

As previously discussed, antimicrobial nanoparticles hold great promise for safeguarding food systems' microbiological quality. Choosing the appropriate antimicrobial agent for a given food is critical in this situation. Packaging film qualities like barrier and transparency can be studied in relation to nanoparticles. Although the evaluation of nanoparticle safety in general (Radusin et al., 2016), as well as statutory limits, is challenging the safety review and permission for the use of nanoparticles as a major ingredient for food packaging remains the most challenging task (Radusin et al., 2016; Rauscher et al., 2017; Amentaet al., 2015).

17.5 Current industrial applications

Nanotechnology may help with the protection and maintenance of quality for the food while it has role in the invention of novel additives as well as supplements and unique flavors for food (Fig. 17.1) (Dimitrijevic et al., 2015). In the food division, this technology can be engaged to create packages that have better heat, mechanical, and safety features. Consumers are notified when foods are about to expire using nanosensors incorporated in food packing arrangements. This technology helps to develop healthier and tasty foods (Gokularaman et al., 2017). Nanostructures with a wide range of qualities are appropriate for usage in foods in addition to wrapping items that improve the nutritious eminence of foods (Aditya et al., 2018).

The current market status of foodstuffs by using this technology has surpassed 1 billion ($US), with the majority of money spent on nanoparticle coverings for wrapping, health-promoting items, and beverages. The world's biggest food companies, comprising Kraft, H.J. Heinz, Nestle, Altria, Unilever, and Hershey, are investing extensively in nanotechnology research and development for a variety of food foodstuffs, containing milk, vegetables, meat, fish, and bakery items (Grumezescu et al., 2018). Nanosensors, tracking devices, targeted delivery of critical constituents, food security, novel produce design, precision handing out, smart wrapping, and other innovative technologies have all made their way into the food business (Huang et al., 2010).

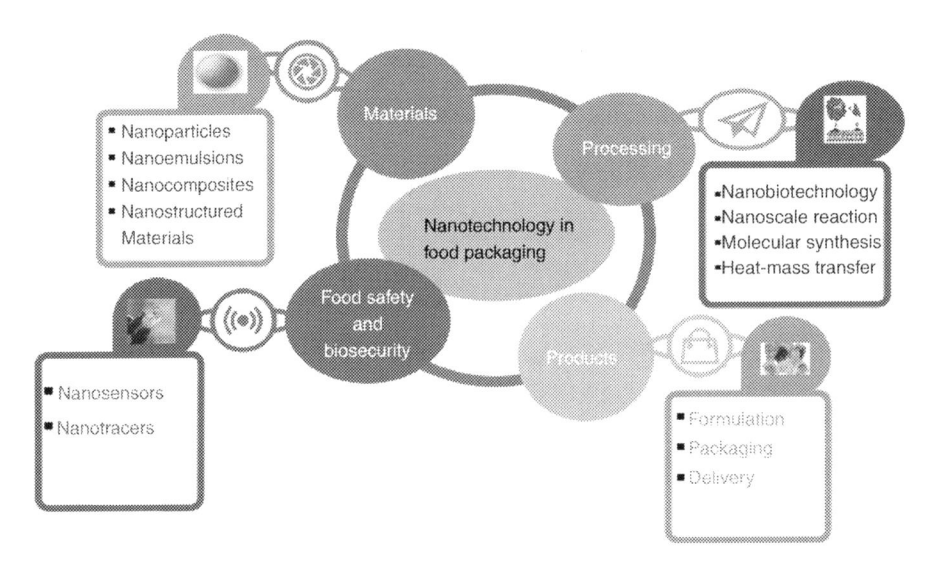

FIGURE 17.1

Nanotechnology current food industry applications.

Nanotechnology is undeniably transforming the food business. Food quality enhancement, bioactive fortification, controlled release of bioactive complexes utilizing nanocarrier encapsulation, food structure, and texture modification, and exposure and neutralization of biochemical, microbiological, and chemical modifications via intelligent packing. Various nanomaterials have a wide variety of applications in the production of food and the enhancement of nutritious properties. Protein nanoparticles, for example, are employed in the production of food goods since protein solubility aids in the construction of protein nanoparticles by desirable efficient qualities in food (Shafiq et al., 2020).

17.5.1 Nanomaterials

Nanomaterials has unique qualities which open up a slew of new possibilities designed for use in the food industry. Various purposeful nanostructures can be utilized as building blocks for developing innovative structures and adding novel functions to foods. Nanoparticles, nanoemulsions, nanocomposites, and nanostructured materials are among them. Numerous of these structures, as well as their current and prospective applications in the food industry, are described here (Sekhon, 2010; Weiss et al., 2006).

17.5.1.1 Nanoparticles

To make nanometer-sized particles, food-grade biopolymers such as proteins or polysaccharides can be employed (Ravichandran, 2010). Nanoparticles are classified into several classes depending on their aptitude to convey various responses by various components and in various environments. Nanoparticles can be categorized into two groups based on their chemical properties: organic and inorganic. When utilized as vehicles for the distribution of health nutrients or medications, organic nanoparticles are frequently referred to as nanocapsules. Moreover, nanoprecipitation, emulsion–diffusion, double emulsification, emulsion–coacervation, polymer–coating, and layer-by-layer are six traditional ways for making nanocapsules. They are more likely to be employed to increase or alter food functionality acceptable to boost the nutritious value of food systems. Organic nanoparticles have been developed to provide vitamins and supplementary nutrients deprived of altering the aroma or appearances of food and beverages. Nanoparticles made of inorganic materials, like organic nanoparticles, have a variety of production methods, for instance, gas phase inorganic nanoparticle production and liquid phase inorganic nanoparticle blend, which are additionally divided into several approaches. When utilized as a nanoparticle, inorganic chemicals synthesized at the nanoscale by deviations of composites and permitted intended for usage in foodstuff, such as titanium dioxide, a food color, can operate as an ultraviolet defense hurdle in foodstuff packing. For the preservation of prepared foods, innovative storing vessels/utensils (food connection constituents) established on surrounded inorganic nanoparticles must remain developed. The usage of silver (Ag) nanoparticles as an antimicrobial agent is the most typical use. Ag nanoparticles are used in refrigerator plates, storing containers, packing outlines, also additional outsides that derive into interaction by means of foodstuff throughout the manufacturing process (Ranjan S et al., 2014).

17.5.1.2 Nanoemulsions

An emulsion is a combination of two or more liquids that are difficult to mix together (such as oil and water). As a result, a nanoemulsion is a mixture in which the distributed droplet diameters are less than 500 nm. Nanoemulsions can enclose constituents inside their droplets, which helps to prevent chemical degradation. In reality, alternative types of nanoemulsions by additional complicated features, such as nanostructured multiple emulsions or nanostructured multilayer emulsions, encapsulate numerous functional components by a single delivery technique (Ravichandran, 2010). Nanoemulsions have been designed for application in food packaging. A nanomicelle-based produce that claims to comprise natural glycerine is a good example. Pesticide residues are removed from fruitlets and vegetables, in addition to oil and grime after flatware. As it has excellent clarity, nanoemulsions ought to newly get loads of consideration from food manufacturing. Nanoemulsions are effective against Gram-negative bacteria as well as other food pathogens. It is also used to decontaminate the surfaces of food processing plants and to minimize the contamination of chicken skin's surface. Moreover, treatment with nanoemulsion inhibited the growth of *S. typhimurium* colonies. Sugar beet pectin should be regarded as a substitute towards milk proteins besides gum Arabic used for the encapsulation of useful foodstuff components, based on the physicochemical properties of microencapsulated fish oil (Sekhon, 2010).

17.5.1.3 Nanocomposites

Nanocomposites are heterogeneous/hybrid ingredients made at the nanometric measure via merging polymers with inorganic solids (clays to oxides). Nanocomposites are hybrids that contain nanoscale morphology in one phase, such as nanoparticles, nanotubes, or lamellar nanostructures. They have multiphases, and multiphasic constituents must have at least one phase with diameters in the 10–100 nm range (Sen, 2020). Montmorillonite (MMT) is the most considered category of polymer-clay nanocomposites, a type of hybrid material made up of an organic polymer matrix and organophilic clay fillers (Weiss et al., 2006).

Nanocomposite materials can be classified based on whether or not polymeric material is present in the composite. Polymer-founded nanocomposites are polymers or copolymers that contain nanoparticles or nanofillers dispersed in the polymer matrix and are further categorized as polymer/ceramic-based, inorganic/organic polymer, inorganic/organic hybrid, and polymer/layered silicate. The compositions of nonpolymer-based nanocomposites do not comprise any polymers or polymer-derived components. Inorganic nanocomposites are made up of nonpolymer-based nanocomposites. The three categories are metal-based nanocomposites, and ceramic-based nanocomposites. Food packaging uses polymer nanocomposites, and specific examples include processed cheese, meat, and dairy items, as well as medicinal vessels for carrying blood collection tubes, infant pacifiers, and drinking water bottles. Clay-based polymer nanocomposites have been employed in plastic bottles to improve the barrier, mechanical potentials, and product life. Nanocomposites are also used in the production of beer bottles to address a variety of issues, including biological and nonbiological features, beer colloid instability, oxygen permeability, and taste changes caused by light exposure (Sen, 2020).

17.5.1.4 Nanostructured materials

Some foods include nano-sized components that are not the same as synthetically created nanomaterials. Many dietary proteins have spherical arrangements between ten and hundred nm in dimension, while the bulk of polysaccharides and lipids are linear polymers with a thickness of less than 1 nm (one-dimensional nanostructures). Natural nanostructures can also be found in milk and milk foodstuffs, such as milk proteins and casein. Polymeric nanoparticles, liposomes, nanoemulsions, and microemulsions are the most important synthetic nanostructured arrangements in foodstuffs. These polymers improve solubility, bioavailability, controlled release, and bioactive component protection throughout manufacturing and storing (Pathakoti et al., 2017).

17.6 Processing of food

The process of transforming raw constituents into marketable foods and other forms with an extended shelf life is recognized as food processing. Toxin removal, disease prevention, preservation, and improving foodstuff uniformity intended for improved promotion and delivery are very illustrations of processing. Processed meals are not as much of possible to spoil rapidly as fresh foodstuffs, and thus are better suitable meant for long-distance transport since the source to the customer. Altogether they have become more effective as an outcome of the usage of nanotechnology. In the processing industry, encapsulating simple solutions, colloids, emulsions, biopolymers, and other materials into foods is crucial, and the functional qualities are preserved. Nanodrops, or nano-sized self-assembled structural lipids, act as liquid transporter for beneficial components that are insoluble in water and fats. They are utilized to keep cholesterol from making its way into the bloodstream from the digestive tract (Abbas et al., 2009).

17.7 **Food products**

Foodstuff nanotechnology is progressively being employed in the development of functional food additives such as foodstuff tastes and antioxidants. The ultimate objective is to increase the functionality of these components in food systems, lowering the concentrations required. We have already discussed supply and controlled release technologies for nutraceutical solubilization in foods. The development of food matrixes is progressively incorporating these new functional components. Commercially accessible food additives include nanoparticulate lycopene and carotenoids. Nanotechnology has made substantial contributions to agriculture and food systems science and engineering, including food security, illness treatment delivery technologies, new molecular and cellular biology tools, innovative pathogen detection materials, and environmental protection are all on the horizon.

The following are some illustrations of nanotechnology as a technique for obtaining supplementary progressions in the food industry:

- Food manufacturing, processing, and transportation have all been made safer thanks to pathogen and contaminant detection sensors.
- Devices that track individual shipments and retain past environmental data for a particular product.
- Food processing and transportation may be made more efficient and secure by using smart/intelligent systems that incorporate sensing, localization, reporting, and remote control of food goods.
- Transport, maintain, and dispense purposeful food components to their intended locations using encapsulation and delivery systems.

Nanotechnology has the prospective to expand food safety and biosecurity in four important areas of food production: the advance of novel functional constituents, microscale and nanoscale handing out, product improvement, and techniques and instrumentation strategy (Weiss et al., 2006).

Gas and moisture penetrability, as well as strength and biodegradability, are all requirements for a suitable packing material. As of well wrapping material through improved mechanical strength, barrier possessions, and antimicrobial films to nanosensing intended for pathogen revealing and alerting customers to the protection grade of foodstuff, nano-based "smart" and "active" food packaging's offers several benefits above old-style wrapping approaches. Foodstuff packing can advantage of the usage of nanocomposites as an active material in wrapping and material covering.

Many researchers were intrigued by the organic substances with antimicrobial properties, such as essential oils, organic acids, and bacteriocins, and also their application in polymeric matrix as antimicrobial wrapping. Nevertheless, since such composites are extremely susceptible to these physical circumstances, they are not suitable for several foodstuff handing out measures that call for high temperatures and pressures. Inorganic nanoparticles can be employed to achieve considerable antibacterial activity at low concentrations while also being more stable in demanding environments. As a consequence, most recently, there has been a surge of attention on using these nanoparticles in antimicrobial food packaging (Sing et al., 2017).

To increase the bioavailability and targeted delivery of natural bioactive chemicals, nanotechnology-based delivery methods are applied. Food includes bioactive chemicals that boost immunity and protect against disease. The potency of the majority of the food products was low, despite the fact that they had larger amounts of bioactive chemicals. Low bioavailability, solubility, and stability in the gut, as well as diminished penetrability and retention time in the digestive system, are all contributing factors. Nanomaterials have a huge surface area per unit mass and a lesser particle size, that expands the biological activity, bioavailability, and solubility of captured food ingredients. The use of nanosized iron and iron/zinc compounds in nutraceutical delivery improved bioavailability and minimized color changes in the finished products. Most vitamins (A, D, and E) and bioactive substances like curcumin, carotenoids, conjugated linoleic acids, coenzyme Q10, and -30 fatty acids have limited or variable bioavailability afterwards ingestion (Nile et al., 2020).

17.8 Food safety and biosecurity

Food fortification (minerals, vitamins, antioxidants, and essential oils), sensory perfection (flavor or color augmentation), shelf-life extension, and antimicrobial food wrapping are all examples of food eminence and food protection (finding of foodborne pathogens or harmful metabolites) have all been successfully implemented using nanotechnology on direct or packaged food products. In the early 2000s, researchers were drawn to nanotechnology because of the application in food safety and quality assurance programs using nanosensors and biochips.

Furthermore, the food business places a high value on food preservation. Food spoilage can be detected using nanosensors, which are made up of hundreds of nanoparticles that glow in diverse colors when they come into touch with pathogens. The main goal of nanosensors is to reduce the time needed to identify pathogens from days to hours or even minutes which is crucial in food microbiology. Nanosensors like these might be entrenched right into packaging constituents, acting as "electronic tongues" or "noses" that detect compounds produced throughout food decomposition. Other nanosensors are focused on microfluidics devices and may be used to swiftly and precisely identify viruses in a timely manner. Microfluidic sensors offer a variety of benefits, notably its compact size and designed to perceive chemicals of interest fast in just µl of sample volume, resulting in widespread application in medical, biological, and chemical research (Ravichandran, 2010).

17.9 Active food packaging by engineered nanoparticles: current difficulties

Emerging food packaging materials that contain modified nanoparticles with active and intelligent capabilities have the probable to help solve several of the world's food delivery problems. These materials have the capability to extend the stability of food goods, enhance food safety, and decrease food waste from rotting. However, worries about the harm presented to people by eating engineered nanoparticles that may move from nanoparticles food packaging into food have slowed the adoption of innovative food packaging materials incorporating designed nanoparticles (Hannon et al., 2015). Because of their heightened toxicity, mobility, and bioaccumulation, engineered nanoparticles have the possibility to destruction individuals and the atmosphere. For consumers to accept nanotechnology, all related hazards must be fully conveyed so that they may make an informed conclusion. Moreover, the level of hazard posed to humans in the worst-case scenario of exposure should be evaluated. If there is an undesirable degree of risk, a risk-managing approach should be devised to reduce it. Risk assessment is a technique for determining the dangers that a chemical or procedure poses to individuals and the environment. When used to design nanoparticle food packaging, migration studies and in vivo toxicity studies are used to establish the level of exposure to humans through nanoparticle ingestion (EFSA, 2012).

Inhalation, ingestion, and dermal penetration are all ways for nanoparticles from manufactured or other nanomaterials to reach the body. Nanotechnology-based medical equipment and medications, as well as the injection and release of nanoparticles from implants, could be a source of nanoparticles. In the food sector, inhalation and skin penetration are nearly exclusively associated with personnel in nanomaterials manufacturing plants, but ingestion is the main source of worry for ultimate consumers. Nanoparticles in food are mostly caused by direct interaction between nanopackaging and food, as well as nanoparticle migration from nanopackaging materials (Dimitrijevic et al., 2015).

17.9.1 Migration of engineered nanoparticles

The release of material from one medium to another is referred to as migration. The substance will migrate owing to a concentration gradient between the two media, according to Fick's first law of diffusion. Any loosely bound designed nanoparticles in the food packing will move from the packaging to the food if there are no modified nanoparticles in the food. This happens because there are less designed nanoparticles in the food, which causes migration. Temperature, time, concentration gradient, material characteristics, migrant position in the material, and migrant-material interaction are all factors that influence migration. The migration potential and diffusion mechanisms for designed nanoparticles from food packaging materials is a field of nanotechnology that has not gotten as much attention as nanoaerosols, nanofluids, and nanomedicines (Hannon et al., 2015).

17.10 Limitations of nanotechnology in food packaging

At the present time, consumers are prepared to pay more for higher quality foods and convenient packaged foods. The development of nanotechnology has led to the introduction of nanomaterials into our lives. While there have been studies on the development of nanomaterials for packaging, there has not been a comprehensive analysis, there are few studies relating to possible toxicity caused to human health. As it pertains to carbon nanotubes, published data suggests that carbon nanotubes may also have cytotoxic effects on human cells, at least when in contact with the skin or lungs (Monteiro-Riviere et al., 2005; Warheit et al., 2003).

Echegoyen and Nerín (2003) studied the migration of silver from three types of nanocomposites into food stimulants, including an analysis of the form in which silver migrates (ions or particles). In their experiments, they found that silver migrated into food stimulants, with acidic food exhibiting the highest levels of migration. Additionally, heating was found to increase migration, with a microwave causing more migration than a classical oven. There are two possible mechanisms through which silver might migrate: detachment of silver nanoparticles from the composites, and oxidative dissolution of silver ions.

The carbon nanotubes may migrate into foods once they are present in the food packaging material, so it is essential to investigate any potential health risks associated with ingested carbon nanotubes. The migration of substances from packaging materials to food can have a detrimental impact on food safety, causing great concern for consumers. All new packaging materials—nanomaterials included or not—are required to undergo migration tests in order to comply with Commission Regulation (EU) No 10/2011 (Agriopoulou, 2016).

The properties of nanomaterials determine how nanoparticles affect the body. Inhalation, ingestion, and skin penetration are all methods for nanoparticle penetration in an organism. Two studies examined clay movement from potato starch films, potato starch polyester blends, and polyethylene terephthalate (PET) bottles. When nanoparticles are positively charged and hydrophilic in nature, they tend to increase circulation time. They have severe effects on microcirculation. The brain is the most affected organ. It is observed to have increased production of ROS in microglial cells. Inhalation and completely related to factory workers (Silvestre et al., 2011; Gokularaman et al., 2017).

The main causes of food nanoparticles are diffusion of nanoparticles from nanopackaging components and the direct interaction of nanopackaging materials with food. In addition to toxicity, genotoxicity and carcinogenicity are other side effects of nanoparticles that have drawn the most attention. ZnO nanoparticles have genotoxic potential in human epidermal cells even though bulk ZnO is nontoxic, suggesting the size of the particles influences their effects on the body (Baltic et al., 2013).

The migration of silver and copper from nanocomposites used for their antimicrobial properties in food packaging was tested. The study revealed the percentage of nanofiller in the nanocomposites was the most critical parameter driving migration, more so than particle size, temperature, or contact time. The model developed for this study also predicted migration of nanosilver and to a lesser degree, nanocopper into food stuffs. This technique, if further developed and validated, could potentially be of benefit to the industry by reducing the time and cost usually involved in migration studies (Cushen et al., 2014).

Recent studies have been examined the migration and toxicological profile of clay polylactic acid nanocomposites migrated in water less than 10 g/dm^2, under these experimental conditions. Additionally, the authors evaluated the potential cytotoxicity of migration extracts both in vitro and in vivo on two cell types representative of the digestive system, as well as their ability to induce DNA mutations. Further, in a 90-day study, the rate exposed to the same migration extracts in drinking water showed no evidence of toxicity in terms of oxidative stress, inflammation, clinical biomarkers, or histopathology (Maisanaba et al., 2014). Deeper migratory and toxicological research is required for nanotechnology to be developed safely in the food packaging sector. Thus, for nano-packaging applications to be performed safely and successfully, three regulations must be applied: (1) food regulations, (2) health regulations, and (3) environmental regulations. Nano-packaging can provide new benefits to society, but it must be maintained at the highest standard of safety, health, and environmental protection. Additionally, these precautions are necessary to guarantee the safety of using nano-packaging, especially in terms of their potential toxicological effects, migration potential, and level of being exposed for both consumers and employees, with a focus on the health effects of the chosen nanomaterials after long-term exposure. As long as the requirements for the utilization of nanomaterials in food packaging are met, then the use of nanomaterials in food packaging will immensely improve the nutrition, flavor, and safety of food supplies as well as the environment. To elucidate the risk related to its employment by the packaging industry, a detailed toxicological analysis is required.

17.11 **Conclusion**

Almost every part of the food industry will likely be impacted by nanotechnology in some way. The rationale of this chapter was to demonstrate to the reader the exciting advantages that nanomaterials can bring to the food industry, including improved packaging materials and safer foods on store shelves that are less likely to be contaminated by chemical adulterants, and potentially lethal microorganisms, insects, etc. Food packs must be easy to handle, able to distribute food, and have a number of other characteristics relating to the physical characteristics of the packing material.

Additionally, unique and effective polymer materials for food packaging based on nanotechnology were reported in this work. In reality, all of the primary features of the package can be extended and implemented through the usage of polymer nanotechnology (containment, protection and preservation, marketing, and communication). Many of the biggest food packaging companies in the world are actively looking into the potential of polymer nanotechnology in order to create novel food packaging materials with enhanced mechanical, barrier, and antimicrobial properties as well as the capacity to record and monitor the state of food during its shipment and storage.

By adding the right kind of nanomaterial to the polymer matrix, the packaging material's mechanical, water-, oxygen-, and microbe-resistance qualities are improved, increasing the life period of the food goods. Additionally, active, IP with extended shelf life is made using nanoparticles. Although the addition of nanoparticles considerably enhances the properties of packing polymer, the procedure is more complicated than it first appears. Certain nanoparticles may cause health problems if they migrate into food and are consumed by a person for an extended period of time. Therefore, it is crucial to examine the migration, toxicity, and permissible limit of nanoparticles when using them in food packaging polymer that comes into close contact with food. The cutting-edge and best method for food packaging, according to the future of food consumption, is nanotechnology.

References

Abbas, K.A., Saleh, A.M., Mohamed, A., MohdAzhan, N., 2009. The recent advances in the nanotechnology and its applications in food processing: a review. J. Food Agric. Environ. 7 (3-4), 14–27.

Abdollahi, M., Alboofetileh, M., Rezaei, M., Behrooz, R., 2013. Comparing physico-mechanical and thermal properties of alginate nanocomposite films reinforced with organic and/or inorganic nanofillers. Food Hydrocoll. 32 (2), 416–424.

Aday, M., Yener, U., 2015. Assessing consumers' adoption of active and intelligent packaging. Br. Food J. 117 (1), 157–177.

Aditya, A., Chattopadhyay, S., Gupta, N., Alam, S., Veedu, A.P., Pal, M., Singh, A., Santhiya, D., Ansari, K.M., Ganguli, M., 2018. ZnO nanoparticles modified with an amphipathic peptide show improved photoprotection in skin. ACS Appl. Mater. Interfaces 11 (1), 56–72.

Agriopoulou, S., 2016. Nanotechnology in food packaging. EC Nutr. 42, 118–142.

Akbar, A., Anal, A.K., 2014. Zinc oxide nanoparticles loaded active packaging, a challenge study against *Salmonella typhimurium* and *Staphylococcus aureus* in ready-to-eat poultry meat. Food Contr. 38, 88–95.

Amenta, V., Aschberger, K., Arena, M., Bouwmeester, H., Moniz, F.B., Brandhoff, P., Gottardo, S., Marvin, H., Mech, A., Pesudeo, L.Q., Reuscher, H., Schoonjaus, R., Vettori, M.V.,

Weigel, S., Peters, R.J., 2015. Regulatory aspects of nanotechnology in the agri/feed/food sector in EU and non-EU countries. Regul. Toxicol. Pharm. 73 (1), 463–476.

Anon, 1995. Scavenger Solution: Packaging News, December ed. p. 20.

Azlin-Hasim., S., Cruz-Romero, M.C., Morris, M.A., Padmanabhan, S.C., Cummins, E., Kerry, J.P., 2016. The potential application of antimicrobial silver polyvinyl chloride nanocomposite films to extend the shelf-life of chicken breast fillets. Food Bioprocess Tech. 9 (10), 1661–1673.

Baltic, M., Boskovic, M., Ćirić, J., Dokmanovic, M., Janjić, J., Djordjevic, J., 2013. Nanotechnology and its potential applications in meat industry. Tehnologija mesa 54, 168–175.

Bilbao-Sainz, C., Bras, J., Williams, T., Sénechal, T., Orts, W., 2011. HPMC reinforced with different cellulose nano-particles. Carbohydr. Polym. 86 (4), 1549–1557.

Bodaghi, H., Mostofi, Y., Oromiehie, A., Zamani, Z., Ghanbarzadeh, B., Costa, C., Conte, A., Del Nobile, M.A., 2013. Evaluation of the photocatalytic antimicrobial effects of a TiO_2 nanocomposite food packaging film by in vitro and in vivo tests. LWT – Food Sci. Technol. 50 (2), 702–706.

Conte, A., Longano, D., Costa, C., Ditaranto, N., Ancona, A., Cioffi, N., Scrocco, C., Sabbatini, L., Conto, F., Del Nobile, M.A., 2013. A novel preservation technique applied to fiordilatte cheese. Innov. Food Sci. Emerg. 19, 158–165.

Cushen, M., Kerry, J., Morris, M., Cruz-Romero, M., Cummins, E., 2014. Evaluation and simulation of silver and copper nanoparticle migration from polyethylene nanocomposites to food and an associated exposure assessment. J. Agric. Food Chem. 62 (6), 1403–1411.

Dainelli, D., Gontard, N., Spyropoulos, D., Zondervan-van den Beuken, E., Tobback, P., 2008. Active and intelligent food packaging: legal aspects and safety concerns. Trends Food Sci. Technol. 19, S103–S112.

De Moura, M.R., Aouada, F.A., Avena-Bustillos, R.J., McHugh, T.H., J.M. Krochta, J.M., Mattoso, L.H.C., 2009. Improved barrier and mechanical properties of novel hydroxypropyl methylcellulose edible films with chitosan/tripolyphosphate nanoparticles. J. Food Eng. 92 (4), 448–453.

Dimitrijevic, M., Karabasil, N., Boskovic, M., Teodorovic, V., Vasilev, D., Djordjevic, V., Kilibarda, N., Cobanovic, N., 2015. Safety aspects of nanotechnology applications in food packaging. Proced. Food Sci. 5, 57–60.

European Commission, 2009. European Commission (E.C.) No. 405/2009 on active and intelligent materials and articles in contact with food.

Echegoyen, Y., Nerín, C., 2013. Nanoparticle release from nano-silver antimicrobial food containers. Food Chem. Toxicol. 62, 16–22.

EFSA (Panel on food contact materials, enzymes, flavourings and processing aids [CEF]), 2012. Scientific opinion on the safety evaluation of the substance, titanium nitride, nanoparticles, for use in food contact materials. EFSA J. 10 (3), 2641.

Emamifar, A., Kadivar, M., Shahedi, M., Soleimanian-Zad, S., 2010. Evaluation of nanocomposite packaging containing Ag and ZnO on shelf life of fresh orange juice. Innov. Food Sci. Emerg. 11 (4), 742–748.

Ganiari, S., Choulitoudi, E., &Oreopoulou, V., 2017. Edible and active films and coatings as carriers of natural antioxidants for lipid food. Trends Food Sci. Technol. 68, 70–82.

Giles, E., Kuznesof, S., Clark, B., Hubbard, C., Frewer, L., 2015. Consumer acceptance of and willingness to pay for food nanotechnology: a systematic review. J. Nanopart. Res. 17 (12), 1–26.

Gokularaman, S., Cruz, S.A., Pragalyaashree, M.M., Nishadh, A., 2017. Nanotechnology approach in food packaging: a review. J. Pharmaceut. Sci. Res. 9 (10), 1743–1749.

Gómez-Estaca, J., López-de-Dicastillo, C., Hernández-Munoz, P., Catalá, R., &Gavara, R., 2014. Advances in antioxidant active food packaging. Trends Food Sci. Technol. 35 (1), 42–51.

Grumezescu, A.M., Holban, A.M., 2018. Impact of Nanoscience in the Food Industry. Academic Press.

Hannon, J.C., Kerry, J., Cruz-Romero, M., Morris, M., Cummins, E., 2015. Advances and challenges for the use of engineered nanoparticles in food contact materials. Trends Food Sci. Technol. 43 (1), 43–62.

Hirvikorpi, T.T., Vähä-Nissi, M., Mustonen, T., Karppinen, M., 2010. Atomic layer deposited aluminium oxide barrier coatings for packaging materials. Thin. Solid. Films 518, 2654–2658.

Huang, Q., Yu, H., Ru, Q., 2010. Bioavailability and delivery of nutraceuticals using nanotechnology. J. Food Sci. 75 (1), R50–R57.

Le Corre, D., Bras, J., Dufresne, A., 2010. Starch nanoparticles: a review. Biomacromolecules 11 (5), 1139–1153.

Li, L., Zhao, C., Zhang, Y., Yao, J., Yang, W., Hu, Q., Wang, C., Cao, C., 2017. Effect of stable antimicrobial nano-silver packaging on inhibiting mildew and in storage of rice. Food Chem. 215, 477–482.

Maisanaba, S., Gutiérrez-Praena, D., Puerto, M., Llana-Ruiz-Cabello, M., Pichardo, S., Moyano, R., 2014. In vivo toxicity evaluation of the migration extract of an organomodified clay-poly(lactic) acid nanocomposite. J. Toxicol. Environ. Health Part A 77 (13), 731–746.

Majeed, K., Jawaid, M., Hassan, A., Abu Bakar, A., Khalil, H.P.S.A., Salema, A.A., Inuwa, I., 2013. Potential materials for food packaging from nanoclay/natural fbresflled hybrid composites. Mater. Des. 46, 391–410.

Majid, I., Nayik, G.A., Dar, S.M., Nanda, V., 2018. Novel food packaging technologies: innovations and future prospective. J. Saudi Society Agricult. Sci. 17 (4), 454–462.

Maxfield, M., Christiani, B.R., 1993. Polymer nanocomposites formed by melt processing of a polymer and an exfoliated layered material derivatized with reactive *Organo silanes*. https://patents.google.com/patent/WO1993011190A1/en.

Mihaly Cozmuta, A., Peter, A., Mihaly Cozmuta, L., Nicula, C., Crisan, L., Baia, L., Turila, A., 2015. Active packaging system based on Ag/TiO$_2$ nanocomposite used for extending the shelf life of bread. Chemical and microbiological investigations. Packag. Technol. Sci 28 (4), 271–284.

Mohanty, A.K., Misra, M., Nalwa, H.S., 2009. Packaging Nanotechnology. American Scientific Publishers. Los Angeles, pp. 341–350.

Monteiro-Riviere, N.A., Nemanich, R.J., Inman, A.O., Wang, YY, Riviere, JE., 2005. Multi-walled carbon nanotube interactions with human epidermal keratinocytes. Toxicol. Lett. 155 (3), 377–384.

Neethirajan, S., Jayas, D.S., 2011. Nanotechnology for the food and bioprocessing industries. Food Bioprocess Technol. 4 (1), 39–47.

Nerín, C., Tovar, L., Djeane, D., Camo, J., Salafranca, J., Beltrán, J., Roncalés, P., 2006. Stabilization of beef meat by a new active packaging containing natural antioxidants. J. Agric. Food Chem. 54, 7840–7846.

Nerin, C., Tovar, L., Salafranca, J., 2008. Behaviour of a new antioxidant active film versus oxidizable model compounds. J. Food Eng. 84, 313–320.

Nile, S.H., Baskar, V., Selvaraj, D., Nile, A., Xiao, J., Kai, G., 2020. Nanotechnologies in food science: applications, recent trends, and future perspectives. Nano-Micro Lett. 12 (1), 1–34.

Okada, Y., Fukushima, M., Kawasumi, I., Shinji, U., Arimitsu, S., Shigetoshi, K., Toshio, K., Osami, 1988. Composite material and process for manufacturing same. https://www.google.com/patents/US4739007.

Paine, F.A., 1991. The Packaging User's Handbook. AVI, Van Nostrand Reinhold, New York.

Panea, B., Ripoll, G., Gonzalez, J., Fernandez-Cuello, A., Alberti, P., 2014. Effect of nanocomposite packaging containing different proportions of ZnO and Ag on chicken breast meat quality. J. Food Eng. 123, 104–112.

Parisi, C., Vigani, M., Rodríguez-Cerezo, E., 2015. Agricultural nanotechnologies: what are the current possibilities? Nano Today 10 (2), 124–127.

Pathakoti, K., Manubolu, M., Hwang, H.M., 2017. Nanostructures: current uses and future applications in food science. J. Food Drug Analy. 25 (2), 245–253.

Potter, L., Campbell, A.J., Cava, D., 2008. Active and Intelligent Packaging: A Review (Review No. 62). Campden BRI, Chipping Campden, Gloucestershire.

Radusin, T.I., Ristic, I.S., Pilić, B.M., Novaković, A.R., 2016. Antimicrobial nanomaterials for food packaging applications. Sci. J. Food Feed Res. 43 (2), 119–126.

Ranjan, S., Dasgupta, N., Chakraborty, A.R., Samuel, S.M., Ramalingam, C., Shanker, R., Kumar, A., 2014. Nanoscience and nanotechnologies in food industries: opportunities and research trends. J. Nanopart. Res. 16 (6), 1–23.

Rauscher, H., Rasmussen, K., Sokull-Kluttgen, B., 2017. Regulatory aspects of nanomaterials in the EU. Chem. Ing. Tech 89 (3), 224–231.

Ravichandran, R., 2010. Nanotechnology applications in food and food processing: innovative green approaches, opportunities and uncertainties for global market. Int. J. Green Nanotechnol. Phys. Chem. 1 (2), 72–96.

Rhim, J., Park, H., Ha, C., 2013. Bio-nanocomposites for food packaging applications. Prog. Polym. Sci. 38 (10-11), 1629–1652.

Robertson, G.L., 1993. Food Packaging: Principles and Practice. Marcel Dekker, New York.

Rooney, M.L. (Ed.), 1995. Active Food Packaging. Chapman and Hall, London, pp. 164–173.

Roosen, J., Bieberstein, A., Blanchemanche, S., Goddard, E., Marette, S., Vandermoere, F., 2015. Trust and willingness to pay for nanotechnology food. Food Policy 52, 75–83.

Rubilar, O., Diez, M.C., Tortella, G.R., Briceno, G., Marcato, P.D., Duran, N., 2014. New strategies and challenges for nano-biotechnology in agriculture. J. Biobased Mater. Bioenergy. 8, 1–12.

Sekhon, B.S., 2010. Food nanotechnology: an overview. Nanotechnol. Sci. Appl. 3 (1), 214.

Sen, M., 2020. Nanocomposite materials. In: Nanotechnology and the Environment. IntechOpen.

Shafiq, M., Anjum, S., Hano, C., Anjum, I., Abbasi, B.H., 2020. An overview of the applications of nanomaterials and nanodevices in the food industry. Foods 9 (2), 148.

Silvestre, C., Duraccio, D., Cimmino, S., 2011. Food packaging based on polymer nanomaterials. Prog. Polym. Sci. 36 (12), 1766–1782.

Sing, T., Shukla, S., Kumar, P., Wahla, V., Bajpai, V.K., Rather, I.A., 2017. Corrigendum: application of nanotechnology in food science: perception and overview. Front. Microbiol. 12 (8), 2517.

Steenis, N.D., Herpen, E.V., 2017. Consumer response to packaging design: the role of packaging materials and graphics in sustainability perceptions and product evaluations. J. Cleaner Prod. 162, 286–298 .

Subramaniam, P.J., 1998. Dairy foods, multi-component products, dried foods and beverages. In: Blakistone, B.A. (Ed.), Principles and Application of Modified Atmosphere Packaging of Foods. Blackie Academic & Professional, London, pp. 158–193.

Vanderroost, M., Ragaert, P., Devlieghere, F., De Meulenaer, B., 2022. Intelligent food packaging: the next generation. Trends Food Sci. Technol. 39 (1), 47–62.

Vermeiren, L., Heirlings, L., Devlieghere, F., Debevere, J., 2003. Oxygen, ethylene and other scavengers. In: Ahvenainen, R. (Ed.), Novel Food Packaging Techniques. Woodhead Publishing Limited, Cambridge, pp. 22–49.

Waite, N., 2003. Active Packaging. PIRA International, Leatherhead.

Warheit, D.B., Laurence, B.R., Reed, K.L., Roach, D.H., Reynolds, G.A., Webb, T.R., 2003. Comparative pulmonary toxicity assessment of single-wall carbon nanotubes in rats. Toxicol. Sci. 77 (1), 117–125.

Weiss., J, Takhistov, P., McClements, D.J., 2006. Functional materials in food nanotechnology. J. Food Sci. 71 (9), R107–R117.

Yildirim, S., Röcker, B., Pettersen, M., Nilsen-Nygaard, J., Ayhan, Z., Rutkaite, R., Radusin, T., Suminska, P., Marcos, B., Coma, V., 2017. Active packaging applications for food. Compr. Rev. Food Sci. Food Saf. 17 (1), 165–199.

Perspectives for polymer-based antimicrobial films in food packaging applications

18

Ashish Tiwari[a], Anurag Tiwari[c], Santosh Kumar[b], Shalinee Singh[b] and PK Dutta[d]

[a] *Department of Chemistry, Government College Dumariya Jarhi, Surajpur, Chhattisgarh, India,*
[b] *Department of Chemistry, Harcourt Butler Technical University, Kanpur, Uttar Pradesh, India,*
[c] *Department of Applied Mechanics, MNNIT Allahabad, Prayagraj, Uttar Pradesh, India,*
[d] *Department of Chemistry, MNNIT Allahabad, Prayagraj, Uttar Pradesh, India*

18.1 Introduction

Food packaging is a high-demanding and crucial aspect of the food business, which uses many petrochemical-derived polymers. After its intended use, a large quantity of trash is produced, resulting in disposal concerns and associated environmental issues (Kanatt and Makwana, 2020). Biopolymers can be used to replace synthetic nondegradable polymers that have a significant environmental effect (Kanatt and Makwana, 2020; Moreno et al., 2015; Shankar and Rhim, 2015). As a result, in the current context, scientists are focusing on the use of bio-based polymers, for example, guar gum (GG), chitosan (CS), moringa extract (ME), cellulose, etc. which are commonly used for packaging food materials (Barbosa et al., 2014; Gatto et al., 2019). Along with providing physical protection, biopolymers-based packaging materials can also act as an interruption to hydrophilicity and O_2, which extend the shelf life of the food (Rhim et al., 2006; Lević et al., 2014). They also serve as a backbone matrix for the incorporation of various plant sap, cross-linker, and antimicrobial agents to improve the biopolymer's functional qualities.

Proteins, lipids, and carbohydrates are all bio-based basic resources that are abundant on the planet. Their nontoxic, renewable, eco-friendly properties made them a viable chemical-free option for the plastic packaging of food. The oxygen interrupter and antibacterial characteristics of bio-based polymers can be increased by adding naturally originating additives and extracts in food packaging (Kumar et al., 2021; Hernandez et al., 2017; Kechichian et al., 2010). These bio-based packaging sheets can be doped with nanomaterials to obtain the requisite strength and increase their barrier properties (Aydogdu et al., 2020). There are many types of naturally occurring (animal and plant-originated) polysaccharides that are generally used for the development of packaging materials, out of which GG, is produced from the seeds of guar or Indian cluster beans (Thombare et al., 2016). Because of the amazing properties of GG, such as ease of availability and biocompatibility, used as an alternative to synthetic polymers. GG is an eco-friendly nonionic polysaccharide derived from GG seed's endosperm. It is made up of galactomannan which has mannose units in it. GG has a high water solubility due to its concentration of hydroxyl groups (El-hefian et al., 2010; Sahariah et al., 2017). It is suited for use in the food, pharmaceutical, paper, textile, and oil sectors because of its thickening and emulsifying properties (Cazón et al., 2020; Das et al., 2015).

In a polar solvent, GG is a highly soluble polysaccharides. It is used as a thickening material in the food packaging industry. Its solubility is affected by temperature, particle size, and pH (Gupta et al., 2018). In comparison to other biopolymers, GG has got a lot of attention because of its higher molecular weight, the longer chain of the polymer, and abundant availability in nature (Das et al., 2011). Another common derivative of cellulose is carboxymethylcellulose (CMC) which has good film-forming properties, biocompatibility, and water solubility as well as being readily accessible and nontoxic.

It also aids in the improvement of packaging film's physical, mechanical, and barrier qualities (Bandyopadhyay et al., 2019). Due to their high hydrophilicity, GG, and CMC are limited in their use in packing films due to weak barrier properties against atmospheric components. To impart hydrophobicity and decrease the solubility of GG in water, different types of cross-linkers are used for example glutaraldehyde, acetic acid, citric acid, sodium sulfate, etc. (Xu et al., 2015; Yun et al., 2006; Olsson et al., 2013). The cross-linking materials that are used in commercial films may migrate into food. Although these biopolymers are good for the environment, they lack mechanical strength and they are easily affected by the environment. Due to this reason, there is a requirement for more research to improve its strength and applications as synthetic polymers. Chitosan (CS) is a multifunctional polysaccharide with a low cost, highly propitious, and biodegradable. Chitosan has an antibacterial and nontoxic nature which protects food against antimicrobial adulterant (Kanatt et al., 2012; Bashir et al., 2018; Grande and Carvalho, 2011). CS films, on the other hand, are limited in their practical application due to their lack of flexibility (Grande and Carvalho, 2011). Nowadays, researchers are mixing CS with other polymers to improve its overall performance as a biopolymer antimicrobial film (Choo et al., 2016).

Recently, it has been revealed that CS and polyvinyl alcohol (PVA) blends have outstanding mechanical strength racwell as robust film-forming synthetic polymers. Because of its distinctive characteristics such as nontoxic, biocompatible, and water-soluble (Yu et al., 2018; Costa et al., 2009), therefore, it is fit for using in a wide range of industries. The carbon skeleton of PVA contains hydroxyl groups which form hydrogen bonding with biopolymers (chitosan, starch, and GG). The extract of moringa leaves contains various types of chemicals (Mbikay, 2012; Gopalakrishnan et al., 2016). It has a high concentration of antioxidant, antibacterial, and antimicrobial properties (Nemade and Sawarkar, 2015). As a consequence, ME can be used to generate a new chitosan-guar gum-polyvinylalcohol-moringa extract (CGPM) active films by adding them in a particular ratio.

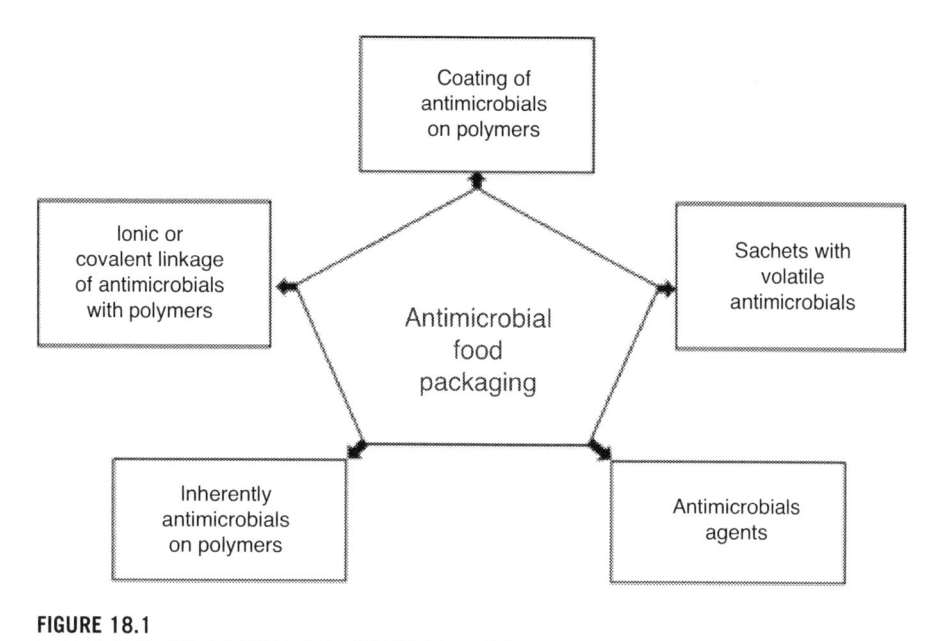

FIGURE 18.1

Types of antimicrobial food packaging.

18.2 Antimicrobial food packaging

Food packaging is required to keep food safe from bacterial infection and deterioration. However, foods that are extremely pathogenic can be protected by antimicrobial substances including carbon dioxide, ethanol, medications, chlorine dioxide, organic acids, essential oils (EOs), and spices. These compounds are being researched for their capacity to prevent bacterial growth that could ruin food (Campillo et al., 2009).

18.2.1 Various kinds of antimicrobial food packaging

Antimicrobial packaging includes volatile antimicrobial sachets and pads, polymers with volatile and nonvolatile antimicrobial agents directly incorporated, antimicrobial coatings on polymer surfaces, ionic, and covalent antimicrobial-polymer linkages resulting from immobilization techniques, and naturally antimicrobial polymers. Packaging packets containing vapor-permeable antimicrobials. Various kinds of packaging of food by using antimicrobial are depicted in Fig. 18.1.

18.2.2 Sachets with antimicrobials

Oxygen barrier, moisture abstracter, and ethanol vapor provider are the three most often used types of sachets in antimicrobial (Appendini et al., 2002). Moisture absorbers inhibit microbiological development by reducing water activity. Sachets with vaporized ethanol are used in backing and other seafood processing to get rid of molding (Smith et al., 1995).

18.2.3 Antimicrobial agents in food packaging

Commercial applications for antimicrobial polymer matrices include pharmaceuticals and household items. By incorporating a small amount of silver into polyethylene (PE), polypropenes, and zeolites, microbial cell enzymatic activity is reduced. Polymers containing antimicrobial enzymes such as lactoferrin, lactoperoxidase, and some antimicrobial peptides such as cecropins, magainins, and others such as natural phenols, antioxidants, and metals such as copper are commonly employed (Appendini et al., 2002). When mixed with Ethylenediaminetetraacetic acid (EDTA), edible films containing lysozymes and nisin inhibit the growth of *Escherichia coli*. Antimicrobials in food packaging slow surface growth.

18.2.4 Coating of antimicrobials to polymer surfaces

Antimicrobial coatings on polymer surfaces are discovered to be loosely or strongly attached to the package inside surface. Oxygen-absorbing, moisture-absorbing, and ethanol vapor-generating sachets are the most common types of antimicrobial sachets (Appendini et al., 2002). These antimicrobials are employed to inhibit oxidative and molding activity, whilst water absorbers reduce water activity and hamper microbiological growth (Rooney, 1995).

18.2.5 Immobilization of antimicrobials by ionic or covalent linkages to polymers

Microbial agent diffusion is less with covalent bonding, while ionic bonding allows the progressive release of antimicrobial agents into food (Appendini et al., 2002). The enzyme glucose oxidase catalyzes the conversion of glucose and oxygen reaction mixtures into antibacterial hydrogen peroxides (Garcia et al., 1990; Wang and Hsiue, 1993), which is covalently bonded to an insoluble substrate that might be used in packaging materials. For milk shelf-life extension, packaging materials containing immobilized beta-galactosidase and glucose oxidase activate the lactoperoxidase system in milk (Garibay et al., 1995).

18.2.6 Inherently antimicrobial polymers

Fungicide-coated waxes were one of the first advances in antimicrobials for vegetables and fruits. Shrink film can be used to wrap cheese and sausages, as well as the addition of quaternary ammonium ion salt to potatoes (Labuza and Breene, 1989). Covered in cellulose casing and sorbic acid-coated wax paper (Food Safety Consortium Newsletter, 2000). The cast film process is used to incorporate antimicrobials into polymers that serve as coatings (Appendini et al., 2002). Nisin-zinc-coated PE films (Food Safety Consortium Newsletter, 2000) and nisin-methylcellulose-coated PE films (Cooksey, 2005) have been employed in the poultry industry. Polyvinyl alcohol, starch, and casein were used as adhesives in enzyme glucose oxidases that were coated on moisture-proof materials (Labuza and Breene, 1989).

18.3 Materials used in antimicrobial food packing

The cast film process is used to incorporate antimicrobials into polymers that serve as coatings (Appendini et al., 2002). In the poultry business, nisin-zinc-coated PE films (Food Safety Consortium Newsletter, 2000) and nisin-methylcellulose-coated PE films (Cooksey, 2005) have been used. In enzyme glucose oxidases coated on moisture-proof materials, polyvinyl alcohol, starch, and casein were employed as adhesives (Labuza and Breene, 1989).

18.3.1 Natural materials used in food packaging

18.3.1.1 Starch

Corn, wheat, rice, and potatoes are the most prevalent sources of starch in industrial applications. Although starches from diverse sources are chemically similar, they may differ in granule size, shape, and molecular content. The most important properties are the ratios of the polysaccharides amylose and amylopectin. Due to the increasing cost and scarcity of traditional polymer film-forming matrices, starch has emerged as a key raw component in the polymeric film-producing area. Starch dissolves into harmless molecules when it comes into touch with soil microbes, making it beneficial for the creation of agricultural mulch films (Chandra and Rustgi, 1998). Starch absorbs water and is thermomechanically resistant. Despite being a polymer, starch does not have a high stress resistance. The glucoside linkages begin to break at temperatures above 150°C, and the starch grain becomes endothermically heated at temperatures above 250°C.

18.3.1.2 Protein

Proteins are polymers made up of 20 distinct amino acids. Proteins range in molecular weight from 6000 to one million. With only 100 amino acid residues, a protein can have about 10,100 distinct conformations (Fried, 2003). Proteins are a valuable source of packing films due to their various molecular properties and chemical activities. Protein films provide good gas barrier qualities as well as appropriate mechanical and optical properties despite their high moisture sensitivity and limited water vapor barrier capabilities.

18.3.1.3 Cellulose

With the exception of the most potent hydrogen bond-breaking solvents, such as N-methylmorpholine-monoxide, cellulose is a high molecular weight crystalline polymer that is insoluble and infusible which is commonly processed into derivatives due to these insolubility characteristics. A cellulose derivative is cellulose acetate with high-volume applications including fibers, films, and thermoplastic injection molding (Chandra and Rustgi, 1998). Composite films of silk fibroin and microcrystalline

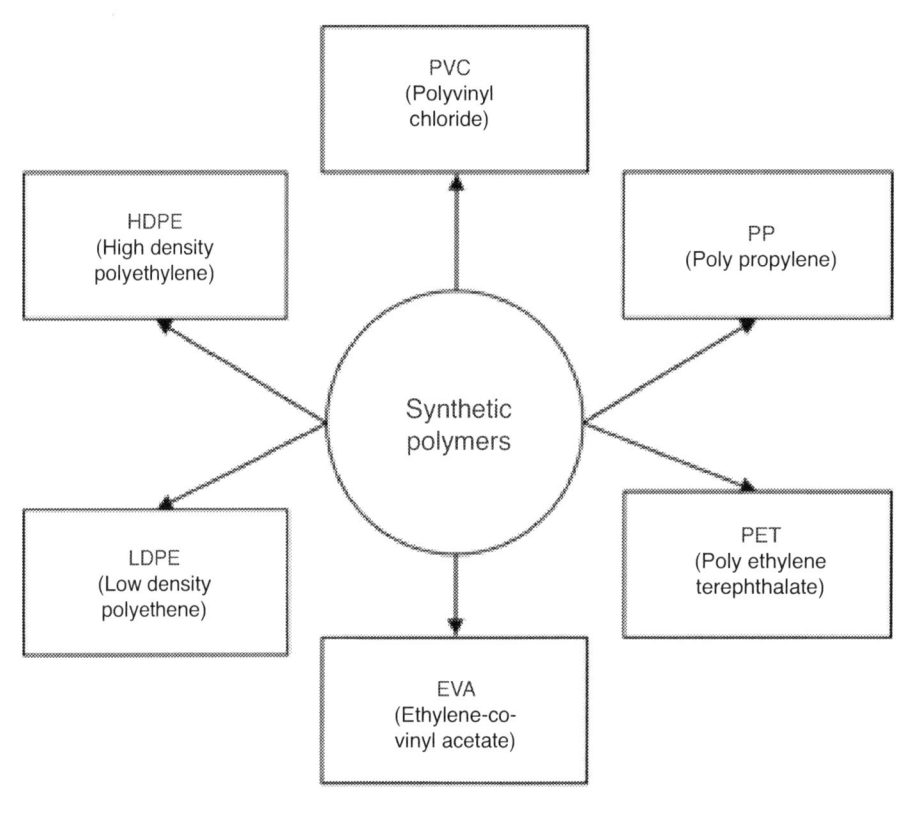

FIGURE 18.2

Different types of synthetic polymers used in food packaging.

cellulose have been explored. In their study, the films' tensile strengths increased as the cellulose content (wt.%) increased.

18.3.1.4 Cellulose acetate

Long strands of glucose molecules make up cellulose, a plant-generated natural polymer. The degree to which acetate groups are replaced affects the properties of cellulose. It can be chemically converted to obtain cellulose acetate. Around 75% of the hydroxyl groups on the polymer's skeleton are replaced by acetate groups. The semi-crystalline material cellulose acetate has good mechanical characteristics.

18.3.2 Antimicrobial synthetic polymer

Plastics are synthetic polymers such as PE and nylon that are commonly referred to as "plastics." Under regulated conditions, synthetically derived polymers or their derivatives (Campbell and Ian, 1994) are generated. High tensile strength and barrier characteristics are common in synthetic polymeric films. Fig. 18.2 offers information about various synthetic polymers.

18.3.2.1 Polyethylene

Polythene is categorized into numerous classes based on its density and branching patterns. The three most common types of PE are high-density (HD) and low-density polyethylene (LDPE), and linear low-density-polyethene (LLDPE) (Baner, 2000). Han and Floros used LDPE resins and potassium sorbate powder to generate an antibacterial coating. To determine its suitability as a packaging material, its tensile characteristics, visibility, and antibacterial activity were assessed. The addition of potassium sorbate did not influence the film's tensile qualities. The film's transparency reduced as the potassium sorbate level increased (Han et al., 1997).

18.3.2.2 Polyvinyl chloride

Polyvinyl chloride (PVC) is a translucent, amorphous polymer that is widely utilized in the production of films and containers. Plasticizers (low-volatility organic liquids) are often added to polymers, resulting in a wide range of types and amounts of the substance utilized, which has different qualities. PVC films that have been plasticized, which are limp, sticky, and flexible are used to package fresh meat. PVC sheets that have not been plasticized are rigid, and they are typically thermoformed into chocolate and biscuit inserts. The transparency and resistant to oils and other barrier properties of polyvinylchloride bottles are excellent. Because of their low thermal processing stability and environmental issues with chlorinated plastics, PVC bottles are rarely used in food packaging.

18.4 **Antimicrobial agents**

Antimicrobial packaging can be manufactured by using chitosan and EOs as antibacterial agents (Ke et al., 2021; Wińska et al., 2019). Antimicrobial compounds can be integrated into or onto the matrix through direct addition, encapsulation, coating, and grafting. Traditional packaging techniques, on the other hand, often use polymeric polymers. These polymers can be found in various areas and are characterized on the basis of their biodegradability. Because of concerns about the environment, legislation requirements, and customer requests for green products, a recent trend has been to replace nonbiodegradable polymers with biodegradable polymers. As a result, natural resources are now used in antimicrobial packaging for antibacterial compounds, as well as the main polymer. The next subtopic will go through the various varieties, as well as their pros and cons.

18.4.1 **Types of antimicrobial agents**

Antimicrobial agents of many types have been investigated for antimicrobial packaging applications. Each antimicrobial agent has a unique method that causes it to react differently to germs. There are some limitations to the antimicrobials used in this case. According to their sources and physiologies, these chemicals are classified as synthetic and natural (Fig. 18.3). Antimicrobials can be organic or inorganic. In the

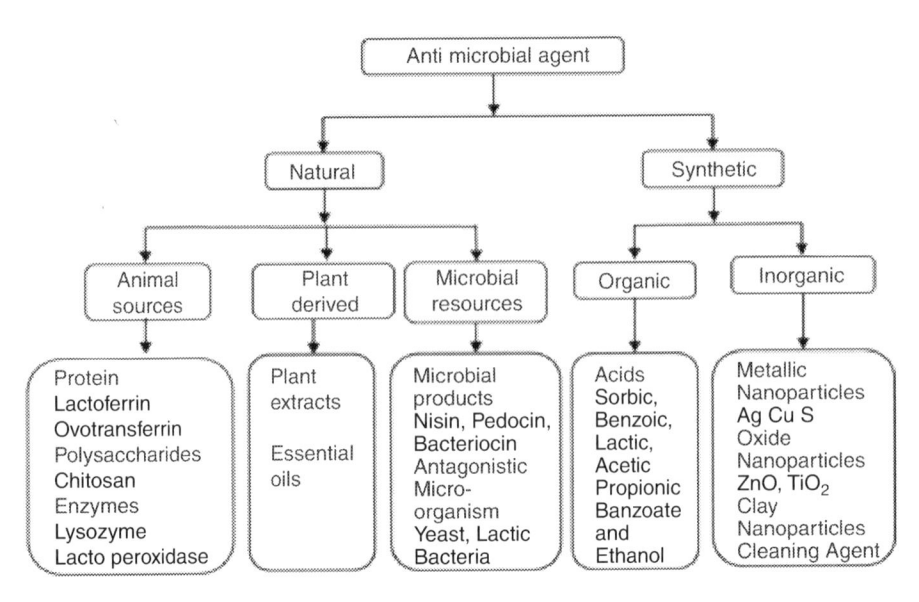

FIGURE 18.3

Sources of antimicrobial agent.

literature, nanoparticles of metals (Ag, Cu, and S) (Sánchez et al., 2020), nanoparticles of metal oxide (ZnO, TiO$_2$, and CuO) (Nguyen et al., 2019), nanoparticles of clay (montmorilonitrile, bentonite, and cloisite) (Ebrahimi et al., 2018), chelating agents (SO$_2$, ClO$_2$, and ethanol) (Chen et al., 2020), organic acids and their salts (García et al., 2019), and other types of synthetic antimicrobial agents have all been reported. The antimicrobial agents used may vary depending on the packing material's intended function. Synthesized antibacterial compounds such as EDTA, fungicides, parabens, and other chemicals are among the most commonly employed substances in the food packaging industry. Metals, especially metal oxides, have antibacterial characteristics, while the long-term effects on the environment and human health are unknown. Silver-based antimicrobial food packaging, on the other hand, is expected to become popular and is already being used in Japan and the United States (González et al., 2020).

18.4.1.1 *Natural antimicrobial agents*

18.4.1.1.1 Essential oils

Essential oils are natural evaporative compounds obtained by either crushing or distillation of plants (Vergis et al., 2015; Vasile et al., 2017). Secondary metabolites are complex lipid oils produced by various parts of the plants root, shoot, stem, flower, and leaf. The antibacterial activities of EOs are due to their oxygenated derivatives, triterpenes, diterpenes phenolic compounds, and monoterpenes (Vickers, 2017). EOs have fungicidal, antioxidative, virocidal, micocidal, and bactericidal characteristics. The Food and Drug Administration (FDA) is in charge of regulating food and drug administration has declared EOs are generally recognized as safe (GRAS). Due to their high biological efficacy, EOs are the most often utilized additives in the food industry.

18.4.1.1.2 Bacteriocins

Bacteriocins are bacteria-generated antibacterial compounds synthesized by ribosomes. Bacteriocins can be found in a variety of sites and have a range of structures, enabling new antibacterial agents to be produced. Popular bacteriocins include nisin and pediocin.

18.4.1.1.3 Nisin

Nisin is used as a most common bacteriocin for the packaging of food, an antimicrobial peptide generated by *Lactococcus lacticus* and lactic acid bacteria. The FDA have approved nisin, which is used in more than 48 nations. Nisin inhibits the growth of a wide spectrum of Gram-positive bacteria, including *Listeria monocytogenes*, by blocking spore germination (Limjaroen et al., 2003; Soto et al., 2016). Because of the cell wall bacteria is lacking affinity with water. The antibiotic (peptide's amino acid) can enter the bacterial cell membrane and impact its permeability due to its hydrophobic nature. Bacteria die when cellular components such as DNA and other cell organelles come out of the cell. The antibacterial characteristics of nisin can be easily added to polymers by coating polymer films with it or by combining nisin with polymers to create food packaging films. Various case studies have shown that smart packing containing nisin helps limit the growth of certain food-borne pathogens. Nisin inhibits microbial development and keeps food fresh at 25°C. Nisin, exhibits antibacterial qualities in seasoned beef, a popular Korean meal (Paik et al., 2006). Seasoned beef is packaged and stored at 4°C–25°C for 60 days. Consequently, it has been observed that seasoned beef with nisin supports a moderate rise in mesophilic microbe growth, whereas in the absence of nisin seasoned beef promotes a significant enhancement in the development of these microorganisms. Therefore, it is observed that even at room temperature, nisin has remarkable antibacterial characteristics. Nisin helps to preserve seasoned beef by improving the mechanical qualities of food containers. Cutting force in packages containing nisin changes minimally after keeping at room temperature, whereas cutting force in packages in the absence of nisin decreases dramatically. As a result, nisin is a potential antibacterial agent that is commonly used in the packaging of food.

18.4.1.1.4 Pediocin

Pediocin is the second most widely used bacteriocin that is mostly produced from the bacterium *Pediococcus acidilactici*. It is a naturally occurring small molecular protein that resists the action of heat, has consistent physicochemical features, and broad antibacterial range. Because of the nature of polypeptides, they can be digested in the human body and may be used to prevent effectively the variety of foodborne pathogenic bacteria. As a result, pediocin has the potential to be exploited in the development of naturally derived food stabilizers of the most recent generation (Mehta et al., 2013; Chikindas et al., 2010).

Silva et al. (Silva et al., 2009) produced a cellulosic film with great antibacterial efficacy based on pediocin, which can prevent pathogenic bacteria from growing. They kept slices of ham for up to 15 days and discovered that cellulosic films have the same antibacterial effect as containing medicines with 25% and 50%. They utilized pediocin and zinc-oxide nanoparticles to make antimicrobial polymeric nanocomposites films that were effective against *L. monocytogens* and *Staphylococcus aureus* (Perez et al., 2013). Elongation at break values rises when pediocin is added. Pediocin forms a yellow color film with nanocomposite, that is counterbalanced by the addition of zinc oxide nanoparticles, giving the films a whitish tint.

18.4.1.1.5 Lysozyme

Lysozyme is a mono peptide chain, hydrophilic nature enzyme. It acts as an antibacterial and protects against Gram-positive bacterial infection. The antibacterial activities of lysozyme are due to its capacity to hydrolyze the 1-4 glycosidic linkage between N-acetylmuramic acid and N-acetylglucosamine in peptidoglycan (Irkin et al., 2015). Lysozymes hydrolyze the peptidoglycans component of the bacterial cell wall, destroying it and enabling intracellular contents to flow out which results in the decay of bacteria (Cha et al., 2017). Several researchers have employed physical mixing or molecular bonding to immobilize antibacterial enzymes in order to manufacture food packaging materials. The sol–gel process is employed to construct an antibacterial active package based on lysozyme (Corradini et al., 2013), for example, using polyethylene terephthalate (PET) polymer base for manufacturing packing polymeric films under rigorous control capabilities. Lysozymes travel from the film to water and become active once absorbed into PET films, making the film antibacterial.

18.4.1.1.6 Grapefruit seed extract

Catechins, epicatechins, gallic acid, and procyanidins are among the several phenolic chemicals found in grapefruit seed extract (GFSE). As a result, GFSE is a natural source of the antibacterial agent that restricts the development of bacterial strains such as Gram-negative and Gram-positive (Cvetnic et al., 2004; Heggers et al., 2002). GFSE contains antigerm, bactericidal, antifungal, and virocidal, antiseptic properties. GFSE used to create antibacterial food packaging with promising qualities using polycaprolactone, chitosan, and PE (Tan et al., 2015; Tong et al., 2018). The crystallinity of biodecomposable polycaprolactone (PCL) and alginic acid increases when GFSE is introduced (Tan et al., 2015). Compression molding is used to create a smooth and uniform film. The film is also effective against *Pseudomonas aeruginosa*. Grapefruit seed extract may be distributed in chitosan polymer at various concentrations (0.5%, 1.0%, and 1.5% v/v) without damaging the clarity of the film. By enhancing GF seed extract improves chitosan-based composite films' tensile strength. It has been observed that chitosan-matrix-based film demonstrates antifungal activity. In addition, Table 18.1 represents different polymeric packaging with various antimicrobial agents and methods of preparation.

Table 18.1 Represent different polymeric packaging with various antimicrobial agents and methods of preparation.

PVC	Ag-nano particles	Solvent casting	*Bacillus subtilis, A. niger* and *F. solani*	Braga et al. (2018)
	PHE-Zn	Milling	*E. coli* and *S. aureus*	Zhang et al. (2020)
	Orange essential oil	Solvent casting	*E. coli* and *S. aureus*	Silva et al. (2018)
PET	Ag-nano particles	Electrospinning process	*E. coli, Pseudomonas aeruginosa,* and *S. aureus*	Sorkhabi et al. (2023)
	LDH-p hydroxybenzoate	Ball milling	*Salmonella spp.* and *Campylobacter*	Bugatti et al. (2019)
	ZnO, TiO$_2$	Melt mixing	*Lactobacillus sakei subsp. E. coli, S. cerevisiae,* and *A. brasiliensis*	Kohannia et al. (2021)
LDPE	ZnO	Extrusion	*S. aureus* and *E. coli*	Dehghani et al., (2019)
	Ag-nano particles	Ball milling	*E. coli*	Olmos et al. (2018)
	Thymol	Melt compounding	*E. coli*	Safakas et al. (2023)
PE	Carvacrol and menthol	Coating	*E. coli, S. aureus, L. innocua,* and *S. cervicea.*	Jahdkaran et al. (2021)
PP	Sorbic acid	Extrusion	*E. coli* and *S. aureus*	Fasihnia et al. (2018)
	Oregano EO	Melt blending	*B. thermosphacta*	Cabello et al. (2018)
	Carvacrol	Melt compounding	*E. coli* and *A. alternata*	Krepker et al. (2018)
PS	GO-p(VBC)	Casting	*B. cereus, P. aeruginosa,* and *C. albicans*	Ghanem et al. (2020)
	ZnO, CaCO$_3$ nanoparticles	Encapsulation	*S. aureus, P. aeruginosa, Candidia albicans,* and *A. niger*	Ibrahim et al. (2019)

18.4.1.2 Synthetic antimicrobial agent

Preservatives are defined as substance that is mixed with food that prevents or delays decomposition; however, common salt, sweeteners, vinegar, spice and spices' oil, and insecticides are not considered under this category.

18.4.1.2.1 Benzoic acid and sodium benzoate

Benzene carboxylic acid (C_6H_5COOH) is one of the most commonly utilized food, cosmetics, and pharmaceutical preservatives. The FDA approved sodium salt of benzoic acid for the first time, as a chemical preservative for food.

18.4.1.2.2 Sorbic acid and sorbateshdienoic acid

Sorbic acid and its salt, particularly potassium salt of sorbic acid, have been widely utilized as preservatives in a variety of foods. In recent years, it has been used in the feed of animals, medicines, and cosmetical things. These are beneficial supplements because they prevent or inhibit the development of a variety of microbes, including yeast strains, molds, and pathogens. In particular conditions, some microbial strains are resistant to sorbates. They are, however, regarded as effective food preservatives when employed in hygienic situations and products made with proper production practices.

18.4.1.2.3 Ethylenediaminetetraacetic acid

Because of restricted cation availability and EDTA's tends to break the cell membranes of bacteria through the formation of bivalent cations, which operate across the membrane as salt bridges and chelators in macromolecules, for example, lipopolysaccharide (Economou et al., 2009; Boziaris and Adams, 1999). EDTA has an antibacterial effect. The interaction of EDTA with many other antibacterial agents has been extensively explored. Adding ethylene diamine tetra acetate to weak acids improves their antibacterial activity concerning Gram-negative strain (Ntzimani et al., 2010). Economou et al. (2009) found that EDTA can help nisin's antibacterial action and food shelf life. The impact of EDTA and oregano essential oil in combination with improved atmosphere-packaged chicken liver flesh has also been investigated (Hasapidou and Savvaidis, 2017). Cannarsi and colleagues (Cannarsi et al., 2008) studied the antibacterial activity of lysozyme and nisin in combination with EDTA on spoilage microorganisms in cooled buffalo meat. Combining 0.5% lysozyme and 2% ethylene diamine tetra acetate produces the best results, inhibiting the development of all bacteria tested, including *Brochotrix thermosphacta*. Incorporating EDTA into several antibacterial treatments to improve their inhibiting effects on deteriorating bacteria.

18.5 Antimicrobial polymeric films

Bags and other forms of plastic containers have been manufactured using hydrocarbons since their inception in 1959, and their environmental pollution is a global issue. There is a simple solution to employ biodegradable materials to create environmentally friendly bags and packaging. By incorporating antimicrobial properties into biodegradable containers gives better protection from food spoiling and hence improves shelf life. In their search for plastic replacements, biopolymers have received special attention from researchers. Naturally originated polymers of carbohydrates and their derivatives such as starch, carboxy-methylated cellulose (CMC), methylated cellulose (MC), chitosans, lignins, poly-lactic acid (PLA), poly-vinyl alcohols (PVA), and others. These materials must be nontoxic, long-lasting, and possess the required qualities before they can be employed in the food sector. Even though many of them are polysaccharides with the distinctive features of each bio-polymer are affected by a similar monomer composition, the polymer chain orientation, substitution, and hydrogen bonding. Polyvinyl alcohol is water-soluble due to its hydrophilic nature, it has good crystal-forming capabilities, and its side chains can establish hydrogen bonds with other polymers via their OH groups or aid in the loading of antimicrobial compounds (Chen et al., 2018). PLA has a high degree of crystallinity, but it can only be dissolved in organic solvents, therefore surface compatibilization is required when using nanoparticles to strengthen it (Baek et al., 2018). Aliginic acid and its salt both are used very effectively. As a result, they are combined with lipidic components (such as EOs, which have antibacterial properties) and can be fortified with nanoparticles (Alboofetileh et al., 2018). Chitosan is a polymer that shows the characteristic of solubility in water due to hydrogen bonding through OH and NH_2 groups at acidic pH. Cellulosic polymer and its various derived polymers are hydrophilic biopolymers with a free hydroxyl group. Combining biopolymers with –OH groups in composite materials, where the latter will aid cross-linking, is one way to increase water resistance.

Numerous hydrogen bonds are created when different polymers are mixed, improving the composite's mechanical properties. Simply combining two polymers does not always result in a superior product. PVA/CMC composite films, for example, have poor water resistance and antimicrobial properties. To improve water vapor permeability (WVP) and antibacterial action, inorganic nanoparticles (such as Ag, ZnO, TiO_2, Fe_3O_4, and others) and EOs can be added to polymeric films (Zhang et al., 2018a; Singh et al., 2018). Oxygen and water vapor-resistant property is very important in the food packaging sector, while vulnerability to CO_2 and aromatic chemicals is also relevant. Composite film engineering typically attempts to enhance the properties of the original biopolymer (usually a polysaccharide) in order to obtain the desired properties. Nanomaterials can also be used as antibacterial agents, oxygen scavengers in the food packaging field. The compatibility of nano components with the polymeric matrix is a major issue in designing antimicrobial packaging (Youssef et al., 2018). In packaging, bio-nano composite materials are a feasible alternative to fossil-based polymers. However, in addition to their evident benefits, polysaccharides have substantial drawbacks. Water resistance and permeability are two mechanical properties that are deficient. Superior thermal and mechanical capabilities, as well as water vapor and oxygen barrier properties, can be produced by incorporating various nanoparticles into the biopolymeric film while keeping the biopolymer's

Table 18.2 Principal film forming biodegradable classes of substances used in packaging.

Film forming category	Substance used	References
Polysaccharides	Cellulose	Khezrian and Shahbazi (2018)
	Chitosans	Kaya et al. (2018)
	Starch	Araújo et al. (2023)
	Hemicellulose	Tedeschi et al. (2020)
	Alginate	Fabra et al. (2018)
	Agarose	Hu et al. (2016)
	Glucomannan	Lie et al. (2017)
	Pullulan	Li et al. (2020)
Proteins	Soya protein	Silva et al. (2018)
	Whey protein	Kouravand et al. (2018)
	Zein	Silva et al. (2018)
	Collagen	Ahmed et al. (2016)
	Gelatin	Lopez et al. (2018)
	Casein	Abdollahzadeh et al. (2018)
Liquid and waxes	Plant oil and animal fat	Akyuz et al. (2018)
	Waxes	Saurabh et al. (2018)
Others substances	PLA_S	Swaroop et al. (2018)
	PVA_S	Shao et al. (2018)

biodegradability and nontoxic nature. The most widely utilized nanoparticles include clay nanoparticles (such as laponites, montmorillonite, and kaolinites), metal oxide nanoparticle (zinc oxide or titanium oxide), and silver nanoparticles. Tables 18.2 and 18.3 contains a list of antimicrobial packaging materials, as well as the polymeric material, antimicrobial agent, and food type examined.

18.5.1 Edible films

The antibacterial nature of edible films makes them useful for the packaging of food. According to the research, there are enormous compounds that may be used in the production of food packaging films, with the qualities of these materials ultimately determining which types of food the film or coating can be applied to (Yun et al., 2017). Nontoxic film coatings must be attached to the surface of the food, be tasteless, and be stable over time to prevent stale formation must have a good appearance and should be acceptable, and economical to the consumers. To increase their impermeability and antibacterial capabilities, edible films made of carbohydrates, polypeptides, or triglycerides are required. Among the various types of polysaccharides are starches, biopolymers, alginate, chitosan, pectin, and GG plant sources including peanuts, gluten, rice bran, soy proteins, and maize zein. Food packing made of edible films is not a new idea. Collagen-based membranes were used to make sausage, hotdogs, and salami, among other meat products. The food sector is currently looking for solutions to increase product shelf life (Yuossef et al., 2018).

Table 18.3 Antimicrobial packages and their applicability.

Product preserved	Packaging material	Antimicrobial agent	References
Icebergs lettuce	Polysaccharide	Oil of clove flowers	Wieczyńska and Cavoski (2018)
Cucumis sativus	Chitosans	Limonene	Maleki et al. (2018)
Tomatoes		Nanoparticles of titanium dioxide	Kaewklin et al. (2018)
Pea and berries		Citric acid ($C_6H_8O_7$)	Yan et al. (2019)
	Polylactic acid	Silver nanoparticles	Zhang et al. (2018b)
	Chitosans	Spearmint oil	Shahbazi et al. (2018)
	Gelatine	Butylated hydroxyanisole	Li et al. (2018a)
Fishes	Polylactic acid	Thymol	Zhu et al. (2018)
Crap fillets	Alginic acid	Ziziphoras oil	Rezaei et al. (2018)
Rainbow trcut fillet	Chitosan	Grapes seed extract	Hassanzadeh et al. (2018)
Salmon	Polylactic acid	Glycerol mono-laurate	Ma et al. (2018)
	Chitosans	Cymophenol,	Wang et al. (2018)
	Gelatine	Oil of clove flowers	Ejaz et al. (2018)

(continued on next page)

Table 18.3 Antimicrobial packages and their applicability—cont'd

Product preserved	Packaging material	Antimicrobial agent	References
Chickens	Chitosans	Acerola residue extract	Zegarra et al. (2018)
Poultry		Zingiber officinale oil	Souza et al. (2018); Pires et al. (2018)
Chickens	Gelatine	Thymus vulgaris oil	Lin et al. (2018)
	Pullulan (polysaccharide polymer)	Nisin	Hassan and Cutter et al. (2020)
	Chitosans	Plantaricin	
Meat of ostrich	Milk kefir	Zataria multiflora oil	Rad et al. (2018)
Lamb meat	Chitosans	Satureja plant oil	Pabast et al. (2018)
minced beef	Polylactic acid	Spearmint oil	Taleb et al. (2018)
Pork	Chitosan and starch	Trihydroxybenzoic acid	Zhao et al. (2018)
Salami (lung meat)	Cheese byproduct	Cinnamomum cassia, Rosmarinus officinalis oils	Ribeiro et al. (2018)
Dairy product (cheese, curd)	Chitosans	Titanium dioxide	Yuossef et al. (2018)
		Monolaurin	Lotfi et al. (2018)
	Starch	Clove leaf oil	Yang et al. (2018)
	Agar-agar	Enterocin	Guitián et al. (2019)
	Prolamine protein	Pomegranate peel extract	Mushtaq et al. (2018)
For roasted Peanut	Bananas flour	Allium sativum oil	Orsuwan and Sothornvit (2018)

It is common to practice making edible films for fruits, cheese, and meat products out of naturally occurring, abundant, and inexpensive edible polymers. These films can improve the quality of food while simultaneously reducing the amount of food that is processed. This type of film is a replacement for the wax coating, which is toxic and banned in an increasing number of nations and is used on a wide variety of foods (apples, oranges, limes, and so on). Edible biopolymers keep food fresher for longer and suit the European Union (EU) market's need for organic, healthier, and long-decaying foods.

18.5.1.1 Polysaccharides-based films

Natural polymer chitosan serves as the backbone of this edible antibacterial packaging innovation. It is a polysaccharide extracted from crab and shrimp shells that possess antibacterial characteristics. It is biodegradable, biocompatible, and most importantly, edible. Chitosan packing films can help to protect food through a variety of processes. The film, like any physical barrier, can block the bacterium from accessing the food. It can suppress respiratory function by limiting oxygen transfer and forming a physical barrier that inhibits bacteria from getting nutrients. Various nutrients can be chelated by chitosan chains, and electrostatic interactions can tear the outer cellular membrane of germs, all of which add to the stress that microorganisms must endure. The antibacterial effect of chitosan could be due to a broken cell membrane, which allows internal electrolytes to escape. Various action mechanisms lead to microbial mortality once the chitosan chains have diffused through the cellular wall. It has the ability to chelate internal nutrients or essential metal ions from cellular plasma, affect gene expression, and enter the nucleus and bind DNA, limiting replication (Wang et al., 2018). It has a number of drawbacks, including higher costs, acidic hydrolysis, and poor mechanical properties. As a result, nanoparticles will typically be employed to make a nanocomposite film to make up for chitosan's shortcomings in certain areas. In order to increase the mechanical properties of polymers, plasticizers are often used as an additive. Sugar alcohols such as glycerol, and sorbitol are compatible with chitosan. The use of chitosan in edible films is objectionable and it raises concerns regarding its use because chitosan is prepared by the exoskeleton of crustaceans, it is not acceptable by vegetarians, however, it can be obtained from a variety of other sources, including mushroom inferior stem (a byproduct that is generally unused) (Singh et al., 2017). Priyadarshi et al. (2018) developed a chitosan-based antimicrobial film that was wrapped directly to the food surface. Citric acid was used, as a binding agent, which increased the film's stability and antioxidant properties. The glycerol was used to make the film very flexible which makes the film elastic. Due to this, the lengthening ability was increased 12 times but the breaking resistance was decreased. The newly acquired packaging material was more water-resistant, resulting in a reduction of WVP of 29%. Furthermore, the films were translucent, which is advantageous from the consumer's standpoint because they will want to see the food within the container at all times. Finally, it was discovered that green chili had a longer shelf life (Sought et al., 2018). By layering more polymers and adding nanoparticles, the properties of chitosan films were improved.

Bilayer films made of chitosans and PCL have been formed by pressing or coating the two substances. Nanocellulose (NC) at concentrations ranging from 2% to 5% and grapes seeds extract at fifteen percent weight-to-weight were injected into both layers. The WVP and film transparency were both significantly improved after the incorporation of NC, but the seeds of the grape extract had the reverse effect. Compared to solution-coated films, pressed films have higher elastic modulus and stretch resistance. Both film forms were bactericidal, and the antioxidant activity of the grape seed extract was preserved when it was put in the chitosan matrix (Sought et al., 2018). Researchers used cellulose nanocrystals and grape pomace extracts to build a composite chitosan-based film (Xu et al., 2018). The tensile strength of grape pomace extract was improved but the elongation and WVP were decreased, whereas NC had the reverse effect. Chitosan, citric acid, and fish gelatin were used by Uranga et al. (2019) to develop antibacterial films that were resistant to germs. The film has antibacterial action against *E. coli*, performs as an ultraviolet barrier, has good mechanical performances, and when exposed to water, the citric acid acts as a swelling inhibitor. This type of film can be used to wrap seafood. Food modification happens in the case of fish or shellfish as a result of enzyme activity present naturally in the flesh, reactions such as lipid oxidation, or microorganism metabolism. Although chitosan films do not have strong antioxidant properties, they can help to reduce lipid oxidation. To address this issue, Uranga et al. (2018) developed a new technique that included the addition of anthocyanins produced from processed food wastes to the chitosan-fish gelatin mixture. The resultant packing demonstrated antibacterial and antioxidant activities, as well as mechanical and WVP properties. Others added procyanidin or other antioxidants to the chitosan gelatin film (Ramziia et al., 2018; Pérez et al., 2018). Adding radicals to chitosan films, such as 2,2,6,6-tetra-methylpyperidine-1-oxyl, also known as TEMPO can increase their antioxidant activity. To obtain the desired mechanical properties, up to 15.5% cellulose nano and microfibers plus a plasticizer such as D-glucitol that can be utilized (25%). Aside from *E. coli*, the films also inhibited the growth of *Salmonella enterica* and *L. monocytogenes* (Soni et al., 2018). Jiwei and his coworkers (Yang et al., 2019) created a chitosan-based edible film for the fresh strawberry. The fruits were coated with a bicomponent film after being immersed in liquids containing 1.5% chitosans and 2.0% carboxymethyl cellulose. When compared to a simple chitosan film, the results showed that the LBL film properties were greatly enhanced. The bicomponent film preserved the fruits for longer, The LBL film lowered the biochemical decomposition of carbs, fatty acid oxidation, and amino acids from fruits, preserving them for 8 days and enabling only minor changes in acidic nature and dissolved material (Yang et al., 2019).

The antibacterial component thymol derived from thyme was also used in the edible chitosan films, which are obtained from thyme essential oil. A thymol nanofluid was applied to a chitin and quinoa whey film. Strawberry with these coatings has been demonstrated to be more fungal resistant. In comparison to the control batch, the edible film's sensory properties began to improve on the fifth day. Furthermore, the life span was increased by four days, losses of mass were reduced, and the pH, titration ability acidity, and percentage of resolvable chemicals were not changed

(Robledo et al., 2018). Yun and his coworkers developed polymeric films that are blended for fruit preservation using chitosan and PVA (Yun et al., 2017). The sorbitol composites have better thermal and mechanical performance. Many samples showed minor WVP as compared to the basic polymeric films. After 70 days of storage, the fruits exhibited no symptoms of degeneration, according to the research. After 220 days, the packaging material disintegrates 40%–65%. Yang et al. (2018) combined chitosan and PVA with 1%–3% lignin nanoparticles to generate a hydrogel.

Because of the aggregation of nanoparticles, the best properties were obtained with 1% lignin. Higher quantities provided no additional benefits. Hydrophilic polymers have a honeycomb-like porous surface that can be filled with antimicrobial chemicals and contain micron-sized pores. The researcher's tests show that chitosan and lignin have a synergistic effect against *E. coli* and *S. aureus*, as well as an increase in antioxidant characteristics (Yang et al., 2018). Vegetables and Fruit that are ripening due to the action of ethylene can also benefit from chitosan-based films. Because such films absorb ethylene, they can cause ripening to be delayed, which is undesirable. According to the literature, 5–6 g chitosan, 6–7 g potassium permanganate, and 2–3 mL polyol were used to preserve tomatoes at room temperature for five times longer (Warsiki et al., 2018). Other nanocomposite chitosan-based films have revealed similar results. TiO_2 nanoparticles can be included in the additives to improve their antibacterial characteristics.

The coating can be applied directly to the packaging from the solvent. These films have been used to keep tomatoes safe until they have been harvested and to prolong ripening. The consistency, and color variations, the authors kept up with the amount of lutein, losses of masses, total soluble material, ascorbic acid, and CO_2 and ethylene concentrations within the package. In comparison to standard chitosan packaging or the control lot, the inclusion of TiO_2 nanoparticles improved packaging quality, with the cherry tomatoes displaying fewer alterations. The behavior was related to TiO_2's photocatalytic activity versus ethylene gas, according to the researchers (Kaewklin et al., 2018). Chitosan and ZnO nanoparticles can be mixed to create an antibacterial composite (Kaewklin et al., 2018). Meat products, in general, have high containing nutritional content and provide a perfect environment used for microbial development, therefore they should only be kept in the refrigerator for a few days. It is no surprise that chitosan-based meat packaging has been obtained by a number of studies (Suo et al., 2017; Kumar et al., 2020). To offer the film antibacterial and antioxidant capabilities, the scientists utilized glycerol as a plasticizer and gallic acid as an antioxidant. Starches, alginate, and gallic acid provided unique hydrogen bonds, electrostatic forces, and esters bonds, reducing the permeability to water vapors. When compared to the control, antimicrobial testing on bacteria associated with foodborne illness revealed that preinoculated ham had a shelf life of 7–25 days. Zheng et al. (2018) studied a variety of edible films based on chitosan and starch finding that raising the starch ratio to 50% enhanced mechanical characteristics by 25% and reduced WVP and oxygen permeability both approximately 2.5 times. The scientists used essential oil from *Litsea cubeba* to promote antibacterial activity while also increasing the packaging's barrier qualities. As previously stated, the antibacterial and antioxidant activity of

Table 18.4 Essential oils and extracts and the usual compatibilities.

Food product	Essential oils or extract	References
Meat	Barbados cherry, *Origanum vulgare* oil, ginger oils, Penang oil, corn mint oil, thyme oil, grape seeds oil, eucalypt leaf oil	Shahbazi et al. (2018); Azadbakht et al. (2018)
Fruit and vegetables	*Litsea cubeba* oil, *Syzygium aromaticum*, and oregano oils, *Mentha spicata*, cinnamon oil, limonene, *Thymus vulgaris* oil	Niu et al. (2018); Dolea et al. (2018)
Bread and pastries	Garlic essential oil, pomegranate peel extract, cinnamon oil, clove oil	Ju et al. (2018)
Dairy product	Oil of clove, *Zingiber officinale* oil	Yang et al. (2018)

chitosan films is usually considerably increased by introducing EOs. While customers will enjoy thymus vulgaris, zingiber officinale, or salvia *Rosmarinus* essential oil in meat-based goods, they will be less enthusiastic about it in fruits. Cinnamon and vanilla are similar in that they pair nicely with pastry but not with meat. Orange or lime EOs, on the other hand, can be utilized in the packing of fruits or seafood (Table 18.4).

Apricot kernels are one of the most common agricultural wastes. Extracting the essential oil with the application of the bitter core is one conceivable application for them. The major fatty acid in this essential oil is oleic acid, but it also contains N-methyl-2-pyrolidone, a powerful antioxidant and antibacterial agent. The WVP was reduced by 41% and the mechanical resistance was raised by 94% when chitosan was combined with apricot essential oil in a 1:1 ratio. On bread slices, the antibacterial activity of the chitosan-essential oil package was examined, and it was found to inhibit fungus development (Priyadarshi et al., 2018a). Zegarra and colleagues investigated chitosan films loaded with acerola extract from wastes to improve the life span of meat, demonstrating the potential for agricultural waste to be exploited. Zhang and his coworkers used EOs from Perilla frutescens Britt. in a chitosan-based film to boost antibacterial activities against *E. coli, S. aureus*, and *Bacillus subtilis*. The essential oil improved mechanical characteristics, the capability of UV blocking, and WVP and inflammation capacity (Zhang et al., 2018a; Ojagh et al., 2010; Xing et al., 2015; Srikandace et al., 2018). Cinnamon EOs is another source of antioxidants and antimicrobials. It increases the antibacterial activities depending on the chitosan-based film while also slowing down lipid oxidation and lowering WVP. Chitosan films have poor WVP performance, which must be addressed through structural changes. The solution is to use hydrophobic lipidic molecules. Fatty acids and EOs, or perhaps chemicals derived from EOs, are the best possibilities. Various chemicals are used to mask some of the unpleasant or excessive aromas of EOs, as well as avoiding the inclusion of potentially harmful components in packaging. According to studies, combining chitosan with carvacrol had a synergistic effect, and the addition of caprylic acid increased the film's antibacterial efficacy even more. Nanoliposomes with a diameter of 90 nm were made from a soy lecithin mixture and exhibited a 46–69% encapsulation efficiency. The release rate of the essential oil was reduced by encapsulating it in nanoliposomes, resulting in a longer antibacterial and antioxidant activity. The meat in this film was kept at 4°C for 20 days without spoiling. (Pabast et al., 2018).

The other most prevalent polysaccharides utilized for edible films and ecological packaging materials is cellulose. It is available from a number of places. Poor mechanical and physical properties and a scarcity of water resistance limit the applicable domains. The added value of packaging materials is increased by coating the cellulose with an antibacterial agent. Many cellulose-based edible films are described in the literature, such as those made with CMC and glycerol with antibacterial essential oil from citrus fruits. Antibacterial activity against *E. coli* or *S. aureus* was significantly increased when up to 2% of essential oil was added (Srikandace et al., 2018). To make biodegradable packaging, Ortiz and his coworkers employed micro-fibrillated cellulose combined with soya protein and glycerin as a plasticizer. Furthermore, clove essential oil was added to the film to boost antibacterial and antioxidant activities against the decay of microbes. Essential oil similarly acted as a plasticizer, improving WVP while simultaneously increasing oxygen permeability. The diameter of the cellulose fibers was 50–60 nm, and the crystallinity was 35% (Ortiz et al., 2018). Yu and his colleagues (2018) used soya protein, cellulose nanocrystals, and pine needle extraction to produce a similar composition. The addition of cellulose nanoparticles and pine extraction to the film lowered WVP and increased mechanical strength by limiting hydrophobic interactions. Other polymers or nanoparticles can also be added to cellulose to improve its mechanical and barrier qualities. Natural, plentiful polymers are the most common choice, and they can be employed as-is or after chemical modification (Fardioui et al., 2018). Cellulose is compatible with other polysaccharides, as well as proteins (Shabanpour et al., 2018; Gilbert et al., 2018) and lipids (Shen et al., 2015; Coma et al., 2001). Zhang coagulated a cellulosic film containing embedded AgCl crystals in one step using a lithium chloride and acetified solution. The presence of PVP plays a significant role in the embedding process. Some of the AgCl was dissolved by light exposure, resulting in Ag@AgCl nanoparticles. PVP concentration can alter the shape and size of these nanoparticles. The test on *E. coli* and *S. aureus* revealed a significant antimicrobial activity, with no live bacteria detected after three hours of treatment (Zhang et al., 2018). Zahedi and colleagues (2018) discuss the creation of a composite film based on CMC containing on average 5% montmorillonite and 1%–4% ZnO. The inclusion of these nanoparticles reduced WVP values by 53% while also improving mechanical resistance. The nanoparticles were also an effective UV blocker. Tyagi and colleagues developed an improved parchment packaging material by covering it with a composite made up of lignocellulose materials, soya protein, and alkyl ketene dimer (Tyagi et al., 2018). When scientists build identical composites that are based on cellulose or chitosan, the intrinsic antibacterial activity of chitosan will always distinguish them (Karami et al., 2021). AgNPs were immobilized on laponite and employed for litchi packing elsewhere (Wu et al., 2018). Converting chitosan to a quaternary ammonium salt or subjecting it to additional additives is an effective way to improve its characteristics. Because the two chemicals are miscible due to hydrogen bonding, CMC and improved chitosan can also be mixed in any proportion. The mechanical and thermal properties of the composite film are superior. In such composites, the WVP is reduced, but the oxygen permeability is raised. The CMC-chitosan compound is antibacterial compared to *S. aureus* and *E. coli*, and a test on fresh ripe bananas revealed that the coated fruits had a longer shelf life (Hu et al., 2016). Acidic hydrolysis can be used to convert bacterial cellulose from *Acetobacter xylinus* into cellulose nanocrystals. These bacterial cellulose nanoparticles, which are 25–35 nm in size, can be placed in chitosan films alongside silver nanoparticles (AgNPs), which are 35–50 nm in size. The embedding procedure has a significant impact on the color and transparency of the films but also improves their mechanical and barrier qualities dramatically. The introduction of new hydrogen bonds between chitosan and BC indicates that embedding represents both a chemical and physical process.

The antibacterial activities of these films against food-borne pathogens show that chitosan and BC-incorporated nanoparticles work in tandem (Salari et al., 2018). Lippia alba extract can be used to incorporate AgNPs into methylcellulose films. The films show lower resistance properties and percentage of elongation than that of the controls lot, but they stretch more easily and are more hydrophobic. The antibacterial and antioxidant activities are increased at the same time as compared to the same control films (Nunes et al., 2018). In addition to AgNPs, cellulose-based nanocomposite films including oxide nanoparticles from transition metals such as an oxide of copper and zinc have been reported in the literature during the production and redevelopment of BC to cotton or nanocrystalline cellulose, nanoparticle oxides can be attached to the fiber surface by multiple hydrogen bonds without any additional surface modification. Regardless, the number of oxide NPs attached to cellulose is lower than the amount of AgNPs that can be added. The thermal stability of nanocomposites comprising AgNPs and ZnONPs is superior to that of bare cellulose film. Antibacterial investigations on *E. coli* and *L. monocytogenes* indicated that they are both strongly inhibited (Shankar et al., 2018). Starch, like cellulose, is a common polysaccharide used in edible films, and it provides nutritious benefits for human beings. Antimicrobial packaging materials can be made from simple starch films combined with diverse natural chemicals (Caetano et al., 2018). Listeria-infected cheese can be preserved for 24 days using starch films infused with clove leaf oil. Clove leaf oil has many functions including increasing tensile properties and reducing elongation rupture. Inhibiting *L. genus Listeria* multiplication, acting as an ultraviolet barrier, and scavenging radicals. Dinika and his colleagues (2019) proposed a new design that depends on milk whey and starch from farming waste materials. In this case, antimicrobial peptide (AMP) from cheese whey is essential for the function of the finished consumable antimicrobial edible films. Trongchuen and colleagues (2018) developed a based on the starch compound which is utilized as an antibacterial packaging substance. The formulation including foam made of starch, 11% extracted spent coffee ground, and essential oil of oregano 9%, showed good effectiveness towards *E. coli* and *S. aureus* in tests. Ounkaew et al. (2018) created a comparable composition with improved antibacterial and antioxidant characteristics by mixing starch with PVA and including spent coffee ground (SCG) and citric acid into the resultant nanocomposite.

18.5.1.2 Other biopolymers-based films

Polylactic acid is another type of polymer that has received a lot of attention, It can be made from easily renewable resources such as maze flour or beet of sugar through the process of bacterial fermentation. PLA goods degrade promptly at the end of their useful lives due to their biodegradable nature. The biggest issue with PLA films being acknowledged as an acceptable alternative for food packaging is their poor barrier qualities, which allow oxygen to easily pass through them. A composite made of PLA and NC infused with plant EOs was utilized to increase the shelf life of ground beef in the refrigerator to 12 days (Talebi et al., 2018). While the authors claim that incorporating NC into PLA has no significant effect on WVP, the values reduce by 50% when EOs are introduced. As a result, some researchers looked at using PLA sheets containing only a few important essential oil components as antimicrobials (Lopez et al., 2018). PLA that has been combined with proteins before being supplemented with vital nutrients has been reported to have better characteristics. Zhu and colleagues (2018) synthesized multilayer films using PLA and gliadin without compromising the surface. Furthermore, the film shows varied antibacterial action on either side, with the gliadin side being more effective than the PLA, implying that researchers have also synthesized PLA films with salicylic acid (SA) added. After harvesting, SA is frequently used to preserve fruits and vegetables (Yin et al., 2018). Wang et al. (2013) developed PLA which is based on three-component composite films which are combined with chitosan and tea polyphenols in several molar proportions. The three-component film offers enhanced mechanical characteristics, decreased WVP, and sealability of heat. The composite film containing PLA-tea polyphenols-chitosan has nearly three times the heat-seal strength of pure PLA. All samples containing tea polyphenols had increased vitamin C concentrations because these chemicals provide antioxidant potential, preventing vitamin C from interacting with free radicals and so degrading more slowly. PLA sheets can be reinforced with antibacterial nanoparticles, as other polymeric materials used in packaged foods (Ahmed et al., 2018). Zhang and colleagues (2018) added up to a 0.1 w/w fraction of AgNPs to PLA to create a film that can keep fruits fresh for longer. The film containing 5% AgNPs produced the best results. Li and colleagues (2017) proposed that TiO_2 or Ag nanoparticles be included in PLA. In comparison to the base polymer, the produced films had lower WVP, transparency, and elastic modulus. Valerini and colleagues (2018) studied about the improvement of antimicrobial efficacy by increasing the concentration of TiO_2 from 1% to 5% and found that the bacterial concentration has been reduced, films based on PLA enhanced with ZnO doped with aluminum nanoparticles showed antibacterial activity against *E. coli*, whereas PLA films loaded with ZnO nanoparticles remained tested for keeping new-cut apple in the refrigerator (Li et al., 2018). Gelatins films were investigated with a number of antimicrobial agents as active materials for packaging. The active component can be enclosed in a co-polymer, beta-cyclodextrin, clay nanoparticles (such as halloysite and others), or added as a nanoemulsion to decrease the release kinetics of antimicrobials (Li et al., 2018). Other polymers such as PCL, alginate (Dou et al., 2018), chitosan (Bonilla et al., 2018), starch (Moreno et al., 2018), cellulose, and its derivatives (Liu et al., 2018) can be used to reinforce gelatin to increase mechanical and barrier qualities. Nanoparticles such as ZnO, TiO_2 (He et al., 2016), and Ag (Kumar et al., 2018) can also increase mechanical characteristics. Most noteworthy, the nanocomposite films had lower oxygen and water vapor permeability than the simple poly(butyleneadipate-co-terephthalate) (PBAT) film. Antimicrobial packaging might be made using nisin incorporated with a PBAT polymer (Zehetmeyer et al., 2017; Khare et al., 2020) which has been represented to work towards *L. monocytogenes*.

18.6 Toxicity of antimicrobial packaging

Natural polymers that are both biodegradable and safe for human consumption are commonly used in the development of innovative antibacterial food packaging films. Most of them are GRAS and can be consumed. For example, at the micron level, chitosan and cellulose are GRAS. While NC is minimal in cytotoxicity, it can impact the gut bacterial community and massively reduce digestion and damage digestive function (Khare et al., 2020; DeLoid et al., 2019; Chen et al., 2020). The application of nanomaterials in packaged foods has numerous benefits, involving active antimicrobial effect, support for further antimicrobial agents, and food contamination sensors. But there is also a need to examine dangers. Desired metals and oxides bound to the nanoparticles interact with bacteria and can limit biofilm formation, but they can enter food as well as interact with human cells (Dimitrijevic et al., 2015; Doskocz et al., 2020; Noori et al., 2020). Consequently, before any new antimicrobial food packaging can be released, a detailed investigation of nanoparticle migration and toxic effects in food is required. According to current research (Dash et al., 2020; Garcia et al., 2018; Zehetmeyer et al., 2017), nanoparticles have toxic effects on living beings and can have significant effects on natural ecosystems. AgNPs are one of the most studied and used antibacterial agents because of the effects and consequences of AgNPs on human health (Istiqola et al., 2020; Motelica et al., 2020; Morais et al., 2020). AgNPs and AuNPs have been shown to increase the production of alanine aminotransferase and aspartate aminotransferase in zebrafish. Increased levels of these enzymes result in increased formation of reactiveness of oxygen species (ROS), which can be resulting in oxidative stresses and immunotoxicity. AgNPs also produced subtle nucleus and nuclear abnormalities (Ramachandran et al., 2018). The increasing use of nanomaterials has resulted in an ever-increasing number of metal nanoparticles in the environment. Nanoparticle accumulation (Zhou et al., 2020; Gong et al., 2020; Shah et al., 2020) has been observed in seafood, as has their both cyto and genotoxic effects in plants (Wu et al., 2020; Becaro et al., 2017; Ma et al., 2020). Copper (Cu) and copper oxide (CuO) nanoparticles are also a source of toxicological concern, with a strong association between nanoparticle size and phytotoxicity (Yang et al., 2020; Tanha et al., 2020; Rajput et al., 2020) being studied. According to recent research, Hordeum sativum distichum (Rajput et al., 2018) and *Abelmoscus esculentus* (Baskar et al., 2021) have an effect on chroroplast, chromoplast, leucoplast, chondriosome, cytoplasm, nucleoplast, and cell membranes. In addition to antimicrobial packaging, ZnO and TiO_2 nanoparticles are often used in cosmetic materials and colourants. Although ZnO is classified as GRAS by the FDA and its harmful effects are still being investigated, some publications on ZnO nanoparticles are already dangerous (Jeon et al., 2020; Vasile et al., 2015). ZnO nanoparticles have the ability to cross from the abdominal and enter into the digestive process in the large bowel (Voss et al., 2019), as well as cause malignant cell hyperproliferation (Meng et al., 2020). In animal studies, TiO_2 nanoparticles have been found to disrupt gastrointestinal homeostasis, induce cytotoxicity, and cause genotoxicity (Musial et al., 2020; Kurtz et al., 2020; Hashem et al., 2020; Cao et al., 2020; Bettencourt et al., 2020). According to research (Najnin et al., 2020; Costa et al., 2020; Bonin et al., 2020), EOs have no risk at all when used in food packaging. There is considerable concern that EOs, at doses inhibiting pathogenic bacteria, may also affect beneficial bacteria (Ingok et al., 2020).

18.7   Antimicrobial packaging systems

Antimicrobial packaged food systems are divided into two categories: package/food systems and packaging/food systems. This packaging system uses headspace material with or without food surfaces or low-viscosity liquid food (Han et al., 2003). There are two main migratory processes in this system diffusion along with the packing substances and the food, as well as partitioning at the interface. This approach is demonstrated by individually packaged dairy products and ready-to-eat beef products (Quintavalla et al., 2002). Some packaging systems are flexible in nature like flasks, cans, saucers, and packets. Because of the migration into the headspace, antimicrobial drugs must be volatile in these environments. Volatile compounds, unlike nonvolatile substances, have the ability to move between the package and the food, into the headspace and gaps with air.

Packaging materials, food products, and the package's headspace make up the majority of food packaging systems. The majority of the food packaging solutions reflect either a package/food system or a headspace/food system where the void volume of solid food products is considered as a headspace. A package/food system is a package that comes into touch with either a solid food product or a low-viscosity/liquid food that does not require head space. Wrapped cheese, deli items, and aseptic meat packages are illustrations of the foods packaged for this method. The major migration processes in this system are diffusion between the packing material and the food, as well as interfacial partition (Talebi et al., 2018). In order to assess a drug's reciprocal distribution, evaporation or balanced distribution between main locations, packing material, and food must be considered as a significant migratory process. Unlike nonvolatile compounds, which can only travel via the contact region between the packaging and the food, volatile elements can move through the headspace and air gap.

18.8 Various engineered properties involved in antimicrobial films for food-packaging

After the incorporation of an antibacterial ingredient, the qualities of the food-packaging sheets are influenced. There are various mechanical, and gas barriers, and thermal and morphological properties of films are described.

18.8.1 Mechanical properties

The addition of antimicrobials results in a dramatic change in the tensile characteristics of polymeric films. Antimicrobial drugs in food significantly alter some important properties, and therefore a significant change in film characteristics is anticipated. When the antibacterial molecule's molecular weight is less than that of the polymeric material, Han and Floros (1997) predict that there will be no significant effect on the tensile properties. In this scenario, the antibacterial should not modify the polymer structure of the packing substance, affecting its tensile qualities. Minor concentrations of antimicrobials can alter the tensile characteristics if relevant antimicrobial drugs interact with the matrix of the packing material (Limjaroen et al., 2003; Pires et al., 2008; Pranoto et al., 2005). The strength and resistance of the film decreased as the concentration of antibacterial contained in the polymer increased. On the other hand, PVOH films with enterosine exhibited better tensile properties, which can be attributed to the fact that antibacterial enterosine acts as a plasticizer for PVOH films, increasing their ductility (Marcos et al., 2010).

18.8.2 Gas barrier properties

The effects of antimicrobials on the gas-resistant characteristics of composite films have not been thoroughly studied. The paucity of such research in this area may be attributable to the fact that the effectiveness of antibacterial for harmful bacteria is identified and optimized before any other features of the films are evaluated. According to Suppakul et al. (2006), this could be due to an increase in the system's hydrophobicity, resulting in a lower permeability to water vapor. The pore effect and the solubility-diffusion effect are two processes by which gases can be transferred through packing materials. In one case, gases cross the substance by bypassing via small pinholes or ruptures in the structure. In another situation, the level of transmission is determined by the difference in the concentration and solubility of the gases in the respective substance between the two sides of the packing material. Antimicrobials appear to modify the structure of food-packaging films and affect gas permeability by changing solubility in the packaging structure, or by producing pinholes. However, generalizations are unattainable, as the final effect is dependent on the type of antimicrobial agent used and the polymer structure.

18.8.3 **Morphological properties**

The surface morphology of antibacterial coatings was examined using scanning electron microscopy (SEM). Micrographs of film surfaces demonstrate the structural alterations caused by antimicrobials in the polymer matrix. Antimicrobials may result in pores forming in the polymer matrix, which may affect the film's compressive and gas resistance. In PBAT films containing nisin, the disruption of link formation caused by the antimicrobial drug's interaction with the polymer led to the production of minute holes and pores (Bastarrachea et al., 2010). Cavities arise in PVOH films when the antimicrobial interacts with the polymer, whereas enterocin interacts with alginate films to generate void spaces (Marcos et al., 2010). As a result, microscope images aid in understanding the variations in film shape caused by the antimicrobial component. The morphological alterations may result in significant changes in the film are other technical properties, but the magnitude of these changes is difficult to predict and is dependent on the specific polymer–antimicrobial interaction.

18.8.4 **Thermal properties**

The thermal characteristics of food-packaging sheets are affected when antimicrobials are used. The thermal characteristics of antimicrobial food-packaging films have only been studied in a few investigations, and no significant changes in Tg and Tm have been detected when antimicrobials have been added (Bastarrachea et al., 2010). However, increasing the nisin content to 5000 IU/cm^2 has been demonstrated to reduce the overall crystallinity of PBAT films. Variations in the extensible characteristics of antibacterial films have been described using this change in crystallinity. As a result, more research into the thermal characteristics of food-packaging films incorporating antimicrobials is urgently needed. This would assist researchers in figuring out how polymers interact with antimicrobials and setting processing conditions for making active films.

18.8.5 **Diffusion**

Examining the movement of antibacterial compounds through packing films is important for evaluating and characterizing food-packaging films with antibacterial properties. With the use of this knowledge, it is possible to assess the likelihood that a packing film would maintain antibacterial properties and release them when it comes into contact with food. The release of antimicrobial components from packing materials is critical because it impacts how successfully germs are eradicated. The rate of discharge should not exceed the rate of microbial development. The antibacterial substance's solubility in specific meals is an essential consideration. The release of antimicrobial components from packing materials is crucial, according to Han et al. (2005), because it impacts how successfully germs are eradicated. The rate of discharge should not exceed that of bacterial development. The solubility of the antibacterial agent in specific meals is an essential consideration. The quantity required to sustain concentration levels above the minimal inhibitory concentration can be calculated using antimicrobial diffusion characteristics (Han et al., 2005). Antimicrobial agent diffusion through packaging materials is influenced by both chemical and physical variables. Hydrophobic interactions, electrostatic interactions, and other factors may influence diffusion. The existence of a tortuous and porous media, as well as the configuration of the films' matrix, can influence the diffusion phenomenon. Antimicrobial agent dispersion in food-packaging films has been studied in several research (Zactiti et al., 2009).

18.9 Conclusion

Due to growing consumer demand for foods that are minimally processed and additive-free, active food packaging films with an antibacterial basis have attracted a lot of interest. Antibacterial packaging of food has been found to inhibit harmful bacteria and hence improve food safety. On the other side, antimicrobial compounds could change the technical characteristics of packing sheets, thereby restricting their usage. The impact of antimicrobials in the polymer matrix on film characteristics was addressed in this chapter. The morphological structure, biomechanical, thermal, and gas barrier qualities are all significantly altered when these antimicrobials are added, according to several research. Antimicrobial substances, on the other hand, can alter the technical properties of packing sheets, limiting their application. The impact of antimicrobials in the polymer matrix on film characteristics has been addressed. When these antimicrobials are added, numerous investigations have revealed significant changes in mechanical, thermal, and gas barrier properties. The advancements made in this sector so far necessitate additional research. Improvements in product quality and identification of foods that function best with antimicrobial packaging solutions will make these solutions more beneficial, cost-effective, and safe for consumers. Cellulose, NC, hemicellulose, chitosan, starch, and pectin are all examples of polysaccharides and a variety of other antimicrobial components such as bacteriocin, EOs, pediocin, nanomaterials of metal or metal oxides have been used as detailed by a number of bibliographic databases. Chitosan and its derivatives which are Akyuz active biomolecules play a crucial role in the food application sector in light of recent food contamination events.

To avoid spoilage and harmful microbe development and increase food safety and shelf life, this chapter verified the need for antimicrobial food packaging systems that may be developed utilizing antimicrobial packaging materials and/or antimicrobial chemicals inside package spaces or foods. Polysaccharide-based composites are proposed for biodegradable food packaging applications because of their challenges and future study. In food packaging, it is envisaged that polysaccharide-based composites will lead to the creation of innovative microbial films for food packaging.

Acknowledgment

Shalinee Singh is grateful to Vice-Chancellor, HBTU Kanpur, India, for financial assistance in the form of University Fellowship.

References

Abdollahzadeh, E., Ojagh, S.M., Fooladi, A.A.I., Shabanpour, B., Gharahei, M., 2018. Effects of probiotic cells on the mechanical and antibacterial properties of sodium-caseinate films. Appl. Food Biotechnol. 5 (3), 155–162.

Ahmad, M., Nirmal, N.P., Danish, M., Chuprom, J., Jafarzedeh, S., 2016. Characterisation of composite films fabricated from collagen/chitosan and collagen/soy protein isolate for food packaging applications. RSC Adv. 6 (85), 82191–82204.

Ahmed, J., Arfat, Y.A., Bher, A., Mulla, M., Jacob, H., Auras, R., 2018. Active chicken meat packaging based on polylactide films and bimetallic Ag–Cu nanoparticles and essential oil. J. Food Sci. 83 (5), 1299–1310.

Akyuz, L., Kaya, M., Ilk, S., Cakmak, Y.S., Salaberria, A.M., Labidi, J., Sargin, I., 2018. Effect of different animal fat and plant oil additives on physicochemical, mechanical, antimicrobial and antioxidant properties of chitosan films. Int. J. Biol. Macromol. 111, 475–484.

Alboofetileh, M., Rezaei, M., Hosseini, H., Abdollahi, M., 2018. Morphological, physico mechanical, and antimicrobial properties of sodium alginate montmorillonite nanocomposite films incorporated with marjoram essential oil. J. Food Process. Preserv. 42 (5), e13596.

Alemdar, A., Sain, M., 2008. Isolation and characterization of nanofibers from agricultural residues–wheat straw and soy hulls. Bioresour. Technol. 99 (6), 1664–1671.

Appendini, P., Hotchkiss, J.H., 2002. Review of antimicrobial food packaging. Innov. Food Sci. Emerg. 3 (2), 113–126.

Araújo, M.N.P., Grisi, C.V.B., Duarte, C.R., Almeida, Y.M.B., Vinhas, G.M., 2023. Active packaging of corn starch with pectin extract and essential oil of *Turmeric Longa Linn:* Preparation, characterization and application in sliced bread. Int. J. Biol. Macromol. 226, 1352–1359.

Aydogdu, A., Radke, C.J., Bezci, S., Kirtil, E., 2020. Characterization of curcumin incorporated guar gum/orange oil antimicrobial emulsion films. Int. J. Biol. Macromol. 148, 110–120.

Azadbakht, E., Maghsoudlou, Y., Khomiri, M., Kashiri, M., 2018. Development and structural characterization of chitosan films containing *Eucalyptus globulus* essential oil: potential as an antimicrobial carrier for packaging of sliced sausage. Food Packag. Shelf Life 17, 65–72.

Baek, N., Kim, Y.T., Marcy, J.E., Duncan, S.E., O'Keefe, S.F., 2018. Physical properties of nanocomposite polylactic acid films prepared with oleic acid modified titanium dioxide. Food Packag. Shelf Life 17, 30–38.

Bandyopadhyay, S., Saha, N., Brodnjak, U.V., Saha, P., 2019. Bacterial cellulose and guar gum based modified PVP-CMC hydrogel films: characterized for packaging fresh berries. Food Packag. Shelf Life 22, 100402.

Baner, A.L., 2000. Plastic Packaging Materials for Food: Barrier Function Mass Transport. Wiley-VCH Weinheim, New York, p. 21.

Barbosa-Pereira, L., Aurrekoetxea, G.P., Angulo, I., Paseiro-Losada, P., Cruz, J.M., 2014. Development of new active packaging films coated with natural phenolic compounds to improve the oxidative stability of beef. Meat Sci. 97 (2), 249–254.

Bashir, A., Jabeen, S., Gull, N., Islam, A., Sultan, M., Ghaffar, A., Khan, S.M., Iqbal, S.S., Jamil, T., 2018. Co-concentration effect of silane with natural extract on biodegradable polymeric films for food packaging. Int. J. Biol. Macromol. 106, 351–359.

Baskar, V., Safia, N., Sree Preethy, K., Dhivya, S., Thiruvengadam, M., Sathishkumar, R., 2021. A comparative study of phytotoxic effects of metal oxide (CuO, ZnO and NiO) nanoparticles on in-vitro grown *Abelmoschus esculentus*. Plant Biosyst. Int. J. Plant Biol. 155 (2), 374–383.

Bastarrachea, L., Dhawan, S., Sablani, S.S., Mah, J.H., Kang, D.H., Zhang, J., Tang, J., 2010. Biodegradable poly (butylene adipate-co-terephthalate) films incorporated with nisin: characterization and effectiveness against *Listeria innocua*. J. Food Sci. 75 (4), 215–224.

Becaro, A.A., Siqueira, M.C., Puti, F.C., de Moura, M.R., Correa, D.S., Marconcini, J.M., Mattoso, L.H.C., Ferreira, M.D., 2017. Cytotoxic and genotoxic effects of silver nanoparticle/carboxymethyl cellulose on Allium cepa. Environ. Monit. Assess. 189 (7), 352.

Bettencourt, A., Gonçalves, L.M., Gramacho, A.C., Vieira, A., Rolo, D., Martins, C., Louro, H., 2020. Analysis of the characteristics and cytotoxicity of titanium dioxide nanomaterials following simulated in vitro digestion. Nanomaterials 10 (8), 1516.

Bonilla, J., Poloni, T., Lourenço, R.V., Sobral, P.J., 2018. Antioxidant potential of eugenol and ginger essential oils with gelatin/chitosan films. Food Biosci. 23, 107–114.

Bonin, E., Carvalho, V.M., Avila, V.D., dos Santos, N.C.A., Benassi-Zanqueta, É., Lancheros, C.A.C., do Prado, I.N., 2020. Baccharis dracunculifolia: chemical constituents, cytotoxicity and antimicrobial activity. LWT–Food Sci. Technol. 120, 108920.

Boziaris, I.S., Adams, M.R., 1999. Effect of chelators and nisin produced in situ on inhibition and inactivation of Gram negatives. Int. J. Food Microbiol. 53 (2-3), 105–113.

Braga, L.R., Rangel, E.T., Suarez, P.A.Z., Machado, F., 2018. Simple synthesis of active films based on PVC incorporated with silver nanoparticles: Evaluation of the thermal, structural and antimicrobial properties. Food Packag. Shelf Life 15, 122–129. doi:10.1016/j.fpsl.2017.12.005.

Bugatti, V., Vertuccio, L., Zuppardi, F., Vittoria, V., Gorrasi, G., 2019. PET and active coating based on a LDH nanofiller hosting p-hydroxybenzoate and food-grade zeolites: Evaluation of antimicrobial activity of packaging and shelf life of red meat. Nanomaterials 9 (12), 1727. doi:10.3390/nano9121727.

Cabello, M.L.R., Pichardo, S., Bermudez, J.M., Baños, A., Ariza, J.J., Guillamón, E., Aucejo, S., Cameán, A.M., 2018. Characterisation and antimicrobial activity of active polypropylene films containing oregano essential oil and Allium extract to be used in packaging for meat products. Food Addit. Contam. Part A 35 (4), 782–791. https://doi.org/10.1080/19440049.2017.1422282.

Campbell, I.M., 1994. Introduction to Synthetic Polymers. Oxford University Press, New York.

Campillo, A.M., Sánchez, I.C., Garai, R.M., Puerta, C.N., 2009. Active packaging that inhibits food pathogens, US Patent, US 2009/0232948 A1.

Cannarsi, M., Baiano, A., Sinigaglia, M., Ferrara, L., Baculo, R., Nobile, M.A.D., 2008. Use of nisin, lysozyme and EDTA for inhibiting microbial growth in chilled buffalo meat. Intl. J. Food Sci. Tech. 43 (4), 573–578.

Cao, X., Han, Y., Gu, M., Du, H., Song, M., Zhu, X., Xiao, H., 2020. Foodborne titanium dioxide nanoparticles induce stronger adverse effects in obese mice than non-obese mice: gut microbiota dysbiosis, colonic inflammation, and proteome alterations. Small 16 (36), 2001858.

Cazón, P., Velazquez, G., Vázquez, M., 2020. Characterization of mechanical and barrier properties of bacterial cellulose, glycerol and polyvinyl alcohol (PVOH) composite films with eco-friendly UV-protective properties. Food Hydrocoll. 99, 105323.

Cha, D.S., Chinnan, M.S., 2004. Biopolymer-based antimicrobial packaging: a review. Crit. Rev. Food Sci. Nutr. 44 (4), 223–237.

Chandra, R., Rustgi, R., 1998. Biodegradable polymers. Prog. Polym. Sci. 23, 1273.

Chen, C., Xu, Z., Ma, Y., Liu, J., Zhang, Q., Tang, Z., … Xie, J., 2018. Properties, vapour-phase antimicrobial and antioxidant activities of active poly (vinyl alcohol) packaging films incorporated with clove oil. Food Control 88, 105–112.

Chen, M., Chen, X., Ray, S., Yam, K., 2020. Stabilization and controlled release of gaseous/volatile active compounds to improve safety and quality of fresh produce. Trends Food Sci. Technol. 95, 33–44.

Chen, Y., Lin, Y.J., Nagy, T., Kong, F., Guo, T.L., 2020. Subchronic exposure to cellulose nanofibrils induces nutritional risk by non-specifically reducing the intestinal absorption. Carbohydr. Polym. 229, 115536.

Chikindas, M., Emond, E., Haandrikman, A.J., Kok, J., Leenhouts, K., Pandian, S., Venema, K., 2010. Heterologous processing and export of the bacteriocins pediocin PA-1 and *lactococcin* a in *Lactococcus lactis*: a study with leader exchange. Probiotics Antimicrob. Proteins 2 (2), 66–76.

Choo, K., Ching, Y.C., Chuah, C.H., Julai, S., Liou, N.S., 2016. Preparation and characterization of polyvinyl alcohol-chitosan composite films reinforced with cellulose nanofiber. Materials (Basel) 9 (8), 644.

Coma, V., Sebti, I., Pardon, P., Deschamps, A., Pichavant, F.H., 2001. Antimicrobial edible packaging based on cellulosic ethers, fatty acids, and nisin incorporation to inhibit *Listeria innocua* and *Staphylococcus aureus*. J. Food Prot. 64 (4), 470–475.

Cooksey, K., 2005. Effectiveness of antimicrobial food packaging materials. Food Addit. Contam. 22, 980–987.

Corradini, C., Alfieri, I., Cavazza, A., Lantano, C., Lorenzi, A., Zucchetto, N., Montenero, A., 2013. Antimicrobial films containing lysozyme for active packaging obtained by sol–gel technique. J. Food Eng. 119 (3), 580–587.

Costa, W.K., de Oliveira, J.R.S., de Oliveira, A.M., da Silva Santos, I.B., da Cunha, R.X., de Freitas, A.F.S., da Silva, M.V., 2020. Essential oil from *Eugenia stipitata* McVaugh leaves has antinociceptive, anti-inflammatory and antipyretic activities without showing toxicity in mice. Ind. Crops Prod. 144, 112059.

Costa-Júnior, E.S., Barbosa-Stancioli, E.F., Mansur, A.A., Vasconcelos, W.L., Mansur, H.S., 2009. Preparation and characterization of chitosan/poly (vinyl alcohol) chemically crosslinked blends for biomedical applications. Carbohydr. Polym. 76 (3), 472–481.

Cvetnic, Z., Vladimir-Knezevic, S., 2004. Antimicrobial activity of grapefruit seed and pulp ethanolic extract. Acta Pharm. 54 (3), 243–250.

da Silva, F.T., da Cunha, K.F., Fonseca, L.M., Antunes, M.D., El Halal, S.L.M., Fiorentini, Â.M., Dias, A.R.G., 2018. Action of ginger essential oil (Zingiber officinale) encapsulated in proteins ultrafine fibers on the antimicrobial control in situ. Int. J. Biol. Macromol. 118, 107–115.

Das, D., Ara, T., Dutta, S., Mukherjee, A., 2011. New water-resistant biomaterial biocide film based on guar gum. Bioresour. Technol. 102, 5878–5883.

Das, T., Yeasmin, S., Khatua, S., Acharya, K., Bandyopadhyay, A., 2015. Influence of a blend of guar gum and poly (vinyl alcohol) on long term stability, and antibacterial and antioxidant efficacies of silver nanoparticles. RSC Adv. 5 (67), 54059–54069.

Dash, S.R., Kundu, C.N., 2020. Promising opportunities and potential risk of nanoparticle on the society. IET Nanobiotechnol. 14, 253–260.

Dehghani, S., Peighambardoust, S.H., Peighambardoust, S.J., Hosseini, S.V., Regenstein, J.M., 2019. Improved mechanical and antibacterial properties of active LDPE films prepared with combination of Ag, ZnO and CuO nanoparticles. Food Packag. Shelf Life 22, 100391. doi:10.1016/j.fpsl.2019.100391.

DeLoid, G.M., Cao, X., Molina, R.M., Silva, D.I., Bhattacharya, K., Ng, K.W., … Demokritou, P., 2019. Toxicological effects of ingested nanocellulose in in vitro intestinal epithelium and in vivo rat models. Environ. Sci. Nano 6 (7), 2105–2115.

Dimitrijevic, M., Karabasil, N., Boskovic, M., Teodorovic, V., Vasilev, D., Djordjevic, V., Cobanovic, N., 2015. Safety aspects of nanotechnology applications in food packaging. Procedia Food Sci. 5, 57–60.

Dinika, I., Utama, G.L., 2019. Cheese whey as potential resource for antimicrobial edible film and active packaging production. Foods Raw Mater. 7 (2), 229–239.

Dolea, D., Rizo, A., Fuentes, A., Barat, J.M., Fernández-Segovia, I., 2018. Effect of thyme and oregano essential oils on the shelf life of salmon and seaweed burgers. Food Sci. Technol. Int. 24 (5), 394–403.

dos Santos Caetano, K., Lopes, N.A., Costa, T.M.H., Brandelli, A., Rodrigues, E., Flôres, S.H., Cladera-Olivera, F., 2018. Characterization of active biodegradable films based on cassava starch and natural compounds. Food Packag. Shelf Life 16, 138–147.

dos Santos Pires, A.C., de Fátima Ferreira Soares, N., de Andrade, N.J., Silva, L., H.M., Camilloto, G.P., Bernardes, P.C, 2008. Development and evaluation of active packaging for sliced mozzarella preservation. Packag. Technol. Sci. Int. J. 21 (7), 375–383.

Doskocz, N., Zaleska-Radziwill, M., Affek, K., Lebkowska, M., 2020. Ecotoxicity of selected nanoparticles in relation to micro-organisms in the water ecosystem. Desalin. Water Treat. 186, 50–55.

Dou, L., Li, B., Zhang, K., Chu, X., Hou, H., 2018. Physical properties and antioxidant activity of gelatin-sodium alginate edible films with tea polyphenols. Int. J. Biol. Macromol. 118, 1377–1383.

Ebrahimi, H., Abedi, B., Bodaghi, H., Davarynejad, G., Haratizadeh, H., Conte, A., 2018. Investigation of developed clay-nanocomposite packaging film on quality of peach fruit (*Prunus persica* Cv. Alberta) during cold storage. J. Food Process. Preserv. 42 (2), e13466.

Economou, T., Pournis, N., Ntzimani, A., Savvaidis, I.N., 2009. Nisin–EDTA treatments and modified atmosphere packaging to increase fresh chicken meat shelf-life. Food Chem. 114 (4), 1470–1476.

Ejaz, M., Arfat, Y.A., Mulla, M., Ahmed, J., 2018. Zinc oxide nanorods/clove essential oil incorporated type B gelatin composite films and its applicability for shrimp packaging. Food Packag. Shelf Life 15, 113–121.

El-hefian, E.A., Nasef, M.M., Yahaya, A.H., 2010. Rheological and morphological studies of chitosan/agar/poly (vinyl alcohol) blends. J. Appl. Sci. Res. 6, 460–468.

Fabra, M.J., Falcó, I., Randazzo, W., Sánchez, G., López-Rubio, A., 2018. Antiviral and antioxidant properties of active alginate edible films containing phenolic extracts. Food Hydrocoll. 81, 96–103.

Fardioui, M., Kadmiri, I.M., Bouhfid, R., 2018. Bio-active nanocomposite films based on nanocrystalline cellulose reinforced styrylquinoxalin-grafted-chitosan: antibacterial and mechanical properties. Int. J. Biol. Macromol. 114, 733–740.

Fasihnia, S.H., Peighambardoust, S.H., Peighambardoust, S.J., Oromiehie, A., 2018. Development of novel active polypropylene based packaging films containing different concentrations of sorbic acid. Food Packag. Shelf Life 18, 87–94. doi:10.1016/j.fpsl.2018.10.001.

Figueroa-Lopez, K.J., Castro-Mayorga, J.L., Andrade-Mahecha, M.M., Cabedo, L., Lagaron, J.M., 2018. Antibacterial and barrier properties of gelatin coated by electrospun polycaprolactone ultrathin fibers containing black pepper oleoresin of interest in active food biopackaging applications. Nanomaterials 8 (4), 199.

Food Safety Consortium Newsletter, 2000. Food Science and Technology. John Wiley & Sons, New York, pp. 1824–1829.

Fried, J.R., 2003. Polymer Science & Technology, second ed. Prentice Hall Professional Technical Reference, pp. 325–353.

Garcia, J., Galindo, E., 1990. An immobilization technique yielding high enzymatic load of nylon nets. Biotechnol. Tech. Biotechnol. 4 (6), 425–428.

Garcia, C.V., Shin, G.H., Kim, J.T., 2018. Metal oxide-based nanocomposites in food packaging: applications, migration, and regulations. Trends Food Sci. Technol. 82, 21–31.

Garibay, G., Luna-Salazar, A., Casas, L., 1995. Antimicrobial effect of the lactoperoxidase system in milk activated by immobilized enzymes. Food Biotechnol. 9 (3), 157–166.

Gatto, M., Ochi, D., Yoshida, C.M.P., Silva, C.F., 2019. Study of chitosan with different degrees of acetylation as cardboard paper coating. Carbohydr. Polym. 210, 56–63.

Ghanem, A.F., Youssef, A.M., Rehim, M.H.A., 2020. Hydrophobically modified graphene oxide as a barrier and antibacterial agent for polystyrene packaging. J. Mater. Sci. 55 (11), 4685–4700. doi:10.1007/s10853-019-04333-7.

Gilbert, J., Cheng, C.J., Jones, O.G., 2018. Vapor barrier properties and mechanical behaviors of composite hydroxypropyl methylcelluose/zein nanoparticle films. Food Biophys. 13 (1), 25–36.

Gómez-García, M., Sol, C., de Nova, P.J., Puyalto, M., Mesas, L., Puente, H., Carvajal, A., 2019. Antimicrobial activity of a selection of organic acids, their salts and essential oils against swine enteropathogenic bacteria. Porc. Health Manag. 5 (1), 1–8.

Gong, Y., Chai, M., Ding, H., Shi, C., Wang, Y., Li, R., 2020. Bioaccumulation and human health risk of shellfish contamination to heavy metals and As in most rapid urbanized Shenzhen, China. Environ. Sci. Pollut. Res. 27 (2), 2096–2106.

González, O.M., Velín, A., García, A., Arroyo, C.R., Barrigas, H.L., Vizuete, K., Debut, A., 2020. Representative hardwood and softwood green tissue-microstructure transitions per age group and their inherent relationships with physical–mechanical properties and potential applications. Forests 11 (5), 569.

Gopalakrishnan, L., Doriya, K., Kumar, D.S., 2016. Moringa oleifera: a review on nutritive importance and its medicinal application. Food Sci. Hum. Wellness 5 (2), 49–56.

Grande, R., Carvalho, A.J., 2011. Compatible ternary blends of chitosan/poly (vinyl alcohol)/poly (lactic acid) produced by oil-in-water emulsion processing. Biomacromolecules 12 (4), 907–914.

Guitián, M.V., Ibarguren, C., Soria, M.C., Hovanyecz, P., Banchio, C., Audisio, M.C., 2019. Anti-*Listeria monocytogenes* effect of bacteriocin-incorporated agar edible coatings applied on cheese. Int. Dairy J. 97, 92–98.

Gupta, S., Variyar, P.S., 2018. Guar gum: a versatile polymer for the food industry. Biopolymers for food design, Hand book of Food Bioengineering. Academic Press, pp. 383–407.

Han, J.H., 2003. Antimicrobial food packaging. Novel Food Packag. Tech. 8, 50–70.

Han, J.H., 2005. Antimicrobial packaging systems. In: Innovations in Food Packaging. Academic Press, pp. 80–107.

Han, J.H., Floros, J.D., 1997. Casting antimicrobial packaging films and measuring their physical properties and antimicrobial activity. J. Plast. Film Sheeting 13 (4), 287–298.

Hasapidou, A., Savvaidis, I.N., 2011. The effects of modified atmosphere packaging, EDTA and oregano oil on the quality of chicken liver meat. Food Res. Int. 44 (9), 2751–2756.

Hashem, M.M., Abo-EL-Sooud, K., Abd-Elhakim, Y.M., Badr, Y.A.H., El-Metwally, A.E., Bahy-El-Dien, A., 2020. The long-term oral exposure to titanium dioxide impaired immune functions and triggered cytotoxic and genotoxic impacts in rats. J. Trace Elem. Med. Biol. 60, 126473.

Hassan, A.H., Cutter, C.N., 2020. Development and evaluation of pullulan-based composite antimicrobial films (CAF) incorporated with nisin, thymol and lauric arginate to reduce foodborne pathogens associated with muscle foods. Int. J. Food Microbiol. 320, 108519.

Hassanzadeh, P., Moradi, M., Vaezi, N., Moosavy, M.H., Mahmoudi, R., 2018. Effects of chitosan edible coating containing grape seed extract on the shelf-life of refrigerated rainbow trout fillet. Vet. Res. Forum 9, 73–79.

He, Q., Zhang, Y., Cai, X., Wang, S., 2016. Fabrication of gelatin–TiO_2 nanocomposite film and its structural, antibacterial and physical properties. Int. J. Biol. Macromol. 84, 153–160.

Heggers, J.P., Cottingham, J., Gusman, J., Reagor, L., McCoy, L., Carino, E., … Zhao, J.G., 2002. The effectiveness of processed grapefruit-seed extract as an antibacterial agent: II. Mechanism of action and in vitro toxicity. J. Alternat. Complement. Med. 8 (3), 333–340.

Hu, Z., Hong, P., Liao, M., Kong, S., Huang, N., Ou, C., Li, S., 2016. Preparation and characterization of chitosan—agarose composite films. Materials 9 (10), 816.

Hu, D., Wang, H., Wang, L., 2016. Physical properties and antibacterial activity of quaternized chitosan/carboxymethyl cellulose blend films. LWT–Food Sci. Technol. 65, 398–405.

Ibrahim, S., El-Naggar, M.E., Youssef, A.M., Abdel-Aziz, M.S., 2019. Functionalization of polystyrene nanocomposite with excellent antimicrobial efficiency for food packaging application. J. Cluster Sci. 31, 1371–1382. doi:10.1007/s10876-019-01748-9.

Irkin, R., Esmer, O.K., 2015. Novel food packaging systems with natural antimicrobial agents. J. Food Sci. Technol. 52, 1–17.

Istiqola, A., Syafiuddin, A., 2020. A review of silver nanoparticles in food packaging technologies: regulation, methods, properties, migration, and future challenges. J. Chin. Chem. Soc. 67 (11), 1942–1956.

Jahdkaran, E., Hosseini, S.E., Nafchi, A.M., Nouri, L., 2021. The effects of methylcellulose coating containing carvacrol or menthol on the physicochemical, mechanical, and antimicrobial activity of polyethylene films. Food Sci. Nutr. 9 (5), 2768–2778. doi:10.1002/fsn3.2240.

Jeon, Y.R., Yu, J., Choi, S.J., 2020. Fate determination of ZnO in commercial foods and human intestinal cells. Int. J. Mol. Sci. 21 (2), 433.

Ju, J., Xu, X., Xie, Y., Guo, Y., Cheng, Y., Qian, H., Yao, W., 2018. Inhibitory effects of cinnamon and clove essential oils on mold growth on baked foods. Food Chem. 240, 850–855.

Kaewklin, P., Siripatrawan, U., Suwanagul, A., Lee, Y.S., 2018. Active packaging from chitosan-titanium dioxide nanocomposite film for prolonging storage life of tomato fruit. Int. J. Biol. Macromol. 112, 523–529.

Kanatt, S.R., Rao, M., Chawla, S., Sharma, A., 2012. Active chitosan–polyvinyl alcohol films with natural extracts. Food Hydrocoll. 29 (2), 290–297.

Kanatt, S.R., Makwana, S.H., 2020. Development of active, water-resistant carboxymethyl cellulose-poly vinyl alcohol-aloe vera packaging film. Carbohydr. Polym. 227, 115303.

Karami, N., Kamkar, A., Shahbazi, Y., Misaghi, A., 2021. Electrospinning of double-layer chitosan-flaxseed mucilage nanofibers for sustained release of *Ziziphora clinopodioides* essential oil and sesame oil. LWT–Food Sci. Technol. 140, 110812.

Kaya, M., Khadem, S., Cakmak, Y.S., Mujtaba, M., Ilk, S., Akyuz, L., Deligöz, E., 2018. Antioxidative and antimicrobial edible chitosan films blended with stem, leaf and seed extracts of *Pistacia terebinthus* for active food packaging. RSC Adv. 8 (8), 3941–3950.

Ke, C.L., Deng, F.S., Chuang, C.Y., Lin, C.H., 2021. Antimicrobial actions and applications of chitosan. Polymers 13 (6), 904.

Kechichian, V., Ditchfield, C., Veiga-Santos, P., Tadini, C.C., 2010. Natural antimicrobial ingredients incorporated in biodegradable films based on cassava. Starch 1088–1094.

Khare, S., DeLoid, G.M., Molina, R.M., Gokulan, K., Couvillion, S.P., Bloodsworth, K.J., Demokritou, P., 2020. Effects of ingested nanocellulose on intestinal microbiota and homeostasis in Wistar Han rats. NanoImpact 18, 100216.

Khezrian, A., Shahbazi, Y., 2018. Application of nanocompostie chitosan and carboxymethyl cellulose films containing natural preservative compounds in minced camel's meat. Int. J. Biol. Macromol. 106, 1146–1158.

Kohannia, N., Beigmohammadi, F., Ghara, A.R., Nayebzadeh, K., 2021. Effect of polyethylene terephthalate incorporated with titanium dioxide and zinc oxide nanoparticles on shelf-life extension of mayonnaise sauce. J. Food Process. Preserv. 45 (5), e15453. doi:10.1111/jfpp.15453.

Kouravand, F., Jooyandeh, H., Barzegar, H., Hojjati, M., 2018. Characterization of cross-linked whey protein isolate-based films containing *Satureja Khuzistanica* Jamzad essential oil. J. Food Process. Preserv. 42 (3), e13557.

Krepker, M., Setter, O.P., Shemesh, R., Vaxman, A., Alperstein, D., Segal, E., 2018. Antimicrobial carvacrol-containing polypropylene films: composition, structure and function. Polymers 10 (1), 79. https://doi.org/10.3390/polym10010079.

Kumar, S., Shukla, A., Baul, P.P., Mitra, A., Halder, D., 2018. Biodegradable hybrid nanocomposites of chitosan/gelatin and silver nanoparticles for active food packaging applications. Food Packag. Shelf Life 16, 178–184.

Kumar, S., Mudai, A., Roy, B., Basumatary, I.B., Mukherjee, A., Dutta, J., 2020. Biodegradable hybrid nanocomposite of chitosan/gelatin and green synthesized zinc oxide nanoparticles for food packaging. Foods 9 (9), 1143.

Kumar, A., Gupta, V., Singh, S., Saini, S., Gaikwad, K.K., 2021. Pine needles lignocellulosic ethylene scavenging paper impregnated with nanozeolite for active packaging applications. Ind. Crops Prod. 170, 113752.

Kurtz, C.C., Mitchell, S., Nielsen, K., Crawford, K.D., Mueller-Spitz, S.R., 2020. Acute high-dose titanium dioxide nanoparticle exposure alters gastrointestinal homeostasis in mice. J. Appl. Toxicol. 40 (10), 1384–1395.

Labuza, T., Breene, W., 1989. Applications of active packaging for improvement of shelf-life and nutritional quality of fresh and extended shelf-life foods. J. Food Process. Preserv. 13, 1–89.

Lei, J., Zhou, L., Tang, Y., Luo, Y., Duan, T., Zhu, W., 2017. High-strength konjac glucomannan/silver nanowires composite films with antibacterial properties. Materials 10 (5), 524.

Lević, S., Obradović, N., Pavlović, V., Isailović, B., Kostić, I., Mitrić, M., Bugarski, B., Nedović, V., 2014. Thermal, morphological, and mechanical properties of ethyl vanillin immobilized in polyvinyl alcohol by electrospinning process. J. Therm. Anal. Calorim. 118 (2), 661–668.

Li, W., Zhang, C., Chi, H., Li, L., Lan, T., Han, P., Qin, Y., 2017. Development of antimicrobial packaging film made from poly (lactic acid) incorporating titanium dioxide and silver nanoparticles. Molecules 22 (7), 1170.

Li, L., Wang, H., Chen, M., Jiang, S., Jiang, S., Li, X., Wang, Q., 2018a. Butylated hydroxyanisole encapsulated in gelatin fiber mats: volatile release kinetics, functional effectiveness and application to strawberry preservation. Food Chem. 269, 142–149.

Li, M., Zhang, F., Liu, Z., Guo, X., Wu, Q., Qiao, L., 2018. Controlled release system by active gelatin film incorporated with β-cyclodextrin-thymol inclusion complexes. Food Bioprocess Technol. 11 (9), 1695–1702.

Li, S., Yi, J., Yu, X., Wang, Z., Wang, L., 2020. Preparation and characterization of pullulan derivative/chitosan composite film for potential antimicrobial applications. Int. J. Biol. Macromol. 148, 258–264.

Limjaroen, P., Ryser, E., Lockhart, H., Harte, B., 2003. Development of a food packaging coating material with antimicrobial properties. J. Plast. Film Sheeting 19 (2), 95–109.

Lin, L., Zhu, Y., Cui, H., 2018. Electrospun thyme essential oil/gelatin nanofibers for active packaging against *Campylobacter jejuni* in chicken. LWT–Food Sci. Technol. 97, 711–718.

Liu, Y., Li, Y., Deng, L., Zou, L., Feng, F., Zhang, H., 2018. Hydrophobic ethylcellulose/gelatin nanofibers containing zinc oxide nanoparticles for antimicrobial packaging. J. Agric. Food Chem. 66 (36), 9498–9506.

Lotfi, M., Tajik, H., Moradi, M., Forough, M., Divsalar, E., Kuswandi, B., 2018. Nanostructured chitosan/monolaurin film: preparation, characterization and antimicrobial activity against *Listeria monocytogenes* on ultrafiltered white cheese. LWT–Food Sci. Technol. 92, 576–583.

Ma, Y., Li, L., Wang, Y., 2018. Development of PLA-PHB-based biodegradable active packaging and its application to salmon. Packag. Technol. Sci. 31 (11), 739–746.

Ma, C., Liu, H., Chen, G., Zhao, Q., Guo, H., Minocha, R., Dhankher, O.P., 2020. Dual roles of glutathione in silver nanoparticle detoxification and enhancement of nitrogen assimilation in soybean (*Glycine max* (L.) Merrill). Environ. Sci. Nano 7 (7), 1954–1966.

Maleki, G., Sedaghat, N., Woltering, E.J., Farhoodi, M., Mohebbi, M., 2018. Chitosan-limonene coating in combination with modified atmosphere packaging preserve postharvest quality of cucumber during storage. J. Food Measur. Character. 12 (3), 1610–1621.

Marcos, B., Aymerich, T., Monfort, J.M., Garriga, M., 2010. Physical performance of biodegradable films intended for antimicrobial food packaging. J. Food Sci. 75 (8), E502–E507.

Mbikay, M., 2012. Therapeutic potential of moringa oleifera leaves in chronic hyperglycemia and dyslipidemia: a review. Front. Pharmacol. 3, 24.

Mehta, R., Arya, R., Goyal, K., Singh, M., K Sharma, A., 2013. Bio-preservative and therapeutic potential of pediocin: recent trends and future perspectives. Recent Pat. Biotechnol. 7 (3), 172–178.

Meng, J., Zhou, X., Yang, J., Qu, X., Cui, S., 2020. Exposure to low dose ZnO nanoparticles induces hyperproliferation and malignant transformation through activating the CXCR2/NF-κB/STAT3/ERK and AKT pathways in colonic mucosal cells. Environ. Pollut. 263, 114578.

Morais, L.D.O., Macedo, E.V., Granjeiro, J.M., Delgado, I.F., 2020. Critical evaluation of migration studies of silver nanoparticles present in food packaging: a systematic review. Crit. Rev. Food Sci. Nutr. 60 (18), 3083–3102.

Moreno, O., Atarés, L., Chiralt, A., 2015. Effect of the incorporation of antimicrobial/antioxidant proteins on the properties of potato starch films. Carbohydr. Polym. 133, 353–364.

Moreno, O., Atarés, L., Chiralt, A., Cruz-Romero, M.C., Kerry, J., 2018. Starch-gelatin antimicrobial packaging materials to extend the shelf life of chicken breast fillets. LWT–Food Sci. Technol. 97, 483–490.

Motelica, L., Ficai, D., Ficai, A., Oprea, O.C., Kaya, D.A., Andronescu, E., 2020. Biodegradable antimicrobial food packaging: trends and perspectives. Foods 9 (10), 1438.

Mushtaq, M., Gani, A., Gani, A., Punoo, H.A., Masoodi, F.A., 2018. Use of pomegranate peel extract incorporated zein film with improved properties for prolonged shelf life of fresh Himalayan cheese (Kalari/kradi). Innovat. Food Sci. Emerg. Technol. 48, 25–32.

Musial, J., Krakowiak, R., Mlynarczyk, D.T., Goslinski, T., Stanisz, B.J., 2020. Titanium dioxide nanoparticles in food and personal care products: what do we know about their safety? Nanomaterials 10 (6), 1110.

Mutlu-Ingok, A., Devecioglu, D., Dikmetas, D.N., Karbancioglu-Guler, F., Capanoglu, E., 2020. Antibacterial, antifungal, antimycotoxigenic, and antioxidant activities of essential oils: an updated review. Molecules 25 (20), 4711.

Najnin, H., Alam, N., Mujeeb, M., Ahsan, H., Siddiqui, W.A., 2020. Biochemical and toxicological analysis of *Cinnamomum tamala* essential oil in Wistar rats. J. Food Process. Preserv. 44 (2), e14328.

Nemade, S., Sawarkar, S., 2015. Recovery and synthesis of guar gum and its derivatives. Int. J. Adv. Res. Chem. Sci. 2, 33–40.

Nguyen, V.T., Vu, V.T., Nguyen, T.H., Nguyen, T.A., Tran, V.K., Nguyen-Tri, P., 2019. Antibacterial activity of TiO_2-and ZnO-decorated with silver nanoparticles. J. Compos. Sci. 3 (2), 61.

Niu, B., Yan, Z., Shao, P., Kang, J., Chen, H., 2018. Encapsulation of cinnamon essential oil for active food packaging film with synergistic antimicrobial activity. Nanomaterials 8 (8), 598.

Noori, A., Ngo, A., Gutierrez, P., Theberge, S., White, J.C., 2020. Silver nanoparticle detection and accumulation in tomato (*Lycopersicon esculentum*). J. Nanopart. Res. 22 (6), 1–16.

Ntzimani, A.G., Giatrakou, V.I., Savvaidis, I.N., 2010. Combined natural antimicrobial treatments (EDTA, lysozyme, rosemary and oregano oil) on semi cooked coated chicken meat stored in vacuum packages at 4 C: microbiological and sensory evaluation. Innovat. Food Sci. Emerg. Technol. 11 (1), 187–196.

Nunes, M.R., Castilho, M.D.S.M., de Lima Veeck, A.P., Rosa, C., G., Noronha, C.M., Maciel, M.V., Barreto, P.M, 2018. Antioxidant and antimicrobial methylcellulose films containing *Lippia alba* extract and silver nanoparticles. Carbohydr. Polym. 192, 37–43.

Odusote, J.K., Oyewo, A.T., 2016. Mechanical properties of pineapple leaf fiber reinforced polymer composites for application as a prosthetic socket. J. Eng. Technol. 7 (1), 125–139.

Ojagh, S.M., Rezaei, M., Razavi, S.H., Hosseini, S.M.H., 2010. Effect of chitosan coatings enriched with cinnamon oil on the quality of refrigerated rainbow trout. Food Chem. 120 (1), 193–198.

Olmos, D., Pontes-Quero, G.M., Corral, A., González-Gaitano, G., González-Benito, J., 2018. Preparation and characterization of antimicrobial films based on LDPE/Ag nanoparticles with potential uses in food and health industries. Nanomaterials 8 (2), 60. doi:10.3390/nano8020060.

Olsson, E., Hedenqvist, M.S., Johansson, C., Jarnstrom, L., 2013. Influence of citric acid and curing on moisture sorption, diffusion and permeability of starch films. Carbohydr. Polym. 94, 765–772.

Orsuwan, A., Sothornvit, R., 2018. Active banana flour nanocomposite films incorporated with garlic essential oil as multifunctional packaging material for food application. Food Bioprocess Technol. 11 (6), 1199–1210.

Ortiz, C.M., Salgado, P.R., Dufresne, A., Mauri, A.N., 2018. Microfibrillated cellulose addition improved the physicochemical and bioactive properties of biodegradable films based on soy protein and clove essential oil. Food Hydrocoll. 79, 416–427.

Ounkaew, A., Kasemsiri, P., Kamwilaisak, K., Saengprachatanarug, K., Mongkolthanaruk, W., Souvanh, M., Chindaprasirt, P., 2018. Polyvinyl alcohol (PVA)/starch bioactive packaging film enriched with antioxidants from spent coffee ground and citric acid. J. Polym. Environ. 26 (9), 3762–3772.

Pabast, M., Shariatifar, N., Beikzadeh, S., Jahed, G., 2018. Effects of chitosan coatings incorporating with free or nano-encapsulated *Satureja* plant essential oil on quality characteristics of lamb meat. Food Control 91, 185–192.

Paik, H.D., Kim, H.J., Nam, K.J., Kim, C.J., Lee, S.E., Lee, D.S., 2006. Effect of nisin on the storage of sous vide processed Korean seasoned beef. Food Control 17 (12), 994–1000.

Perez Espitia, P.J., Ferreira Soares, N.d.F., Teofilo, R.F., dos Reis Coimbra, J.S., Vitor, D.M., Batista, R.A., Ferreira, S.O., de Andrade, N.J., Alves Medeiros, E.A., 2013. Physical-mechanical and antimicrobial properties of nanocomposite films with pediocin and ZnO nanoparticles. Carbohydr. Polym. 94, 199–208.

Pérez-Córdoba, L.J., Norton, I.T., Batchelor, H.K., Gkatzionis, K., Spyropoulos, F., Sobral, P.J., 2018. Physico-chemical, antimicrobial and antioxidant properties of gelatin-chitosan based films loaded with nanoemulsions encapsulating active compounds. Food Hydrocoll. 79, 544–559.

Piñeros-Hernandez, D., Medina-Jaramillo, C., López-Córdoba, A., Goyanes, S., 2017. Edible cassava starch films carrying rosemary antioxidant extracts for potential use as active food packaging. Food Hydrocoll. 63, 488–495.

Pires, J.R.A., de Souza, V.G.L., Fernando, A.L., 2018. Chitosan/montmorillonite bionanocomposites incorporated with rosemary and ginger essential oil as packaging for fresh poultry meat. Food Packag. Shelf Life 17, 142–149.

Portugal Zegarra, M.D.C.C., Santos, A.M.P., Silva, A.M.A.D., Melo, E.D.A., 2018. Chitosan films incorporated with antioxidant extract of acerola agroindustrial residue applied in chicken thigh. J. Food Process. Preserv. 42 (4), e13578.

Pranoto, Y., Rakshit, S.K., Salokhe, V.M., 2005. Enhancing antimicrobial activity of chitosan films by incorporating garlic oil, potassium sorbate and nisin. LWT–Food Sci. Technol. 38 (8), 859–865.

Priyadarshi, R., Kumar, B., Negi, Y.S., 2018. Chitosan film incorporated with citric acid and glycerol as an active packaging material for extension of green chilli shelf life. Carbohydr. Polym. 195, 329–338.

Priyadarshi, R., Kumar, B., Deeba, F., Kulshreshtha, A., Negi, Y.S., 2018a. Chitosan films incorporated with apricot (*Prunus armeniaca*) kernel essential oil as active food packaging material. Food Hydrocoll. 85, 158–166.

Quintavalla, S., Vicini, L., 2002. Antimicrobial food packaging in meat industry. Meat Sci. 62 (3), 373–380.

Rad, F.H., Sharifan, A., Asadi, G., 2018. Physicochemical and antimicrobial properties of kefiran/waterborne polyurethane film incorporated with essential oils on refrigerated ostrich meat. LWT–Food Sci. Technol. 97, 794–801.

Rajput, V., Minkina, T., Fedorenko, A., Sushkova, S., Mandzhieva, S., Lysenko, V., Ghazaryan, K., 2018. Toxicity of copper oxide nanoparticles on spring barley (*Hordeum sativum distichum*). Sci. Total Environ. 645, 1103–1113.

Rajput, V., Minkina, T., Sushkova, S., Behal, A., Maksimov, A., Blicharska, E., Barsova, N., 2020. ZnO and CuO nanoparticles: a threat to soil organisms, plants, and human health. Environ. Geochem. Health 42 (1), 147–158.

Ramachandran, R., Krishnaraj, C., Kumar, V.K., Harper, S.L., Kalaichelvan, T.P., Yun, S.I., 2018. In vivo toxicity evaluation of biologically synthesized silver nanoparticles and gold nanoparticles on adult zebrafish: a comparative study. 3 Biotech 8 (10), 1–12.

Ramziia, S., Ma, H., Yao, Y., Wei, K., Huang, Y., 2018. Enhanced antioxidant activity of fish gelatin–chitosan edible films incorporated with procyanidin. J. Appl. Polym. Sci. 135 (10), 45781.

Rezaei, F., Shahbazi, Y., 2018. Shelf-life extension and quality attributes of sauced silver carp fillet: a comparison among direct addition, edible coating and biodegradable film. LWT–Food Sci. Technol. 87, 122–133.

Rhim, J.W., Hong, S.I., Park, H.M., Ng, P.K., 2006. Preparation and characterization of chitosan-based nanocomposite films with antimicrobial activity. J. Agric. Food Chem. 54 (16), 5814–5822.

Ribeiro Santos, R., de Melo, N.R., Andrade, M., Azevedo, G., Machado, A.V., Carvalho Costa, D., Sanches Silva, A., 2018. Whey protein active films incorporated with a blend of essential oils: characterization and effectiveness. Packag. Technol. Sci. 31 (1), 27–40.

Robledo, N., López, L., Bunger, A., Tapia, C., Abugoch, L., 2018. Effects of antimicrobial edible coating of thymol nanoemulsion/quinoa protein/chitosan on the safety, sensorial properties, and quality of refrigerated strawberries (Fragaria×ananassa) under commercial storage environment. Food Bioprocess Technol. 11 (8), 1566–1574.

Rooney, M., 1995. Active Food Packaging. Blackie Academic and Professional, Glasgow.

Safakas, K., Giotopoulou, I., Giannakopoulou, A., Katerinopoulou, K., Lainioti, G.C., Stamatis, H., Barkoula, N.M., Ladavos, A., 2023. Designing antioxidant and antimicrobial polyethylene films with bioactive compounds/clay nanohybrids for potential packaging applications. Molecules 28 (7), 2945. https://doi.org/10.3390/molecules28072945.

Sahariah, P., Masson, M., 2017. Antimicrobial chitosan and chitosan derivatives: a review of the structure–activity relationship. Biomacromolecules 18 (11), 3846–3868.

Salari, M., Khiabani, M.S., Mokarram, R.R., Ghanbarzadeh, B., Kafil, H.S., 2018. Development and evaluation of chitosan based active nanocomposite films containing bacterial cellulose nanocrystals and silver nanoparticles. Food Hydrocoll. 84, 414–423.

Sánchez-López, E., Gomes, D., Esteruelas, G., Bonilla, L., Lopez-Machado, A.L., Galindo, R., Souto, E.B., 2020. Metal-based nanoparticles as antimicrobial agents: an overview. Nanomaterials 10 (2), 292.

Santiago-Silva, P., Soares, N.F., Nóbrega, J.E., Júnior, M.A., Barbosa, K.B., Volp, A.C.P., Würlitzer, N.J., 2009. Antimicrobial efficiency of film incorporated with pediocin (ALTA® 2351) on preservation of sliced ham. Food Control 20 (1), 85–89.

Saurabh, C.K., Gupta, S., Variyar, P.S., 2018. Development of guar gum based active packaging films using grape pomace. J. Food Sci. Technol. 55 (6), 1982–1992.

Shabanpour, B., Kazemi, M., Ojagh, S.M., Pourashouri, P., 2018. Bacterial cellulose nanofibers as reinforce in edible fish myofibrillar protein nanocomposite films. Int. J. Biol. Macromol. 117, 742–751.

Shah, N., Khan, A., Ali, R., Marimuthu, K., Uddin, M.N., Rizwan, M., … Khisroon, M., 2020. Monitoring bioaccumulation (in gills and muscle tissues), hematology, and genotoxic alteration in Ctenopharyngodon idella exposed to selected heavy metals. Biomed. Res. Int. 2020.

Shahbazi, Y., Karami, N., Shavisi, N., 2018. Effect of Mentha spicata essential oil on chemical, microbial, and sensory properties of minced camel meat during refrigerated storage. J. Food Saf. 38 (1), e12375.

Shankar, S., Rhim, J.W., 2015. Amino acid mediated synthesis of silver nanoparticles and preparation of antimicrobial agar/silver nanoparticles composite films. Carbohydr. Polym. 130, 353–363.

Shankar, S., Oun, A.A., Rhim, J.W., 2018. Preparation of antimicrobial hybrid nano-materials using regenerated cellulose and metallic nanoparticles. Int. J. Biol. Macromol. 107, 17–27.

Shao, P., Yan, Z., Chen, H., Xiao, J., 2018. Electrospun poly (vinyl alcohol)/permutite fibrous film loaded with cinnamaldehyde for active food packaging. J. Appl. Polym. Sci. 135 (16), 46117.

Shen, Z., Kamdem, D.P., 2015. Antimicrobial activity of sugar beet lignocellulose films containing tung oil and cedarwood essential oil. Cellulose 22 (4), 2703–2715.

Silva, C.F., Oliveira, F.S.M., Caetano, V.F., Vinhas, G.M., Cardoso, S.A., 2018. Orange essential oil as antimicrobial additives in poly(vinyl chloride) films. Polímeros 28 (4), 332–338. doi:10.1590/0104-1428.16216.

Simoes, M., Simões, L.C., Vieira, M.J., 2010. A review of current and emergent biofilm control strategies. LWT–Food Sci. Technol. 43 (4), 573–583.

Singh, A., Dutta, P.K., 2017. Extraction of chitin-glucan complex from *Agaricus bisporus*: characterization and antibacterial activity. J. Polym. Mater. 34 (1), 1.

Singh, S., Gaikwad, K.K., Lee, Y.S., 2018. Antimicrobial and antioxidant properties of polyvinyl alcohol bio composite films containing seaweed extracted cellulose nano-crystal and basil leaves extract. Int. J. Biol. Macromol. 107, 1879–1887.

Sinha, A.K., Narang, H.K., Bhattacharya, S., 2018. Evaluation of bending strength of abaca reinforced polymer composites. Mater. Today: Proc. 5 (2), 7284–7288.

Smith, J., Hoshino, J., Abe, Y., 1995. Interactive packaging involving sachet technology. In: Rooney, M.L. (Ed.), Active Food Packaging. Blackie Academic and Professional, Glasgow, p. 143.

Sogut, E., Seydim, A.C., 2018. Development of Chitosan and polycaprolactone based active bilayer films enhanced with nanocellulose and grape seed extract. Carbohydr. Polym. 195, 180–188.

Soni, B., Mahmoud, B., Chang, S., El-Giar, E.M., 2018. Physicochemical, antimicrobial and antioxidant properties of chitosan/TEMPO biocomposite packaging films. Food Packag. Shelf Life 17, 73–79.

Sorkhabi, T.S., Samberan, M.F., Kumaravel, V., 2023. Antimicrobial activities of polyethylene terephthalate-waste-derived nanofibrous membranes decorated with green synthesized Ag nanoparticles. Molecules (Basel, Switzerland) 28 (14), 5439. https://doi.org/10.3390/molecules28145439.

Soto, K.M., Hernández-Iturriaga, M., Loarca-Piña, G., Luna-Bárcenas, G., Gómez-Aldapa, C.A., Mendoza, S., 2016. Stable nisin food-grade electrospun fibers. J. Food Sci. Technol. 53 (10), 3787–3794.

Souza, V.G., Pires, J.R., Vieira, É.T., Coelhoso, I.M., Duarte, M.P., Fernando, A.L., 2018. Shelf life assessment of fresh poultry meat packaged in novel bionanocomposite of chitosan/montmorillonite incorporated with ginger essential oil. Coatings 8 (5), 177.

Srikandace, Y., Indrarti, L., Sancoyorini, M.K., 2018, June. Antibacterial activity of bacterial cellulose-based edible film incorporated with *Citrus* spp essential oil. IOP Conf. Ser.: Earth Environ. Sci., 160. IOP Publishing.

Suo, B., Li, H., Wang, Y., Li, Z., Pan, Z., Ai, Z., 2017. Effects of ZnO nanoparticle-coated packaging film on pork meat quality during cold storage. J. Sci. Food Agric. 97 (7), 2023–2029.

Suppakul, P., Miltz, J., Sonneveld, K., Bigger, S.W., 2006. Characterization of antimicrobial films containing basil extracts. Packag. Technol. Sci. Int. J. 19 (5), 259–268.

Swaroop, C., Shukla, M., 2018. Nano-magnesium oxide reinforced polylactic acid biofilms for food packaging applications. Int. J. Biol. Macromol. 113, 729–736.

Talebi, F., Misaghi, A., Khanjari, A., Kamkar, A., Gandomi, H., Rezaeigolestani, M., 2018. Incorporation of spice essential oils into poly-lactic acid film matrix with the aim of extending microbiological and sensorial shelf life of ground beef. LWT–Food Sci. Technol. 96, 482–490.

Tan, Y.M., Lim, S.H., Tay, B.Y., Lee, M.W., Thian, E.S., 2015. Functional chitosan-based grapefruit seed extract composite films for applications in food packaging technology. Mater. Res. Bull. 69, 142–146.

Tanha, E.Y., Fallah, S., Rostamnejadi, A., Pokhrel, L.R., 2020. Particle size and concentration dependent toxicity of copper oxide nanoparticles (CuONPs) on seed yield and antioxidant defense system in soil grown soybean (Glycine max cv. Kowsar). Sci. Total Envir. 715, 136994.

Tedeschi, G., Guzman-Puyol, S., Ceseracciu, L., Paul, U.C., Picone, P., Di Carlo, M., Heredia-Guerrero, J.A., 2020. Multifunctional bioplastics inspired by wood composition: effect of hydrolyzed lignin addition to xylan–cellulose matrices. Biomacromolecules 21 (2), 910–920.

Thombare, N., Jha, U., Mishra, S., Siddiqui, M.Z., 2016. Guar gum as a promising starting material for diverse applications: a review. Int. J. Biol. Macromol. 88, 361–372.

Todkar, S.S., Patil, S.A., 2019. Review on mechanical properties evaluation of pineapple leaf fibre (PALF) reinforced polymer composites. Compos. Part B: Eng. 174, 106927.

Tong, S.Y., Lim, P.N., Wang, K., Thian, E.S., 2018. Development of a functional biodegradable composite with antibacterial properties. Mater. Technol. 33 (11), 754–759.

Trongchuen, K., Ounkaew, A., Kasemsiri, P., Hiziroglu, S., Mongkolthanaruk, W., Wanna-sutta, R., Chindaprasirt, P., 2018. Bioactive starch foam composite enriched with natural antioxidants from spent coffee ground and essential oil. Starch 70 (7-8), 1700238.

Tyagi, P., Hubbe, M.A., Lucia, L., Pal, L., 2018. High performance nanocellulose-based composite coatings for oil and grease resistance. Cellulose 25 (6), 3377–3391.

Uranga, J., Etxabide, A., Guerrero, P., de la Caba, K., 2018. Development of active fish gelatin films with anthocyanins by compression molding. Food Hydrocoll. 84, 313–320.

Uranga, J., Puertas, A.I., Etxabide, A., Dueñas, M.T., Guerrero, P., De La Caba, K., 2019. Citric acid-incorporated fish gelatin/chitosan composite films. Food Hydrocoll. 86, 95–103.

Valerini, D., Tammaro, L., Di Benedetto, F., Vigliotta, G., Capodieci, L., Terzi, R., Rizzo, A., 2018. Aluminum-doped zinc oxide coatings on polylactic acid films for antimicrobial food packaging. Thin Solid Films 645, 187–192.

Vasile, O.R., Serdaru, I., Andronescu, E., Truşcă, R., Surdu, V.A., Oprea, O., Vasile, B.Ş., 2015. Influence of the size and the morphology of ZnO nanoparticles on cell viability. C.R. Chim. 18 (12), 1335–1343.

Vasile, C., Sivertsvik, M., Miteluţ, A.C., Brebu, M.A., Stoleru, E., Rosnes, J.T., Popa, M.E., 2017. Comparative analysis of the composition and active property evaluation of certain essential oils to assess their potential applications in active food packaging. Materials 10 (1), 45.

Vergis, J., Gokulakrishnan, P., Agarwal, R.K., Kumar, A., 2015. Essential oils as natural food antimicrobial agents: a review. Crit. Rev. Food Sci. Nutr. 55 (10), 1320–1323.

Vickers, N.J., 2017. Animal communication: when I'm calling you, will you answer too? Curr. Biol. 27 (14), R713–R715.

Voss, L., Saloga, P.E., Stock, V., Böhmert, L., Braeuning, A., Thünemann, A.F., Sieg, H., 2019. Environmental impact of ZnO nanoparticles evaluated by in vitro simulated digestion. ACS Appl. Nano Mater. 3 (1), 724–733.

Wang, C., Hsiue, G., 1993. Glucose oxidase immobilization onto plasma induced graft copolymerized polymeric membrane modified by polyethylene oxide as a spacer. J. Appl. Polym. Sci. 50, 1141–1149.

Wang, L., Dong, Y., Men, H., Tong, J., Zhou, J., 2013. Preparation and characterization of active films based on chitosan incorporated tea polyphenols. Food Hydrocoll. 32 (1), 35–41.

Wang, H., Qian, J., Ding, F., 2018. Emerging chitosan-based films for food packaging applications. J. Agric. Food Chem. 66 (2), 395–413.

Wang, Q., Lei, J., Ma, J., Yuan, G., Sun, H., 2018. Effect of chitosan-carvacrol coating on the quality of Pacific white shrimp during iced storage as affected by caprylic acid. Int. J. Biol. Macromol. 106, 123–129.

Warsiki, E., 2018. Application of chitosan as biomaterial for active packaging of ethylene absorber. IOP Conf. Ser.: Earth Environ. Sci. 141, 012036.

Wieczyńska, J., Cavoski, I., 2018. Antimicrobial, antioxidant and sensory features of eugenol, carvacrol and trans-anethole in active packaging for organic ready-to-eat iceberg lettuce. Food Chem. 259, 251–260.

Wińska, K., Mączka, W., Łyczko, J., Grabarczyk, M., Czubaszek, A., Szumny, A., 2019. Essential oils as antimicrobial agents-myth or real alternative? Molecules 24 (11), 2130.

Wu, Z., Huang, X., Li, Y.C., Xiao, H., Wang, X., 2018. Novel chitosan films with laponite immobilized Ag nanoparticles for active food packaging. Carbohydr. Polym. 199, 210–218.

Wu, J., Wang, G., Vijver, M.G., Bosker, T., Peijnenburg, W.J., 2020. Foliar versus root exposure of AgNPs to lettuce: phytotoxicity, antioxidant responses and internal translocation. Environ. Pollut. 261, 114117.

Xing, Y., Lin, H., Cao, D., Xu, Q., Han, W., Wang, R., Li, X., 2015. Effect of chitosan coating with cinnamon oil on the quality and physiological attributes of China jujube fruits. Biomed. Res. Int. 2015.

Xu, H., Canisag, H., Mu, B., Yang, Y., 2015. Robust and flexible films from 100% starch crosslinked by biobased disaccharide derivative. ACS Sustain. Chem. Eng. 3, 2631–2639.

Xu, Y., Willis, S., Jordan, K., Sismour, E., 2018. Chitosan nanocomposite films incorporating cellulose nanocrystals and grape pomace extracts. Packag. Technol. Sci. 31 (9), 631–638.

Yan, J., Luo, Z., Ban, Z., Lu, H., Li, D., Yang, D., Aghdam, M.S., Li, L., 2019. The effect of the layer-by-layer (LBL) edible coating on strawberry quality and metabolites during storage. Postharvest Biol. Technol. 147, 29–38.

Yang, S.Y., Cao, L., Kim, H., Beak, S.E., Song, K.B., 2018. Utilization of foxtail millet starch film incorporated with clove leaf oil for the packaging of queso blanco cheese as a model food. Starch 70 (3-4), 1700171.

Yang, W., Fortunati, E., Bertoglio, F., Owczarek, J.S., Bruni, G., Kozanecki, M., Puglia, D., 2018. Polyvinyl alcohol/chitosan hydrogels with enhanced antioxidant and antibacterial properties induced by lignin nanoparticles. Carbohydr. Polym. 181, 275–284.

Yang, W., Xie, Y., Jin, J., Liu, H., Zhang, H., 2019. Development and application of an active plastic multilayer film by coating a plantaricin BM-1 for chilled meat preservation. J. Food Sci. 84 (7), 1864–1870.

Yang, Z., Xiao, Y., Jiao, T., Zhang, Y., Chen, J., Gao, Y., 2020. Effects of copper oxide nanoparticles on the growth of rice (*Oryza Sativa* L.) seedlings and the relevant physiological responses. Int. J. Environ. Res. Public Health 17 (4), 1260.

Ye, J., Wang, S., Lan, W., Qin, W., Liu, Y., 2018. Preparation and properties of polylactic acid-tea polyphenol-chitosan composite membranes. Int. J. Biol. Macromol. 117, 632–639.

Yin, X., Sun, C., Tian, M., Wang, Y., 2018. Preparation and characterization of salicylic acid/polylactic acid composite packaging materials. In: Applied Sciences in Graphic Communication and Packaging. Springer, Singapore, pp. 811–818.

Youssef, A.M., El-Sayed, S.M., 2018. Bionanocomposites materials for food packaging applications: concepts and future outlook. Carbohydr. Polym. 193, 19–27.

Yu, M., Huang, R., He, C., Wu, Q., Zhao, X., 2016. Hybrid composites from wheat straw, inorganic filler, and recycled polypropylene: morphology and mechanical and thermal expansion performance. Int. J. Polym. Sci. 2016, 2520670.

Yu, Z., Li, B., Chu, J., Zhang, P., 2018. Silica in situ enhanced PVA/chitosan biodegradable films for food packages. Carbohydr. Polym. 184, 214–220.

Yu, Z., Sun, L., Wang, W., Zeng, W., Mustapha, A., Lin, M., 2018. Soy protein-based films incorporated with cellulose nanocrystals and pine needle extract for active packaging. Ind. Crops Prod. 112, 412–419.

Yun, Y.H., Na, Y.H., Yoon, S.D., 2006. Mechanical properties with the functional group of additives for starch/PVA blend film. J. Polym. Environ. 14, 71–78.

Yun, Y.H., Lee, C.M., Kim, Y.S., Yoon, S.D., 2017. Preparation of chitosan/polyvinyl alcohol blended films containing sulfosuccinic acid as the crosslinking agent using UV curing process. Food Res. Int. 100, 377–386.

Zactiti, E.M., Kieckbusch, T.G., 2009. Release of potassium sorbate from active films of sodium alginate crosslinked with calcium chloride. Packag. Technol. Sci. Int. J. 22 (6), 349–358.

Zahedi, Y., Fathi-Achachlouei, B., Yousefi, A.R., 2018. Physical and mechanical properties of hybrid montmorillonite/zinc oxide reinforced carboxymethyl cellulose nanocomposites. Int. J. Biol. Macromol. 108, 863–873.

Zehetmeyer, G., Meira, S.M.M., Scheibel, J.M., de Brito da Silva, C., Rodembusch, F.S., Brandelli, A., Soares, R.M.D, 2017. Biodegradable and antimicrobial films based on poly (butylene adipate-co-terephthalate) electrospun fibers. Polym. Bull. 74 (8), 3243–3268.

Zhang, C., Li, W., Zhu, B., Chen, H., Chi, H., Li, L., Xue, J., 2018. The quality evaluation of postharvest strawberries stored in Nano-Ag packages at refrigeration temperature. Polymers 10 (8), 894.

Zhang, Z.J., Li, N., Li, H.Z., Li, X.J., Cao, J.M., Zhang, G.P., He, D.L., 2018a. Preparation and characterization of biocomposite chitosan film containing *Perilla frutescens* (L.) Britt. essential oil. Ind. Crops Prod. 112, 660–667.

Zhang, X., Shu, Y., Su, S., Zhu, J., 2018b. One-step coagulation to construct durable antifouling and antibacterial cellulose film exploiting Ag@ AgCl nanoparticle-triggered photocatalytic degradation. Carbohydr. Polym. 181, 499–505.

Zhang, M., Han, W., Hu, X., Li, D., Ma, X., Liu, H., Liu, L., Lu, W., Liu, S., 2020. Pentaerythritol p-hydroxybenzoate ester-based zinc metal alkoxides as multifunctional antimicrobial thermal stabilizer for PVC. Polym. Degrad. Stab. 181, 109340. https://doi.org/10.1016/j.polymdegradstab.2020.109340.

Zhao, Y., Teixeira, J.S., Gänzle, M.M., Saldaña, M.D., 2018. Development of antimicrobial films based on cassava starch, chitosan and gallic acid using subcritical water technology. J. Supercrit. Fluids 137, 101–110.

Zheng, K., Li, W., Fu, B., Fu, M., Ren, Q., Yang, F., Qin, C., 2018. Physical, antibacterial and antioxidant properties of chitosan films containing hardleaf oatchestnut starch and *Litsea cubeba* oil. Int. J. Biol. Macromol. 118, 707–715.

Zhou, Q., Liu, L., Liu, N., He, B., Hu, L., Wang, L., 2020. Determination and characterization of metal nanoparticles in clams and oysters. Ecotoxicol. Environ. Saf. 198, 110670.

Zhu, J.Y., Tang, C.H., Yin, S.W., Yang, X.Q., 2018. Development and characterisation of polylactic acid–gliadin bilayer/trilayer films as carriers of thymol. Int. J. Food Sci. Technol. 53 (3), 608–618.

Further readings

Alemdar, A., Sain, M., 2008. Biocomposites from wheat straw nanofibers: Morphology, thermal and mechanical properties. Compos. Sci. Technol. 68 (2), 557–565.

Costa, L.A., Assis, D.D.J., Gomes, G.V., da Silva, J.B., Fonsêca, A.F., Druzian, J.I., 2015. Extraction and characterization of nanocellulose from corn stover. Mater. Today Proc. 2 (1), 287–294.

De Andrade, M.R., Nery, T.B.R., de Santana e Santana, T.I., Leal, I.L., Rodrigues, L.A.P., de Oliveira Reis, J.H., Machado, B.A.S., 2019. Effect of cellulose nanocrystals from different lignocellulosic residues to chitosan/glycerol films. Polymers 11 (4), 658.

Krepker, M., Shemesh, R., Danin Poleg, Y., Kashi, Y., Vaxman, A., Segal, E., 2017. Active food packaging films with synergistic antimicrobial activity. Food Contr. 76, 117–126.

Morais, J.P.S., Rosa, M.D.F., De Souza Filho, M.D.S.M., Nascimento, L.D., Do Nascimento, D.M., Cassales, A.R., 2013. Extraction and characterization of nanocellulose structures from raw cotton linter. Carbohydr. Polym. 91, 229–235.

Panthapulakkal, S., Zereshkian, A., Sain, M., 2006. Preparation and characterization of wheat straw fibers for reinforcing application in injection molded thermoplastic composites. Bioresour. Technol. 97 (2), 265–272.

Ranum, P., Peña-Rosas, J.P., Garcia-Casal, M.N., 2014. Global maize production, utilization, and consumption. Ann. N.Y. Acad. Sci. 1312 (1), 105–112.

Salmah, H., Marliza, M.Z., Selvi, E., 2014. Biocomposites from polypropylene and corn cob: effect maleic anhydride grafted polypropylene. Adv. Mater. Res. 3 (3), 129.

Perspectives for carbon-based nanomaterial and its antimicrobial films in food applications

19

**Eli José Miranda Ribeiro Júnior[a], Marcos Túlio da Silva[b],
Alexandre Gonçalves Pinheiro[c,d] and Stephen Rathinaraj Benjamin[e]**

[a] *Department of Pharmacy, Faculty of CGESP (Centro Goiano de Ensino Superior), Goiânia,
Goiás, Brazil,* [b] *Instituto de Saúde e Biotecnologia, Universidade Federal do Amazonas, Espírito
Santo, Coari – Amazonas, AM, Brazil,* [c] *Department of Physics, Faculty of Education, State
University of Ceará-UECE, Planalto Universitário, Quixadá, Ceará, Brazil,* [d] *Department of
Physics, The University of Texas at Dallas, Texas, United States,* [e] *Drug Research and Development
Center (NPDM), Department of Physiology and Pharmacology, Federal University of Ceará-UFC,
Fortaleza, Ceará, Brazil*

19.1 Introduction

Nanotechnology is a multidisciplinary field that includes physics, chemistry, biology, and engineering. The term refers to nanomaterials with structures ranging from 1 nm to 100 nm in size. Materials take on new features at these nanoscale scales that were not present in their original forms. Nanoscientists worldwide are working to uncover these distinct features to develop new and better goods using environmentally friendly methods. The term "nanobiotechnology" refers to the fusion of nanotechnology with biology, resulting in a new class of devices and systems that serve many purposes. Nanomaterials have unique chemical and physical features that make them ideal for formulating novel food products. The innovative features of nanomaterials open a variety of new possibilities for the food and agricultural sectors, such as developing more intense food colorings, flavors, nutritious additives, antimicrobial compounds for food packaging, and various other applications. Furthermore, nanobiotechnology may aid in producing healthier foods that include less fat, sugar, and salt, assisting in the prevention and treatment of many food-related disorders.

Conversely, nanomaterials in food items may generate concerns about their health effects on consumers. Consequently, the presence of nanoscale-sized objects in food demands a thorough evaluation of any possible harmful consequences. In addition to excellent electrical properties, superior thermal conductivity, and high mechanical strength, CBN has gained considerable interest from the scientific community. Low-dimensional carbon nanomaterials (CNTs) are often classified as either zero-dimensions (0D) such as fullerene, carbon quantum dots (CQDs), and nanodiamonds (NM), or one-dimension (1D) such as carbon nanotubes (CNTs) and carbon nanohorns or two-dimensions (2D) including graphene/graphene oxide/reduced graphene oxide (G/GO/rGO), carbon (graphene) nanoribbons, and three-dimensions (3D) nanomaterials in the 1 nm–100 nm range based on their unique dimensionality and nanoscale ranges (Pathakoti et al., 2017). Integrating nanostructured components with other polymers, biomolecules, and nanostructured materials or existing in integrated form may lead to the formation of nanocomposite materials with larger particle sizes (>100 nm). Moreover, nanomaterials exhibit large surface volume fractions with significant solubility, strength, solid magnetism, low scattering, and high optical and thermodynamic characteristics (Singh et al., 2017). The potential fields of food chemistry, food analysis, and food safety are quickly increasing with carbon-based nanomaterials.

The unique physiochemical and antimicrobial capabilities of nanoparticles make them extensively utilized against a broad range of pathogenic microorganisms and in healthcare, crop protection, water treatment, food safety, and food preservation, among other applications (Fu, 2014; Baranwal et al., 2018). The antimicrobial effect of carbon nanomaterials (CNMs) influenced by their content, surface functionalization, targeted microorganisms, and interactive environment. In addition to the physical process of biological separation of microbial cells from their supportive environment, the antibacterial techniques of CNMs based on penetrating the microbial cell wall/membrane and causing structure damage (Ji et al., 2016). The third set of processes includes the interaction of CNMs with microbes and the generation of hazardous materials like reactive oxygen species (ROS). CNM-microbial interactions lead to ROS-independent oxidative stress and mortality (Ji et al., 2016; Chong et al., 2017). Food nanotechnology refers to using nanostructures and nanotechnologies in the food industry and related subjects. It has tremendous potential applications in many areas connected to the food sector and the food chain. According to the research aims, the application of CNMs as antimicrobial agents, either alone or in nanocomposites, has advanced significantly in recent years. In this session, we will critically review antimicrobial research (in vitro/in situ studies) and the potential applications of nanostructures as antimicrobials. CNMs are a unique family of antibiotics/antimicrobial drugs with food applications. In this investigation of nanosensors and electrochemical sensors' potential applications in the food sector, examines the antibacterial characteristics of nanocomposites, a class of polymers and carbon nanostructures.

19.2 Types of carbon-based nanomaterials

19.2.1 Graphene

Graphene is a 2D sheet of sp^2-bonded carbon atoms arranged in a hexagonal lattice configuration. The term "graphene" refers to a layer of carbon atoms with 3–10 layers. Graphene with more than 10 layers is referred to as "multilayer graphene," while graphene with more than 30 layers is referred to as "thick graphene" or "thin graphite nanocrystals." The graphene carbon atoms do not use any of the material's potential energy. The extra electron vacancy might be used in a chemical reaction, perhaps resulting in a nonplanar atomic structure. This free or suspended capacity is available to interact with nearby organic compounds or radical atoms. Several methods have produced graphene, including mechanical exfoliation, GO reduction synthesis, electrochemical exfoliation, chemical vapor deposition (CVD), and vacuum arc discharge (Georgakilas et al., 2015). The most important members of this category are GO, rGO, and graphene quantum dots (GQDs). Hummer's technique, which involves the oxidative exfoliation of graphite with the help of $KMnO_4$ and H_2SO_4 (Bagheri et al., 2019), is most often used to synthesize GO. In addition to the traditional hydrothermal approach (Ghorbani et al., 2015), reducing agents like hydrazine and l-ascorbic acid may be used to convert GO into rGO. GOs or carbon precursors are thermally oxidized GQDs (Shen et al., 2012).

Electrochemical sensing can benefit from graphene's excellent low-noise characteristics due to its exceptional electrical, optical, mechanical, and thermal capabilities. Graphene has many functional groups, including carboxyl (-COOH) and hydroxyl (-OH) groups (Sundramoorthy and Gunasekaran, 2014).

Aside from its large specific surface area of 2630 m^2/g, graphene is also highly conductive (200,000 $cm^2/V/s$), has remarkable mechanical strength, and has attractive optical properties. Compared to CNTs, its effective surface area is about twice as large and cost-efficient. The homogeneous surface functionalization is also due to its exceptionally uniform surface. Graphene has been studied in electroanalytical chemistry because of its remarkable physical and electrochemical characteristics.

19.2.2 Carbon nanotubes

Carbon nanotubes are cylindrical molecules with high mechanical properties and electrochemical, optical, thermal, and chemical capabilities. The strong covalent bond between carbon atoms and their smaller size is responsible for the characteristics of CNTs. In terms of CNTs, SWCNT (single-wall carbon nanotube), DWCNT (double-walled carbon nanotube), and MWCNT (multiwalled carbon nanotube) are the three most common forms. CNTs are often produced from carbonaceous materials that are either solid or gaseous, such as HOPG and hydrocarbons. The arc discharge technique, CVD procedure, and laser deposition method are the three primary synthesis methods for the fabrication of CNTs, respectively (Ying et al., 2011).

19.2.3 Fullerene

The remarkable three-dimensional nanostructure and exceptional electrochemical photocatalytic characteristics of fullerene, the third carbon allotrope discovered in 1985, have lately drawn a lot of interest from physicists and chemists (Kroto, 1987). Fullerene, in particular, has a high electroactive surface area, broad electrical conductivity, and chemical resistance. Allotropes of carbon with numerous redox states and the capacity to receive and transport electrons, allowing for functionalization, signal mediation, and other unique properties, have been identified by researchers (Pan et al., 2020; Pilehvar and De Wael, 2015). Despite its relatively new usage as an electrode modifier material, fullerene is a suitable material; it is regarded as a novel and appealing aspect of the construction of sensors and biosensors.

19.3 Antimicrobial carbon nanoparticles and nanocomposites

Furthermore, CNTs' surface area and size have also been shown to be essential parameters that can influence their antibacterial activity. Nanoparticles' remarkable capacity to interact with bacteria is enhanced by their size reduction as their surface area increases (Kang et al., 2008; Shi et al., 2016). The antibacterial efficacy of nanoparticles is often dependent on the kind of microorganisms that are employed as well as their composition, surface functionalization, and intrinsic properties (Hajipour et al., 2012). Carbon-based nanomaterials may harm bacterial membranes by causing oxidative stress (Shvedova et al., 2012). Researchers have shown that the predominant antibacterial effect of carbon-based nanomaterials is due to their physical contact with bacteria rather than oxidative stress (Manke et al., 2013). Indeed, the interactions between bacteria and carbon-based nanoparticles (CBN) play a crucial role in their antibacterial function. Carbon nanoparticles and bacteria have been shown to aggregate, and studies have shown that this leads to cell death when bacteria and CNTs interact (Kang et al., 2007; Noreen et al., 2019), synthesized nanocomposites GO-Ag-TiO_2 via the hydrothermal technique, and tested their ability to inhibit *Campylobacter jejuni*, a foodborne pathogen. It was shown that the composites developed reduced the growth of *C. jejuni* and prevented the development of biofilms by decreasing motility, hydrophobicity, and aggregation.

19.3.1 Antimicrobial characteristics of carbon nanotubes

Recent studies have shown that SWCNTs have the potential to exhibit antibacterial activity. The surface-to-volume ratio of CNMs increases with decreasing size, resulting in a stronger connection with the microorganisms' cell wall or membrane and as a result, more efficacy in their function. CNTs are linked to microorganisms and damage their cell membrane, metabolic operations, and morphology using CNTs as a catalyst (Khan et al., 2016). According to the study, the bacteriostatic properties of CNTs are produced by direct contact-induced disruption to microbe cell membranes. SEM patterns showed that incubation with CNTs caused morphological alterations in microorganisms linked to a loss of cellular integrity. Additionally, exposure to nanoscale CNTs increased plasmid DNA by fivefold, RNA by twofold, and cytoplasmic material outflow by fivefold during the following exposure. CNTs' high surface-to-volume ratio and significant interior volume are credited with their bacteriostatic qualities, which are becoming widely recognized. By using CNTs as antibiotic transporters, researchers hope to improve antibiotic bioavailability while enabling more precise dosing (Mohammed et al., 2019).

19.3.2 Antimicrobial properties of graphene

Antibacterial effects of the graphene family have been linked to physical damage caused by the interaction of graphene's sharp edges with the membranes of bacteria, as well as to photon-thermal combustion and wrapping, which are also caused by physical damage. Oxidative stress is associated with chemical processes due to charge transfer and ROS. The antibacterial activity of graphene is enhanced in polymers with well-dispersed graphene sheets because of the improved dispersion, which provides bacteria with a larger surface area to interact with and causes mechanical disruption of its membranes by the sharp and intense edge graphene flakes (Carvalho et al., 2020).

Deng et al. (2017) examined the effects of GO and Ag_3PO_4 composite on *Escherichia coli* and *Staphylococcus aureus* growth by 78% and 96%, respectively. Chen et al. (2017) investigated the excellent antibacterial activity using GQD/silver nanoparticles (AgNP) hybrid concentrations (1 mg/mL) with dual enzyme activity equal to peroxidases and oxidases and effective killing bacteria (80%*E. coli* and 70%*S. aureus*). Recently, various graphene and polymer composite materials have improved the antibacterial efficacy of various graphene metal and polymer composite materials. These materials include graphene copper (Deng et al., 2017), zinc oxide (Zanni et al., 2017), iron oxide (Pan et al., 2016), composite, polyvinylalchol (PVA) (Hu et al., 2017; Cao et al., 2015), and polylactide (PLA) film with GO and thermally reduced graphene oxide (TrGO) (Arriagada et al., 2018) compositions. Table 19.1 lists and summarizes the bactericidal properties of graphene-based metal and polymer composition.

19.4 Food packaging

The active packaging system uses a material that endorses the packed product and removes elements contributing to food degradation. Active packaging materials improve food's taste and nutritional content and increase the product's shelf life. Biopolymer packaging (mainly polysaccharides and proteins) was developed to compete with synthetic materials and products. Food packaging is an essential and integral aspect of the food industry, the food supply chain (in its entirety), food safety, and public health. Packaged foods are significant in several parts of the food supply chain. Mechanical, barrier, and antibacterial qualities are critical in food packaging creation and evaluation. The advancement of food packaging technology has focused on and used cost-effective, novel, innovative, and eco-friendly materials. Nanomaterials are interesting for food packaging due to their unique features (such as the mechanical qualities of graphene). Aromatic unsaturated aldehydes like cinnamaldehyde and stacked graphene are used in food packaging for their antibacterial qualities against various bacteria, yeasts, and molds. This investigation used the antibacterial cinnamaldehyde

Table 19.1 The antibacterial performance of graphene/metal and polymer composite.

Type	Material	Pathogen	Killing rate	Cells/mL	References
Graphene metal composition	GO/Ag$_3$PO$_4$[a]	E. coli	78% (25 min)	10^6–10^7	Deng et al. (2017)
		S. aureus	96% (25 min)	10^6–10^7	
	GQD/Ag[b]	E. coli	80% (15 min)	10^7–10^8	Chen et al. (2017)
		S. aureus	70% (15 min)	10^7–10^8	
	Graphene/copper	E. coli	99.9% (15 min)	10^7	Deng et al. (2017)
	Graphene/zinc oxide	S. aureus	80%–90% (4 h)	6×10^5	Zanni et al. (2017)
	rGO/iron oxide	S. aureus	81%	10^7	Pan et al. (2016)
Graphene polymer composition	GO fiber/PVA[c]	E. coli	84% (plate counting)	–	Hu et al. (2017)
		S. aureus	90% (plate counting)		
	Graphene/PVA/ -biocide (10%)	E. coli	97.1% (24 h)	2×10^4	Cao et al. (2015)
		S. aureus	99.7% (24 h)	2×10^4	
	TrGO[d] (10%)/PLA	E. coli	44.9% (24 h)	2×10^6	Arriagada et al. (2018)
		S. aureus	35.5% (24 h)	2×10^6	
	GO film/PLLA[e]	E. coli	90% (12 h)	10^9	Yang et al. (2017)
		S. aureus	92% (12 h)	10^9	
		B. subtilis	90% (12 h)	10^9	
	GO/PSPH[f]	S. aureus	95% (30 min)	5.5×10^6	Pan et al. (2017)
		E. coli	97% (30 min)	10^8	
	GO(5%)/PLA	E. coli	100% (24 h)	8×10^7	Arriagada et al. (2018)
		S. aureus	100% (24 h)	8×10^7	

[a] Ag$_3$PO$_4$: Silver orthophosphate
[b] GQD: Graphene quantum dot
[c] PVA: Polyvinylalchol
[d] TrGO: Thermally reduced graphene oxide
[e] PLLA: Poly(lactic acid)
[f] PSPH: Poly[5,5-dimethyl-3-(30-triethoxysilylpropyl)hydantoin]

schiff base formation to functionalize chitosan (CS). The active bioactive components (cinnamaldehyde) in CS (the carrier) may be released slowly and gradually using schiff base compounds since they are reversible (Higueras et al., 2015).

19.5 Food deterioration and food born pathogen contamination

Food safety continues to be a key concern in public health, particularly given the contemporary environment in which biosafety is given considerable attention, CNTs (like SWCNTs and graphene) have emerged as sensor materials for the detection of microorganisms (Muniandy et al., 2019). Today, spoilage and dangerous bacteria are widely recognized as among the leading causes of food degradation and waste. Various microorganisms, such as *E. coli, Bacillus, Listeria*, and *Salmonella* are involved in the spoilage of food products. Table 19.2 summarizes carbon nanomaterial-based pathogen detection techniques.

19.5.1 Carbon dots

Carbon dots (CDs) are a prospective new category of fluorescent CNTs which have been intensively explored in recent decades because of their great fluorescent features, simple synthesis procedures, high biocompatibility, and a wide range of detecting applications. It is projected that CDs will eventually replace the classic semiquantum dots. In recent years, many novel synthetic methods have been designed that employ small chemicals and renewable technologies as precursors, as well as heating, microwave irradiation, and ultrasonic techniques to accomplish the transformation.

Carbon dots are particularly well suited for diagnostic and therapeutic applications because of their small particle size (less than 10 nm) and remarkable biological characteristics such as low toxicity and high biocompatibility (Devi et al., 2019). CDs can be employed in bioimaging and biosensors, catalysts, and photoelectric conversion because of their exceptional adjustable luminous range, excellent light stability, low toxicity, and general biocompatibility. CDs are currently composed of CQD and GQDs (Li et al., 2017). *Clostridium sporogenous, Pseudomonas aeruginosa, S. aureus, Bacillus subtilis*, and *Mycobacterium smegmatis* are among the microorganisms resistant to CD entrance (Zhang et al., 2016). CDs enter bacteria cells via an unknown mechanism, although studies have speculated that eukaryotic cells take up CDs through endocytosis (Cao et al., 2007). Recently, Zhao et al. (2021) employed CDs-microsphere immunosensors for assaying the *E. coli* O157:H7 in milk samples, with an LOD of 2.4×10^2 CFU/mL.

19.5.2 Carbon nanotubes

Carbon nanotubes have recently acquired interest because of their distinctive features, including chemical, thermal, electrical, and mechanical capabilities. Due to their large surface area to volume ratio and quick physical or chemical interactions made

Table 19.2 List of different nanosensors applications in the food industry.

Pathogen	Type of sensor/material	Linear range	Detection limit	Recovery (%)	Food analysis	References
Salmonella enteritidis	Electrochemical immunosensor[a], Fe_3O_4/graphene	2.4×10^2–2.4×10^7 CFU/mL	2.4×10^2 CFU/mL	90.8–102.6	Milk	Song et al. (2021)
E. coli O157:H7	Lateral flow immune assay[b], GO/rGO	$\sim10^8$–$\sim10^3$ CFU/mL	$\sim10^5$ CFU/mL	55–103.25	Milk/water	Shirshahi et al. (2019)
Vibrio parahaemolyticus	Ecetrochemiluminescence Osensor, GO	10–10^8 CFU/mL	5 CFU/mL	94.4–112.0	Seafood and water	Sha et al. (2016)
Listeria monocytogenes	Electrochemiluminescence sensor[c], NCDs	2.0–1.0×10^6 CFU/mL	1.0×10^{-1} CFU/mL	105–121	Milk, sausage, and ham	Jampasa et al. (2021)
Salmonella typhimurium	Aptasensor, rGO-CS[d]	10^1–10^6 CFU/mL	10^1 CFU/mL	-	Raw chicken	Dinshaw et al. (2017)
Yersinia enterocolitica	Biosensor, SWCNT[e] Linear sweep voltammetry (LSV)	10^6–10^4 CFU/mL	10^4 CFU/mL	-	Kimchi	Sobhan et al. (2019)

[a] *Fe_3O_4: Ferric oxide*
[b] *GO/rGO: Graphene oxide/reduced graphene oxide*
[c] *NCDs: Nitrogen-decorated carbon dots*
[d] *CS: Chitosan*
[e] *SWCNT: Single wall carbon nanotubes*

them detect target molecules in minor quantities (Yang et al., 2015; Benjamin et al., 2018). *Salmonella enterica* can be detected directly using a label-free impedimetric aptasensor constructed by electrodepositing rGO-MWCNT nanocomposite in a step technique (Jia et al., 2016). This aptasensor can view *S. enterica* in the range of 75–7.5–10^5 CFU/mL, and the low limit of detection is 25 CFU/mL and applied in chicken samples. Appaturi et al. (2020) designed a biosensor based on GCE utilizing an rGO-CNT composite. They improved the electrochemical activity and immobilized the aptamer on a broad surface area of CNT by adding an amino-modified DNA aptamer that efficiently identified *Salmonella typhimurium* in raw chicken meat samples at concentrations of 10^1–10^8 CFU/mL and LOD value of 10^1 CFU/mL.

Foodborne pathogens such as *E. coli K12* (Yamada et al., 2014), *Salmonella infantis* (Villamizar et al., 2008), and *S. aureus* (Choi et al., 2017), can be detected by SWCNT-based biosensors in peanut allergy Ara h1 (Sobhan et al., 2018), and Ara h2 (Sobhan et al., 2018). SWCNT-based biosensors can identify target bacteria with high sensitivity without labeling. However, this report (Sobhan et al., 2019) revealed that enrichment could be required for food samples with bacterial counts below 10^4 CFU/mL. The biosensor based on functionalized SWCNTs was effectively developed to detect *Yersinia enterocolitis* in commercial Kimchi food products. The appropriate concentration range for the SWCNT-based biosensors was 10^6–10^4 CFU/mL, and the LOD was 10^4 CFU/mL.

19.5.3 Ordered mesoporous carbon

Ordered mesoporous carbon (OMC) is a novel form of nanostructured carbon material that has been extensively employed in electrochemical sensing applications. MC materials may be divided into two categories based on pore regularity: disordered MC and ordered MC. Mesoporous carbon is replaced with macroporous carbon nanorods in OMC materials, which are highly organized and macroporous. The excellent electrochemical capacitance of OMC nanoparticles makes them ideal for electrochemical capacitors. The OMC-modified electrode demonstrated much enhanced electrochemical performance against RAC oxidation, which resulted in greater detection selectivity and sensitivity owing to the increased electrochemical current. Zhang et al. (2021) incorporated the OMC-modified SPE to detect trans-resveratrol (TRA) in red wine samples. OMC has a large specific area, well-ordered porosity structures, and extremely strong electrical conductivity on the surface of the SPE, which improves electron and mass transport. The authors discovered that OMC/SPE-based electrochemical sensors for TRA molecules had a strong linear range from 5 μM to 50 μM, with a LOD of 0.473 μM, and excellent selectivity and sensitivity using the amperometry technique.

19.5.4 Nanopolymer composite

Nanopolymer composite (NPC) is utilized in packaging soft drinks and food products because of its enhanced thermal, strength, and conductivity properties due to its superior physical, mechanical, and chemical qualities. Polymer nanocomposites with a nanoscale improver are among the most influential groups of composites. Many polymer nanocomposites have excellent mechanical properties, such as strong strength and minimal weight, excellent thermal and electrical stability, and excellent chemical resistance. Polymer nanocomposites can make various products, including coatings resistant to wear and corrosion, gas and vapor-permeable packaging, conductive polymers, strong textile fibers, and industrial tools. In a recent study, researchers examined the antibacterial efficacy of a PLA-based composite film, including GO nanosheets and clove essential oil (CLO), against gram-positive (*S. aureus*) and gram-negative (*E. coli*) (Arfat et al., 2018). The study showed that PLA composite is an excellent active packaging for preserving food against microorganisms. While incorporating CQDs into fullerene films was able to limit the growth of *S. aureus* and *E. coli*. The results demonstrated that films containing nanoparticles have antibacterial properties without releasing active substances into the environment and are ideal for employing films as active food packaging. In addition to improving the thermomechanical characteristics of PLA, Poly(3-hydroxybutyrate-co-3-hydroxy valerate) (PHBV) and diversity of nanocomposites have been used to date, with CNTs (Liu et al., 2019), GO (Li et al., 2019) (Arfat et al., 2018), and rGO (Gouvêa et al., 2018) that are further highlighted in Table 19.3.

For example, various efforts have been made to produce CNTs based on multiple composition CNTs/PLA nanocomposites with better mechanical and electrical characteristics. Recently, it was shown that GO and rGO have potent antibacterial activity. CS is a bioactive macromolecule having antibacterial characteristics. For example, a self-assembled film of GO/CS/TiO$_2$ nanocomposites produced by Xu et al. (2017) displayed substantial antibacterial properties against the development of biofilms by *B. subtilis* and *Aspergillus niger*. Furthermore, this study found that coating can be used as an efficient cling film that successfully delays moisture loss in fruits and vegetables, suppresses PPO activity, and improves antioxidant enzyme activity, particularly SOD activity. The new GO-Fe$_3$O$_4$@NPVP-Ag nanocomposites with numerous antibacterial modes and properties have been developed. *E. coli* and *S. aureus* were used to test the antibacterial characteristics of the developed nanocomposites. *E. coli* exhibited MICs of 31.25 g/mL for GO-Ag and GO-Fe$_3$O$_4$@NPVP-Ag; *S. aureus* showed a MIC of 62.5 g/mL (Li et al., 2018). The anti-bacterial properties of graphene-based polymer nanocomposites and multiple composites are described in Table 19.4.

The authors Liu et al. (2019) used electrospinning to develop polylactic acid/CNTs/CS composite fibers with varying CS concentrations and tested their capacity to preserve strawberries. The fiber films' mechanical characteristics, solubility, and swelling ratio improved with increasing CS concentration, reaching a peak at 7 wt%. The composite fiber films' antibacterial activity against four pathogens improved with increasing CS concentration, peaking at 7 wt%. The composite fibers were also more effective against *S. aureus* than *E. coli*. The nanocomposite fibers with varying CS levels demonstrated excellent preservation benefits for strawberries, with the fiber of 7 wt% CS exhibiting the best results. However, these fibers may be helpful

Table 19.3 Antibacterial properties of graphene-based polymer nanocomposites.

Polymer	Nanocarbon	Pathogens	Method	Load	Growth inhibition	References
PLA/PEG[h]	GO-CLO release	S. aureus E. coli	Plate count	~10^8 CFU/mL	7 log CFU/mL 6 log CFU/mL	Arfat et al. (2018)
PLA/CS[a]	MWCNTs	S. aureus	PDA[b]	$1–2 \times 10^6$ CFU/mL	94.6%	Liu et al. (2019)
PHBV[f]	CNCs[g]-g-GO hybrids	S. aureus	ADA	1×10^6 CFU/mL	99.9%	Li et al. (2019)
Plasticized PHBV	rGO-ZnO hybrids	E. coli	Plastic surface	1.3×10^6 CFU/mL	100%	Gouvêa et al. (2018)
PLA[c]	MWCNTs-AgNPs[d]	S. haemolyticus	ADA[e]	1×10^5 CFU/mL	1.5 mm	Gan et al. (2020)
PEG, AGAR	Ag–ZnO-rGO hybrids	S. aureus P. aeruginosa	Antibiofilm assay	$1–5 \times 10^5$ CFU/mL	97% 80%	Naskar et al. (2018)

[a] PLA: Polylactide CS: chitosan
[b] PDA: Nutrient agar and potato dextrose agar
[c] PLA: Polylactide
[d] AgNPs: Silver nanoparticles
[e] ADA: Agar diffusion assay
[f] PHBV: Poly(3-hydroxybutyrate-co-3-hydroxy valerate)
[g] CNCs: Cellulose nanocrystals
[h] PEG: Polyethylene glycol

Table 19.4 The antibacterial performance of graphene nanomaterial-based multiple composites.

Type	Material	Bacteria	Antibacterial performance	References
Graphene nanomaterial-based multiple composition	GO/CS/TiO$_2$	B. subtilis A. niger	100%, 40 mg/mL, OD600 = 0.79, 12 h 95%, 200 mg/mL, OD600 = 0.7, 12 h	Xu et al. (2017)
	GO/Fe$_3$O$_4$/NPVP[a]/Ag	E. coli S. aureus	MIC[b] = 31.25 MIC = 62.5	Li et al. (2018)
	Graphene/Fe/Ag	E. coli S. aureus B. subtili	90.2%, 100 mg/L, ~10^6 CFU/mL, 2 h 95.1%, 100 mg/L, ~10^6 CFU/mL, 2 h 99.6%, 2 mg/L, ~10^6 CFU/mL, 2 h	Ahmad et al. (2016)
	Graphene/Ag/Au	E. coli	100%, 5 × 10^{-3} M, 6 h	Perdikaki et al. (2016)
	rGO/Au/PdO[c]	E. coli S. aureus	99%, OD600 = 1.2, 12 h 99.5%, OD600 = 1.2, 12 h	Fakhri and Naji (2017)
	Graphene/ZnFe$_2$O$_4$/PANI[d]	E. coli S. aureus C. albicans	MIC = 12.5 mg/L, sunlight MIC = 12.5 mg/L, sunlight MIC = 12.5 mg/L, sunlight	Ma et al. (2014)
	GO/TiO$_2$[e]/BC	S. aureus	91.3%, 5% wt, OD$_{600}$ = 0.5, 2 h, UV light	Liu et al. (2017)
	GO/PLA/ZnO	E. coli S. aureus	97.6%, 24 h, UV–visible light 99.2%, 24 h, UV–visible light	Huang et al. (2015)
	Graphene/Ag/PVA	E. coli S. aureus	99%, 104 –106, 3 h 99%, 104 –106, 24 h	Surudžić et al. (2016)

[a] NPVP: N-alkylated poly (4-vinylpyridine)
[b] MIC: Minimum inhibitory concentration
[c] PbO: Palladium oxide
[d] decorated polyaniline-coupled-graphene aerogels
[e] TiO$_2$: Titanium oxide

in the future for preserving fruits and vegetables. Demitri and colleagues (Demitri et al., 2016) suggested that research showed that cinnamaldehyde with graphite stacks improves antifungal activity and mechanical properties in CS substrates. The fungicidal action of cinnamaldehyde was effectively examined *in vitro* against the fungus *Rhizopus stolonifer*, demonstrating that raising the concentration of cinnamaldehyde enhanced the inhibitory effect. In order to improve mechanical properties, nanometric graphene stacks were incorporated into cinnamaldehyde-functionalized CS films. Finally, antifungal capabilities were evaluated on slices of bread against a mold line.

19.6 Analytical and biosensor instruments for food analysis and safety

In addition to food testing, chemistry, and food security, CBNs have been intensively explored as analytical instruments and biosensors owing to substantial nanoscale phenomenon studies and innovative nanofabrication technologies. Due to extensive research, CBNs have found a wide variety of uses. In terms of nanosensors, these CBNs stand out because of their ability to detect heavy metal ions and gas molecules, as well as antibodies, food additives, and harmful toxins due to considerable research into nanoscale phenomena and new nanofabrication methods. Despite the availability of numerous classical biosensors, we focused on nanozymes-based biosensors for food additives/adulterants detection, as shown in Table 19.5.

19.6.1 Food additives/adulterants

This study developed and evaluated the novel electrochemical sensor for vanilla (VAN) commercialized in actual samples of commercialized vanilla sugar. The authors Taouri et al. (2021) discussed the facile construction and effectiveness of nanostructured and responsive f-MWCNTs-FNTs modified carbon paste electrode (CPE) that was used to compare to a ultra performance liquid chromatography (UPLC)/UV control technique. VAN determination with the newly manufactured electrode has several advantages over conventional approaches, including ease of sensor preparation, low cost, speed, and excellent analytical features such as selectivity (high sensitivity), sensitivity (high), and low detection limits (3.5×10^{-8} mol/L). Furthermore, the most significant benefit is the simplicity with which the actual VAN concentration in real matrices can be determined without prior pretreatment. Recent publications mainly in the literature have described numerous electrodes, and modified electrodes for detecting VAN based on carbon materials, such as graphite (Giray Dilgin, 2019), carbon black (Kutty et al., 2019), CNT-AuNPs (Karabiberoğlu and Koçak, 2018; Chen et al., 2019), Gr (including rGO, GO) containing metal nanoparticles (Gao et al., 2018; Ning et al., 2018), NiO-SWCNTs and ionic liquid (Gupta et al., 2018), and graphene-ionic surfactant (Raril and Manjunatha, 2020).

The accurate and sensitive detection of toxins in food products recently attracted significant interest from the analytical and food science communities. In addition, the synthetic azo-colorant dye known as sudan I (1-Phenylazo-2-naphthol) has been

Table 19.5 Nanomaterial composite-based biosensors for food contamination detection.

Target analyte	Nanomaterial composite	Type of sensor/ technique	Linear range	Limit of detection	Recovery (%)	Food matrix	References
Sudan I	rGO	Electrochemical sensor, LSV	0.04–8.0 µM/L	0.01 µM/L	84–110	Chili sauce/tomato sauce	Zhang et al. (2013)
	Gr/β-CD/PtNPs[a]	Electrochemical sensor, DPV	0.005–66.68 µM	1.6 nM	96–99.3	Chili powder/chili sauce/tomato sauce/ketchup	Palanisamy et al. (2017)
Tartrazine	CTAB[b]-GO/MWCNT	Electrochemical sensor, DPV[c]	0.03–0.6 µM	0.005 µM	93–111	Soft drinks	Yang and Li (2015)
	CQDs	Fluorescence analysis	0.25–32.5 µM	73 nM	87.3–106.6	Steamed buns/honey/candy	Xu et al. (2015)
	Gr/PLPA[d]	Electrochemical sensor, DPV	1.54–5.14 µM	1.54 µmol/L	98.71–104.44	Orange juice	Tahtaisleyen et al. (2020)
Melamine	AuNP/rGO	Electrochemical sensor, DPV	5.0–50 nM	1.0 nM	–	Food contact materials (plate and fruit tray)	Chen et al. (2015)
	CS/GO	Electrochemical sensor, DPV	4–2300 ppb	0.83 ppb	–	Milk	Feng et al. (2019)
Urea	NF/Ag-N-SWCNT[f]	Non-enzymatic electrochemical detection, CV	66 nM–20.6 mM	4.7 nM	95.3	Milk/water	Kumar and Sundramoorthy (2018)
	Graphene-polyaniline	Non-enzymatic electrochemical detection, current-potential (i-t)	10-200 µM	5.88 µM	98.4–104.7	Milk/Tap water	Sha et al. (2017)

(continued on next page)

Analyte	Material	Method	Linear range	LOD	Recovery (%)	Sample	Reference
Urea, uric acid	rGO-gold nanoparticles	Electrochemical sensor, LSV	10–500 μM	10.95 μM	–	Milk and fruit juice and urine	Mazzara et al. (2021)
Sodium ion	AgNPs/GO	Electrochemical sensor, CV	0–100 mM	9.344 mM	94.48–100.80	Fish sauce and seasoning powder of instant noodle.	Traiwatcharanon et al. (2020)
Sunset yellow	GO/MWCNT	Electrochemical sensor, DPV	10^{-7}–2×10^{-5} M	10^{-8} M	90–110	Soft drink	Yang and Li (2015)
Sucrose	GOx[e]-PtNPs-MWCNT	Electrochemical biosensor, Chronoamperometry	1×10^{-4}–1×10^{-9} mol/L	1×10^{-9} mol/L	–	Commercial fruit, vegetable, and mix juice	Bagal-Kestwal and Chiang (2019)
BHA Synthetic antioxidant	POC[h]/MWCNT	Electrochemical sensor, DPV	0.33–110 μM	0.11 μM	97.9–104.6	Potato chips	Manoranjitham and Narayanan (2021)
Bisphenol-A	Chitosan/MWCNTs-Au	Electrochemiluminescent sensor, DPV	0.25–100 μM	0.083 μM	95.6–105	Milk	Guo et al. (2016)
	rGO-Fc[g]-NH$_2$/AuNP	Electrochemical sensor, DPV	5.0×10^{-9}–1.0×10^{-5} M	2.0×10^{-9} M	96–106	Milk	Huang et al. (2015)
	COOH-MWCNNT-Au-Pd	Electrochemical sensor, DPV	0.18–18 μM	60 nM	99.8–100.4	Milk	Mo et al. (2019)
H_2O_2	MWCNTs-FeC[i]	Electrochemical sensor, amperometry	1–1000 μM	0.49 μM	94.33–97.62	Milk, apple juice	Wu et al. (2022)
	RGO/CNTs-Pt	Electrochemical sensor, chronoamperometry	0.0003–0.018 μM and 0.01–4.0 μM	0.31 μM	96.1–109.5	Milk	Zhang et al. (2020)

(continued on next page)

Table 19.5 Nanomaterial composite-based biosensors for food contamination detection—cont'd

Target analyte	Nanomaterial composite	Type of sensor/ technique	Linear range	Limit of detection	Recovery (%)	Food matrix	References
Rhodamine B	Graphene nanoplatelets (GPLs) and AgNPs	Electrochemical sensor, SWV	2–100 µM	1.94 µM	105.40	Crackers	Kartika et al. (2021)
	Exfoliated graphene nanosheets	Electrochemical sensor, DPV	5–120 nM	1.5 nM	–	Soy sauce,	Sun and Yang (2017)
	rGO-Cu$_2$O	Electrochemical sensor, second derivative linear sweep voltammetry	0.01–20 µM	0.006 µM	96.3–103.0	Tomato juice, chili sauce, chili powder, and soy sauce	He et al. (2019)
Aflatoxin	rGO-AuNPs	Electrochemical aptasensor, EIS	0.5–800 ng/mL	0.3 ng/mL	92–108	Raw milk, pasteurized full-fat milk and low-fat milk	Ahmadi et al. (2022)
	SPCE/FGO^j	Apta sensor DPV	0.05–6.0 ng/mL	0.05 ng/mL	84–89	Beer and wine	Goud et al. (2017)
Nitrite	Prussian blue analogue/MWCNT	Amperometric sensor, DPV	10–400, 400–2100 mM	0.5 mM	99.52–101.8	Ham sausage and mustard	Liu H-Y et al. (2019)
	Cetyltrimethylammonium bromide-chitosan functionalized carbon nanotube	Electrochemical nonenzymatic sensor, LSV	30–800 µM	9.6 µM	93.2–102.8	Yuba Wheat flour Corn starch	Wu et al. (2018)

(*continued on next page*)

Methyl parathion	ERGO-CS/Hb[k]	Electrochemical biosensor, SWV	20–260 ng/mL	20.98 ng/mL	94–101	Onion and lettuce leaves	Kaur et al. (2020)
Chlorpyrifos	AChE/AuNCs/GO-CS		1×10^{-2} to 5×10^{2} µg/L	3×10^{-3} µg/L	96.12–101.68	Spinach, lettuces	Yao et al. (2019)
Ara h1	SWCNTs -pAb[l]	Nano biosensor, linear sweep voltammetry	1–10^{5} ng/L	10^{3} ng/L	101–109	Milk, peanut	Sobhan et al. (2018)

[a] Gr/β-CD/PtNPs: Graphene-β-cyclodextrin platinum nanoparticles
[b] CTAB: Cetyltrimethylammonium bromide
[c] DPV: Diffrential pulse voltammetry
[d] PLPA: Poly(L–phenylalanine)
[e] Gox: Glucose oxidase
[f] NF/Ag-N-SWCNT: Nafion/silvernanoparticles-nitrogen-doped single-walled carbon nanotube
[g] Fc: Ferrocene
[h] POC: Poly octanediol-co-citrate
[i] MWCNT-FeC: Ferrocene-functionalized multi-walled carbon nanotubes
[j] SPCE/FGO: Screen printed electrode, functionalized graphene oxide
[k] ERGO-CS/Hb: Electrochemically reduced graphene oxide-chitosan-hemoglobin
[l] pAb: Ara h1 antibody

found in high levels in polluted chili powder, curry, and sauces. Recently, graphene/-CD/platinum nanoparticles (PtNPs) nanocomposites are used for the purpose of electrochemically detecting sudan I. The analytical sensitivity of the electrode modified by the produced nanocomposite was enhanced (2.82 μA/μM) and LOD (1.6 nM). Sudan I is electrochemically detected using a graphene/-CD/PtNPs nanocomposite. The fabricated sensor determined that the food samples, including red chili powder, chili and tomato sauce, found good recovery values of 96.0%–99.3% (Palanisamy et al., 2017).

19.6.2 Sweeteners

Carbon nanotubes have great potential as chemical sensors because of their large aspect ratio, wide surface area, remarkable chemical and thermal durability, and ability to be controlled electrically. Lee and his colleagues (Lee and Kim, 2020) examined the Au-CNT-FET sensor that used a specifically produced CNT in a nano template to detect sucrose selectively with sensitivity to measure (0.64 m/M) up to 2.5 m/M, but extremely sensitive to glucose (0.28 m/M) and fructose (0.05 m/M) at lower concentrations. According to this theory, an enhanced potential barrier between conductive CNTs due to bulky sucrose adsorption on AuNP surfaces would be the predicted mechanism for selective sensing. The built Au-CNT-FET biosensor will easily measure the sucrose concentration in a food's concentrated sugar combination.

19.6.3 Food colorants

Artificially flavored drinks and meals, which are becoming more popular due to the world's fast growth, attract more attention due to their vibrant look and sweet taste. Synthetic food colorants are commonly used as additives in many items, including food. Tartrazine (TZ, E102) is a monoazo pyrazolone dye and synthetic color used in sweeteners, beverages, breads, dairy, drinks, and fast-food products. The electrochemical identification of TZ was used in a novel graphene-Gr/poly(L-phenylalanine)-PLPA-modified pencil graphite electrode (PGE). The developed novel sensor platform's detection and quantitation limits were 1.54 μM and 5.14 μM, respectively. The sensor platform's recovery values ranged from 98.71% to 104.44% on actual samples of orange juice. For instance, a fluorescent CQD was fabricated by Xu et al. (2015) for the fluorescence detection of TZ in food samples. In this method, fluorescence quenching may enable C-dots a great sensor for sensitive and precise measurement. The linearity of TZ concentrations ranging from 0.25 μM to 32.50 μM could be determined due to the reduction in fluorescence intensity. The proposed method detected trace levels of TZ in food, including honey, steamed buns, and candy.

19.6.4 Dyes

Commercially available azoic dye rhodamine B (RhB) has found widespread use as a fluorescent labeling and food coloring agent due to its solubility in water.

Screen-printed carbon electrodes-graphene nanoplatelets (SPCE-GPLs)/AgNPs were used to build an electrochemical RhB sensing platform in recent work. The high conductivity of GPLs and AgNPs improves RhB's current responsiveness. RhB has a linear response range of 2–100 µM, with an LOD of 1.94 µM. It is also possible to detect RhB in crackers food samples (Kartika et al., 2021).

19.6.5 Antioxidants

Antioxidants are groups of active substances that are often found in food. Antioxidants in drinks, foods, and bodily fluids must be identified, evaluated, and studied quantitatively using standard methodologies (Benjamin et al., 2015; Garcia et al., 2016). The team led by José Palacios has developed a simple electrochemical sensor employing a sonogel glassy carbon electrode to detect ascorbic acid in commercial baby apple juice samples (Abdelrahim et al., 2013). The electrochemical sensor demonstrated a sensitive detection value between 1.59 µM and 2.93 µM. As a synthetic antioxidant, butylated hydroxyanisole (BHA) is utilized as a preservative in many different types of products, including packaged foods, cosmetics, biofuels, and pharmaceuticals. The authors designed a novel, sensitive, and easy-to-use electrochemical sensor for detecting BHA (Manoranjitham and Narayanan, 2021). The electropolymerized poly octanediol-co-citrate (POC) film containing MWCNTs greatly increases the sensitivity of BHA because of the film's wide surface area and fast electron transport rate. According to the results, the linear range for the POC/MWCNTs electrode was between 0.33 µM and 110 µM, with a LOD of 0.11 µM. The electrochemical sensing of BHA in samples of potato chips using a modified electrode worked well in practice.

19.6.6 Hydrogen peroxide

Over the past several decades, hydrogen peroxide (H_2O_2) residue detection has been imperative to the food industry and environmental studies. Hydrogen peroxide is widely used as a sterilizing agent in aseptically packed foods and as a preservative in milk owing to its intrinsic microbiological and sporicidal characteristics. Recently, a simple, sensitive, and nonenzymatic H_2O_2 sensors are based on MWCNT-FeC nanocomposites was reported by Wu et al. (2022). This sensor is capable of measuring the exact concentration of H_2O_2 in aseptically packaged foods, with a detection value as low as 0.49 µM and a linear range from 1 µM to 1 mM. Additionally, CNT, rGO, and PtNP nanocomposites were used to construct an enzyme-free H_2O_2 electrochemical sensor using modified glassy carbon electrodes (GCE). In the optimal settings, RGO/CNTs-Pt/GCE achieves excellent sensitivity (347 5A m/M/cm^2) and linear responses of 0.0003–0.018 mM, 0.01–4.00 mM were observed with a LOD of 0.31 mM in milk samples (Zhang et al., 2020). According to recent research, electrostatic interactions with copper oxide (CuO)-SWCNT-PDDA(poly-(di allyl dimethyl ammonium chloride) nanocomposites were used to fabricate an electrochemical sensor with GCE to detect H_2O_2 in milk and chicken paw samples (Liu et al., 2021). The linear response range of the biosensor was 1–1150 µM with a LOD of 0.39 µM.

Furthermore, the biosensor above was shown to have 96%–104% greater analytical recoveries.

19.6.7 Aflatoxin B1

Methylene blue (MB) was used as the signaling molecule in the development of the electrochemical aptasensor detecting aflatoxin B1 (AFB1), and graphene was used as the platform to enable signal amplification. The response of the sensor was found to be linear between 0.05 ng/mL and 6.0 ng/mL, with the LOD of 0.05 ng/mL. For the samples of alcoholic beverages (beer and wine), good recoveries ranging from 84.0% to 89.0% were obtained (Goud et al., 2017). rGO and AuNPs-based PGE were used to construct an aptasensor to detect AFM1. The label-free impedimetric assay of AFM1 in milk samples has been developed. The proposed sensor has great promise in AFM1 by using the electrochemical impedance spectroscopy (EIS) technique, with a linearity of 0.5–800 ng/L and LOD of 0.3 ng/L, adequate selectivity, and high stability (Ahmadi et al., 2022).

19.6.8 Bisphenol-A

The epoxy resins and polycarbonate plastics used in many common food and drink containers, infant bottles, and dental fillings are made using bisphenol A (BPA, 2, 2-bis (4-hydroxyphenyl) propane), a crucial chemical processing reactant. BPA has been measured using a range of methodologies, including MWCNTs-Au nanocomposites (Guo et al., 2016) and rGO-AuNPs (Huang et al., 2015). The recently suggested electrochemical biosensor is a simple, economical, and powerful option with excellent performance for detecting BPA. It may be utilized in the analytical food industry to identify novel electroactive chemicals and real samples. Bimetallic alloys of Au and Pd were used to include carboxylic MWCNTs to fabricate a novel sensor for BPA. The sensor revealed a linear range from 0.18 μM to 18 μM, with an LOD of 0.06 μM. The sensitive electrochemical sensor response showed a linear dynamic relationship from 0.18 μM to 18 μM, and 60 nM for the detection, respectively. The sensor has excellent stability and anti-interference properties. Moreover, the proposed sensor detects BPA in water and milk samples with excellent sensitivity (Mo et al., 2019).

19.6.9 Pesticides

Pesticide residue detection has become a key concern for food safety experts, and a quick technique for detecting pesticide residues is required to minimize human exposure to pesticides. Organophosphorus pesticides (OPs) are the most often used because of their low persistence. Kaur et al. (2020) constructed a new bioelectrode using biocompatible ERGO-CS/Hb/FTO as the electrode material. Hb-redox active protein hemoglobin was successfully immobilized, which led to a fast charge transfer rate that didn't change the structure of the cell itself. The designed bioelectrode showed an excellent linear range (20–260 ng/mL) and LOD value (20.98 ng/mL) in vegetable samples to detect methyl parathion.

19.6.10 **Food allergy**

Food allergies are one of the most dangerous health problems people face today. Peanuts are believed to be the most prevalent cause of allergy sickness compared to other familiar sources such as milk, tree nuts, eggs, fish, grain, seafood, and soy. The proposed biosensor (Sobhan et al., 2018) employs a linker consisting of 1-pyrenebutanoic acid succinimidyl ester (1-PBSE) and SWCNTs to facilitate signal transfer from the target material to the SWCNTs. Additionally, SWCNTs were capable of transforming the electrical responses produced by the biological interaction between antibodies and antigens. The newly designed SWCNT-based biosensor's sensor capability includes the LOD, 1 ng/mL, the linear range (1–1000 ng/mL), and the detection sensitivity was determined. In addition, a nanobiosensor based on SWCNTs conjugated with polyclonal anti-Ara h1 antibody was used to detect the peanut allergen Ara h1 in processed meals (Sobhan et al., 2018).

Additionally, Gómez-Arribas et al. (2018) determined that biosensors based on nanoparticles and CNTs for detecting food allergens are effective for food allergen analysis. Recent research has shown that the label-free electrochemical immunosensor proposed by Lim and Ahmed (2016) can detect porcine serum albumin (PSA) in raw meat. In this research, carbon nanofiber (CN)-modified SPCE was electro-grafted with a 4-carboxyphenyl layer for covalent antibody attachment and PSA detection. The proposed immunosensor displayed a linear range of 0.5–500 pg/mL for PSA measurement in a buffer solution with LOD of 0.5 pg/mL.

19.7 **Conclusion**

Antimicrobial composites utilizing CBN have the potential to revolutionize healthcare. CNTs (CNT and graphene derivatives) embedded into or on the surface of a polymeric matrix were the focus of this study. Despite their popularity, these forms of CNTs continue to be among the most widely studied. In most studies, the antimicrobial properties of CNTs are attributed to direct particle-microbe interactions. For this reason, materials' surfaces, which serve as the contact between CNMs and microbes, must be the focus of attention. Nanocomposites including nanoparticles at their surfaces tend to be more effective for antibacterial qualities, allowing for maximum efficiency via targeted localization of the nanoparticles. Physical damage and oxidative stress can still produce antimicrobial effects despite being obstructed. Once the carbon nanoparticles have been securely encased in a polymeric substance, the next difficulty is ensuring that the nanoparticles are not released into the environment and hence, do not threaten human health.

Acknowledgment

The author would like to acknowledge the support provided by CAPES—Coordenação de Aperfeiçoamento de Pessoal de Nível Superior, Brazil.

References

Abdelrahim, M.Y.M., Benjamin, S.R., Cubillana-Aguilera, L.M., Naranjo-Rodríguez, I., de Cisneros, J., Delgado, J.J., et al., 2013. Study of the electrocatalytic activity of cerium oxide and gold-studded cerium oxide nanoparticles using a sonogel-carbon material as supporting electrode: electroanalytical study in apple juice for babies. Sensors. 13 (4), 4979–5007. http://www.mdpi.com/1424-8220/13/4/4979.

Ahmad, A., Qureshi, A.S., Li, L., Bao, J., Jia, X., Xu, Y., et al., 2016. Antibacterial activity of graphene supported FeAg bimetallic nanocomposites. Colloids Surf. B Biointerfaces 143, 490–498. https://linkinghub.elsevier.com/retrieve/pii/S092777651630220X.

Ahmadi, S.F., Hojjatoleslamy, M., Kiani, H., Molavi, H., 2022. Monitoring of aflatoxin M1 in milk using a novel electrochemical aptasensor based on reduced graphene oxide and gold nanoparticles. Food Chem. 373, 131321. https://linkinghub.elsevier.com/retrieve/pii/S030881462102327X.

Appaturi, J.N., Pulingam, T., Thong, K.L., Muniandy, S., Ahmad, N., Leo, B.F., 2020. Rapid and sensitive detection of *Salmonella* with reduced graphene oxide-carbon nanotube based electrochemical aptasensor. Anal. Biochem. 589, 113489. https://linkinghub.elsevier.com/retrieve/pii/S0003269719306578.

Arfat, Y.A., Ahmed, J., Ejaz, M., Mullah, M., 2018. Polylactide/graphene oxide nanosheets/clove essential oil composite films for potential food packaging applications. Int. J. Biol. Macromol. 107, 194–203. https://linkinghub.elsevier.com/retrieve/pii/S0141813017328325.

Arriagada, P., Palza, H., Palma, P., Flores, M., Caviedes, P., 2018. Poly(lactic acid) composites based on graphene oxide particles with antibacterial behavior enhanced by electrical stimulus and biocompatibility. J. Biomed. Mater. Res. Part A 106 (4), 1051–1060.

Bagal-Kestwal, D.R., Chiang, B.H., 2019. Platinum nanoparticle-carbon nanotubes dispersed in gum Arabic-corn flour composite-enzymes for an electrochemical sucrose sensing in commercial juice. Ionics (Kiel) 25, 5551–5564.

Bagheri, M., Jafari, S.M., Eikani, M.H., 2019. Development of ternary nanoadsorbent composites of graphene oxide, activated carbon, and zero-valent iron nanoparticles for food applications. Food Sci. Nutr. 7 (9), 2827–2835. https://onlinelibrary.wiley.com/doi/10.1002/fsn3.1080.

Baranwal, A., Srivastava, A., Kumar, P., Bajpai, V.K., Maurya, P.K., Chandra, P., 2018. Prospects of nanostructure materials and their composites as antimicrobial agents. Front. Microbiol. 9, 422.

Benjamin, S., de Oliveira Neto, J.R., de Macedo, I., Bara, M., da Cunha, L., de Faria Carvalho, L., et al., 2015. Electroanalysis for quality control of acerola (*Malpighia emarginata*) fruits and their commercial products. Food Anal. Methods. 8 (1), 86–92. http://link.springer.com/10.1007/s12161-014-9872-0.

Benjamin, S., Vilela, R., de Camargo, H., Guedes, M., Fernandes, K., Colmati, F., 2018. Enzymatic electrochemical biosensor based on multiwall carbon nanotubes and cerium dioxide nanoparticles for rutin detection. Int. J. Electrochem. Sci. 13 (1), 563–586.

Cao, L., Wang, X., Meziani, M.J., Lu, F., Wang, H., Luo, P.G., et al., 2007. Carbon dots for multiphoton bioimaging. J. Am. Chem. Soc. 129 (37), 11318–11319. https://pubs.acs.org/doi/10.1021/ja073527l.

Cao, Y.-C., Wei, W., Liu, J., You, Q., Liu, F., Lan, Q., et al., 2015. The preparation of graphene reinforced poly(vinyl alcohol) antibacterial nanocomposite thin film. Int. J. Polym. Sci. 2015, 1–7. http://www.hindawi.com/journals/ijps/2015/407043/.

Chen, L., Chaisiwamongkhol, K., Chen, Y., Compton, R.G., 2019. Rapid electrochemical detection of vanillin in natural vanilla. Electroanalysis. 31 (6), 1067–1074. https://onlinelibrary.wiley.com/doi/10.1002/elan.201900037.

Chen, N., Cheng, Y., Li, C., Zhang, C., Zhao, K., Xian, Y., 2015. Determination of melamine in food contact materials using an electrode modified with gold nanoparticles and reduced graphene oxide. Microchim. Acta 182 (11–12), 1967–1975. http://link.springer.com/10.1007/s00604-015-1533-5.

Chen, S., Quan, Y., Yu, Y.-L., Wang, J.-H., 2017. Graphene quantum dot/silver nanoparticle hybrids with oxidase activities for antibacterial application. ACS Biomater. Sci. Eng. 3 (3), 313–321. https://pubs.acs.org/doi/10.1021/acsbiomaterials.6b00644.

Choi, H.-K., Lee, J., Park, M.-K., Oh, J.-H., 2017. Development of single-walled carbon nanotube-based biosensor for the detection of *Staphylococcus aureus*. J. Food Qual. 2017, 1–8. https://www.hindawi.com/journals/jfq/2017/5239487/.

Chong, Y., Ge, C., Fang, G., Wu, R., Zhang, H., Chai, Z., et al., 2017. Light-enhanced antibacterial activity of graphene oxide, mainly via accelerated electron transfer. Environ. Sci. Technol. 51 (17), 10154–10161.

de Carvalho, A.P.A., Junior, C.A.C., 2020. Green strategies for active food packagings: a systematic review on active properties of graphene-based nanomaterials and biodegradable polymers. Trends Food Sci. Technol. 103, 130–143. https://linkinghub.elsevier.com/retrieve/pii/S0924224420305410.

Demitri, C., De Benedictis, V.M., Madaghiele, M., Corcione, C.E., Maffezzoli, A., 2016. Nanostructured active chitosan-based films for food packaging applications: effect of graphene stacks on mechanical properties. Measurement 90, 418–423. https://linkinghub.elsevier.com/retrieve/pii/S0263224116301531.

Deng, C.-H., Gong, J.-L., Ma, L.-L., Zeng, G.-M., Song, B., Zhang, P., et al., 2017. Synthesis, characterization and antibacterial performance of visible light-responsive Ag_3PO_4 particles deposited on graphene nanosheets. Process Saf. Environ. Prot. 106, 246–255. https://linkinghub.elsevier.com/retrieve/pii/S0957582017300095.

Deng, C.H., Gong, J.L., Zeng, G.M., Zhang, P., Song, B., Zhang, X.G., et al., 2017. Graphene sponge decorated with copper nanoparticles as a novel bactericidal filter for inactivation of *Escherichia coli*. Chemosphere 184, 347–357.

Devi, P., Saini, S., Kim, K.-H., 2019. The advanced role of carbon quantum dots in nanomedical applications. Biosens. Bioelectron. 141, 111158. https://linkinghub.elsevier.com/retrieve/pii/S0956566319301794.

Dinshaw, I.J., Muniandy, S., Teh, S.J., Ibrahim, F., Leo, B.F., Thong, K.L., 2017. Development of an aptasensor using reduced graphene oxide chitosan complex to detect *Salmonella*. J. Electroanal. Chem. 806, 88–96. https://linkinghub.elsevier.com/retrieve/pii/S1572665717307622.

Fakhri, A., Naji, M., 2017. Degradation photocatalysis of tetrodotoxin as a poison by gold doped PdO nanoparticles supported on reduced graphene oxide nanocomposites and evaluation of its antibacterial activity. J. Photochem. Photobiol. B 167, 58–63.

Feng, N., Zhang, J., Li, W., 2019. Chitosan/graphene oxide nanocomposite-based electrochemical sensor for ppb level detection of melamine. J. Electrochem. Soc. 166 (14), B1364–B1369.

Fu, P.P., 2014. Introduction to the special issue: nanomaterials - toxicology and medical applications. J. Food Drug Anal. 22 (1), 1–2.

Gan, L., Geng, A., Jin, L., Zhong, Q., Wang, L., Xu, L., et al., 2020. Antibacterial nanocomposite based on carbon nanotubes–silver nanoparticles-co-doped polylactic acid. Polym. Bull. 77 (2), 793–804. http://link.springer.com/10.1007/s00289-019-02776-1.

Gao, J., Yuan, Q., Ye, C., Guo, P., Du, S., Lai, G., et al., 2018. Label-free electrochemical detection of vanillin through low-defect graphene electrodes modified with au nanoparticles. Materials (Basel) 11 (4), 489.

Garcia, L., Benjamin, S., Antunes, R., Lopes, F., Somerset, V., Gil E de, S., 2016. *Solanum melongena* polyphenol oxidase biosensor for the electrochemical analysis of paracetamol. Prep. Biochem. Biotechnol. 46 (8), 850–855. www.tandfonline.com/doi/full/10.1080/10826068.2016.1155060.

Georgakilas, V., Perman, J.A., Tucek, J., Zboril, R., 2015. Broad family of carbon nanoallotropes: classification, chemistry, and applications of fullerenes, carbon dots, nanotubes, graphene, nanodiamonds, and combined superstructures. Chem. Rev. 115 (11), 4744–4822. https://pubs.acs.org/doi/10.1021/cr500304f.

Ghorbani, M., Abdizadeh, H., Golobostanfard, M.R., 2015. Reduction of graphene oxide via modified hydrothermal method. Procedia Mater. Sci. 11, 326–330.

Giray Dilgin, D., 2019. Koşullandırılmış kalem grafit elektrot kullanilarak vanilinin voltammetrik tayini. Akad Gıda 17, 1–8. https://dergipark.org.tr/en/doi/10.24323/akademik-gida.543981.

Gómez-Arribas, L., Benito-Peña, E., Hurtado-Sánchez, M., Moreno-Bondi, M., 2018. Biosensing based on nanoparticles for food allergens detection. Sensors. 18 (4), 1087. www.mdpi.com/1424-8220/18/4/1087.

Goud, K.Y., Hayat, A., Catanante, G., M., S., Gobi, K.V., Marty, J.L., 2017. An electrochemical aptasensor based on functionalized graphene oxide assisted electrocatalytic signal amplification of methylene blue for aflatoxin B1 detection. Electrochim. Acta 244, 96–103. https://linkinghub.elsevier.com/retrieve/pii/S001346861731085X.

Gouvêa, R.F., Del Aguila, E.M., Paschoalin, V.M.F., Andrade, C.T., 2018. Extruded hybrids based on poly(3-hydroxybutyrate-co-3-hydroxyvalerate) and reduced graphene oxide composite for active food packaging. Food Packag. Shelf Life 16, 77–85. https://linkinghub.elsevier.com/retrieve/pii/S2214289417303010.

Guo, W., Zhang, A., Zhang, X., Huang, C., Yang, D., Jia, N., 2016. Multiwalled carbon nanotubes/gold nanocomposites-based electrochemiluminescent sensor for sensitive determination of bisphenol A. Anal. Bioanal. Chem. 408 (25), 7173–7180. http://link.springer.com/10.1007/s00216-016-9746-y.

Gupta, V.K., Karimi-Maleh, H., Agarwal, S., Karimi, F., Bijad, M., Farsi, M., et al., 2018. Fabrication of a food nano-platform sensor for determination of vanillin in food samples. Sensors (Switzerland) 18 (9), 2817.

Hajipour, M.J., Fromm, K.M., A, A.A., D, J.A., de, L.I.R., Rojo, T., et al., 2012. Antibacterial properties of nanoparticles. Trends Biotechnol. 30 (10), 499–511. https://linkinghub.elsevier.com/retrieve/pii/S0167779912000959.

He, Q., Liu, J., Tian, Y., Wu, Y., Magesa, F., Deng, P., et al., 2019. Facile preparation of Cu_2O nanoparticles and reduced graphene oxide nanocomposite for electrochemical sensing of rhodamine B. Nanomaterials 9 (7), 958. www.mdpi.com/2079-4991/9/7/958.

Higueras, L., López-Carballo, G., Gavara, R., Hernández-Muñoz, P., 2015. Reversible covalent immobilization of cinnamaldehyde on chitosan films via schiff base formation and their application in active food packaging. Food Bioproc. Technol. 8 (3), 526–538. http://link.springer.com/10.1007/s11947-014-1421-8.

Hu, X., Ren, N., Chao, Y., Lan, H., Yan, X., Sha, Y., et al., 2017. Highly aligned graphene oxide/poly(vinyl alcohol) nanocomposite fibers with high-strength, antiultraviolet and antibacterial properties. Compos. Part A Appl. Sci. Manuf. 102, 297–304.

Huang, N., Liu, M., Li, H., Zhang, Y., Yao, S., 2015. Synergetic signal amplification based on electrochemical reduced graphene oxide-ferrocene derivative hybrid and gold nanoparticles as an ultra-sensitive detection platform for bisphenol A. Anal. Chim. Acta 853, 249–257. https://linkinghub.elsevier.com/retrieve/pii/S0003267014012495.

Huang, Y., Wang, T., Zhao, X., Wang, X., Zhou, L., Yang, Y., et al., 2015. Poly(lactic acid)/graphene oxide-ZnO nanocomposite films with good mechanical, dynamic mechanical, anti-UV and antibacterial properties. J. Chem. Technol. Biotechnol. 90 (9), 1677–1684. https://onlinelibrary.wiley.com/doi/10.1002/jctb.4476.

Jampasa, S., Ngamrojanavanich, N., Rengpipat, S., Chailapakul, O., Kalcher, K., Chaiyo, S., 2021. Ultrasensitive electrochemiluminescence sensor based on nitrogen-decorated carbon dots for *Listeria* monocytogenes determination using a screen-printed carbon electrode. Biosens. Bioelectron. 188, 113323. https://linkinghub.elsevier.com/retrieve/pii/S0956566321003602.

Ji, H., Sun, H., Qu, X., 2016. Antibacterial applications of graphene-based nanomaterials: recent achievements and challenges. Adv. Drug. Deliv. Rev. 105, 176–189.

Jia, F., Duan, N., Wu, S., Dai, R., Wang, Z., Li, X., 2016. Impedimetric *Salmonella aptasensor* using a glassy carbon electrode modified with an electrodeposited composite consisting of reduced graphene oxide and carbon nanotubes. Microchim. Acta 183 (1), 337–344. http://link.springer.com/10.1007/s00604-015-1649-7.

Kang, S., Herzberg, M., Rodrigues, D.F., Elimelech, M., 2008. Antibacterial effects of carbon nanotubes: size does matter!. Langmuir 24 (13), 6409–6413. https://pubs.acs.org/doi/10.1021/la800951v.

Kang, S., Pinault, M., Pfefferle, L.D., Elimelech, M., 2007. Single-walled carbon nanotubes exhibit strong antimicrobial activity. Langmuir 23 (17), 8670–8673.

Karabiberoğlu, Ş., Koçak, ÇC., 2018. Electrochemical vanillin determination on gold nanoparticles modified multiwalled carbon nanotube electrode. Deu Muhendis Fak Fen ve Muhendis 20 (59), 461–470.

Kartika, A.E., Setiyanto, H., Manurung, R.V., Jenie, S.N.A., Saraswaty, V., 2021. Silver nanoparticles coupled with graphene nanoplatelets modified screen-printed carbon electrodes for rhodamine B detection in food products. ACS Omega 6 (47), 31477–31484. https://pubs.acs.org/doi/10.1021/acsomega.1c03414.

Kaur, R., Rana, S., Lalit, K., Singh, P., Kaur, K., 2020. Electrochemical detection of methyl parathion via a novel biosensor tailored on highly biocompatible electrochemically reduced graphene oxide-chitosan-hemoglobin coatings. Biosens. Bioelectron. 167, 112486. https://linkinghub.elsevier.com/retrieve/pii/S0956566320304796.

Khan, A.A.P., Khan, A., Rahman, M.M., Asiri, A.M., Oves, M., 2016. Lead sensors development and antimicrobial activities based on graphene oxide/carbon nanotube/poly(O-toluidine) nanocomposite. Int. J. Biol. Macromol. 89, 198–205.

Kroto, H.W., 1987. The stability of the fullerenes Cn, with n = 24, 28, 32, 36, 50, 60 and 70. Nature 329 (6139), 529–531. http://www.nature.com/articles/329529a0.

Kumar, T.H.V., Sundramoorthy, A.K., 2018. Non-enzymatic electrochemical detection of urea on silver nanoparticles anchored nitrogen-doped single-walled carbon nanotube modified electrode. J. Electrochem. Soc. 165 (8), 21–28.

Kutty, M., Settu, R., Chen, S.M., Chen, T.W., Tseng, T.W., Hatamleh, A.A., et al., 2019. An electrochemical detection of vanillin based on carbon black nanoparticles modified screen printed carbon electrode. Int. J. Electrochem. Sci. 14, 5972–5983.

Lee, M., Kim, D., 2020. Exotic carbon nanotube based field effect transistor for the selective detection of sucrose. Mater. Lett. 268, 127571.

Li, F., Yu, H.-Y., Wang, Y.-Y., Zhou, Y., Zhang, H., Yao, J.-M., et al., 2019. Natural biodegradable poly(3-hydroxybutyrate- co -3-hydroxyvalerate) nanocomposites with multifunctional cellulose nanocrystals/graphene oxide hybrids for high-performance food packaging. J. Agric. Food Chem. 67 (39), 10954–10967. https://pubs.acs.org/doi/10.1021/acs.jafc.9b03110.

Li, K., Liu, W., Ni, Y., Li, D., Lin, D., Su, Z., et al., 2017. Technical synthesis and biomedical applications of graphene quantum dots. J. Mater. Chem. B 5, 4811–4826.

Li, Q., Yong, C., Cao, W., Wang, X., Wang, L., Zhou, J., et al., 2018. Fabrication of charge reversible graphene oxide-based nanocomposite with multiple antibacterial modes and magnetic recyclability. J. Colloid Interface Sci. 511, 285–295. https://linkinghub.elsevier.com/retrieve/pii/S002197971731158X.

Lim, S.A., Ahmed, M.U., 2016. A label free electrochemical immunosensor for sensitive detection of porcine serum albumin as a marker for pork adulteration in raw meat. Food Chem. 206, 197–203. https://linkinghub.elsevier.com/retrieve/pii/S030881461630437X.

Liu, H.-Y., Wen, J.-J., Huang, Z.-H., Ma, H., Xu, H.-X., Qiu, Y.-B., et al., 2019. Prussian blue analogue of copper-cobalt decorated with multi-walled carbon nanotubes based electrochemical sensor for sensitive determination of nitrite in food samples. Chin. J. Anal. Chem. 47 (6), e19066–e19072. https://linkinghub.elsevier.com/retrieve/pii/S1872204019611680.

Liu, L., Wang, L., Liang, Q., Guo, T., Guo, F., 2021. Hydrogen peroxide residue determination in food samples by a glassy carbon electrode modified with CuO-SWCNT-PDDA nanocomposites. Microchem. J. 167, 106327. https://linkinghub.elsevier.com/retrieve/pii/S0026265×21004112.

Liu, L.P., Yang, X.N., Ye, L., Xue, D.D., Liu, M., Jia, S.R., et al., 2017. Preparation and characterization of a photocatalytic antibacterial material: graphene oxide/TiO$_2$/bacterial cellulose nanocomposite. Carbohydr. Polym. 174, 1078–1086.

Liu, Y., Wang, S., Lan, W., Qin, W., 2019. Fabrication of polylactic acid/carbon nanotubes/chitosan composite fibers by electrospinning for strawberry preservation. Int. J. Biol. Macromol. 121, 1329–1336. https://linkinghub.elsevier.com/retrieve/pii/S0141813018331027.

Ma, G., Chen, Y., Li, L., Jiang, D., Qiao, R., Zhu, Y., 2014. An attractive photocatalytic inorganic antibacterial agent: preparation and property of graphene/zinc ferrite/polyaniline composites. Mater. Lett. 131, 38–41.

Manke, A., Wang, L., Rojanasakul, Y., 2013. Mechanisms of nanoparticle-induced oxidative stress and toxicity. Biomed. Res. Int. 2013, 94291.

Manoranjitham, J.J., Narayanan, S.S., 2021. Electrochemical sensor for determination of butylated hydroxyanisole (BHA) in food products using poly O-cresolphthalein complexone coated multiwalled carbon nanotubes electrode. Food Chem. 342, 128246. https://linkinghub.elsevier.com/retrieve/pii/S0308814620321087.

Mazzara, F., Patella, B., Aiello, G., O'Riordan, A., Torino, C., Vilasi, A., et al., 2021. Electrochemical detection of uric acid and ascorbic acid using r-GO/NPs based sensors. Electrochim. Acta 388, 138652. https://linkinghub.elsevier.com/retrieve/pii/S0013468621009427.

Mo, F., Xie, J., Wu, T., Liu, M., Zhang, Y., Yao, S., 2019. A sensitive electrochemical sensor for bisphenol A on the basis of the AuPd incorporated carboxylic multi-walled carbon nanotubes. Food Chem. 292, 253–259. https://linkinghub.elsevier.com/retrieve/pii/S0308814619306831.

Mohammed, M.K.A., Ahmed, D.S., Mohammad, M.R., 2019. Studying antimicrobial activity of carbon nanotubes decorated with metal-doped ZnO hybrid materials. Mater. Res. Express 6 (5), 12–18.

Muniandy, S., Teh, S.J., Thong, K.L., Thiha, A., Dinshaw, I.J., Lai, C.W., et al., 2019. Carbon nanomaterial-based electrochemical biosensors for foodborne bacterial detection. Crit. Rev. Anal. Chem. 49 (6), 510–533.

Naskar, A., Khan, H., Sarkar, R., Kumar, S., Halder, D., Jana, S., 2018. Anti-biofilm activity and food packaging application of room temperature solution process based polyethylene glycol capped Ag-ZnO-graphene nanocomposite. Mater. Sci. Eng. C 91, 743–753. https://linkinghub.elsevier.com/retrieve/pii/S0928493117334033.

Ning, J., He, Q., Luo, X., Wang, M., Liu, D., Wang, J., et al., 2018. Rapid and sensitive determination of vanillin based on a glassy carbon electrode modified with Cu2O-electrochemically reduced graphene oxide nanocomposite film. Sensors (Switzerland) 18 (9), 2762.

Noreen, Z., Khalid, N.R., Abbasi, R., Javed, S., Ahmad, I., Bokhari, H., 2019. Visible light sensitive Ag/TiO$_2$/graphene composite as a potential coating material for control of *Campylobacter jejuni*. Mater. Sci. Eng. C 98, 125–133. https://linkinghub.elsevier.com/retrieve/pii/S0928493118307082.

Palanisamy, S., Kokulnathan, T., Chen, S.-M., Velusamy, V., Ramaraj, S.K., 2017. Voltammetric determination of Sudan I in food samples based on platinum nanoparticles decorated on graphene-β-cyclodextrin modified electrode. J. Electroanal. Chem. 794, 64–70. https://linkinghub.elsevier.com/retrieve/pii/S1572665717302254.

Pan, N., Liu, Y., Fan, X., Jiang, Z., Ren, X., Liang, J., 2017. Preparation and characterization of antibacterial graphene oxide functionalized with polymeric N-halamine. J. Mater. Sci. 52, 1996–2006.

Pan, W.Y., Huang, C.C., Lin, T.T., Hu, H.Y., Lin, W.C., Li, M.J., et al., 2016. Synergistic antibacterial effects of localized heat and oxidative stress caused by hydroxyl radicals mediated by graphene/iron oxide-based nanocomposites. Nanomed. Nanotechnol. Biol. Med. 12 (2), 431–438.

Pan, Y., Liu, X., Zhang, W., Liu, Z., Zeng, G., Shao, B., et al., 2020. Advances in photocatalysis based on fullerene C60 and its derivatives: properties, mechanism, synthesis, and applications. Appl. Catal. B 265, 118579. https://linkinghub.elsevier.com/retrieve/pii/S0926337319313256.

Pathakoti, K., Manubolu, M., Hwang, H.M., 2017. Nanostructures: current uses and future applications in food science. J. Food Drug Anal. 25, pp. 245–253.

Perdikaki, A., Galeou, A., Pilatos, G., Karatasios, I., Kanellopoulos, N.K., Prombona, A., et al., 2016. Ag and Cu monometallic and Ag/Cu bimetallic nanoparticle–graphene composites with enhanced antibacterial performance. ACS Appl. Mater. Interfaces 8 (41), 27498–27510. https://pubs.acs.org/doi/10.1021/acsami.6b08403.

Pilehvar, S., De Wael, K., 2015. Recent advances in electrochemical biosensors based on fullerene-C60 nano-structured platforms. Biosensors 5 (4), 712–735. http://www.mdpi.com/2079-6374/5/4/712.

Raril, C., Manjunatha, J.G., 2020. A simple approach for the electrochemical determination of vanillin at ionic surfactant modified graphene paste electrode. Microchem. J. 154, 104575.

Sha, R., Komori, K., Badhulika, S., 2017. Graphene–polyaniline composite based ultra-sensitive electrochemical sensor for non-enzymatic detection of urea. Electrochim. Acta 233, 44–51. https://linkinghub.elsevier.com/retrieve/pii/S0013468617305091.

Sha, Y., Zhang, X., Li, W., Wu, W., Wang, S., Guo, Z., et al., 2016. A label-free multi-functionalized graphene oxide based electrochemiluminscence immunosensor for ultrasen-sitive and rapid detection of Vibrio parahaemolyticus in seawater and seafood. Talanta 147, 220–225. https://linkinghub.elsevier.com/retrieve/pii/S0039914015303490.

Shen, J., Zhu, Y., Yang, X., Li, C., 2012. Graphene quantum dots: emergent nanolights for bioimaging, sensors, catalysis and photovoltaic devices. Chem. Commun. 48 (31), 3686. http://xlink.rsc.org/?DOI=c2cc00110a.

Shi, L., Chen, J., Teng, L., Wang, L., Zhu, G., Liu, S., et al., 2016. The antibacterial applications of graphene and its derivatives. Small 12 (31), 4165–4184. https://onlinelibrary.wiley.com/doi/10.1002/smll.201601841.

Shirshahi, V., Tabatabaei, S.N., Hatamie, S., Saber, R., 2019. Functionalized reduced graphene oxide as a lateral flow immuneassay label for one-step detection of *Escherichia coli* O157:H7. J. Pharm. Biomed. Anal. 164, 104–111. https://linkinghub.elsevier.com/retrieve/pii/S0731708518316856.

Shvedova, A.A., Pietroiusti, A., Fadeel, B., Kagan, V.E., 2012. Mechanisms of carbon nanotube-induced toxicity: focus on oxidative stress. Toxicol. Appl. Pharmacol. 261 (2), 121–133.

Singh, T., Shukla, S., Kumar, P., Wahla, V., Bajpai, V.K., 2017. Application of nanotechnology in food science: perception and overview. Front. Microbiol. 8, 1501.

Sobhan, A., Lee, J., Park, M.-K., Oh, J.-H., 2019. Rapid detection of *Yersinia enterocolitica* using a single–walled carbon nanotube-based biosensor for Kimchi product. LWT 108, 48–54. https://linkinghub.elsevier.com/retrieve/pii/S0023643819302324.

Sobhan, A., Oh, J.-H., Park, M.-K., Kim, S.W., Park, C., Lee, J., 2018. Assessment of peanut allergen Ara h1 in processed foods using a SWCNTs-based nanobiosensor. Biosci. Biotechnol. Biochem. 82 (7), 1134–1142. https://academic.oup.com/bbb/article/82/7/1134/5938710.

Sobhan, A., Oh, J.-H., Park, M.-K., Kim, S.W., Park, C., Lee, J., 2018. Single walled carbon nanotube based biosensor for detection of peanut allergy-inducing protein ara h1. Korean J. Chem. Eng. 35 (1), 172–178. http://link.springer.com/10.1007/s11814-017-0259-y.

Sobhan, A., Oh, J.H., Lee, J.Y., 2018. Rapid detection of Ara h2 using single walled carbon nanotube based biosensor for peanut allergen control. Appl. Mech. Mater. 878, 286–290. https://www.scientific.net/AMM.878.286.

Song, Y., Ostermeyer, G.P., Du, D., Lin, Y., 2021. Carbon nanodot-hybridized sil-ica nanospheres assisted immunoassay for sensitive detection of *Escherichia coli*. Sens. Actuators B Chem. 349, 130730. https://linkinghub.elsevier.com/retrieve/pii/S0925400521012983.

Sun, D., Yang, X., 2017. Rapid determination of toxic rhodamine B in food samples us-ing exfoliated graphene-modified electrode. Food Anal. Methods. 10 (6), 2046–2052. http://link.springer.com/10.1007/s12161-016-0773-2.

Sundramoorthy, A.K., Gunasekaran, S., 2014. Applications of graphene in quality assurance and safety of food. Trends Analyt. Chem. 60, 36–53. https://linkinghub.elsevier.com/retrieve/pii/S0165993614001113.

Surudžić, R., Janković, A., Bibić, N., Vukašinović-Sekulić, M., Perić-Grujić, A., Mišković-Stanković, V., et al., 2016. Physico–chemical and mechanical properties and antibacterial activity of silver/poly(vinyl alcohol)/graphene nanocomposites obtained by electrochemical method. Compos. Part B Eng. 85, 102–112.

Tahtaisleyen, S., Gorduk, O., Sahin, Y., 2020. Electrochemical determination of tartrazine using a graphene/poly(L-phenylalanine) modified pencil graphite electrode. Anal. Lett. 53 (11) 1683-1170.

Taouri, L., Bourouina, M., Bourouina-Bacha, S., Hauchard, D., 2021. Fullerene-MWCNT nanostructured-based electrochemical sensor for the detection of vanillin as food additive. J. Food Compost. Anal. 100, 103811. https://linkinghub.elsevier.com/retrieve/pii/S0889157521000119.

Traiwatcharanon, P., Siriwatcharapiboon, W., Wongchoosuk, C., 2020. Electrochemical sodium ion sensor based on silver nanoparticles/graphene oxide nanocomposite for food application. Chemosensors 8 (3), 58.

Villamizar, R.A., Maroto, A., Rius, F.X., Inza, I., Figueras, M.J., 2008. Fast detection of *Salmonella infaintis* with carbon nanotube field effect transistors. Biosens. Bioelectron. 24 (2), 279–283. https://linkinghub.elsevier.com/retrieve/pii/S0956566308001814.

Wu, B., Yeasmin, S., Liu, Y., Cheng, L.-J., 2022. Ferrocene-grafted carbon nanotubes for sensitive non-enzymatic electrochemical detection of hydrogen peroxide. J. Electroanal. Chem. 908, 116101. https://linkinghub.elsevier.com/retrieve/pii/S1572665722000935.

Wu, Z., Guo, F., Huang, L., Wang, L., 2018. Electrochemical nonenzymatic sensor based on cetyltrimethylammonium bromide and chitosan functionalized carbon nanotube modified glassy carbon electrode for the determination of hydroxymethanesulfinate in the presence of sulfite in foods. Food Chem. 259, 213–218. https://linkinghub.elsevier.com/retrieve/pii/S0308814618305132.

Xu, H., Yang, X., Li, G., Zhao, C., Liao, X., 2015. Green synthesis of fluorescent carbon dots for selective detection of tartrazine in food samples. J. Agric. Food Chem. 63 (30), 31–38.

Xu, W., Xie, W., Huang, X., Chen, X., Huang, N., Wang, X., et al., 2017. The graphene oxide and chitosan biopolymer loads TiO_2 for antibacterial and preservative research. Food Chem. 221, 267–277. https://linkinghub.elsevier.com/retrieve/pii/S0308814616316715.

Yamada, K., Kim, C.-T., Kim, J.-H., Chung, J.-H., Lee, H.G., Jun, S., 2014. Single walled carbon nanotube-based junction biosensor for detection of *Escherichia coli*. Docoslis A., (Ed.). PLoS One 9 (9), e105767. https://dx.plos.org/10.1371/journal.pone.0105767.

Yang, N., Chen, X., Ren, T., Zhang, P., Yang, D., 2015. Carbon nanotube based biosensors. Sens. Actuators B Chem. 207, 690–715. http://www.sciencedirect.com/science/article/pii/S0925400514012507.

Yang, Y.J., Li, W., 2015. CTAB functionalized graphene oxide/multiwalled carbon nanotube composite modified electrode for the simultaneous determination of sunset yellow and tartrazine. Russ. J. Electrochem. 51 (3), 21–24.

Yang, Z., Sun, C., Wang, L., Chen, H., He, J., Chen, Y., 2017. Novel poly(1-lactide)/graphene oxide films with improved mechanical flexibility and antibacterial activity. J. Colloid Interface Sci. 507, 344–352.

Yao, Y., Wang, G., Chu, G., An, X., Guo, Y., Sun, X., 2019. The development of a novel biosensor based on gold nanocages/graphene oxide–chitosan modified acetylcholinesterase for organophosphorus pesticide detection. New J. Chem. 43 (35), 13816–13826. http://xlink.rsc.org/?DOI=C9NJ02556A.

Ying, L.S., bin Mohd Salleh, M.A., Mohamed Yusoff, H.B., Abdul Rashid, S.B., Razak, J.b.Abd., 2011. Continuous production of carbon nanotubes: a review. J. Ind. Eng. Chem. 17 (3), 367–376 https://linkinghub.elsevier.com/retrieve/pii/S1226086×11001110.

Zanni, E., Bruni, E., Chandraiahgari, C.R., De Bellis, G., Santangelo, M.G., Leone, M., et al., 2017. Evaluation of the antibacterial power and biocompatibility of zinc oxide nanorods decorated graphene nanoplatelets: new perspectives for antibiodeteriorative approaches. J. Nanobiotechnol. 15 (1), 12–14.

Zhang, L., Zhang, X., Li, X., Peng, Y., Shen, H., Zhang, Y., 2013. Determination of Sudan I using electrochemically reduced graphene oxide. Anal. Lett. 46 (6), 923–935. http://www.tandfonline.com/doi/abs/10.1080/00032719.2012.747096.

Zhang, Q., Zhang, C., Ying, Y., Ping, J., 2021. An easy-fabricated ordered mesoporous carbon-based electrochemical sensor for the analysis of trans-resveratrol in red wines. Food Control 129, 108203. https://linkinghub.elsevier.com/retrieve/pii/S0956713521003418.

Zhang, Y., Cao, Q., Zhu, F., Xu, H., Zhang, Y., Xu, W., et al., 2020. An amperometric hydrogen peroxide sensor based on reduced graphene oxide/carbon nanotubes/Pt NPs modified glassy carbon electrode. Int. J. Electrochem. Sci. 15 (9), 8771–8785.

Zhang, Y., Li, C., Fan, Y., Wang, C., Yang, R., Liu, X., et al., 2016. A self-quenching-resistant carbon nanodot powder with multicolored solid-state fluorescence for ultra-fast staining of various representative bacterial species within one minute. Nanoscale 8 (47), 19744–19753. http://xlink.rsc.org/?DOI=C6NR06553H.

Zhao, Y., Li, Y., Zhang, P., Yan, Z., Zhou, Y., Du, Y., et al., 2021. Cell-based fluorescent microsphere incorporated with carbon dots as a sensitive immunosensor for the rapid detection of *Escherichia coli* O157 in milk. Biosens. Bioelectron. 179, 113057. https://linkinghub.elsevier.com/retrieve/pii/S0956566321000944.

Perspectives for binary and ternary composites films for food applications

20

Shashank T. Mhaske, Jyoti Darsan Mohanty and Umesh R. Mahajan

Department of Polymer and Surface Engineering, Institute of Chemical Technology, Mumbai, Maharashtra, India

20.1 Introduction

In the never-ending quest for better performance, commonly used polymeric materials have reached the end of their usefulness. Thus, material sector analysts always look to produce new materials with the required properties. A composite is a material made up of two or more phases that are joined to generate an advanced material with unique properties. Two or more distinct phases are separated by a specific interface in a composite material. Due to advantages like low weight, faster assembly, high fatigue strength, and high barrier characteristics, composites are becoming an essential part of the material industry. The key distinction between a blend and a composite is that in composites, the main components can be easily visualized, whereas, in blends, they are not distinguishable. Ancient civilizations established composite technology by making bricks out of mud strengthened with straw. Polymer-based composite demand, production, and property change have all increased dramatically during the previous 50 years. According to the forecast, this need will continue in the packing sector in the following years, contributing more significantly. Fiber-reinforced and particulate composites (FRC) are the most popular in the market.

Composite comprises two distinct phases: polymer matrix and the other one is reinforcement. The components of composite material are shown in Fig. 20.1. Polymer matrix serves as a continuous phase, whereas reinforcement serves as the discontinuous or dispersed phase. Matrix is generally a more flexible and pliable material, comprised of flexible polymeric material. The matrix gives shape, surface appearance, environmental protection, and durability to the composite where reinforcement material is well integrated into the matrix to provide macroscopic stiffness and strength.

The reinforcement material is also called the dispersed phase, which is more complicated than the continuous phase. The reinforcement adds unique properties to the composite material, such as chemical resistance, wear resistance, strength, and stiffness. A reduction in tooling and assembly costs is one of the many possible benefits of composites. Except for temperature resistance, the composite material can

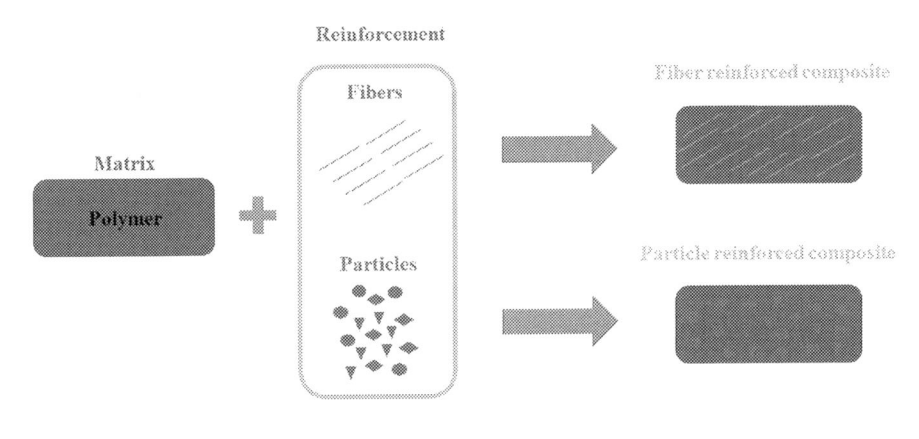

FIGURE 20.1

Representation of matrix and fiber phase in composite system.

outperform any engineering material. Material can be classified as composite if it meets three characteristics given below:

1. The mixing of two phases is possible to yield the final product.
2. There must be an interphase that separates both phases by a phase boundary.
3. The final material should have a unique property that is not governed by any phase.

The addition of distinct reinforcement classifies composites into two categories binary and ternary composites. A single type of reinforcement addition can make binary composites. Two or more types of reinforcement addition can make ternary composites. There is a slight misconception about the kind of reinforcement addition into the polymer matrix. Suppose two similar types of reinforcing material are added. In that case, a particle with two different particle sizes is also referred to as a ternary composite instead of a binary composite rather than having a similar form of reinforcing material. The polymeric material used in the composite fabrication is shown in Fig. 20.2. The most commonly used polymer matrices are polyethylene (PE), polypropylene (PP), and polyethylene terephthalate (PET) (Pereira de Abreu et al., 2007; Novák et al., 2016; Attaran et al., 2017; Leila et al., 2021), aamongst them trending into the market concerning the packaging sector. The reinforcement can be of particles, fibers, and structurally aligned materials. Amongst them, nanoparticle (NP) addition into the matrix is one of the most widespread techniques used to improve the physical properties of the composite material. The current trend in the packaging industry is to develop composites with high performance and quality to overcome the hurdles for special applications. The reinforcement in the matrix is added to increase composite qualities. The binary and ternary composite can be developed by adding both particulate fillers in nano and micro dimensional sizes. The incorporation of NPs improved the performance of base polymers. The NP has a higher surface area. The interaction with the matrix is very effective and it can be homogeneously

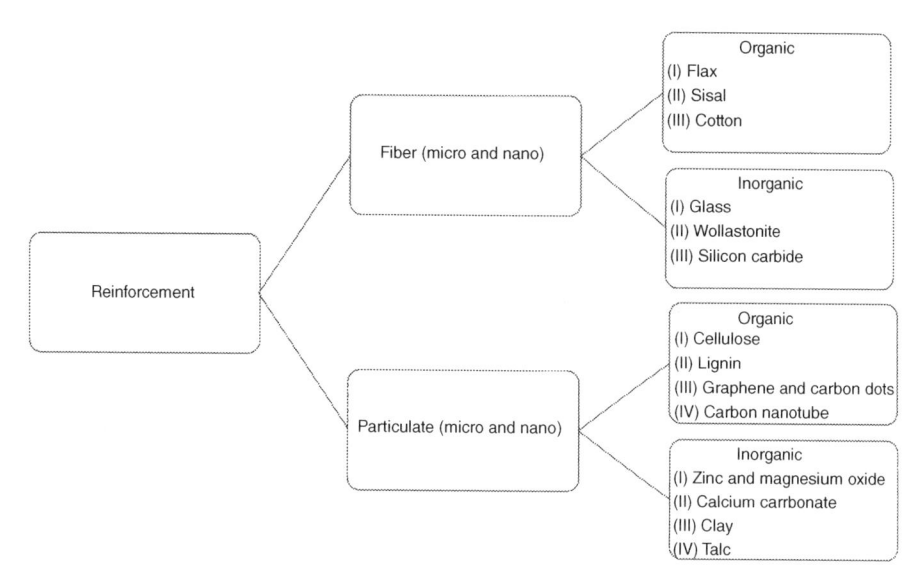

FIGURE 20.2

Different types of reinforcement.

dispersed throughout the matrix. Whereas the microform is mostly dispersed in both homogenous and heterogeneous phases and is less effective at low amounts of loading.

This chapter focuses on the structure-property relationship of composite concerning matrix and reinforcement. The impact of the type of reinforcement addition on the composite property was discussed by taking reference to different research and review journals.

20.1.1 **Matrix phase**

As we discussed, the composite material has two phases, the polymer, i.e., the matrix phase, and reinforcement. There are several matrix phases used for manufacturing food packaging, either synthetic polymer or biopolymer. The different polymers were selected according to their distinct properties for different packaging applications.

The polymer section can be divided into four categories:

1. Synthetic polymer (nonbiodegradable): The polymer is derived from petrochemical sources mainly PE, PP, polyvinyl chloride (PVC), and PET.
2. Synthetic polymer (biodegradable): The polymer is derived from synthetic sources, but they are biodegradable in the presence of a typical weathering environment. A few examples are polycaprolactone (PCL), polybutylene adipate trepthalate (PBAT), and polybutylene succinate (PBS).
3. Biopolymer (nonbiodegradable): The polymer is derived from biosource but nonbiodegradable bio PET, bio PE, and bio polyamide (PA). Terephthalic acid

is derived from furfural and has the potential raw material for bio-based PET. Bioethanol is produced from sugarcane and starch molasse. The bioethanol is hydrolyzed to produce ethylene which is further polymerized to produce PE.

4. Biopolymer (biodegradable): The polymer is derived from bio-sourced and biodegradable products. A few examples are polyhydroxy alkonate (PHA), PBS, and polylactic acid (PLA).

20.1.1.1 Modification of matrix

The matrix phase is modified through physical compatibilization or chemical compatibilization. The physical modification includes the mixing of polymer with matrix by using different random copolymers or coupling agents. It increases the phase compatibility between hydrophobic polymers and hydrophilic NPs. Different coupling agents are silane-grafted maleic anhydride, PP-grafted maleic anhydride, and PP-grafted acrylic acid. It increases the phase compatibility of a matrix with the reinforcement. Biofillers such as nano cellulose, lignin, flax, and sisal fiber are hydrophilic. It required a coupling agent or dispersing agent which improves the compatibility of polymer PP, PLA, and PE are a few examples that improve the compatibility with organic and inorganic filler through physical compatibilization. Similarly, biopolymers like PLA, PHA, and polyhydroxybutyrate (PHB) are grafted with ethylene and propylene glycol to stabilize the melt viscosity while processing (Attaran et al., 2017; Li et al., 2019; Videira-Quintela, 2021).

20.1.2 Reinforcement

The reinforcement is used to improve the matrix properties as necessary for application. Reinforcement is divided into two types based on chemical structure and atomic element composition, i.e., organic and inorganic. Both reinforcements are available in nano and micro in fiber and particulate form. Also, depending on the cost and performance, it is further divided into inert and active reinforcement. Inert reinforcement is used to reduce the cost without any significant improvement in composite properties, whereas, active reinforcement is used to improve the properties of composites. It can be cost-effective depending on the application. Nano form has a higher surface area as compared to microform. Thus, nanomaterial is used as active reinforcement to improve the physical, thermal, chemical, antimicrobial, and barrier properties. Nanomaterial is used in a very small amount in the matrix phase ideally below or equal to 5 wt%. Different organic reinforcements are micro and nanocellulose, nanolignin, graphene, CNT, and carbon dots in particulate form with organic fibers like flax, sisal, and cotton used to manufacture bio-based organic composites. Similarly, nano zinc and magnesium oxide, silver, iron, clay, and micro size talc, mica, and calcium carbonate are used for manufacturing inorganic composite. The different forms of nano reinforcements are nanotubes, nanofibers, and nanocrystals (Mohanty et al., 2018; Videira-Quintela, 2021).

20.1.2.1 Inorganic reinforcement

The inorganic reinforcements are of different shapes and sizes that can be added to the base polymer matrix to improve its physical properties. The shape of reinforcement can be in particulate or fiber form and the size of reinforcement ranges from a few micrometers to nanometers. The common inorganic reinforcements are talc, mica, calcium carbonate, titanium oxide, zinc, and magnesium oxide. The methods for synthesizing from nano to micro size can be produced through different mechanical and chemical modification methods. Mechanical methods involve attrition, shearing, and reduction in the dimension of the particle in nano-size but it can destroy the original crystal structure which impacts the physical properties of the NP. The common mechanical methods used are ball mill, planetary ball mill, vibratory mill, and horizontal ball mill. The other miscellaneous methods used are the collodial method, RF plasma, chemical vapor deposition method, pulsed laser, and thermal decomposition method.

Talc: The talcum material is a composition of clay mineral composed of hydrated magnesium silicate with different compositional elements like magnesium, silicon, oxygen, and hydrogen with the chemical formula $Mg_3Si_4O_{10}(OH)_2$. It has a density between 2.58 g/cc and 2.83 g/cc and is most commonly used for reinforcement in the polymer matrix. The incorporation of talc improves the coefficient of friction of film and can be used as an antiblocking agent. It improves the mechanical strength and higher insulation of polymer composites.

Mica: Mica is a group of silicate minerals with outstanding physical properties with a chemical formula of $X_2Y_{4-6}Z_8O_{20}(OH, F)_4$ where X is K, Na, or Ca, Y is Al, Mg, or Fe, Z is Si or Al. The mica crystals can be easily split into thin elastic phases. The density of mica is in the range between 2.7 g/cc and 3 g/cc. The mica crystals used in the polymer are mostly phlogopite and biotite and it has higher iron and aluminum content. The incorporation of mica provides exceptional surface, gloss, dielectric, stiffness, and strength to polymer composites. It also helps in reducing the thermal expansion coefficient and wrinkling of the polymer.

Calcium Carbonate: The calcium carbonate ($CaCO_3$) is in the form of limestone, chalkboard, and marble. The density of calcium carbonate is in the range between 2.7 g/cc and 2.9 g/cc. It is widely used in polymers to provide balanced properties, i.e., strength and stiffness. It is considered to be a soft and nonabrasive filler. It is cost-effective and widely used in masterbatch industries for producing reinforced polymer.

20.1.2.2 Organic reinforcement

The organic reinforcement consists of fillers, most of them are naturally occurring and chemically synthesized with carbon as the main constituent. The derived source of natural filler is the carbohydrate, and the derived source of carbon-based filler is a different form of carbon compound. Cellulose, lignin, and other biopolymers that can be used to make NPs for composite reinforcement fall under the category of sustainable natural resources.

20.1.2.2.1 Natural filler

The synthesis and development of lignin and cellulose-based nano and microparticles improve the physical properties of the polymer at a low percentage of loading without causing a severe threat to the environment. This type of filler has the potential to act as a replacement for many metals and nonmetal oxide-based compounds in packaging applications (Fortunati et al., 2014; Wang and Rhim, 2015; Yang et al., 2016a; Dunlop, 2021).

Polymer nanocomposites have gained a lot of attention because of their potential for low-cost industrial-scale production, minimum environmental effect, low safety risks, and renewable nature due to their advantages such as nanoscale dimensions, high aspect ratios, low densities, low manufacturing costs, and superior biodegradability. For the development of biocomposites, cellulosic nanomaterials have the potential to be used as the next generation of renewable reinforcements for the development of high-performance biocomposites. The interconnected nanocomposites also contribute to these exceptional mechanical capabilities.

The justifications for employing nanocellulose as reinforcement are its high crystallinity and the possibility of utilizing its high stiffness and strength in composite applications. Nanocrystalline cellulose (NCC) or CNCs, nanofibrillated cellulose (NFC), microfibrillated cellulose (MFC), and cellulose produced by bacteria are a few of the various kinds of nanocellulose (BC). These kinds of nanocellulose can be used to create a variety of reinforcing structures for continuous networked structures, planar reinforcement, or scattered reinforcement. PP, polystyrene, and high-density PE are just a few of the polymers that have been strengthened with nanocellulose. Composites provide the improved properties necessary for several industrial and technological applications by the inclusion of trace amounts of nanometre-sized fillers. The main problem of CNC is the dispersion in a hydrophobic polymer matrix. The chemical modification of CNC through esterification, etherification, and sialylation with fatty acids and alcohols leads to better dispersion in the polymer matrix and also improves the water vapor transmission (WVT) properties (Brinchi et al., 2013).

Modern chemistry describes lignin as an organic complex biopolymer containing a significant number of polyphenols. It has potential use in environmentally friendly technology and is the second-most prevalent biologically generated polymeric material in nature. The extremely heterogeneous polymers that comprise lignin were produced using many distinct precursors of lignols. Heterogeneity results from the different types and intensities of crosslinking between these lignols. Paracoumaryl alcohol (4-hydroxyphenylpropane), sinapyl alcohol (3,5-dimethoxy-4-hydroxyphenylpropane), and coniferyl alcohol (4-Hydroxy-3-methoxy phenyl propane) depending on the plant source, different levels of the precursor "monomers" (lignols) are present. Typically, the syringyl/guaiacyl ratio is used to classify lignins. NPs are generated from renewable resources and have potential use in several industries, including agriculture. Numerous methods, including solvent exchange, dialysis, CO_2 saturation, flash precipitation, and precipitation procedures, can be used to create lignin nanoparticles (LNPs). To enhance biodegradability and reduce production costs, lignin is being used as a filler in commercial polymers in composite development. Depending on the source and method of separation, the natural polymer lignin may have different properties that have an impact on composites made of polymers. It is most commonly used in bio packaging in combination with nano cellulose to improve the ultraviolet (UV) and antimicrobial properties (Brinchi et al., 2013).

20.1.2.2.2 Carbon-sourced reinforcement

The production of binary and ternary composites for commercial usage uses a variety of carbon-derived reinforcing sources such as carbon nanotubes (CNTs) (SWNT and MWNT), graphene oxide (GO), fullerenes, and carbon dots. The use of carbon-based allotrope is widely used for the manufacturing of conducting composites and improves the thermal and mechanical properties. The drawback associated with the carbon-based filler is that it reduces the optical clarity because of its dark color and higher cost as compared to other commercial fillers. However, the specialized application required the use of different forms of carbon-derived allotrope to attain certain desired properties (Díez-Pascual, 2020; Kausar, 2020).

Carbon nanotubes are cylindrical molecules made of sheets of single-layer carbon atoms that have been coiled up (graphene). They come in two varieties: single-walled (SWCNT), with a diameter of less than 1 nanometer (nm), and multiwalled (MWCNT), with diameters up to more than 100 nm and made up of multiple concentrically interconnected nanotubes. They can be as long as a millimeter or even several micrometers.

Fullerenes are molecular allotropes of carbon that exhibit a range of unique behaviors because of their nature as electron molecules and the simplicity of chemical manipulation. A fullerene family is a diverse group of materials with important applications in physics, chemistry, and biology due to the large curvature of these hollow spheres' conjugated electron systems, which has permitted complicated chemical behavior and the production of a wide variety of derivatives. We first provide a quick summary of the fullerene family before describing the molecular and crystal structure of C60, the most widespread and thoroughly studied member of the family. It has 12 pentagonal and 20 hexagonal faces and it is joined together through a single or double bond to form a football-like structure. The derivatives, heterofullerenes, and fullerene polymers are also briefly mentioned.

The carbon allotrope known as graphene is made up of a single sheet of atoms arranged in a two-dimensional (2D) honeycomb lattice nanostructure. The name comes from the fact that the carbon allotrope in graphite has a lot of double bonds. It has a hexagonal network of carbon atoms that resembles a honeycomb that is 1 nm thick and it has a 2D planar surface in the same vein as GO, which is composed of a single sheet of carbon with areas of disrupted aromaticity where carbon atoms are oxidized. Reduced GO is GO that has more carbon from graphene and fewer oxygen functional groups than conventional GO. The graphene is added to the polymer matrix to improve the key properties like mechanical, thermal, and electrical conductivity, and neglect the development of static charges (Tiwari et al., 2016).

20.2 **Preparation method**

The development route for binary and ternary composites is generally through the conventional method of solution blending and melt blending. Both methods are practically used in research laboratories. However, the melt blending method is commercially adopted in the industry because of the process' feasibility and large volume of production. Both methods have certain advantages and drawbacks. In the melt blending technique, the selection of processing temperature range and other constraints like screw revolutions per minute (RPM) and torque can impact the mixing of polymer. Whereas in solution blending, the selection of appropriate solvent and dissolution temperature also plays an important role. The addition of NPs through the solution blending method required the use of a high-speed physical dispenser for dispersing the NP in the polymer solution. So that the added NP should not coagulate and give better dispersion when casted into a film or sheet. Similarly, the melt blending process required certain engineering actions while adding NPs into the molten polymer. A better shearing path is to be provided through two intermeshing screws for homogeneous mixing of polymer and NP. For enhancing the dispersion, NP can be added to the polymer through the masterbatch route. The preparation of micro composites required the addition of a filler main hopper if the filler is particulate with a low amount of loading up to 5 wt%, but a greater than 5 wt% side feeder is needed to reduce the screw torque and to prevent material from degradation with better homogenous dispersion. There are other techniques for mixing polymer through the counter-rotating extruder, single screw extruder, Banbury mixer, and two-roll mills. However, this process has certain limitations in homogenous mixing as compared to twin-screw co-rotating extruders.

Different processing techniques for binary and ternary composites:

1. Solution blending
2. Melt blending
   a. Extrusion
   b. Two-roll mill
   c. Batch mixer

20.2.1 **Solution blending**

The solution blending (Fig. 20.3) technique is commonly used for mixing polymer in a suitable system through a mixer with a rotating blade connected to the motor. Suitability of solvent is required to achieve a complete dissolution process. Different parameters like temperature, time, and rotor speed are also taken into consideration for achieving better dispersion of organic and inorganic reinforcement particles in a polymer solution. Then the obtained solution is cast in a closed mold and kept for drying or it can be passed through a hot and cold roller to produce a film of desired thickness. There are many examples when NPs are thoroughly dispersed and cast into a mold for producing nanocomposites. Solution blending is appropriate to process

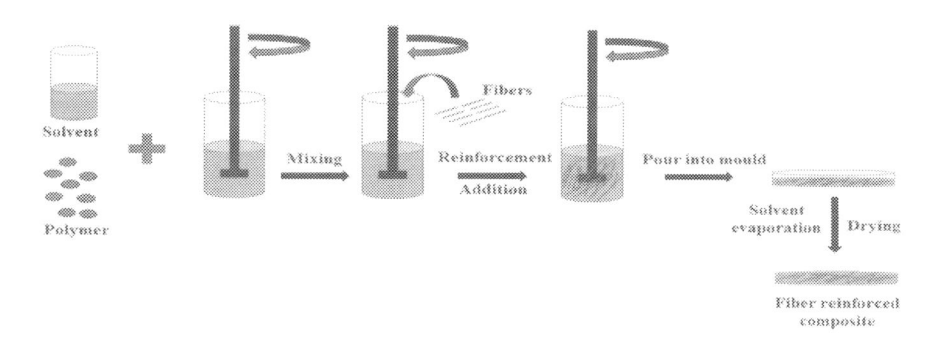

FIGURE 20.3

Schematic diagram of solution blending.

a heat-sensitive polymer with a nanofiller to achieve better homogeneity. However, this process is not commercially viable and only can be used for laboratory research purposes to compare the properties with melted blended polymer.

20.2.2 **Melt blending**

Melt blending is a commercially viable and industrially adopted technique for producing polymer composites. There are many processes used for melt blending of a polymer such as extrusion, two-roll mill, and batch mixer. The different process is to be selected based on the shear rate required for mixing the polymer with organic and inorganic filler. Moisture and shear-sensitive polymers required special precautions while processing to control the melt viscosity and prevent thermolytic degradation. To set the processing parameter screw RPM, torque, mixing time, and residence time are to be studied through torque and capillary rheometer. The rheometry method is scientifically used to understand the melt rheology for the shear rate inside a closed-loop adiabatic system. Polyester, polyamide, and polyvinylidene chloride (PVDC) polymer show a drop in melt viscosity at the low shearing region as compared to polyolefin. So, the screw is specially designed with a low helix angle and high pitch depth to reduce the shear rate. The mixing quality of melt depends on the quality of material discharged from the extruder. A batch mixer and two-roll mill are used for blending the polymer and producing homogenous compounded mass at a low shear rate.

20.2.2.1 *Extrusion process*

The Extrusion process (Fig. 20.4) involved mixing filler with synthetic and biopolymer to prepare binary and ternary composites. The mixing is performed using an intermeshing co-rotating twin screw extruder. The co-rotating provides highly intensive mixing due to the crossflow path and leads to better dispersion of micro and nanofiller in the polymer matrix. The addition of nanofiller can be fed through the side feeder

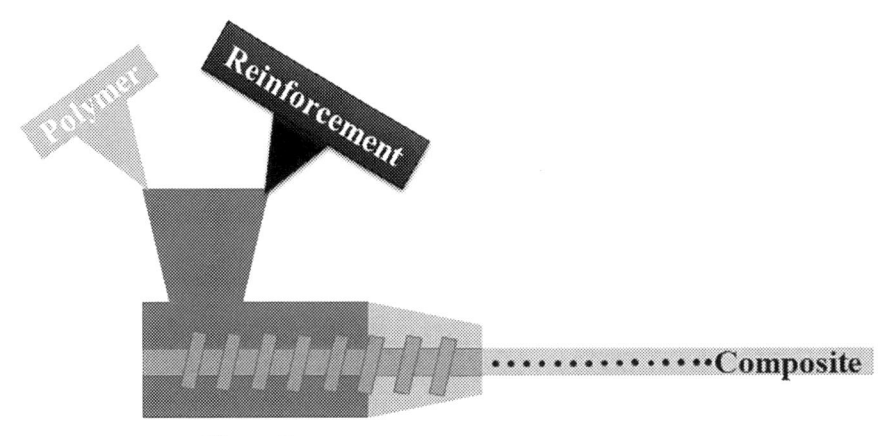

Extruder

FIGURE 20.4

Schematic diagram of extrusion process.

or can be directly fed through the premix with the main feeder. The final product is obtained in the form of granules which are conventionally made into the desired product through different processing techniques. A moderate shear rate and proper processing temperature are recommended for better dispersive mixing of nanofillers because a high shear rate and improper processing temperature lead to low viscosity, degradation, charring, improper mixing, and less dispersion of micro and NP in the polymer matrix. The design of mixing elements and conveying elements is also vital when we consider mixing the polymer with micro or nanofiller to obtain good physical properties of binary and ternary composites.

20.2.2.2 Two-roll mill

Two-roll mills (Fig. 20.5) are conventionally used for low-shear mixing of polymer with additives and fillers. The two identically cylindrical rolls rotate in both clockwise and counterclockwise directions for the mixing of polymer and filler. The low shear mixing provides distributive mixing rather than dispersive mixing. The mixing process is handful for the shear-sensitive polymer to be utilized for the preparation of binary and ternary composites. However, the process is only useful for observing the primary processing of polymer composites. The composite film thickness can be controlled by adjusting the nip gap to the produced sheet of the desired thickness.

20.2.2.3 Batch mixer

The batch mixer (Fig. 20.6) is designed with kneading, blending, and mixing different polymers with fillers inside the enclosed chamber. The design is based on counter-rotating pairs of rotors with different kneading element designs. The polymer is extensively mixed inside the chamber jacketed by different heating and cooling

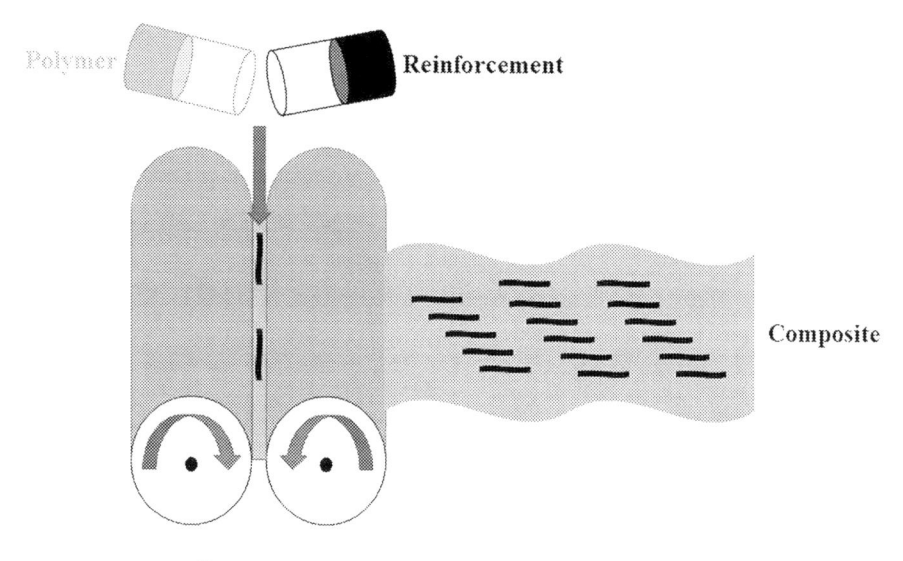

FIGURE 20.5

Schematic diagram of the two-roll mill.

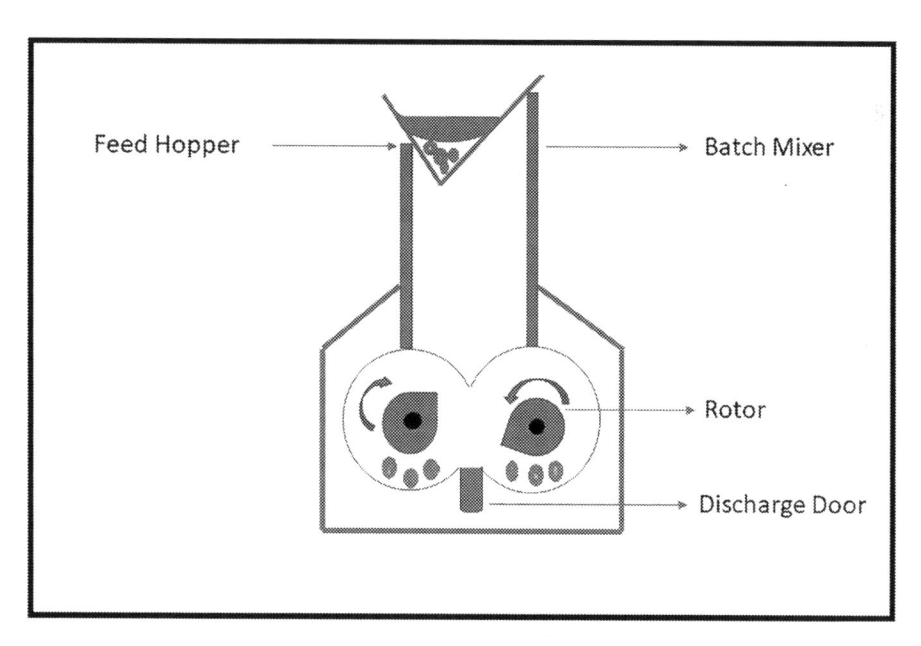

FIGURE 20.6

Schematic diagram of batch mixer.

elements to control the temperature. The batch mixer is the modified version of the two-roll mill with better shear and temperature control. It is extensively used for measuring the mixing torque, viscosity, and degradation time of polymer mixture at a given shear rate of time. It helps produce small batches and the final product is in the form of a rolled sheet. It is more extensively used in the research and development of biopolymers, biopolymer composites, mixing of elastomers, and more importantly, studying the thermal and processing behavior of synthetic and biopolymers.

20.3 Properties of binary and ternary composites

20.3.1 Functional group analysis

Microstructure analysis of polymer is done to understand the functional group and compositional change in polymer by the addition of the functional additives. It is suitable to understand the synergistic effect of additives with polymer. Also, the FT-IR analysis is suitable for understanding deterioration in properties through thermal, UV, and hydrolytic degradation of polymer by comparing it with the base polymer functional group and helps to understand the conformational change by the addition of nanofiller into the polymer matrix. It can be used for both qualitative and quantitative analysis. The degradation of polymer over time is studied through structural analysis as the polymer is exposed to environmental conditions. The degradation can be analyzed through a change in the intensity of spectra and a change in the chemical identity of spectra. Fourier transform infrared spectroscopy (FT-IR) spectroscopy tool was used to analyze the intermolecular interactions between polyvinyl alcohol (PVA), chitosan (CH), and lignin NPs. Accordingly, the outcomes for PVA/3LNP, CH/3LNP, and PVA/CH/3LNP systems are given. The stretching of the OH group, C-O bonds, and bands at 3262/cm and 1412/cm are all seen in the clean PVA spectrum. The interaction between the OH group on the LNP surface and the hydroxyl groups of the PVA matrix by lignin NPs in the matrix causes the intensity change of the bands indicated. Additionally, modifications in the shape of the bands at 846/cm and 2900–2950/cm may be attributed to PVA chain conformational variations. The intensity at wavenumbers greater than 1700/cm demonstrates PVA has been highly deacetylated. Some peak intensities showed modest alterations, as a result of the presence of lignin NPs. The Peaks can also be detected at 1566/cm (the aromatic skeleton stretched C-C) and 1648/cm (water absorption). The FT-IR spectra of pure N-H exhibit the characteristic peak of NHCOCH3 at 1648/cm and NH peak at 1546/cm, and the addition of LNP has no impact on the CH's basic structure. The two main characteristic peaks associated with CH in the PVA/CH blend were shifted towards higher wavenumbers (at 1637/cm and 1550/cm, respectively), as a result of the interaction between OH and NH bending vibration in PVA/CH, whereas no shift has been detected for the PVA peak measured at 1316/cm and 1412/cm compared to the spectra of the PVA/CH film. Although there are minor changes, the bands at 3261/cm and 1086/cm are moved towards lower wavenumber values (3251/cm and 1080/cm, respectively) in the PVA/CH/3LNP ternary system. The groups -OH of PVA, -OH, -NH of CH, and -OH of LNP modify the spectra through intermolecular hydrogen interactions. In addition, the FT-IR spectra of the ternary system (PVA/CH/LNP) show an increase in the band at 1559/cm absorbance. The decreased intensity ratio of this band and 1423/cm band with the presence of LNP indicates that interactions are established between LNP and PVA -OH groups (Yang et al., 2016b). Thymol (8 wt%) and commercial montmorillonite (D43B) at various concentrations of 2.5 wt% and 5 wt% were added to the PLA-based nanocomposite for study and over time, structural change is noticed. In addition to three peaks at 1123/cm, 1082/cm, and 1055/cm associated with the C-C-O groups, PLA displayed distinctive bands at 1750/cm ($C = O$), 1440/cm (CH-CH3), and 1267/cm (C-O-C). The C-O-C peak's intensity starts to decline after 7 days, and it continues to decline after 21 days. The chain scission of PLA by the hydrolysis reaction is what causes the peak's strength to diminish (Ramos et al., 2020). Through FT-IR analysis, the chemical structure of pure lignin and lignin NPs is determined. The positions 1610/cm, 1525/cm, and 1426/cm of aromatic backbone vibration in the lignin structure are there. The bands at 1112/cm, 836/cm, and the shoulder at 1152/cm, all showed signs of HGS-type lignin (C-H) out-of-plane. Further evidence that the lignin is of the HGS type comes from the absorption band at 1161/cm, which is attributable to $C = O$ in conjugated ester groups. Whereas the beta-unsaturated carbonyl group present as a result of the alkaline treatment induces lignin NPs to exhibit an absorption band at 1654/cm. Whereas the peak of 1370/cm is OH in the form of phenol and alcohol deformation. The vibration of C-O in primary alcohol is indicated by the peak at 1205/cm. Following acidolysis, the existence of $C = O$ denotes-carbonyl extending between 1700/cm and 1660/cm. The aromatic vibration of around 1600/cm, C-C stretching is around 1540/cm, C-H stretching is around 1480/cm, C-H vibration of the methyl group is around 1440/cm, and -OH stretching of primary alcohol is around 1040/cm. The IR spectra show an increase in the intensity of the peak at 1120/cm assigned to the condensed aromatic unit and a decrease in intensity at 1700/cm of vibration of the $C = O$ group of hemicellulose (Yang et al., 2015). The characterization of CNC and acetylated CNC structure was confirmed through FT-IR. The O-H stretching vibration and the C-H stretching vibrations are both indicated by bands around 3338/cm and 2901/cm, respectively. Two unique functional group peaks, at 1732/cm and 1248/cm in comparison to CNC, are related to the stretching of the $C = O$ and C-O functional groups as a result of the acetylation process (Yu et al., 2021). To comprehend the synergistic effect of ZnO on PET/ carboxymethyl cellulose (CMC), a ternary blend of PET/CMC and zinc oxide is subjected to FT-IR analysis. Due to the presence of carbohydrates, the CMC/PET possesses an absorption band around 3430/cm that refers to the symmetric and asymmetric vibration of the -OH group. The stretching of the C-H aliphatic group is indicated by the peak at about 2800/cm. The carboxyl $C = O$ group is present when there is a functional group at 1650/cm. The flexural vibration of the hydroxyl group and the vibration of the methyl group are responsible for the absorption bands around 1250/cm and 1400/cm. The nanocomposite films FT-IR spectra also revealed two new peaks at 1100/cm and 1650/cm, and with the increased inclusion of ZnO, the peak changed from 522/cm to 516/cm (Leila et al., 2021).

20.3.2 **Mechanical properties**

Mechanical properties (Fig. 20.7) such as tensile strength, tear strength, and coefficient of friction are important parameters to evaluate the performance of packaging material. The increment in the mechanical properties of packaging film by the addition of micro and nanofillers both organic and inorganic like CNC, LNP, talc, AgO, etc. The addition of functionalized filler improves the dispersion of NPs in the polymer matrix, as a result of this, there is an improvement in crystallinity and the formation of a more uniform crystal size. Smaller crystals improve the geometrical array of packing through the symmetric arrangement. It leads to better load transfer and lower crack propagation rate, so eventually the improvement in mechanical properties can be observed. Nanofiller has a higher surface area as compared to microfiller. So, it helps in the improvement of properties at a lower percentage of loading. The different binary composite system has been developed and successfully commercialized for food packaging applications but achieving all the desired properties required certain modifications and the addition of optional filler. It leads to the formation of the ternary composite system. To produce binary or ternary formulations, cellulose nanocrystals were placed into a PLA matrix or a PLA/PBS mixture both unmodified CNC and surfactant-modified CNC (s-CNC). The mechanical qualities are improved by the incorporation of CNC and s-CNC. Surfactants with CNC modifications are employed in both clean PLA and PLA-PBS blends at the compositions of 1 wt% and 3 wt%. Improvement of tensile strength is observed from 21 MPa to 23 MPa. There is no significant change in tensile strength from 1 wt% to 3 wt% loading but tensile modulus certainly increased from 1300 MPa to 1550 MPa. The obtained mechanical properties of CNC and s-CNC are relatively similar in tensile strength. However, the tensile modulus of s-CNC composite is lower at 3 wt% compared to CNC over all the compositions for the PLA-based composite system. The same loading percentage has been tried with 80% and 20% PLA/PBS blend, there is an increment in tensile yield strength observed for loading s-CNC at 3 wt%, whereas, unmodified CNC shows the yield strength lower than the neat PLA/PBS (80/20) blend. It may be due to the presence of NPs that induced different nucleation effects in the PLA and PBS phases. Due to thermodynamic incompatibility, it creates phase separation and hence, lowers the value of tensile strength (Luzi et al., 2016). But a different study has also been reported where cellulose nanocrystal and cellulose microcrystal has been added to PLA/PHB blend with the addition of plasticizer and different composition of nanocomposite and micro composite has been prepared and it was observed that the addition of plasticizer with PHB improves the flexibility of PLA through reducing tensile strength and improving the elongation at break. By adding microcrystalline cellulose (MCC), tensile strength is reduced from 25 MPa to 14 MPa, and by adding CNC, tensile strength is reduced from 25 MPa to 10 MPa. Also, the strain value increases higher from 22%

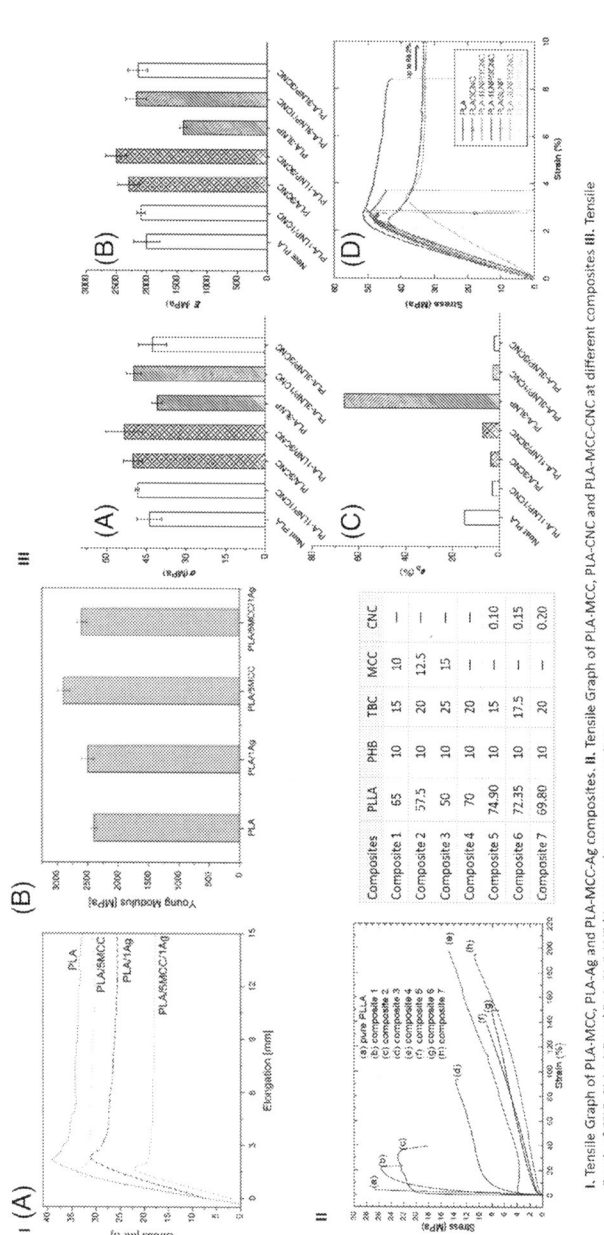

Composites	PLLA	PHB	TBC	MCC	CNC
Composite 1	65	10	15	10	—
Composite 2	57.5	10	20	12.5	—
Composite 3	50	10	25	15	—
Composite 4	70	10	20	—	—
Composite 5	74.90	10	15	—	0.10
Composite 6	72.35	10	17.5	—	0.15
Composite 7	69.80	10	20	—	0.20

I. Tensile Graph of PLA-MCC, PLA-Ag and PLA-MCC-Ag composites. **II.** Tensile Graph of PLA-MCC, PLA-CNC and PLA-MCC-CNC at different composites. **III.** Tensile Graph of CNC, PLA-LNP and PLA-CNC-LNP binary and ternary composites.

FIGURE 20.7

Mechanical properties of binary and ternary composites (Yang et al., 2016a), tensile data of PLA/CNC and LNP (El-Hadi, 2017) tensile data of PLA-PHB and CNC.

to 80% and from 140% to 190%, respectively. Similarly, the structure of CNC and MCC has been studied, and it was found that the CNC is well dispersed in the PLA matrix, and MCC formed as a less dispersed and agglomerate structure. The amount of CNC added is much less compared to the MCC used (El-Hadi, 2017). Starch-based nanocomposites are prepared by dispersing montmorillonite and cellulose nanofibre (CNF) nanofibre through solution blending. Mechanical properties of binary and ternary nanocomposites were studied in the ternary composite, CS/3MMT/3CNF shows higher tensile strength and modulus than CS/3MMT and CS/3CNF. This is possible because of the better dispersion of montmorillonite (MMT) and synergistic effect with CNF. 2D MMT and one-dimensional (1D) CNF are stacked alternately to provide a better-layered ternary system. The addition of only MMT up to 5 wt% also leads to an increase in tensile strength and modulus but a slight reduction in elongation at break can be observed. Whereas keeping fixed MMT quantity at 5 wt% and altering the CNF composition from 1 wt% to 7 wt%, leads to improvement in strength and modulus with improved elongation at break. It confirms that the addition of CNF improves the mechanical properties through a better synergistic effect (Li et al., 2019). To evaluate the mechanical properties in both machine direction (MD) and transverse direction, PP random copolymer with clay (modified MMT) and poly-β-pinene (PβP) is studied transverse direction (TD). In TD, an increase in y was shown by the addition of PP and clay/PP compared to pure polypropylene random (PPR). However, in MD there was no noticeable difference in the strain at a yield of the films. For all samples, the strain at break does not significantly alter in either direction. However, the yield stress is larger for the PPR/clay and PPR/PβP/clay films than for plain PPR. (Hurley et al., 2013). The addition of sepiolite to PET in 0 wt% and 3 wt% leads to an increase in the tensile strength of PET sheets by 28% as compared to pure PET. This can be because of the good dispersion of sepiolite clay in PET and the larger specific area of sepiolite that influenced the properties through physical and chemical interaction. But in the range between 1 wt% and 3 wt%, there is only a 4% increase in strength observed. Also, the elongation of break is decreased up to approximately 50%. It suggests there is a good interfacial bond strength between nano clay and matrix (Fernández-Menéndez et al., 2020). PLA/CNC binary composite is incorporated with LNP to observe the change in mechanical parameter tensile strength and elongation at break. The observation shows a reduction in tensile strength and elongation at the break due to incompatibility and the nonsynergistic effect of CNC and LNP NPs (Yang et al., 2016a). The addition of neat halloysite in the PET matrix does not affect much the mechanical strength and modulus. However, functionalized halloysite at 2% loading improves the ultimate tensile strength and reduces the ultimate tensile strain of binary composites (Garcia-Escobar et al., 2021).

20.3.3 Thermal properties

Thermal properties of binary and ternary composite (Fig. 20.8) are vital because of their stability under the service condition. Growing demand for polymer in food packaging leads to the innovation of synthetic as well as bio-based binary and ternary

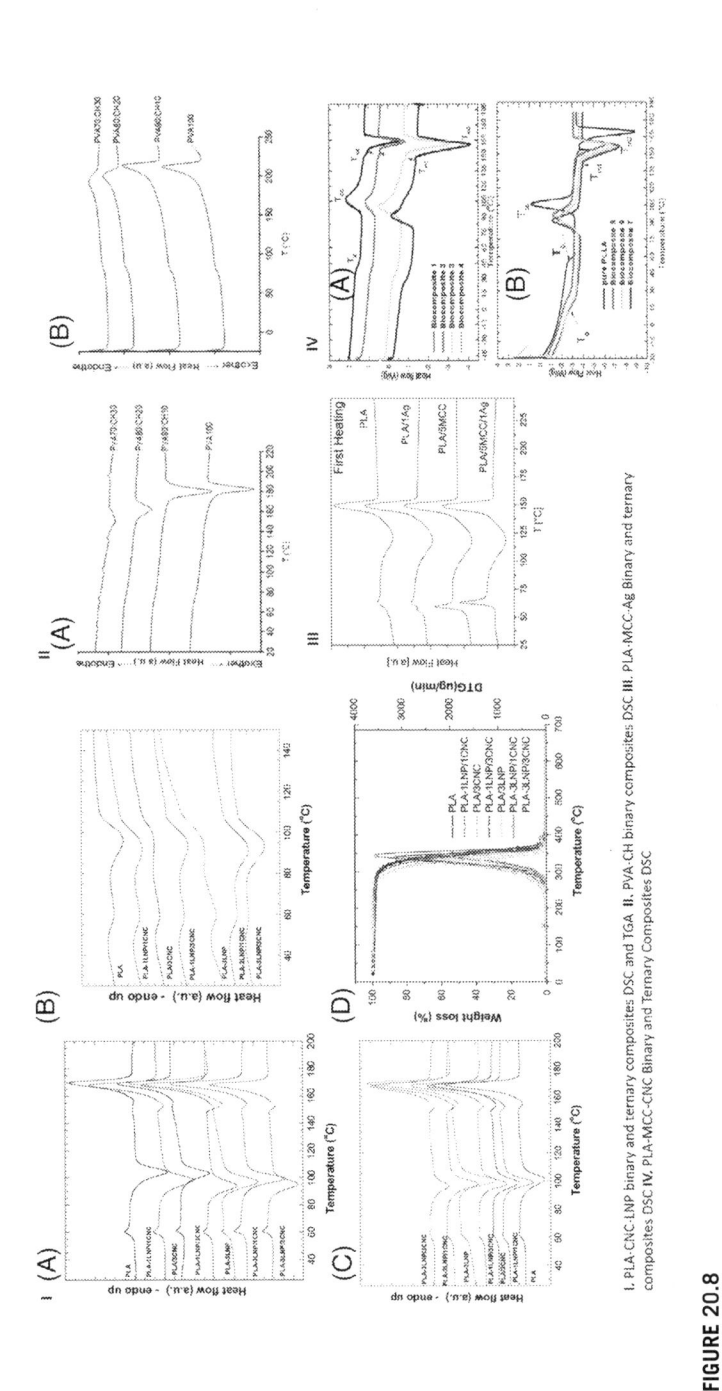

FIGURE 20.8

Thermal properties of binary and ternary composites (Yang et al., 2016a), DSC graph of PLA-CNC and LNP (El-Hadi, 2017), DSC graph of PLA/PHB-MCC and CNC.

composites. The improvement in thermal properties with the addition of micro and nanofillers was examined through different analytical techniques. The crystallization behavior of the polymer should get influenced by the addition of nanofiller. It provides nucleation points during the crystallization and recrystallization process. As an effect of this, the primary crystal structure increases with a more uniform crystal size. Microfiller also influenced the crystallization process but not up to the limit of nano-sized filler. The thermal behavior of PLA/acrylated crystalline nanocellulose (ACNC) nanocomposites significantly improved in comparison to pure PLA with the addition of 1 wt% ACNC. Because PLA chains could not move about as freely due to the homogeneous distribution of ACNC in the PLA matrix, the phase transition in the thermal behavior of composite film needed more energy. Also, by addition of zinc oxide with CNC improved the crystallinity up to 5 wt% but there is no such improvement with 1 wt% loading of CNC and the thermal stability of neat PLA and composite film slightly decreased by the addition of ZnO NP with a larger amount of up to 7 wt%. This indicates the catalyzed intermolecular transesterification reaction with low molecular weight and unzipping of the polymer chain, also reduces the degradation peak from 379°C to 292°C and onset temperature from 320°C to 208°C (Yu et al., 2021). The thermal behavior of PLA is studied with thymol and silver NPs. It does not show apprehensive changes in Tg for PLA-Ag-filled composites. However, the thymol-based binary and ternary composite shows a reduction in Tg up to 10°C from 53°C to 43°C. It confirms the presence of the thymol-induced plasticization effect in PLA. The thermal stability of PLA film and plasticized composite PLA was studied to understand the effect of thymol and silver NPs on PLA. It was observed that the addition of thymol slightly decreased the thermal stability of PLA with a shifting of degradation peak from 363°C to 334°C and onset temperature with 5 wt% weight loss from 320°C to 284°C (Ramos et al., 2016). The addition of CNC, s-CNC, and Ag NP was incorporated in PLA to study the improvement in crystallinity. The s-CNC shows improvement in the percentage of crystallinity from 9.2% to 18%, whereas, the addition of CNC and Ag NPs only managed to increase crystallinity up to 12% each at 1 wt% loading for each. Whereas an increase in crystallinity can be observed in the ternary composite of PLA/s-CNC/Ag up to 17% but PLA/CNC/Ag composite shows a reduction in percentage in crystallinity to 6.4%. It can be predicted that s-CNC has better dispersion characteristics and compatibility with the binary composite of PLA/Ag (Fortunati et al., 2012a; Fortunati et al., 2012c; Fortunati et al., 2013). The addition of CNC and LNP with PLA improves the crystallization temperature but does not affect the percentage of crystallinity at a low percentage of loading and an increase in crystallinity observe at a loading of 3 wt% CNC/LNP (Yang et al., 2016a). The incorporation of CH in polyvinyl alcohol structure improves the thermal stability of PVA-CH film as the thermal stability of CH is better than PVA. Similarly, there is an improvement in Tg as the addition of CH increases. It can be predicted that there is strong hydrogen bonding between PVA and CH due to the presence of hydroxyl and amine functional groups. Also, the CH content reduces the melting temperature and enthalpy, as well as reduces the crystallization temperature and rate of PVA-CH film (Bonilla et al., 2014). A similar phenomenon was observed by the addition of MCC and CNC with PLA/PHB/tributyl citrate (TBC) blend. The addition of the CNC-PLA/PHB/TBC ternary physical blend provides a synergistic effect, improved elongation, tensile strength at break, and shows improved results as compared to MCC-PLA/PHB/TBC ternary blend-based composites (El-Hadi, 2017). Halloysite nanotube (HNT) and low-density polyethylene (LDPE) composite do not show an improvement in melting and crystallization temperature up to 5 wt% loading. There is less improvement in properties due to the absence of compatibilizer. It results in phase separation and agglomeration of nanoclay in LDPE. There is a slight jump in crystallinity from 15% to 18% at 1 wt% loading. It may be because of the nucleation effect induced by a low concentration of HNT in PE. The thermal stability of neat PE and PE/HNT composites was studied to understand the influence of HNT on LDPE but there is no such improvement observed from the degradation peak as the thermal stability decreased from 473°C to 463°C and the onset point decreased from 441°C to 435°C (Tas et al., 2017).

20.3.4 Structural and morphological properties

The morphological properties play an important role in determining the other performance properties of composites such as mechanical, thermal, and barrier properties. A better-developed morphology can enhance properties at a low percentage of loading and vice-versa. Different functionalized NPs are doped into the polymer matrix to overcome certain barriers in properties. Two developed morphologies that can be observed during nanocomposite preparation are intercalation and exfoliation. The more the exfoliated structure, the better is the dispersion of NPs within the polymer chain. Whereas the intercalated structure does not show better dispersion, an agglomeration of NPs within the polymer chain can be observed. The higher loading of NPs depending upon their surface area and surface energy can lead to agglomeration and less dispersed structure. For preparing HNT and PE composites, the structural analysis of HNT is done through TEM, it was observed that HNT has a hollow and open-ended structure with a diameter from 15 nm to 40 nm and a length of 100–200 nm with a mostly agglomerate form with a size of 65 μm. In the scanning electron microscopy (SEM) image on the scale of 20 μm and 2 μm, there is a proper dispersion observed, it can be easily observed the low concentration of HNT loading (1 wt%) and less dispersed with aggregate structure can be observed in high concentration (5 wt%) loading of HNT (Tas et al., 2017). The high-density polyethylene (HDPE) and kaolinite and linear low-density polyethylene (LLDPE) and kaolinite composite are characterized through optical microscopy and X-ray diffraction (XRD) to determine the morphology and structural identity. The kaolinite, an average iron particle diameter of 115 nm is observed with particle diameters varying from 78 nm to 220 nm. This XRD, the base peak reflection is visible at 12.4°C on the intact kaolinite diffractogram, while the residual illite reflection is evident at 8.9°C. The loss of kaolinite and illite reflections following iron modification was most likely caused by the structural irregularity brought on by the intercalation of the active metallic component. The HDPE kaolinite and LLDPE kaolinite nanocomposites have been characterized in accordance to determine their structure and morphology with the addition of kaolinite. Both the nanostructures of the composites remain aggregated, which may be due to thermodynamic incompatibility that can be observed through optical microscopy and the structural analysis suggest no peak of modified kaolinite in the diffractogram of HDPE and LLDPE composite. It implies a less dispersed and exfoliated structure. The exfoliation is not possible unless there is a possibility of affinity of the polymer matrix toward the nanoclay (Busolo and Lagaron, 2012). The addition of cellulose NPs (CNC) and LNPs in PLA for ternary composite shows mixed developed morphology. The addition of lignin LNP 3 wt% shows smooth morphology like neat PLA. However, the addition of CNC 3 wt% shows rough morphology and a more brittle tendency. The combination of lignin and CNC at 3LNP/1CNC and 3LNP/3CNC maintains good morphology as compared to 1LNP/3CNC:1LNP/1CNC ratio. The LNP shows better dispersion of NPs and creates an exfoliated structure even with the synergistic combination with CNC (Fortunati et al., 2012b; Yang et al., 2016a). The nanocomposite morphology of PLA, PLA/CNC, PLA/Ag, PLA/s-CNC, and PLA/Ag/s-CNC was determined through SEM and TEM. The s-CNC shows better dispersion as compared to unmodified CNC. The s-CNC dispersion was deeply investigated through SEM analysis. But the transmission electron microscopy (TEM) is not able to identify the single-crystal structure due to the low-contrast image of both cellulose and PLA. The addition of Ag NPs shows better crystal growth and a distinct single crystal structure can be visible through TEM analysis. TEM analysis shows higher homogeneity of dispersion of Ag NPs with less agglomeration of the metal NP. The wide-angle X-ray analysis was performed to understand the effect of NPs on the crystal structure of PLA nanocomposites. PLA wide angle X-ray scattering (WAXS) pattern only showed a wide diffraction band at 2θ angle of 16.5°convert all °C to ° in XRD section. It gives a clear indication that PLA is amorphous. PLA ternary systems with 1 wt% of CNC or s-CNC and PLA with 5 wt% of CNC showed two main peaks at 16.5°C and 22.4°C. It represents the crystalline peak of PLA and cellulose nanocrystals, respectively. The addition of both 5 wt% CNC and s-CNC improves the crystallinity which is observed through an increase in the intensity of the primary peak of PLA at 16.5°C and an additional peak can be observed at 38.3°C, 44.2°C, 65.0°C, and 77.3°C. It represents the crystal structure of Ag NPs in a crystal plane of (1 1 1), (2 0 0), (2 2 0), and (3 1 1), respectively (Fortunati et al., 2013). TEM images of PET sheets containing 3 weight percent of cloisite 20A NPs reveal a substantially exfoliated/intercalated structure with distributed tactoids. The PET/3 wt% NC sample's intercalated nanoclay structure is shown to have a width of around 72 nm. The distinct delaminated silicate layers exhibit additional exfoliated structures with a width of about 25–27 nm. The image shows that the PET nanocomposite structures' nanoclay particle lengths range from 130 nm to 200 nm. The dispersion of nanocolsite particles in the PET matrix must be ascertained. After intense interaction with the clay surface, colsite NPs that resemble plates were dispersed throughout the polymer chain. Better interaction can result in improved exfoliated structure and dispersion (Hurley et al., 2013). The XRD graph halloysite and functional halloysite shows a clear presence of quartz impurities in the system and the spacing at the (0 0 1) plane of the nanoclay is 0.72 nm. The addition of HNT and functionalized HNT does not alter the inherited crystallinity of PET (Garcia-Escobar et al., 2021).

20.3.5 **Antimicrobial properties**

Because of the random, nonenzymatic chain scissions of the ester groups, the PLA's molecular weight is reduced, which results in the production of oligomers and lactic acid when it reaches a molecular weight of 10,000–20,000 g/mol. They can be metabolized and transformed into carbon dioxide, water, and humus by microorganisms like fungi and bacteria. The PLA-limonene-CNC ternary composite, PLA with 20 wt% limonene, and CNC binary composites from 1 wt% to 3 wt% were also studied. It was observed that the binary PLA, 20 wt% limonene, and the PLA-limonene-CNC systems reached about 25% weight loss disintegration on the 7th day, while 20% of weight loss was measured for the PLA-CNC binary systems. The inclusion of cellulose nanocrystals increased the systems' crystallinity, which in turn affected the pace at which water diffused through the PLA matrix and resulted in a decreased disintegration rate (Fortunati et al., 2014). The antibacterial activities of the neat PLA film and PLA nanocomposite were investigated, and it was discovered that they were unable to contrast the progression of bacterial cell multiplication over time because they appeared to support a postgrowth with bacterial values that increased from 1×10^6 to 8.4×10^7 CFU/mL in just 24 hours. It was discovered that PLA/3LNP has better antimicrobial capabilities than PLA/3CNC after measuring the antibacterial activity of PLA nanocomposite film. The antimicrobial activity of lignin is due to the presence of the phenolic compound and different functional groups like (-OH, -CO, and -COOH) containing oxygen in its structure. Also, the presence of the combined effect of 3LNP/1CNC and 3LNP/3CNC improves crystallinity. It reduces the diffusion mechanism and hence, better antimicrobial action as compared to CNC-based binary nanocomposites (Yang et al., 2016a). Many antibacterial activities have been reported by the addition of LNP, same study has reported the addition of LNP in PVA and CH. The same bacterial growth (PCO-115) and (Xap-3894) were observed as a control in PVA and CH film. Differently, bacterial development was significantly reduced with the addition of LNP in PVA/3LNP, PVA/CH/3LNP, and CH/3LNP films (Yang et al., 2016b). The binary blend of PVA/CH film has also been tested for antimicrobial properties and it shows the improvement in antimicrobial activity of PVA 80% and CH 20% film is better as compared to 100% PVA film (Bonilla et al., 2014). The microbiological analysis is done for virgin PET tray and PET tray nanocomposite filled with sepiolite tested in the presence of aerobic mesophilic and enterobacteriaceae. The test is carried out over 14 days. The average value of PET trays analyzed is 3.83 log (cfu/g) (8.25×10^3 ufc/g) and the nanocomposite tray number is 2.98 log (cfu/g) (9.67×10^2 ufc/g). The bacterial growth count in the nanocomposite tray is from 2.83 log (cfu/g) to 2.98 log (cfu/g), whereas the PET tray has a growth count from 2.43 log (cfu/g) to 3.80 log (cfu/g). The nanocomposite tray shows better results due to the better retention of CO_2 and also the inhabiting nature of N_2 (Fernández-Menéndez et al., 2021a). The modification of PET by silicon oxide (SiOx) through a silane coupling agent for compatibility with CH/nano-ZnO particles leads to an improvement in antibacterial properties. The OD600 test was conducted to understand the effect of bacteria (*Staphylococcus aureus* and *Escherichia coli*). The addition of ZnO shows 100% growth inhibition of bacteria within 2 hours (Li et al., 2021). By considering the initial counts of *S. aureus*, it is also possible to examine how ZnO NPs affect the antibacterial qualities of PET-CNC composite. Bacterium and *E. coli* were 5 × 105 cfu per milliliter and 6 × 105 cfu per antimicrobial count, respectively. Following the setting up of cultural media, *E. coli* and *S. aureus* were counted after *S. aureus* was exposed to the nano biocomposite films to gauge their level of antimicrobial protection. *E. coli* and *S. aureus* assessing the outcomes from colonies of *S. aureus*. The findings demonstrated that the antibacterial capabilities of the film were enhanced as the microbiological count considerably decreased with increasing ZnO NP content. As can be seen, a greater reduction in the cell viability of the gram-positive bacterium (*S. aureus*) than the Gram-negative bacterium (*E. coli*) strain was observed. The *S. aureus* strain decreased from $5.00 \times 10^5 \pm 0.01$ to $2.49 \times 10^5 \pm 0.014$ from 0 wt% to 4 wt% loading of ZnO. Similarly, *E. coli* strain reduces from $5.00 \times 10^5 \pm 0.01$ to $4.12 \times 10^5 \pm 0.02$ loading from 0 wt% to 2 wt% loading of ZnO NP (Leila et al., 2021).

20.3.6 Barrier properties

Barrier and antibacterial properties (Fig. 20.9) are very much vital in the case of food packaging applications. Barrier properties of synthetic and biopolymers depend on the molecular structure and surface properties, such as hydrophilicity and hydrophobicity. Most of the biopolymers are hydrophilic such as PLA, PBAT, PVA, and many more synthetic polymers like polyolefin are hydrophobic and some of them are partially hydrophilic such as PET, PBT, and polycarbonate (PC). To cover such issues, polymers are incorporated with nanofillers and other active additives for surface and structural modification. The barrier properties for nanocomposites depend on the structural path of diffusion, aspect ratio, and physical nature of the nanofiller. By considering the example of PE and PET, in comparison to PET, PE has better WVT rate (WVTR) and lower oxygen transmission rate (OTR) characteristics. In binary systems with 1 wt% of Ag NPs, the OTR values for PLA and s-CNC/CNC/nano Ag nanobio composites decreased by about 22%. In the case of ternary systems, the greatest OTR reductions were found. PLA/1CNC/1 Ag demonstrated a 46% reduction and PLA/5CNC/1Ag demonstrated a similar reduction. As previously reported for binary nanobiocomposites reinforced with cellulose nanocrystals, even greater drops in OTR were seen for the s-CNC-based systems, highlighting the beneficial effect of the cellulose modification in the barrier properties of ternary PLA systems. The addition of 5 wt% s-CNC imparted the best result with a more positive effect on barrier properties (Fortunati et al., 2013). In PLA-CH-PVOH composites, the barrier properties of PLA and CH decrease, and WVTR increases, this is because of the hydrophilic nature of CH but the addition of polyvinyl alcohol (PVOH) reduces the WVTR from

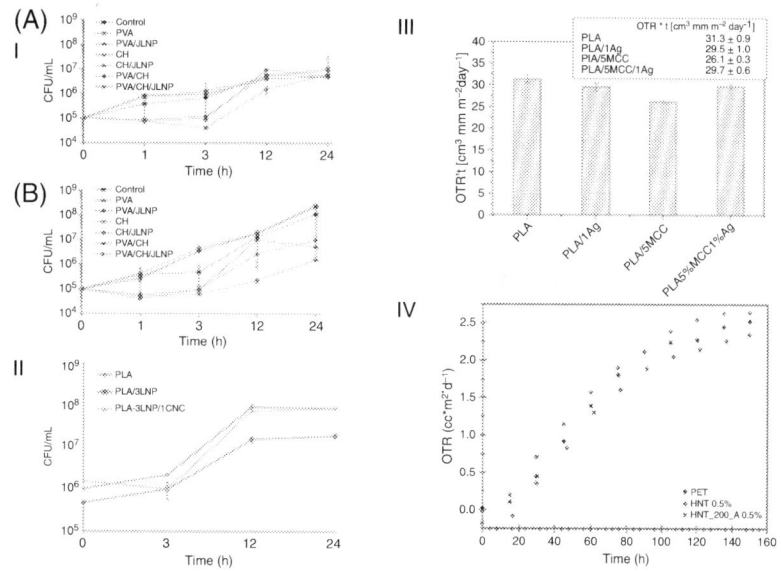

I. **Antimicrobial results** of PVA/CH, PVA/LNP, and PVA/CH/LNP composites. II. **Antimicrobial results** of PLA, PLA/3LNP, and PLA/3LNP/1CNC properties. III. **OTR and WVTR** results of PE film loaded with HNT. IV. **OTR results** of PET/HNT nanocomposites.

FIGURE 20.9

Antimicrobial and barrier properties of binary and ternary composites.

26% to 31%. The increase in WVTR is because of the good phase compatibilization of PLA and CH in the presence of PVOH. The ternary composites of poly(3-hydroxy butyrate-co-3-hydroxyvalerate) (PHBV) with GO and CNC were evaluated for barrier properties like WVTR, the data suggest the higher loading of CNC at 3 wt% reduces the WVT and GO shows the same result at 0.7 wt% loading but in the case of synergistic effect, hybrid NP of CNC grafted GO at an amount ratio 1:0.5 wt% to 1 wt%. By adding 1 wt% and 3 wt% of the CNC, the WVTR reduces by 23.1% and 43.4%, and with the addition of 0.5 wt% and 0.7 wt% of GO, WVTR reduces by 20.3% and 46.8%. With the addition of 1 wt% CNC-GO, the WVT increases from 70.2% to 72.6%. It may be due to the presence of a well-dispersed structure of hybrid NPs with a better symmetrical arrangement (Li et al., 2019). PET is surface-modified with SiOx with the help of a silane-based coupling agent. PET/SiOx, PET/SiOx/CH, and PET/SiOx/CH-nano-zno composites are there for barrier properties and antimicrobial action. The WVTR properties of PET are improved and WVT is reduced due to the presence of SiOx as a protective layer to the PET but the presence of CH and metallic NP again decreases the WVTR properties of PET. It is because of the hydrophilic nature of CH and metallic NP. The oxygen barrier properties of PET improve with the addition of SiOx and further improved with the addition of CH (PET/SiOx/CH) and (PET/SiOx/CH-nano ZnO) coating shows an improvement near to 11 times higher in barrier properties as compared to neat PET film (Li et al., 2021). It observes that the addition of functionalized nano halloysite with silane (BOPTMS) improved the barrier properties of nanocomposites as compared to neat halloysite by blocking the porous structure of halloysite (Garcia-Escobar et al., 2021). Measurements of the WVTR for CMC/PET bilayer films with various ZnO NP content levels. According to the statistical findings, WVTR did not substantially alter across samples containing various amounts of ZnO NPs ($p > 0.05$). By combining ZnO NPs with PET and CNC binary system, the WVTR decreased from nearly 0.32 g/dm 2 KP to approximately 0.22 g/dm 2 KP. It is confirmed that the moisture-inhibitory properties improve with the addition of ZnO NPs, although they remain consistent from 1 wt% to 4 wt% ZnO loading. The addition of fillers, which naturally have poorer permeability qualities than the polymer matrix, can be blamed for the decrease in WVTR. The enhancement in dispersion and synergistic action of ZnO NPs is credited with the decrease in WVTR (Leila et al., 2021).

20.3.7 Migration properties

The addition of NPs creates the possible risk of migration to the food stimulant. It is important to create a nanocomposite film with low portability NPs with the least migration properties. The migration problem occurs through various phenomena such as desorption, diffusion, dissolution, and disintegration phenomenon. According to the European regulation, the acceptable migration limit is 10 mg/dm^2 which roughly translates to 60 mg/kg by considering 1 kg of food. The use of nanotechnology in food packaging application is still a new field to explore, hence, very limited attempt has been made in the study of the nanocomposite. Also, the limitation in the characterization method makes it difficult for qualitative and quantitative analysis. Titanium nitride NPs are added to PET and they have a migration in the concentration of 20 mg/kg (Fernández-Menéndez et al., 2021b).

20.4 Recent development and future scope

In the recent development scenario, the polymer used for food packaging needs to meet the standard properties required for the particular application. Many synthetic polymers have good performance properties with a low migration rate, high stability, and inertness toward food items. Because of the minimal recycling of synthetic polymer and segregation issues, it creates landfill problems and environmental concerns. It leads to the use of biopolymers in food packaging. The biopolymer has some limitations in properties like poor WVTR, high migration, and low stability. To enhance such properties modifications of bio and synthetic polymers are required for the application into food packaging. Biocomposite and bio-based binary and the ternary phase system have gained attention for the use of food packaging for single-use disposable items. The binary and ternary composite system can be developed by adding different functionalized filler nanolignin and modifying cellulose nanocrystals. Also, inorganic nanofiller GO, zinc oxide, titanium oxide, and Ag NPs improve physical, thermal, barrier, and antimicrobial properties. The results show improvement in the properties and performance of nanocomposite film. Also, similarly, synthetic polymer can be blended with biodegradable polymers like thermoplastic starch (TPS), guar gum, bio-based additives, and with an active nanofiller to prepare binary and ternary composite with an increase in properties. The TPS has been commercially blended with PE to manufacture biodegradable carry bags. Similarly, the starch has been modified through chemical grafting and physical mixing of NPs to improve the processability and performance for end-use applications. Some of the typical examples are the addition of silver, copper, and lignin inhabited the microbial growth, as they are costly, they can be used with CNC to provide a synergistic effect at a low percentage of loading.

20.5 Conclusion

The emerging research area of binary and ternary composites is gaining attention for achieving the potential needs of the customer and fulfilling the market demand. The use of different organic nanofillers and microfillers is added to the polymer matrix to enhance the physical, chemical, and biological properties. This book's chapter focuses on the recent development in binary and ternary composite systems in packaging applications. It discusses the effect of the addition of reinforcement in the polymer matrix and the alteration in the overall properties of composites. The most preferable method for preparing binary and ternary composites is through melt blending of polymer with reinforcement. However, the selection of reinforcement plays an important role in achieving certain distinct properties. The current issue in food packaging is the degradability of synthetic polymer due to its adverse impact on the environment. Now, the focus of research is moving toward the development of biocomposites for food packaging. But most biopolymers are moisture-sensitive and possess certain limitations in barrier properties. To overcome the issue, different forms of nanofillers are added to the polymers but they should be environmentally friendly, nonhazardous, and follow the food safety norms in packaging.

References

Attaran, S.A., Hassan, A., Wahit, M.U., 2017. Materials for food packaging applications based on bio-based polymer nanocomposites. J. Thermoplast. Compos. Mater. 30 (2), 143–173. https://doi.org/10.1177/0892705715588801.

Bonilla, J., et al., 2014. Physical, structural and antimicrobial properties of poly vinyl alcohol–chitosan biodegradable films. Food Hydrocoll. 35, 463–470. https://doi.org/10.1016/j.foodhyd.2013.07.002.

Brinchi, L., et al., 2013. Production of nanocrystalline cellulose from lignocellulosic biomass: technology and applications. Carbohydr. Polym. 94 (1), 154–169. https://doi.org/10.1016/j.carbpol.2013.01.033.

Busolo, M.A., Lagaron, J.M., 2012. Oxygen scavenging polyolefin nanocomposite films containing an iron modified kaolinite of interest in active food packaging applications. Innov. Food Sci. Emerg. Technol. 16, 211–217. https://doi.org/10.1016/j.ifset.2012.06.008.

Díez-Pascual, A.M, 2020. Carbon-based polymer nanocomposites for high-performance applications. Polymers 12 (4), 1–6. https://doi.org/10.3390/polym12040872.

Dunlop, M.J., et al., 2021. Polylactic acid cellulose nanocomposite films comprised of wood and tunicate CNCs modified with tannic acid and octadecylamine. Polymer 13 (21), 3661. doi:10.3390/polym13213661.

El-Hadi, A.M., 2017. Increase the elongation at break of poly (lactic acid) composites for use in food packaging films. Sci. Rep. 7, 1–14. https://doi.org/10.1038/srep46767.

Fernández-Menéndez, T., et al., 2020. Industrially produced PET nanocomposites with enhanced properties for food packaging applications. Polym. Test. 90, 106729. https://doi.org/10.1016/j.polymertesting.2020.106729.

Fernández-Menéndez, T., et al., 2021a. Application of PET/sepiolite nanocomposite trays to improve food quality. Foods 10 (6), 1188. https://doi.org/10.3390/foods10061188.

Fernández-Menéndez, T., et al., 2021b. Shelf life of fresh sliced sea bream pack in PET nanocomposite trays. Polymers 13 (12), 1974. https://doi.org/10.3390/polym13121974.

Fortunati, E., et al., 2013. Combined effects of cellulose nanocrystals and silver nanoparticles on the barrier and migration properties of PLA nano-biocomposites. J. Food Eng. 118, 1–8. https://doi.org/10.1016/j.jfoodeng.2013.03.025.

Fortunati, E., et al., 2014. Investigation of thermo-mechanical, chemical and degradative properties of PLA-limonene films reinforced with cellulose nanocrystals extracted from *Phormium tenax* leaves. Eur. Polym. J. 56, 77–91. https://doi.org/10.1016/j.eurpolymj.2014.03.030.

Fortunati, E., Armentano, I., Zhou, Q., Iannoni, A., et al., 2012a. Multifunctional bionanocomposite films of poly (lactic acid), cellulose nanocrystals and silver nanoparticles. Carbohydr. Polym. 87 (2), 1596–1605. https://doi.org/10.1016/j.carbpol.2011.09.066.

Fortunati, E., Armentano, I., Zhou, Q., Puglia, D., et al., 2012b. Microstructure and nonisothermal cold crystallization of PLA composites based on silver nanoparticles and nanocrystalline cellulose. Polym. Degrad. Stab. 97 (10), 2027–2036. https://doi.org/10.1016/j.polymdegradstab.2012.03.027.

Fortunati, E., Peltzer, M., et al., 2012c. Effects of modified cellulose nanocrystals on the barrier and migration properties of PLA nano-biocomposites. Carbohydr. Polym. 90 (2), 948–956. https://doi.org/10.1016/j.carbpol.2012.06.025.

Garcia-Escobar, F., et al., 2021. Halloysite silanization in polyethylene terephthalate composites for bottling and packaging applications. J. Mater. Sci. 56 (29), 16376–16386. https://doi.org/10.1007/s10853-021-06337-8.

Hurley, B.R.A., et al., 2013. Effects of private and public label packaging on consumer purchase patterns. Packag. Technol. Sci. 29, 399–412. https://doi.org/10.1002/pts.

Kausar, A., 2020. A review of fundamental principles and applications of polymer nanocomposites filled with both nanoclay and nano-sized carbon allotropes: graphene and carbon nanotubes. J. Plast. Film Sheeting 36 (2), 209–228. https://doi.org/10.1177/8756087919884607.

Leila, S., et al., 2021. Potential perspectives of CMC–PET/ZnO bilayer nanocomposite films for food packaging applications: physical, mechanical and antimicrobial properties. J. Food Meas. Charact. 15 (4), 1–10. https://doi.org/10.1007/s11694-021-00880-3.

Li, J., et al., 2019. Fabrication and characterization of starch-based nanocomposites reinforced with montmorillonite and cellulose nano fibers. Carbohydr. Polym. 210, 429–436. https://doi.org/10.1016/j.carbpol.2019.01.051.

Li, Y., et al., 2021. High-barrier and antibacterial films based on PET/SiOx for food packaging applications. Food Sci. Technol. 41 (3), 763–767. https://doi.org/10.1590/fst.37720.

Luzi, F., et al., 2016. Production and characterization of PLA PBS biodegradable blends reinforced with cellulose nanocrystals extracted from hemp fibre. Ind. Crops Prod. 93, 1–14. https://doi.org/10.1016/j.indcrop.2016.01.045.

Mohanty, A.K., et al., 2018. Composites from renewable and sustainable resources: challenges and innovations. Science 362 (6414), 536–542. https://doi.org/10.1126/science.aat9072.

Novák, I. et al., 2016. Polyolefin in packaging and food industry. In: Polyolefin Compounds and Materials Fundamentals and Industrial Applications, 1st edition, pp. 181–199. https://doi.org/10.1007/978-3-319-25982-6_7.

Pereira de Abreu, D.A., et al., 2007. Development of new polyolefin films with nanoclays for application in food packaging. Eur. Polym. J. 43 (6), 2229–2243. https://doi.org/10.1016/j.eurpolymj.2007.01.021.

Ramos, M., et al., 2016. Characterization and disintegrability under composting conditions of PLA-based nanocomposite films with thymol and silver nanoparticles. Polym. Degrad. Stab. 132, 2–10. https://doi.org/10.1016/j.polymdegradstab.2016.05.015.

Ramos, M., et al., 2020. Controlled release, disintegration, antioxidant, and antimicrobial properties of poly (lactic acid)/thymol/nanoclay composites. Polymers 12 (9), 1878. https://doi.org/10.3390/polym12091878.

Tas, C.E., et al., 2017. Halloysite nanotubes/polyethylene nanocomposites for active food packaging materials with ethylene scavenging and gas barrier properties. Food Bioproc. Technol. 10 (4), 789–798. https://doi.org/10.1007/s11947-017-1860-0.

Tiwari, S.K., et al., 2016. Magical allotropes of carbon: prospects and applications. Crit. Rev. Solid State Mater. Sci. 41 (4), 257–317. https://doi.org/10.1080/10408436.2015.1127206.

Videira-Quintela, D., Martin, O., Montalvo, G., 2021. Recent advances in polymer-metallic composites for food packaging applications. Trends Food Sci. Technol. 109, 230–244. https://doi.org/10.1016/j.tifs.2021.01.020.

Wang, L.F., Rhim, J.W., 2015. Preparation and application of agar/alginate/collagen ternary blend functional food packaging films. Int. J. Biol. Macromol. 80, 460–468. https://doi.org/10.1016/j.ijbiomac.2015.07.007.

Yang, W., Fortunati, E., et al., 2016a. Synergic effect of cellulose and lignin nanostructures in PLA based systems for food antibacterial packaging. Eur. Polym. J. 79, 1–12. https://doi.org/10.1016/j.eurpolymj.2016.04.003.

Yang, W., Kenny, J.M., Puglia, D, 2015. Structure and properties of biodegradable wheat gluten bionanocomposites containing lignin nanoparticles. Ind. Crops Prod. 74, 348–356. https://doi.org/10.1016/j.indcrop.2015.05.032.

Yang, W., Owczarek, J.S., et al., 2016b. Antioxidant and antibacterial lignin nanoparticles in polyvinyl alcohol/chitosan films for active packaging. Ind. Crops Prod. 94, 800–811. https://doi.org/10.1016/j.indcrop.2016.09.061.

Yu, F., et al., 2021. Poly(lactic acid)-based composite film reinforced with acetylated cellulose nanocrystals and zno nanoparticles for active food packaging. Int. J. Biol. Macromol. 186, 770–779. https://doi.org/10.1016/j.ijbiomac.2021.07.097.

CHAPTER

Perspective for electrospinning polymeric nanofibers in food processing and packaging

21

Jhansi Lakshmi Parimi[a,b], Pawan Prabhakar[c], Padmavati Manchikanti[a], Santanu Dhara[b] and Mamoni Banerjee[c]

[a] *Plant Metabolic Pathway Laboratory, Rajiv Gandhi School of Intellectual Property Law (RGSOIPL), Indian Institute of Technology Kharagpur, Kharagpur, West Bengal, India,* [b] *Biomaterial and Tissue Engineering Laboratory (BMTE), School of Medical Sciences & Technology (SMST), Indian Institute of Technology Kharagpur, Kharagpur, West Bengal, India,* [c] *Bio-Research Laboratory, Rajendra Mishra School of Engineering Entrepreneurship, Indian Institute of Technology Kharagpur, Kharagpur, West Bengal, India*

21.1 Introduction

Swift developments of nanotechnology, first and foremost those addressing production, packaging, preservation, distribution, performance, conservation, and food hygiene (Wang et al., 2021). According to food packaging guidelines, the packaging is "a coordinated system of preparing goods and ensuring safe delivery to the customers." Emerging challenges posed by urbanization and the problems posed to the environment are the pressing stones for food technology. The prime goal of food packing is undoubtedly to provide mechanical support and protection from external contamination (Vilela et al., 2018). Conventional food packaging modalities failed to provide enough protection and created the necessity for multifunctional packaging systems as shown in Fig. 21.1.

Innovative packaging approaches emerged, ensuring flexibility throughout the handling of food, safeguarding the quality of the food, extending shelf life, and minimizing decomposition and contamination (Lai, 2021). Advances in technology brought life to active food packaging strategies; which were designed to release oxygen-scavenging substances to provide an alternative method of preventing oxidation (Terra et al., 2021). These active packaging systems not only act as barriers between the environment and food material but also provide the release of active substances that possess the capability to stop the growth of micro-organisms and prevent oxidation of the products, etc. as shown in Fig. 21.2. These systems possess the capacity to protect the activity of the stored substances by protecting them from

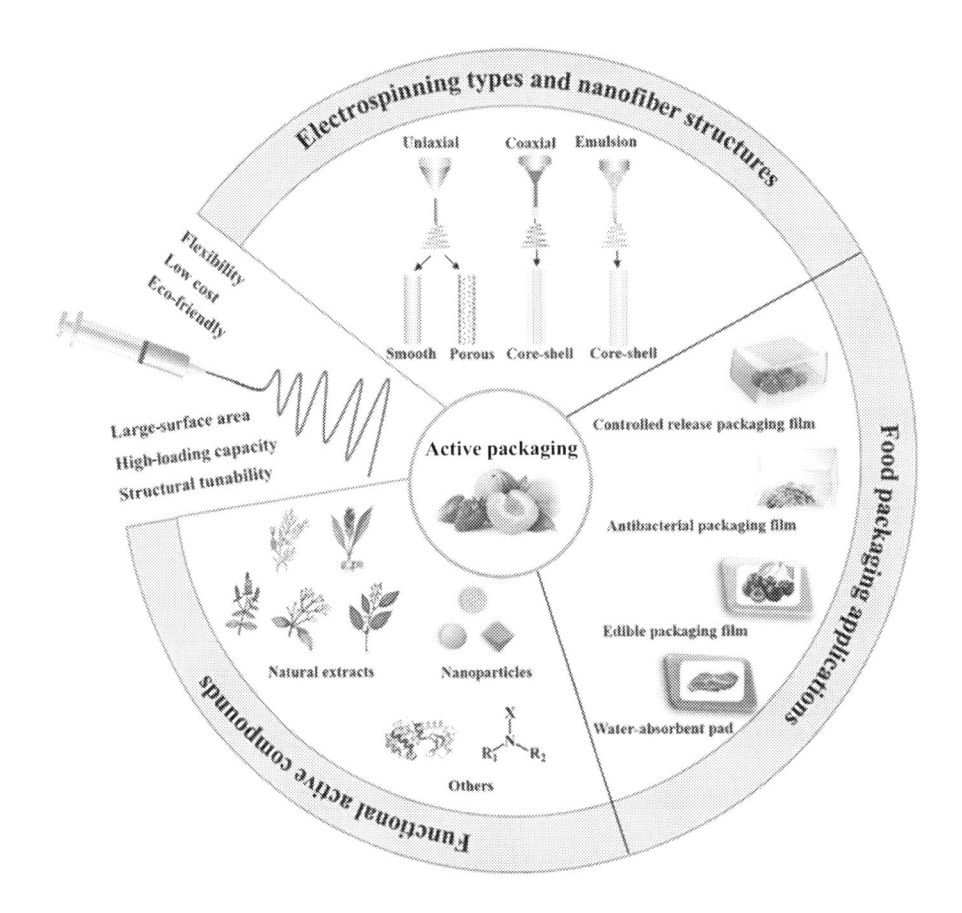

FIGURE 21.1

Schematic representation of electrospinning types for food packaging: functional active components, food package applications (reprinted with permission from Min et al. [2021]).

released gases too, which in turn stops the degradation occurred due to physical, chemical, or microbial degradation (Vilela et al., 2017).

The way forward for active packaging systems is the design and development of indicator systems, which otherwise are smart and intelligent performers to indicate the status of the stored food during transportation and storage. It is an emerging area in food packaging (Shavisi and Shahbazi, 2022). Among various types in this category, are time-temperature indicator packing materials, and color change indicator materials. These materials show the change in the status of the food materials as the reaction to temperature changes over time and changes in the pH of the food products, respectively. These are the most useful and commendable advancements brought about by the research outputs of many scientists working in this field. These indicators contain integrated dyes within the carrier polymer matrix to show the parameters of

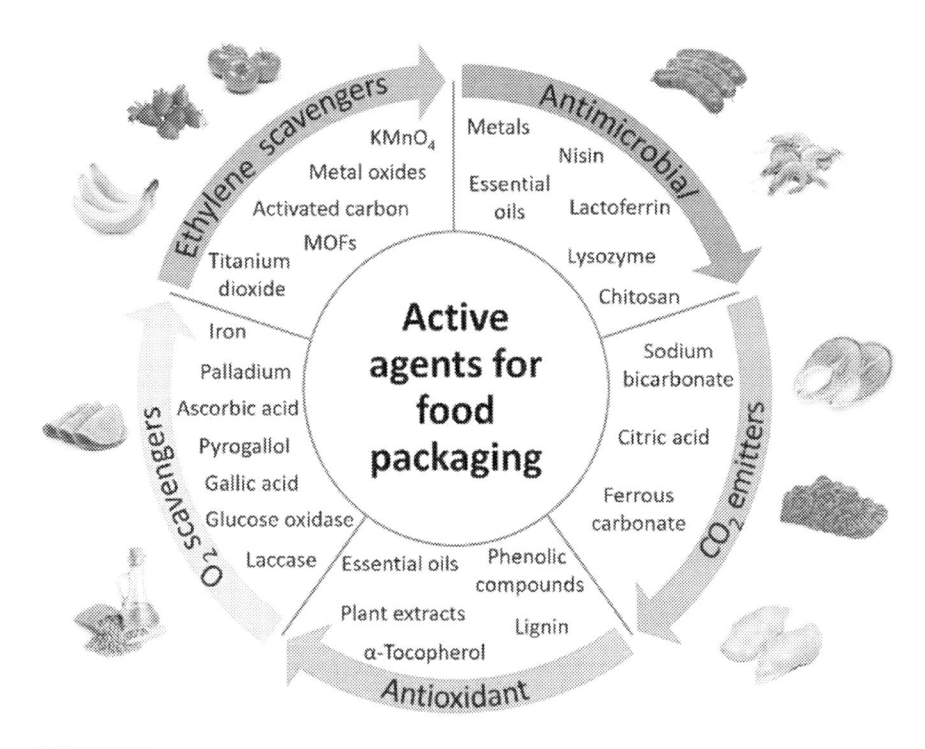

FIGURE 21.2

Active agents for active food packaging (reprinted with permission from Vilela et al. [2017]).

variables such as temperature, pH, etc. Over the use of chemical dyes natural dyes such as curcumin and quercetin, and polycyanin extracted from Spirulina (microalgae) are being in use over chemical dyes offers promising biocompatibility, nontoxicity, and reduces the bioaccumulation of dyes and pigments (Alizadeh-Sani et al., 2020).

Food packaging with nanofibers matrices or scaffolds made of electrospinning has been explored for a decade as they offer many advantageous merits to packaging strategies because of their large surface area, highly porous network, water permeability and the ability to encapsulate sensitive active materials, etc. (Mihindukulasuriya and Lim, 2014). Electrospun nanofiber membrane is gradually employed in food packaging in the preservation of some perishable foods, including fruits, meat, fish, mushrooms, and so on. After being processed into commodities or purchased by consumers, this kind of food has strict storage requirements (Doğan et al., 2022).

Polymeric materials are trendy materials used for food packaging due to their easy formability, lightweight, and low cost (Rhim et al., 2013). Rapid development of a range of new materials from renewable resources, including natural biopolymers, such as polysaccharide and proteinous biopolymers and synthetic polymers such as polycaprolactone (PCL), polyethylene oxide (PEO), polyvinyl alcohol (PVA), and polylactic acid (PLA), etc. These polymers with suitable functionalization strategies of the nanofiber surface with or without added bioactive components are proposed by an extensive number of research efforts (Wen et al., 2017).

Technological advancements in the packaging industry brought the focus to indicator systems in the packaging membranes. These upon observing the temperature change, degradation products release, gas accumulation shown color, etc. indicate the freshness of the stored food materials. These show, product and package interaction, and its impact on the product-package environment. When integrated with the food packages, the advanced active food packaging technologies proved useful to indicate the changes as and when happened during the supply chain, thereby offering convenience to the consumers (Shao et al., 2021).

This chapter is a comprehensive review of the strategies involved with electrospun nanofibers being under exploration for food processing and packaging for active and smart indicating functions. This list is not only limited to the summary of the applications but also covers the list of functional and active packaging strategies available so far. More to the point, the encapsulation of active substances and various functionalization strategies explored were discussed in detail.

21.2 Nanofiber-based food packaging strategies

Modern packaging systems gave birth to active packaging systems, which possess activity control functionalities within (Yildirim, 2017). Depending upon the activity desired, functional/active components will be included in the polymer matrix or film that is in use for packing.

The exploitation of electrospun nanofibers as active packaging systems' intended purpose is to provide barrier properties against H_2O, O_2 (Aydogdu et al., 2019), and CO_2 (Restuccia et al., 2010). They not only act as barriers but also possess the ability to protect water vapor, a key parameter for maintaining the quality of packaged food by protecting the food from environmental contamination.

Polymers and active agents (protecting agents) are the key functional components of nanofiber compositions. Polymer alone or in composite form with or without functionalization or cross-linking forms the structural support for the encapsulation of the loaded active materials to achieve the desired function.

21.2.1 Polymers

In the present scenario, the selection of packaging material is crucial for the food and beverage industries. Adaptation of biodegradable and environment-friendly food packaging materials became a dire need. In this direction, the use of biodegradable polymers as novel materials and attracting sustainable development strategies. Several biodegradable natural (starch, cellulose, and protein) and synthetic (polyhydroxyalkanoates, polyglycolide, and polylactide) like biodegradable polymers have been used for packaging (Mangaraj et al., 2019) as shown in Fig. 21.3. These biodegradable polymers are divided into two basic classes based on their mechanism of action: antioxidant and antimicrobial (antibacterial and antifungal).

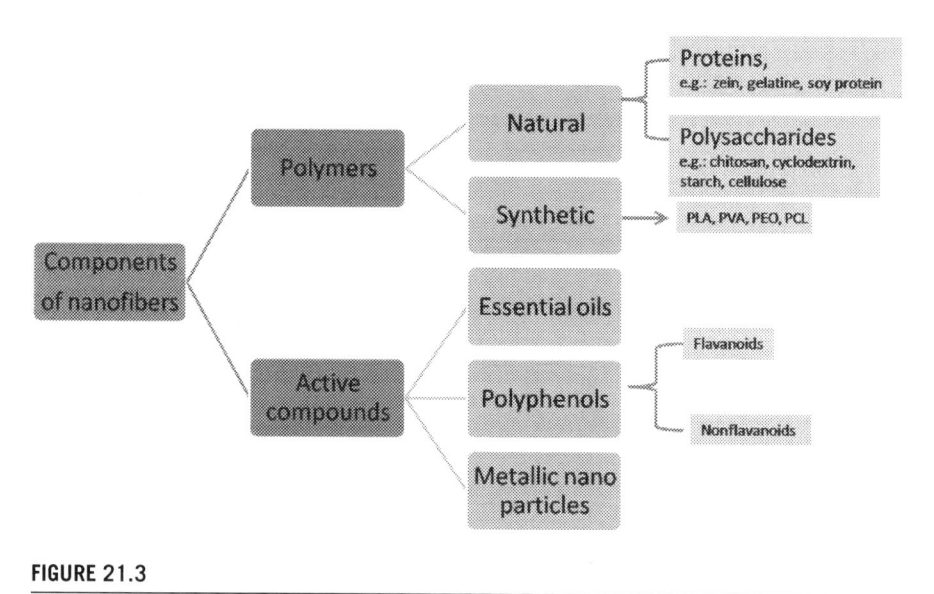

FIGURE 21.3

Schematic illustration of the components involved in nanofibers used for food processing and packaging.

21.2.1.1 Antioxidant biodegradable polymers

All the food items that contain lipids are susceptible to lipid oxidation which may result in loss of flavor or odd flavor, color, functionality, and nutritional value. This may lead to toxicity and be detrimental to the health of consumers. Food items like eggs, meat cheese, fried foods, etc. are highly susceptible to lipid peroxidation (Ahmed et al., 2016). The reason behind lipid peroxidation is the presence of oxygen in the food material which is dissolved during food packaging and is very difficult to eliminate. Therefore, the food industry has recognized the use of antioxidants for scavenging these free radicals. As per the directives of European Union Legislation (Directive 2006/52/EC), few artificial and natural antioxidants have been permitted to be used in food items such as tertbutylhydroquinone, butylated hydroxytoluene, butylated hydroxyanisole, iron or ferrous oxide, catechol, glucose oxidase, and ascorbic acid (De et al., 2016).

Various biodegradable polymer materials have been reported that work as delivery materials for natural antioxidants (Lai, 2021). Corn starch/coconut shell extract/Sepiolite clay results in improved antioxidant activities, elongation break, and water vapor permeability properties (Tongdeesoontorn et al., 2020). Gallic acid grafted on modified silica nanoparticles that are incorporated in chitosan fibers has shown a barrier to ultraviolet (UV) light, and water vapor and significantly increased antioxidant activity (Dong et al., 2022). A few of them along with their intended use are shown in Table 21.1.

Table 21.1 Electrospun nanofibers in active packaging for the food stability and safety.

Function/s	Polymer/s	Active component/s	Food/s	References
Antioxidant and antimicrobial	Zein	Thyme essential oil	Strawberry	Ansarifar and Moradinezhad (2021)
Antimicrobial	Polyvinyl alcohol (PVA)	Ag nanoparticles	Lemon and strawberry	Kowsalya et al. (2019)
Antioxidant	Zein	Allyl isothiocyanate	Strawberry	Colussi et al. (2021)
Antimicrobial	Polycaprolactone	Polycaprolactone *Acalypha indica* leaf extract	Carrot pieces	Colussi et al. (2021)
Ethylene scavenger	Zein	TiO_2 nanoparticles	Cherry and tomatoes	Böhmer-Maas et al. (2020)
Ethylene scavenger	Polyacrylonitrile	TiO_2 nanoparticles	Bananas	Min et al. (2021)
Antioxidant	Lentil flour/polyethylene oxide (PEO)	Gallic acid	Walnut	Aydogdu et al. (2019)
Scavenger	Ethylene	Alumina/carbon potassium permanganate loaded into alumina nanoparticle	Bananas and avocados	Tirgar et al. (2018)
Antimicrobial	Zein	*Laurus nobilis* and *Rosmarinus officinalis* essential oils	Cheese slices	Göksen et al. (2020)
Antibacterial	Gelatin	Thyme	Chicken	Lin et al. (2018)
Antioxidant and antimicrobial	Polyvinyl alcohol	*Laurus nobilis* and *Rosmarinus officinalis* essential oils	Chicken fillets	Göksen et al. (2020)
Antibacterial	PVA	Poly (hexamethylene biguanide)	Japanese sea bass fillets	Zhang et al. (2021)
Antibacterial	PVA	Cinnamon	Shrimp	Nazari et al. (2019)
Antibacterial	Phospholipid	Cinnamaldehyde/hydroxypropyl-β-cyclodextrin	Fresh-cut cucumber	Surendhiran et al. (2020)
Antibacterial	Carboxymethyl cellulose-gelatin	*Mentha longifolia* L. essential oil	Peeled giant freshwater prawn	Shahbazi et al. (2021)

21.2.1.2 Antimicrobial biodegradable polymers

Controlling bacterial and fungal growth and their proliferation is the great challenge of food packing. Reduction in the growth of microorganisms enhances food quality and safety with an improved shelf life of food. Using antimicrobial materials for food packaging maintains the quality of food items like vegetables, meat, and dairy products (Han, 2020). These microbes include varieties of bacterial and fungal species, yeasts, molds, etc. which have their deteriorative effect on different food commodities. The mechanism of spoilage also depends on several factors like water activity, temperature, pH, and the partial pressures of gases like oxygen and carbon dioxide (Chawla et al., 2021).

The composite film of lignin was incorporated into chitosan and found to be an effective broad spectrum of microorganisms (*Pseudomonas aeruginosa* and *Bacillus subtilis*) (Rai et al., 2017). Limolene-chitosan-based biodegradable antibacterial nanofiber matrices showed efficient delivery of cucumber (Bahrami et al., 2020). Essential oil of Satureja khuzistanica Jamzadon cross-linked whey-protein which inhibits the growth of *P. aeruginosa* and *S. aureus* (Maleki et al., 2018). UV-A activated TiO_2-embedded biodegradable polymers, which were found to be very sensitive for *Escherichia coli* bacterial species. When clove essential oil and nisin were incorporated in chitosan (CS) singly or in combination, they were found to inhibit *Staphylococcus aureus, Salmonella typhimurium, E. coli, Listeria monocytogenes* and maintained the shelf life of pork patties (Kouravand et al., 2018). Artemisia fragrance essential oil coated on chitosan film which demonstrated significant antimicrobial activity against coliforms, molds, and yeasts and enhances the nutritional value and shelf life of chicken fillets refrigerated storage (Xie and Hung, 2018). The biodegradable polymers like polybutylene adipate terephthalate (PBAT), polylactic acid (PLA), and carboxymethyl cellulose (CMC) films were coated with silver which was found to inhibit the growth of two prominent bacterial species, e.g., *S. aureus* and *E. coli* thus capable of killing 100% of bacteria under suitable conditions (Venkatachalam and Lekjing, 2020; Yaghoubi et al., 2021). A few of the successful attempts are shown in Tables 21.1 and 21.2.

21.2.2 Active agents

A diverse variety of active agents as shown in Fig. 21.2 to achieve antimicrobial, antioxidant, oxygen-scavenging carbon dioxide emission, and ethylene scavenging are available as performing materials in functional food packaging systems (Vilela et al., 2018).

The addition of active agents will be done either by coating on the surface of nanofibers or blending into the spinning solution used for the electrospinning to obtain nanofibers (Bastarrachea et al., 2015). A schematic illustration of this process and a list of active agents were shown in Figs. 21.1 and 21.2.

21.2.2.1 Antimicrobial agents

A diverse list of antimicrobial agents is available to be used in antimicrobial packaging. These packing systems protect the packed food material from contamination

Table 21.2 Nanofiber-based smart packaging systems.

Function/s	Polymer/s	Component/s	Food/s	References
Oxygen indicator	Polyvinyl alcohol covered with polystyrene	......	Meatball	Yılmaz and Altan (2021)
Food-freshness indicator	Polylactic acid	Naringin and bromocresol purple	Salmon	Xu et al. (2021)
Food-freshness indicator	PCL/PEO	Hibiscus rosa sinensis extract and silver nanoparticles	Shrimp	Jovanska et al. (2022)
Temperature–time indicator	Laccase and guaiacol	Chitosan/polyvinyl alcohol/tetra ethyl ortho silicate	Milk	Tsai et al. (2021)
Food-freshness indicator	Polyvinylidene fluoride	Anthocyanin and cinnamon essential oil	Pork	Zhang et al. (2022)
Food-freshness indicator	Zein	Alizarin	Rainbow trout	Aghaei et al. (2020)
Food-freshness indicator	Chitosan and polyethylene oxide	Curcumin	Chicken breast	Aydogdu et al. (2019)
pH biosensor	Polyvinyl alcohol	Red cabbage (*Brassica oleracea* L.) extract	Fresh date fruit (rutab)	Maftoonazad and Ramaswamy (2019)
Food-freshness indicator	Double layer pullulan and zein	Purple sweet potato extract, glycerol, and carvacrol	Pork	Guo et al. (2020)

occurring by the attack of bacteria, fungi, etc. Antimicrobial agents include metal ions. Natural plant extracts and synthetic antimicrobial compounds, etc.

For the control of fungal spoilage of food materials; natural products like natamycin and essential oils and preservatives like potassium sorbate showed high inhibition of fungal growth (Paster et al., 1990). Fungal growth can be easily controlled by the use of antifungal nanofiber membranes as packaging materials, as mold/fungal growth generally develops on the surfaces. Incorporation of cinnamaldehyde into the PEO/chitosan nanofibers (Rieger and Schiffman, 2014) and green tea extract/PVA/chitosan-glucose oxidase nanofibers to achieve deoxidation for the inhibition of microbial growth under oxygen-deprived conditions (Ge et al., 2012). A comprehensive list of active antimicrobial substances incorporated into the nanofibers and their subtle benefits are listed in Table 21.1.

21.2.2.2 Antioxidant agents

Oxidation, of the bioactive components from protein, lipids, and polyunsaturated fatty acids occurs due to chain reactions that occur in the presence of oxygen deteriorating the stored food and resulting in off-flavors and off-odors. Oxidation will be generally affected by the processing, packaging, and storage conditions; which in turn affects the food quality and degradation. Conventional methods of addition of antioxidants into packaging materials and structures tend to delay the oxidation process. Recent developments in the technology involving electro-spun nanofibers-based antioxidant-releasing packaging systems provide sustained release of antioxidant agents during storage to achieve efficient protection of stored foods as shown in Table 21.1.

Generally, antioxidants are categorized by the way they show their effects: (1) by radical scavenging (carotenoids, flavonoids, and tocopherols), (2) chelating (lactoferrin, citric acid, and EDTA), (3) oxygen scavenging (iron, ascorbic acids, and curcumin), and (4) oxidative stress reduction of nutritional value (tocopherols and essential oils) and synthetic agents like butylated hydroxyanisole (BHA) tert–butylhydroquinone (TBHQ) (Alehosseini et al., 2019) as shown in the Fig. 21.2.

21.2.2.3 Oxygen scavenging agents

Oxygen scavengers are substances used to protect packaged foods from deterioration or damage caused by to accumulation of excessive residual oxygen levels within the package. The said residual oxygen concentration in the headspace of the package is the key determinant of the oxidation status of the foodstuffs. It may cause a change in color, flavor, or nutritional composition as a result of the growth of aerobic microbes.

The oxygen scavenging capacity of gallic acid upon inclusion into hydroxypropyl methylcellulose/PEO nanofibers (Aydogdu et al., 2019) and poly(3-hydroxybutyrate) (PHB)/palladium np fiber films (Cherpinski et al., 2018) for walnut storage paved the way for novel antioxidant-oxygen scavenging packaging materials.

21.2.2.4 Carbon dioxide emitting agents

Carbon dioxide emitters are the substances that release CO_2, which is then absorbed by the food products with carbonic acid formation due to changes in the bacterial cell wall and its internal components (Yildirim, 2017). Hence, as a result, the life and quality of the stored food will be extended due to preservation from degradation. Substances such as sodium bicarbonate, ferrous carbonate ($FeCO_3$) (Restuccia et al., 2010), and citric acid. The use of these substances as nanofiber materials is in its infancy.

21.2.2.5 Ethylene scavenging agents

Control of ethylene (C_2H_4) levels, which is a phytohormone regulator used for the ripening and senescence of the harvested fruits and vegetables plays a crucial role in food packaging and processing. One such example of this type is potassium permanganate. These substances are toxic hence demanding the non-contact utility. For this reason, immobilization or encapsulation will be the ideal method, hence

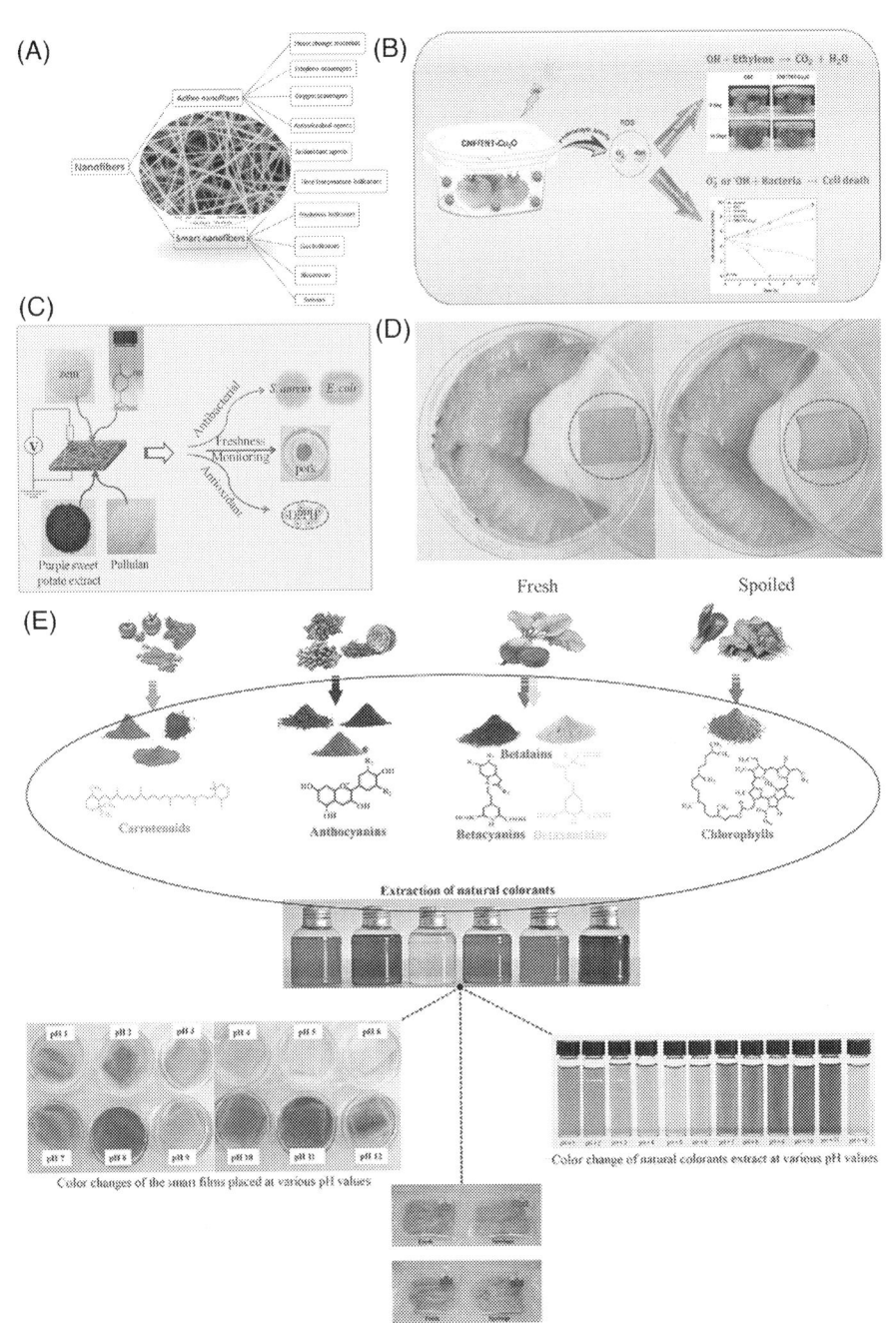

electrospun nanofibers are going to be a gist of hope for ethylene scavenging. Use of scavengers like nanosized clays, activated alumina, and activated carbon (Kaewklin et al., 2018). Cellulose nanofiber (CNF)/TiO_2 nanotubes (TNT)-Cu_2O films showed excellent ethylene oxide scavenging and antimicrobial functions for the active packaging application of fruits and vegetables (Riahi et al., 2022).

21.2.3 Functionalization of nanofibers

Electro-spun nanofibrous matrices perform the function of interlaying between the external environment and stored food materials; hence they can also be called coatings. Their optical, transparent properties will be made functional by various functionalization techniques depending upon the intended use (Fabra et al., 2013). They include antibacterial, antioxidant, scavenging, etc. The surfaces of the nanofiber matrices functionalization will be done by either of the following or a combination of one or two methods: (1) encapsulation (Yao et al., 2016; Altan et al., 2018), (2) thermal post-treatment (Kim et al., 2018; Sagitha et al., 2018), and (3) Surface absorption (Kim et al., 2018; Cherpinski et al., 2018).

21.3 New-generation food packaging

New-generation packaging systems are smart in their action, which means along with the active packaging properties they also contain embedded sensing indicators to show the changes in the product or environment or both (Schaefer and Cheung, 2018). The indicators explored so far are indicators showing changes in temperature-based sensing, degradation product-based sensing (pH), and evolution of released gases or biosensors. Unlike active packaging materials, plastics, and biopolymers are widely used. Biodegradable, biocompatible, and renewable low-cost materials replaced the place of synthetic materials for smart packaging systems as shown in Fig. 21.4 (Ghoshal, 2018).

In the recent past, upholding features of the nanofibrous membranes attracted the attention of the intelligent or smart packaging industry for the monitoring and maintenance of packaged goods throughout the length of the supply chain

FIGURE 21.4

Nanofibersa strategies for active and smart packaging strategies; (A) types of nanofiber packaging strategies, (B) cellulose nanofiber-based ethylene scavenging antimicrobial films (reprinted with permission from Riahi et al. [2022]); (C) intelligent double-layer fiber mats with high colorimetric response sensitivity for food freshness monitoring and preservation (reprinted with permission from Li et al. [2020]); (D) pullulan/chitin nanofibers containing curcumin and anthocyanins for active-intelligent packaging (reprinted with permission from Duan et al. [2021]); (E) pH-sensitive (halochromic) smart packaging films based on natural food colorants for the monitoring of food quality and safety (reprinted with permission from Alizadeh-Sani et al. [2020]).

(Aman Mohammadi et al., 2022) as listed in Table 21.2. The main difference between active packaging systems and smart systems is the indication of the change without releasing their components or migration of the functional components and an interrupted environment of the food (Drago et al., 2020) as shown in Fig. 21.4A.

21.3.1 Time-temperature indicators

These are the devices, which show irreversible changes in the form of color or opacity changes in response to certain thermal events during storage and transportation. Temperature change occurs due to changes in the kinetics due to the physical-chemical or biological decomposition of the packed food material (Forghani et al., 2021) as shown in Table 21.3.

Time-temperature indicators (TTIs) are available in different types; photochemical, microbial, polymer-based, and enzymatic. Among them, enzymatic TTIs are highly sensitive to environmental variations. Hence, immobilization into nanofibers will be a commendable alternative for this type. Laccase enzyme immobilization on zein and chitosan/PVA nanofibers are a few examples in this category (Jhuang et al., 2020).

21.3.2 pH indicators

Food freshness indicators (FFIs) are also called freshness indicators. Generally, these are halochromic materials, which are sensitive to pH variations. Changes in the pH can happen at any time during the supply chain, either by chemical reactions or microbial attacks within the package (Balbinot-Alfaro et al., 2019). These indicators are generally dyes that are sensitive to pH variations. They have been immobilized onto nanofibers and showed great stability as shown in the Fig. 21.4E. Natural pigment from red cabbage (*Brassica oleracea* L.)-PVA nanofibers (Maftoonazad and Ramaswamy, 2019). PCL, PEO nanofibers immobilized with curcumin, quercetin, and phycocyanin showed pH variations in the range of 3–6 (Terra et al., 2021). Anthocyanin–PLA nanofibers for ferric ions detection in the pH range from 3.75 to 6.0 (El-Naggar et al., 2021) and PLA-PEO and microalga spirulina biomass were used as pH indicators successfully (Kuntzler et al., 2020).

21.3.3 Gas indicators

These are the indicator sensors for the released gases through the packaged foods during storage and transportation. Gases like O_2 and CO_2 are released due to changes in the state of the food materials, i.e., physical, chemical, or biological (ZabihzadehKhajavi et al., 2020). Oxygen indicator material coated onto the polystyrene nanofibers showed improved dye leakage (Yılmaz and Altan, 2021).

21.3.4 Sensors and biosensors

These are the sensors developed to detect contaminants/allergens in foods by detecting the target analyte molecules (Mustafa and Andreescu, 2018). Widely used sensing

Table 21.3 Applications of electrospun nanofibers for food packaging.

Category	Food	Form	Applications	References
Meat products	Chicken	Wrapper film	Antimicrobial, antioxidant, and sensory properties	Göksen et al. (2020); Surendhiran et al. (2020); Lin et al. (2018)
	Meat	Film and wrapper film		Altan et al. (2018); Cui et al. (2018); Altan et al. (2018)
	Pork	Wrapper film		Li et al. (2020); Yang et al. (2021); Wen et al. (2017)
Sea products	Fish	Wrapper film	Antimicrobial, antioxidant, and sensory properties	Eghbalian et al. (2021); Zhang et al. (2021); Ozogul et al. (2017)
	Seafood	Cover film		Shahbazi et al. (2021)
Fruits and vegetables	Fruits	Bag, wrapper film, and sachet	Weight loss, firmness, antioxidant, antimicrobial, sensory properties, ripening rate, and ethylene production	Colussi et al. (2021); Min et al. (2021)
	Vegetables	Film and wrapper film		Ansarifar and Moradinezhad (2021); Pan et al. (2019)
Bakery products	Bread	Sachet	Antimicrobial	Altan et al. (2018); Göksen et al. (2020)
Dairy products	Cheese	Wrapper film	Antimicrobial and sensory properties	Soto et al. (2019); Lin et al. (2018); Tayebi-Moghaddam et al. (2021)

materials are based on graphene nanocomposite, metallic nanoparticles, and metal nanocomposites (Manikandan et al., 2018). $LaMnO_3$ nanofibers used for the detection of fructose, a highly performing electrochemical sensor (Xu et al., 2014), and other types of electrochemical sensors are used for the identification microbial attack of foods (Silva et al., 2020) and identification of biphenol in drinking water are few reports under this category. Sensors based on marker compounds, for example, ThermoLux violet or light PhotoLux purple (Mancipe et al., 2021). Sodium alginate/PVA sensors used for ammonia detection of packaged fish (Saraf and Vigneshwaran, 2021) and curcumin/PVP/ethyl cellulose/PEO nanofibers for the monitoring of kinds of seafood (Luo and Lim, 2020) and others in the food industry (Mustafa and Andreescu, 2018) are a few examples.

21.4 Nanofibers applications in food packaging

Up-to-date research inputs of electrospun nanofibers show their immense potential in the food industry. The detailed list of materials explored and their respective functional properties are listed in the following Table 21.2 (active packaging strategies) and Table 21.3 (smart packaging strategies).

A concise overview of various types of food products and the research efforts made in the specific food materials packaging gives us information about the available packaging materials and their efficient performance status as shown in Fig. 21.4B–E.

21.4.1 Dairy products

Monitoring of dairy products is an important area, where packaging materials made of zein/*Laurus nobili*-rosmarinus officinalis essential oil (Göksen et al., 2020) and gelatine/allyl isothiocyanate nanofibers are used for the monitoring and control of the microbial growth on cheese slices during storage (Al-Moghazy et al., 2020).

For the improvement of the shelf life of the dairy products like cheese and paneer. In addition to the activity nanofiber membranes developed by encapsulating laccase enzyme within the nanofiber matrix made up of chitosan/polyvinyl alcohol/tetraethyl ortho-silicate nanofibers colored with guaiacol (Tsai et al., 2021) a TTI, is used to detect bacterial growth in milk during storage at different temperatures.

21.4.2 Fruits and vegetables

Fruits and vegetables undoubtedly require larger support and protection as they tend to degrade fast due to transpiration, ethylene production, respiration, and microbial growth active packaging, and smart packaging materials showed tremendous benefits for the fruits and vegetable packaging industry.

21.4.2.1 Strawberries

Strawberries are sensitive fruits and demand careful and proper storage and transportation. As they will be prone to the attack of molds and temperature, protection is the key to improving shelf life. For this, the cellulose acetate-PVA-silver NP's nanofibers matrix showed a reduction in mold growth up to 7 days (Tarus et al., 2019) PLA/Thyme oil NF's (Min et al., 2021). Pullulan/SCMC/tea polyphenol extract NF's (Shao et al., 2018) showed a reduction in weight loss and firmness on 5 day storage period zein-thyme oil NF's protected them from microbial attack and preserved the antioxidant capacity, total phenolic content, etc. at 4°C (Ansarifar and Moradinezhad, 2021).

21.4.2.2 Lemons

Polyvinyl alcohol/Ag active nanofibers showed an enhanced shelf life of about 10 days owing to their antimicrobial activity (Kowsalya et al., 2019). Photocatalytic nanofibers prepared with Zein/TiO$_2$-NPs showed commendable ethylene scavenging properties and protected lemons from discoloration/color change from yellow to brown.

21.4.2.3 Bananas

Polyacrylonitrile-TiO$_2$ nanofibers improved the stability of bananas during storage by preventing photo-oxidation (Zhu et al., 2019). Similarly, Zein/TiO$_2$ nanofibers showed extended shelf life of tomatoes due to ethylene absorption capacity (Böhmer-Maas et al., 2020).

21.4.3 Meat and seafood

Spoilage of meat is a common problem due to biochemical activities and microbial attacks. Nanofiber-based packing has been utilized to reduce and detect the level of degradation. For this purpose, thyme oil/β-cyclodextrin ε-polylysine nanoparticles (TCPNs)/gelatin nanofibers, coated directly onto chicken showed good antimicrobial action against *Campylobacter jejuni* (Lin et al., 2018).

In another report, PVA/cinnamon nanophytosomes loaded nanofibers have shown an increase in the shelf life of shrimp (Nazari et al., 2019). Pomegranate peel extract/chitosan/PEO nanofiber matrices showed inhibition of the attack of bacteria on beef (Surendhiran et al., 2020). Gallic acid-loaded collagen/zein nanofibers for tilapia stability (Song et al., 2022), PCL/PEO/*Hibiscus rosa* Sinensis extract for shrimp (Jovanska et al., 2022) and PLA/blueberry-derived anthocyanins for the freshness monitoring of mutton are some of the examples in this category (Sun et al., 2021).

21.4.4 Food grains and other foods

Food grains and cereals are comparatively less fragile and stable food products, wherein electrospun nanofibers find their use to protect them from microbial attack and oxidation. Gelatin/essential oil NF's is one such example that showed protection from microbes and oxidation (Mahmood et al., 2022). Gallic acid-HPMC/PEO nanofibers (Aydogdu et al., 2019) reduced the oxidation of oils and fats in walnuts.

21.5 **Conclusion and future scope**

The crucial role played by packaging in the quality and stability of many food materials gave birth to many innovative packaging and processing strategies. One among them is the use of electrospun nanofibers and nanomaterials for food processing and packaging. They are proven to be highly performing materials and structures as presented in the chapter. The advances and developments of the e-spun nanofibers in active and smart packaging products. They are not only mechanically robust but also provide better barrier and optical performance. The incorporation of active functional agents and their release to respond to the environmental interactions and their performance to protect and monitor the food status is shown in Fig. 21.4.

The high loading capacity and surface interaction capacity of the nanofibers open up many features for the innovative active and intelligent packaging systems development. They not only provide mechanical support but also provide interactive release of the active substances for the control of degradation products during transport and storage. The porous structure of the nanofibers offers water vapor permeability, which is an advantageous feature for gas-releasing food product stability. Facilitation of renewable natural/synthetic and degradable/nondegradable materials for the fabrication of various types of nanofibers is the commendable advantage of the electro-spun nanofibers. It is in its infancy and there is more room for future explorations.

The cost-effectiveness, safety, and efficacy of these nanomaterials to develop economic, scalable products to be used in the food industry are the future challenges.

Challenges include mass production for a wide range of applications. The low fluidity nature of the natural polymers, which demands the addition of supporting polymers to improve the viscosity, is yet another main drawback and determinant of the stability of these nanofibers. More research efforts are needed to explore sustainable natural polymer fabrication as nanofibers to avoid toxic interaction with stored food products. Furthermore, strict validation is required for these systems' utility for a wide range of food product packaging. Structural stability of the nanofibers throughout the supply chain is a critical research parameter so far for the commercial applicability of active and smart packaging systems.

The future directions of these systems are not only limited to the improvements in performance but also have great scope toward the edible packaging and processing systems sector. This might pave the way for nontoxic, degradable packaging alternatives while providing active and smart packing strategies.

Acknowledgment

The author, Dr. Jhansi Lakshmi Parimi, wishes to thank the Department of Science and Technology (DST) for the WOS-B research fellowship (DST/WOS-B/2017/477-HFN) and IIT Kharagpur for the generous support and P. Prabhakar wishes to thank Ministry of Education, Government of India for fellowship and to IIT Kharagpur, Bio-Research Laboratory, Rajendra Mishra School of Engineering Entrepreneurship, Indian Institute of Technology Kharagpur, Kharagpur, India for the generous support of the studies which contributed to this chapter.

References

Aghaei, Z., Ghorani, B., Emadzadeh, B., Kadkhodaee, R., Tucker, N., 2020. Protein-based halochromic electrospun nanosensor for monitoring trout fish freshness. Food Control 111, 107065.

Ahmed, M., Pickova, J., Ahmad, T., Liaquat, M., Farid, A., Jahangir, M., 2016. Oxidation of lipids in foods. Sarhad J. Agric. 32 (3), 18–24. https://doi.org/10.17582/journal.sja/2016.32.3.230.238.

Alehosseini, A., Gómez-Mascaraque, L.G., Martínez-Sanz, M., López-Rubio, A., 2019. Electrospun curcumin-loaded protein nanofiber mats as active/bioactive coatings for food packaging applications. Food Hydrocoll. 87, 758–771. https://doi.org/10.1016/j.foodhyd.2018.08.056.

Alizadeh-Sani, M., Mohammadian, E., Rhim, J.W., Jafari, S.M., 2020. pH-sensitive (halochromic) smart packaging films based on natural food colorants for the monitoring of food quality and safety. Trends Food Sci. Technol. 105, 93–144. https://doi.org/10.1016/j.tifs.2020.08.014.

Al-Moghazy, M., Mahmoud, M., Nada, A.A., 2020. Fabrication of cellulose-based adhesive composite as an active packaging material to extend the shelf life of cheese. Int. J. Biol. Macromol. 160, 264–275. https://doi.org/10.1016/j.ijbiomac.2020.05.217.

Altan, A., Aytac, Z., Uyar, T., 2018. Carvacrol loaded electrospun fibrous films from zein and poly (lactic acid) for active food packaging. Food Hydrocoll. 81, 48–59. https://doi.org/10.1016/j.foodhyd.2018.02.028.

Aman Mohammadi, M., Dakhili, S., Mirza Alizadeh, A., Kooki, S., Hassanzadazar, H., Alizadeh-Sani, M., McClements, D.J., 2022. New perspectives on electrospun nanofiber applications in smart and active food packaging materials. Crit. Rev. Food Sci. Nutr. 22, 1–7. https://doi.org/10.1080/10408398.2022.2124506.

Ansarifar, E., Moradinezhad, F., 2021. Preservation of strawberry fruit quality via the use of active packaging with encapsulated thyme essential oil in zein nanofiber film. Int. J. Food Sci. Technol. 56 (9), 4239–4247. https://doi.org/10.1111/ijfs.15130.

Aydogdu, A., Yildiz, E., Aydogdu, Y., Sumnu, G., Sahin, S., Ayhan, Z., 2019. Enhancing oxidative stability of walnuts by using gallic acid loaded lentil flour based electrospun nanofibers as active packaging material. Food Hydrocoll. 95, 245–255. https://doi.org/10.1016/j.food-hyd.2019.04.020.

Aydogdu, A., Sumnu, G., Sahin, S., 2019. Fabrication of gallic acid loaded hydroxypropyl methylcellulose nanofibers by electrospinning technique as active packaging material. Carbohydr. Polym. 208, 241–250. https://doi.org/10.1016/j.carbpol.2018.12.065.

Bahrami, A., Delshadi, R., Assadpour, E., Jafari, S.M., Williams, L., 2020. Antimicrobial-loaded nanocarriers for food packaging applications. Adv. Colloid. Interface Sci. 278, 102140. https://doi.org/10.1016/j.cis.2020.102140.

Balbinot-Alfaro, E., Craveiro, D.V., Lima, K.O., Costa, H.L., Lopes, D.R., Prentice, C., 2019. Intelligent packaging with pH indicator potential. Food Eng. Rev. 11 (4), 235–244. https://doi.org/10.1007/s12393-019-09198-9.

Bastarrachea, L.J., Wong, D.E., Roman, M.J., Lin, Z., Goddard, J.M., 2015. Active packaging coatings. Shanghai Tuliao 5 (4), 771–791. https://doi.org/10.3390/coatings5040771.

Böhmer-Maas, B.W., Fonseca, L.M., Otero, D.M., da Rosa Zavareze, E., Zambiazi, R.C., 2020. Photocatalytic zein-TiO$_2$ nanofibers as ethylene absorbers for storage of cherry tomatoes. Food Packag. Shelf Life 24, 100508. https://doi.org/10.1016/j.fpsl.2020.100508.

Chawla, R., Sivakumar, S., Kaur, H., 2021. Antimicrobial edible films in food packaging: current scenario and recent nanotechnological advancements-a review. Carbohydr. Polym. Technol. Appl. 2, 100024. https://doi.org/10.1016/j.carpta.2020.100024.

Cherpinski, A., Gozutok, M., Sasmazel, H.T., Torres-Giner, S., Lagaron, J.M., 2018. Electrospun oxygen scavenging films of poly (3-hydroxybutyrate) containing palladium nanoparticles for active packaging applications. Nanomater. 8 (7), 469. https://doi.org/10.3390/nano8070469.

Colussi, R., Ferreira da Silva, W.M., Biduski, B., Mello El Halal, S.L., da Rosa Zavareze, E., Guerra Dias, A.R., 2021. Postharvest quality and antioxidant activity extension of strawberry fruit using allyl isothiocyanate encapsulated by electrospun zein ultrafine fibers. LWT 143, 111087. https://doi.org/10.1016/j.lwt.2021.111087.

Cui, H., Bai, M., Lin, L., 2018. Plasma-treated poly(ethylene oxide) nanofibers containing tea tree oil/beta-cyclodextrin inclusion complex for antibacterial packaging. Carbohydr. Polym. 179, 360–369. https://doi.org/10.1016/j.carbpol.2017.10.011.

de Silva, F.M., Lopes, P.S., da Silva, C.F., Yoshida, C.M., 2016. Active packaging material based on buriti oil—*Mauritia flexuosa L.f.* (Arecaceae) incorporated into chitosan films. J. Appl. Polym. Sci. 133 (12), 12–18. https://doi.org/10.1002/app.43210.

Doğan, C., Doğan, N., Gungor, M., Eticha, A.K., Akgul, Y., 2022. Novel active food packaging based on centrifugally spun nanofibers containing lavender essential oil: rapid fabrication, characterization, and application to preserve of minced lamb meat. Food Packag. Shelf Life 34, 100942. https://doi.org/10.1016/j.fpsl.2022.100942.

Dong, W., Su, J., Chen, Y., Xu, D., Cheng, L., Mao, L., Gao, Y., Yuan, F., 2022. Characterization and antioxidant properties of chitosan film incorporated with modified silica nanoparticles as an active food packaging. Food Chem. 373, 131414. https://doi.org/10.1016/j.foodchem.2021.131414.

Drago, E., Campardelli, R., Pettinato, M., Perego, P., 2020. Innovations in smart packaging concepts for food: an extensive review. Nutrafoods 9 (11), 1628. https://doi.org/10.3390/foods9111628.

Duan, M., Yu, S., Sun, J., Jiang, H., Zhao, J., Tong, C., Hu, Y., Pang, J., Wu, C., 2021. Development and characterization of electrospun nanofibers based on pullulan/chitin nanofibers containing curcumin and anthocyanins for active-intelligent food packaging. Int. J. Biol. Macromol. 187, 332–340. https://doi.org/10.1016/j.ijbiomac.2021.07.140.

Eghbalian, M., Shavisi, N., Shahbazi, Y., Dabirian, F., 2021. Active packaging based on sodium caseinate-gelatin nanofiber mats encapsulated with *Mentha spicata* L. essential oil and MgO nanoparticles: preparation, properties, and food application. Food Packag. Shelf Life 29, 100737. https://doi.org/10.1016/j.fpsl.2021.100737.

El-Naggar, M.E., El-Newehy, M.H., Aldalbahi, A., Salem, W.M., Khattab, T.A., 2021. Immobilization of anthocyanin extract from red-cabbage into electrospun polyvinyl alcohol nanofibers for colorimetric selective detection of ferric ions. J. Environ. Chem. Eng. 9 (2), 105072. https://doi.org/10.1016/j.jece.2021.105072.

Fabra, M.J., Busolo, M.A., Lopez-Rubio, A., Lagaron, J.M., 2013. Nanostructured biolayers in food packaging. Trends Food Sci. Technol. 31 (1), 79–87. https://doi.org/10.1016/j.tifs.2013.01.004.

Forghani, S., Almasi, H., Moradi, M., 2021. Electrospun nanofibers as food freshness and time-temperature indicators: a new approach in food intelligent packaging. Innovat. Food Sci. Emerg. Technol. 73, 102804. https://doi.org/10.1016/j.ifset.2021.102804.

Ge, L., Zhao, Y.S., Mo, T., Li, J.R., Li, P., 2012. Immobilization of glucose oxidase in electrospun nanofibrous membranes for food preservation. Food Control 26 (1), 188–193. https://doi.org/10.1016/j.foodcont.2012.01.022.

Ghoshal, G., 2018. Recent trends in active, smart, and intelligent packaging for food products. In: Food Packaging and Preservation. Academic Press, pp. 343–374. https://doi.org/10.1016/b978-0-12-811516-9.00010-5.

Göksen, G., Fabra, M.J., Ekiz, H.I., López-Rubio, A., 2020. Phytochemical-loaded electrospun nanofibers as novel active edible films: characterization and antibacterial efficiency in cheese slices. Food Control 112, 107133. https://doi.org/10.1016/j.foodcont.2020.107133.

Guo, M., Wang, H., Wang, Q., Chen, M., Li, L., Li, X., Jiang, S., 2020. Intelligent double-layer fiber mats with high colorimetric response sensitivity for food freshness monitoring and preservation. Food Hydrocoll. 101, 105468.

Han, G., Guo, R., Yu, Z., Chen, G., 2020. Progress on biodegradable films for antibacterial food packaging. In: E3S Web of Conferences, 145. EDP Sciences, p. 01036. https://doi.org/10.1051/e3sconf/202014501036.

Jhuang, J.R., Lin, S.B., Chen, L.C., Lou, S.N., Chen, S.H., Chen, H.H., 2020. Development of immobilized laccase-based time temperature indicator by electrospinning zein fiber. Food Packag. Shelf Life 23, 100436. https://doi.org/10.1016/j.fpsl.2019.100436.

Jovanska, L., Chiu, C.H., Yeh, Y.C., Chiang, W.D., Hsieh, C.C., Wang, R., 2022. Development of a PCL-PEO double network colorimetric pH sensor using electrospun fibers containing *Hibiscus rosa sinensis* extract and silver nanoparticles for food monitoring. Food Chem. 368, 130813. https://doi.org/10.1016/j.foodchem.2021.130813.

Kaewklin, P., Siripatrawan, U., Suwanagul, A., Lee, Y.S., 2018. Active packaging from chitosan-titanium dioxide nanocomposite film for prolonging storage life of tomato fruit. Int. J. Biol. Macromol. 112, 523–529. https://doi.org/10.1016/j.ijbiomac.2018.01.124.

Kim, J.H., Joshi, M.K., Lee, J., Park, C.H., Kim, C.S., 2018. Polydopamine-assisted immobilization of hierarchical zinc oxide nanostructures on electrospun nanofibrous membrane for photocatalysis and antimicrobial activity. J. Colloid Interface Sci. 513, 566–574. https://doi.org/10.1016/j.jcis.2017.11.061.

Kouravand, F., Jooyandeh, H., Barzegar, H., Hojjati, M., 2018. Characterization of cross-linked whey protein isolate-based films containing *Satureja khuzistanica* Jamzad essential oil. J. Food Process. Preserv. 42 (3), e13557. https://doi.org/10.1111/jfpp.13557.

Kowsalya, E., MosaChristas, K., Balashanmugam, P., Rani, J.C., 2019. Biocompatible silver nanoparticles/poly (vinyl alcohol) electrospun nanofibers for potential antimicrobial food packaging applications. Food Packag. Shelf Life Life 21, 100379. https://doi.org/10.1016/j.fpsl.2019.100379.

Kuntzler, S.G., Costa, J.A., Brizio, A.P., de Morais, M.G., 2020. Development of a colorimetric pH indicator using nanofibers containing *Spirulina* sp. LEB 18. Food Chem. 328, 126768. https://doi.org/10.1016/j.foodchem.2020.126768.

Lai, W.F., 2021. Design of polymeric films for antioxidant active food packaging. Int. J. Mol. Sci. 23 (1), 12. https://doi.org/10.3390/ijms23010012.

Li, L., Wang, H., Chen, M., Jiang, S., Cheng, J., Li, X., et al., 2020. Gelatin/zein fiber mats encapsulated with resveratrol: kinetics, antibacterial activity and application for pork preservation. Food Hydrocoll. 101, 105577. https://doi.org/10.1016/j.foodhyd.2019.105577.

Lin, L., Zhu, Y., Cui, H., 2018. Electrospun thyme essential oil/gelatin nanofibers for active packaging against *Campylobacter jejuni* in chicken. LWT 97, 711–718. https://doi.org/10.1016/j.lwt.2018.08.015.

Luo, X., Lim, L.T., 2020. Curcumin-loaded electrospun nonwoven as a colorimetric indicator for volatile amines. LWT 128, 109493. https://doi.org/10.1016/j.lwt.2020.109493.

Maftoonazad, N., Ramaswamy, H., 2019. Design and testing of an electrospun nanofiber mat as a pH biosensor and monitor the pH associated quality in fresh date fruit (Rutab). Polym. Test. 75, 76–84. https://doi.org/10.1016/j.polymertesting.2019.01.011.

Mahmood, K., Kamilah, H., Alias, A.K., Ariffin, F., Mohammadi Nafchi, A., 2022. Functionalization of electrospun fish gelatin mats with bioactive agents: comparative effect on morphology, thermo-mechanical, antioxidant, antimicrobial properties, and bread shelf stability. Food Sci. Nutr. 10 (2), 584–596. https://doi.org/10.1002/fsn3.2676.

Maleki, G., Sedaghat, N., Woltering, E.J., 2018. Chitosan-limonene coating in combination with modified atmosphere packaging preserve postharvest quality of cucumber during storage. Food Measure 12, 1610–1621. https://doi.org/10.1007/s11694-018-9776-6.

Mancipe, J.M., Nista, S.V., Caballero, G.E., Mei, L.H., 2021. Thermochromic and/or photochromic properties of electrospun cellulose acetate microfibers for application as sensors in smart packing. J. Appl. Polym. Sci. 138 (11), 50039. https://doi.org/10.1002/app.50039.

Mangaraj, S., Yadav, A., Bal, L.M., Dash, S.K., Mahanti, N.K., 2019. Application of biodegradable polymers in food packaging industry: a comprehensive review. J. Packag. Technol. Res. 3 (1), 77–96. https://doi.org/10.1007/s41783-018-0049-y.

Manikandan, V.S., Adhikari, B., Chen, A., 2018. Nanomaterial based electrochemical sensors for the safety and quality control of food and beverages. Analyst 143 (19), 4537–4554. https://doi.org/10.1039/c8an00497h.

Mihindukulasuriya, S.D.F., Lim, L.T., 2014. Nanotechnology development in food packaging: a review. Trends Food Sci. Technol. 40, 149–167. https://doi.org/10.1016/j.tifs.2014.09.009.

Min, T., Sun, X., Yuan, Z., Zhou, L., Jiao, X., Zha, J., Zhu, Z., Wen, Y., 2021. Novel antimicrobial packaging film based on porous poly (lactic acid) nanofiber and polymeric coating for humidity-controlled release of thyme essential oil. LWT-Food Sci. Technol. 135, 110034. https://doi.org/10.1016/j.lwt.2020.110034.

Mustafa, F., Andreescu, S., 2018. Chemical and biological sensors for food-quality monitoring and smart packaging. Foods 7 (10), 168. https://doi.org/10.3390/foods7100168.

Nazari, M., Majdi, H., Milani, M., Abbaspour-Ravasjani, S., Hamishehkar, H., Lim, L.T., 2019. Cinnamon nanophytosomes embedded electrospun nanofiber: its effects on microbial quality and shelf-life of shrimp as a novel packaging. Food Packag. Shelf Life 21, 100349. https://doi.org/10.1016/j.fpsl.2019.100349.

Ozogul, Y., Yuvka, İ., Ucar, Y., Durmus, M., Kösker, A.R., Öz, M., et al., 2017. Evaluation of effects of nanoemulsion based on herb essential oils (rosemary, laurel, thyme and sage) on sensory, chemical and microbiological quality of rainbow trout (*Oncorhynchus mykiss*) fillets during ice storage. LWT-Food Sci. Technol. 75, 677–684. https://doi.org/10.1016/j.lwt.2016.10.009.

Pan, J., Ai, F., Shao, P., Chen, H., Gao, H., 2019. Development of polyvinyl alcohol/β-cyclodextrin antimicrobial nanofibers for fresh mushroom packaging. Food Chem. 300, 125249. https://doi.org/10.1016/j.foodchem.2019.125249.

Paster, N., Juven, B.J., Shaaya, E., Menasherov, M., Nitzan, R., Weisslowicz, H., Ravid, U., 1990. Inhibitory effect of oregano and thyme essential oils on moulds and foodborne bacteria. Lett. Appl. Microbiol. 11 (1), 33–37. doi.org/10.1111/j.1472-765x.1990.tb00130.x.

Rai, S., Dutta, P.K., Mehrotra, G.K., 2017. Lignin incorporated antimicrobial chitosan film for food packaging application. J. Polym. Mater. 34 (1), 171.

Restuccia, D., Spizzirri, U.G., Parisi, O.I., Cirillo, G., Curcio, M., Iemma, F., et al., 2010. New EU regulation aspects and global market of active and intelligent packaging for food industry applications. Food Control 21 (11), 1425–1435. https://doi.org/10.1016/j.foodcont.2010.04.028.

Rhim, J.W., Park, H.M., Ha, C.S., 2013. Bio-nanocomposites for food packaging applications. Progr. Polym. Sci., 38, 1629–1652. https://doi.org/10.1016/j.progpolymsci.2013.05.008.

Riahi, Z., Ezati, P., Rhim, J.W., Bagheri, R., Pircheraghi, G., 2022. Cellulose nanofiber-based ethylene scavenging antimicrobial films incorporated with various types of titanium dioxide nanoparticles to extend the shelf life of fruits. ACS Appl. Polym. Mater. 4 (7), 4765–4773. https://doi.org/10.1021/acsapm.2c00338.

Rieger, K.A., Schiffman, J.D., 2014. Electrospinning an essential oil: cinnamaldehyde enhances the antimicrobial efficacy of chitosan/poly (ethylene oxide) nanofibers. Carbohydr. Polym. 113, 561–568. https://doi.org/10.1016/j.carbpol.2014.06.075.

Sagitha, P., Reshmi, C.R., Sundaran, S.P., Sujith, A., 2018. Recent advances in postmodification strategies of polymeric electrospun membranes. Eur. Polym. J. 105, 227–249. https://doi.org/10.1016/j.eurpolymj.2018.05.033.

Saraf, K., Vigneshwaran, N., 2021. Paper from cotton linters as substrate for ammonia nanosensor using electrospun alginate nanofibers. Cott. Res. J. 10 (1).

Schaefer, D., Cheung, W.M., 2018. Smart packaging: opportunities and challenges. Procedia Cirp. 72, 1022-1027. https://doi.org/10.1016/j.procir.2018.03.240

Shahbazi, Y., Shavisi, N., Karami, N., Lorestani, R., Dabirian, F., 2021. Electrospun carboxymethyl cellulose-gelatin nanofibrous films encapsulated with *Mentha longifolia* L. essential oil for active packaging of peeled giant freshwater prawn. LWT-Food Sci. Technol. 152, 112322. https://doi.org/10.1016/j.lwt.2021.112322.

Shao, P., Liu, L., Yu, J., Lin, Y., Gao, H., Chen, H., Sun, P., 2021. An overview of intelligent freshness indicator packaging for food quality and safety monitoring. Trends Food Sci. Technol. 118, 285-296. https://doi.org/10.1016/j.tifs.2021.10.012

Shao, P., Niu, B., Chen, H., Sun, P., 2018. Fabrication and characterization of tea polyphenols loaded pullulan-CMC electrospun nanofiber for fruit preservation. Int. J. Biol. Macromol. 107, 1908–1914. https://doi.org/10.1016/j.ijbiomac.2017.10.054.

Shavisi, N., Shahbazi, Y., 2022. Chitosan-gum Arabic nanofiber mats encapsulated with pH-sensitive *Rosa damascena* anthocyanins for freshness monitoring of chicken fillets. Food Packag. Shelf Life 32, 100827. https://doi.org/10.1016/j.fpsl.2022.100827.

Silva, N.F., Neves, M.M., Magalhães, J.M., Freire, C., Delerue-Matos, C., 2020. Emerging electrochemical biosensing approaches for detection of *Listeria monocytogenes* in food samples: an overview. Trends Food Sci. Technol. 99, 621–633. https://doi.org/10.1016/j.tifs.2020.03.031.

Song, Z., Liu, H., Huang, A., Zhou, C., Hong, P., Deng, C., 2022. Collagen/zein electrospun films incorporated with gallic acid for tilapia (*Oreochromis niloticus*) muscle preservation. J. Food Eng. 317, 110860. https://doi.org/10.1016/j.jfoodeng.2021.110860.

Soto, K.M., Hernández-Iturriaga, M., Loarca-Piña, G., Luna-Bárcenas, G., Mendoza, S., 2019. Antimicrobial effect of nisin electrospun amaranth: pullulan nanofibers in apple juice and fresh cheese. Int. J. Food Microbiol. 295, 25–32. https://doi.org/10.1016/j.ijfoodmicro.2019.02.001.

Sun, W., Liu, Y., Jia, L., Saldaña, M.D., Dong, T., Jin, Y., Sun, W., 2021. A smart nanofibre sensor based on anthocyanin/poly-l-lactic acid for mutton freshness monitoring. Int. J. Food Sci. Technol. 56 (1), 342–351. https://doi.org/10.1111/ijfs.14648.

Surendhiran, D., Li, C., Cui, H., Lin, L., 2020. Fabrication of high stability active nanofibers encapsulated with pomegranate peel extract using chitosan/PEO for meat preservation. Food Packag. Shelf Life 23, 100439. https://doi.org/10.1016/j.fpsl.2019.100439.

Tarus, B.K., Mwasiagi, J.I., Fadel, N., Al-Oufy, A., Elmessiry, M., 2019. Electrospun cellulose acetate and poly (vinyl chloride) nanofiber mats containing silver nanoparticles for antifungi packaging. SN Appl. Sci. 1 (3), 1–2. https://doi.org/10.1007/s42452-019-0271-4.

Tayebi-Moghaddam, S., Khatibi, R., Taklavi, S., Hosseini-Isfahani, M., Rezaeinia, H., 2021. Sustained-release modeling of clove essential oil in brine to improve the shelf life of Iranian white cheese by bioactive electrospun zein. Int. J. Food Microbiol. 355, 109337. https://doi.org/10.1016/j.ijfoodmicro.2021.109337.

Terra, A.L., Moreira, J.B., Costa, J.A., de Morais, M.G., 2021. Development of time-pH indicator nanofibers from natural pigments: an emerging processing technology to monitor the quality of foods. LWT-Food Sci. Technol. 142, 111020. https://doi.org/10.1016/j.lwt.2021.111020.

Tirgar, A., Han, D., Steckl, A.J., 2018. Absorption of ethylene on membranes containing potassium permanganate loaded into alumina-nanoparticle-incorporated alumina/carbon nanofibers. J. Agric. Food Chem. 66 (22), 5635–5643.

Tongdeesoontorn, W., Mauer, L.J., Wongruong, S., Sriburi, P., Rachtanapun, P., 2020. Physical and antioxidant properties of cassava starch–carboxymethyl cellulose incorporated with quercetin and TBHQ as active food packaging. Polymers 12 (2), 366. https://doi.org/10.3390/polym12020366.

Tsai, T.Y., Chen, S.H., Chen, L.C., Lin, S.B., Lou, S.N., Chen, Y.H., Chen, H.H., 2021. Enzymatic time-temperature indicator prototype developed by immobilizing laccase on electrospunfibers to predict lactic acid bacterial growth in milk during storage. Nanomaterials 11 (5), 1160. https://doi.org/10.3390/nano11051160.

Venkatachalam, K., Lekjing, S., 2020. A chitosan-based edible film with clove essential oil and nisin for improving the quality and shelf life of pork patties in cold storage. RSC Adv. 10 (30), 17777–17786. https://doi.org/10.1039/d0ra02986f.

Vilela, C., Kurek, M., Hayouka, Z., Röcker, B., Yildirim, S., Antunes, M.D.C., Freire, C.S., 2018. A concise guide to active agents for active food packaging. Trends Food Sci. Technol. 80, 212–222. https://doi.org/10.1016/j.tifs.2018.08.006.

Vilela, C., Pinto, R.J.B., Coelho, J., Domingues, M.R.M., Daina, S., Sadocco, P., Freire, C.S.R., 2017. Bioactive chitosan/ellagic acid films with UV-light protection for active food packaging. Food Hydrocoll. 73, 120–128. https://doi.org/10.1016/j.foodhyd.2017.06.037.

Wang, Y., Xu, H., Wu, M., Yu, D.G., 2021. Nanofibers-based food packaging. ES Food Agrofor. 7, 1–24. https://doi.org/10.30919/esfaf598.

Wen, P., Wen, Y., Zong, M.H., Linhardt, R.J., Wu, H., 2017. Encapsulation of bioactive compound in electrospun fibers and its potential application. J. Agric. Food Chem. 65, 9161–9179. https://doi.org/10.1021/acs.jafc.7b02956.

Xie, J., Hung, Y.C., 2018. UV-A activated TiO_2 embedded biodegradable polymer film for antimicrobial food packaging application. LWT-Food Sci. Technol. 96, 307–314. https://doi.org/10.1016/j.lwt.2018.05.050.

Xu, D., Luo, L., Ding, Y., Jiang, L., Zhang, Y., Ouyang, X., Liu, B., 2014. A novel nonenzymatic fructose sensor based on electrospun $LaMnO_3$ fibers. J. Electroanal. Chem. 727, 21–26. https://doi.org/10.1016/j.jelechem.2014.05.010.

Xu, Y., Yang, D., Huo, S., Ren, J., Gao, N., Chen, Z., Liu, Y., Xie, Z., Zhou, S., Qu, X., 2021. Carbon dots and ruthenium doped oxygen sensitive nanofibrous membranes for monitoring the respiration of agricultural products. Polym. Test. 93, 106957.

Yaghoubi, M., Ayaseh, A., Alirezalu, K., Nemati, Z., Pateiro, M., Lorenzo, J.M., 2021. Effect of chitosan coating incorporated with *Artemisia fragrans* essential oil on fresh chicken meat during refrigerated storage. Polymers 13 (5), 716. https://doi.org/10.3390/polym13050716.

Yang, Y., Shi, Y., Cao, X., Liu, Q., Wang, H., Kong, B., 2021. Preparation and functional properties of poly(vinyl alcohol)/ethyl cellulose/tea polyphenol electrospun nanofibrous films for active packaging material. Food Control 130, 108331. https://doi.org/10.1016/J.foodcont.2021.108331.

Yao, Z.C., Chang, M.W., Ahmad, Z., Li, J.S., 2016. Encapsulation of rose hip seed oil into fibrous zein films for ambient and on demand food preservation via coaxial electrospinning. J. Food Eng. 191, 115–123. https://doi.org/10.1016/j.jfoodeng.2016.07.012.

Yildirim, S., Röcker, B., Pettersen, M.K., Nilsen-Nygaard, J., Ayhan, Z., Rutkaite, R., Radusin, T., Suminska, P., Marcos, B., Coma, V., 2017. Active packaging applications for food. Compr. Rev. Food Sci. Food Saf. 17 (1), 165–199. https://doi.org/10.1111/1541-4337.12322.

Yılmaz, M., Altan, A., 2021. Optimization of functionalized electrospun fibers for the development of colorimetric oxygen indicator as an intelligent food packaging system. Food Packag. Shelf Life 28, 100651. https://doi.org/10.1016/j.fpsl.2021.100651.

ZabihzadehKhajavi, M., Ebrahimi, A., Yousefi, M., Ahmadi, S., Farhoodi, M., Mirza Alizadeh, A., Taslikh, M., 2020. Strategies for producing improved oxygen barrier materials appropriate for the food packaging sector. Food Eng. Rev. 12 (3), 346–363. https://doi.org/10.1007/s12393-020-09235-y.

Zhang, J., Huang, X., Zhang, J., Liu, L., Shi, J., Muhammad, A., Zhai, X., Zou, X., Xiao, J., Li, Z., Li, Y., 2022. Development of nanofiber indicator with high sensitivity for pork preservation and freshness monitoring. Food Chem. 381, 132224.

Zhang, Y., Yang, L., Dong, Q., Li, L., 2021. Fabrication of antibacterial fibrous films by electrospinning and their application for Japanese sea bass (*Lateolabrax japonicus*) preservation. LWT-Food Sci. Technol. 149, 111870. https://doi.org/10.1016/j.lwt.2021.111870.

Zhu, Z., Zhang, Y., Shang, Y., Wen, Y., 2019. Electrospun nanofibers containing TiO_2 for the photocatalytic degradation of ethylene and delaying postharvest ripening of bananas. Food Bioprocess Technol. 12 (2), 281–287. https://doi.org/10.1007/s11947-018-2207-1.

Use of nanobio-technological methods for the analysis and stability of food antimicrobials and antioxidants

22

Megha Pant[a], Kumai Kiran[a], Veena Pande[a], Biswajit Mishra[b] and Anirban Dandapat[c]

[a]*Department of Biotechnology, Sir J.C. Bose Technical Campus Bhimtal, Kumaun University, Nainital, Uttarakhand, India,* [b]*Department of Medicine, Warren Alpert Medical School of Brown University, Providence, RI, United States,* [c]*University School of Automation and Robotics, Guru Gobind Singh Indraprastha University, East Delhi Campus, Delhi, India*

22.1 Role of nanobiotechnology in food sector

Nanobiotechnology has been constantly revolutionizing the food industry. It is the technology where nanoparticles (NPs) within a size range of approximately 1–100 nm were developed for further use in different areas. These particles possess different properties like surface-to-volume ratio, color, optical property magnetic properties, biocompatibility, stability, solubility, etc. (Rai et al., 2009). To develop materials and structures with extraordinary properties nanotechnology has been applied in different fields like food, agriculture, and medicine. In the area of food sciences, innovative and impactful solutions were offered by nanotechnology from processing to packaging. The researchers are compelled to find a way that will not only enhance the food quality but will also preserve the nutritional content of the product. Nanomaterials thus ensure that food delivers the health benefits that it needs to provide by maintaining nutritional value and quality. The food sector now comes up with new innovational ideas, novel methods, and techniques that have been developed and integrated by many industries, organizations, and research laboratories (Dasgupta et al., 2015). From processing to food packaging, it can be used to improve and evaluate the quality and safety of food (Ezhilarasi et al., 2013). Nanostructures for food processing are employed to be used as additives, antimicrobial agents, nutrient delivery, and for sustaining the mechanical properties, and robustness of packing material. One such nanomaterial is nanopolymers which have now been employed to replace conventional food packaging material nanostructures improve the text, consistency, and texture, and improve the effectiveness of numerous kinds of food components c resulting in lesser wastage which would otherwise be abused by microbial contamination (Pradhan et al., 2015).

Nanobiotechnology for Food Processing and Packaging. DOI: https://doi.org/10.1016/B978-0-323-91749-0.00003-4

In modern times, there has been a rapid spread of food contamination mainly due to the globalization of food production and trade. Most of these diseases are the direct result of food and drinking water contamination (Smith et al., 2005). It is difficult to estimate outbreaks, but there have been reports that every year 30% population of industrialized nations suffers from food-borne infections. The rate of microbial infection through food is projected at around 76 million cases annually, receiving treatment in hospitals about 325,000 cases, and about 5000 deaths. Most of the time, the main sufferers of these food-borne infections are the developing countries. These infections carry a serious threat to society. In 1994, a salmonella infection was reported in the United States which was caused by contaminated ice cream resulting in 224,000 individuals being infected. Medical costs and production losses for diseases caused by nurses in the United States during 1997 were estimated at US$ 35 billion (Smith et al., 2005). Around 1998, there was a serious outbreak of hepatitis A in China, which resulted from contaminated molluscs, resulting in 300,000 people being infected (Mead et al., 1999). Thus, overall social and economic pressure got built in the communities to be now vulnerable to invasion and outbreaks. Based on this, the researchers emphasize developing sensitive, fast, and accurate monitoring systems to detect bacteria in food and the environment. There are different conventional techniques in the market to detect and isolate these foods-borne pathogenic bacteria (Pitcher and Fry, 2000). There is a diagnostic kit that relies on antibodies to detect microorganisms. Although these tests are sensitive and less time-consuming but the growth of these organisms is first required in the preparation of 10^6 cells/mL (Siragusa et al., 1995). Modern advances in nanotechnology make it easy to detect the microorganisms that complement the gene of the microorganism using nanomaterials. In addition, the analysis of food antimicrobials and antioxidants that maintain the stabilities and shelf-life of the food items are more important factors. Nanobiotechnology utilizing NPs and suitable biomolecules are highly influential in detecting and analyzing pathogenic microbes, food antimicrobials, and antioxidants. The advantages of these nanomaterials compared to other materials can be ascribed to their smaller size, higher stability, penetrability, less reactivity, and more compatibility (Fakhouri et al., 2014; Shi et al., 2013; Yu et al., 2012; Tang et al., 2012). This chapter deals with the efficiency of nanomaterials to be used as potential devices for analyzing microbial contamination and evaluating antimicrobes and antioxidants in food. Sensors made of nano-size granules are the simplest and most inexpensive species to detect the status and safety of a packaged food product. These sensors detect any changes that occur due to storage and refrigeration during supply or distribution.

Food stability and nutritional content are now been preserved by nanocarriers which are high in demand due to their efficacy in delivering food flavors in foodstuffs by not compromising the overall morphology. An ideal delivery method is required where the primary aim is to effectively provide the important ingredient to its target site. A second important goal is to maintain the shelf life of the compound and make sure that it will not be compromised in the long run. The traditional encapsulation system was slow and less efficient, but NPs are fast, efficient, and possess high release efficiency. These encapsulated NPs have several advantages including the controlled

Table 22.1 Food coatings to increase the overall texture and quality of food products.

Nanotechnique	Characteristic features	Examples	References
Edible coatings	Preservation of food quality during long-term storage	Gelatin-based coatings having cellulose nanocrystal chitosan Chitosan coating with nanosilica alginate lysozyme nanolaminate coatings	Fakhouri et al. (2014) Shi et al. (2013) Yu et al. (2012)
Inorganic NPs	Encapsulation capability with controlled functionalization	Mesoporous silica nanoparticles	Tang et al. (2012)
Liposomes	Liposome surrounds an aqueous solution inside a hydrophobic membrane and has potentials as delivery vehicles for hydrophobic molecules	Cationic lipid incorporated liposomes modified with an acid-labile polymer hyper-branched poly(glycidol)	Yoshizaki et al. (2014)
Hydrogels	Easy encapsulation potential that prevents drug degradation by external factors such as pH and temperature	Protein hydrogels	Qiu and Park (2001)
Polymeric micelles	High water solubility and low toxicity	Poly(ethylene glycol)block-poly(caprolactone) polymeric micelles	Ma et al. (2008)

release of active ingredients, also these delivery systems maintain effectual delivery by taking advantage of the small size of the NPs which due to their size can easily penetrate the tissues and thus release the active compound to the desired area or site in the body (Lamprecht et al., 2004). Other advantages of these nanomaterials are masking any bad odor, shielding the food components from any kind of degradation during processing, storage, and utilization also they protect the food from humidity and high temperature (Ubbinik and Kruger, 2006; Weiss et al., 2006). Food component preservation is a must which has been maintained by developing different natural and synthetic polymer-based delivery systems. These systems encapsulate the food component thus ensuring its proper bioavailability and shelf life preservation. Further, the role of nanotechnology is discussed where it provides novel qualities to food products by improving the overall food taste, appearance, texture, and nutritional value. Different techniques can be followed to coat the food materials to sustain the overall quality of the food items, as given in Table 22.1.

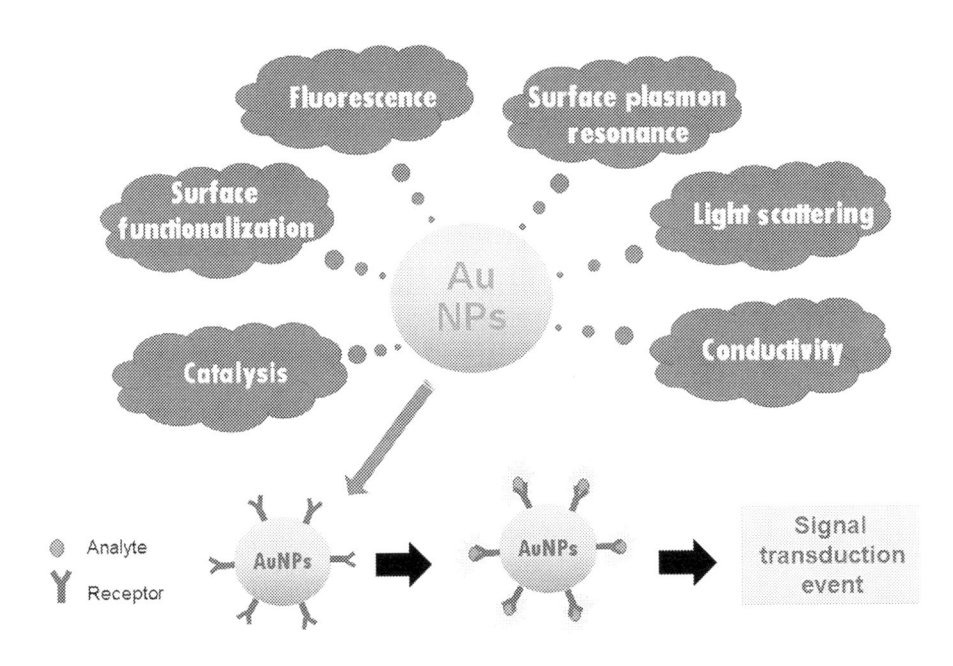

FIGURE 22.1

Physical and chemical properties of gold nanoparticles (AuNPs) and schematic illustration of AuNPs-based detection systems.

22.2 Common nanomaterials used in the detection and analysis

22.2.1 Gold nanoparticles

One frequently used nanoparticle in nanosensing technology is the gold (Au) NPs. These AuNPs have advantages over other NPs as they are more biocompatible and the ratio of volume to surface area is more and it possess remarkable conductance and optical properties (Guo and Wang, 2007). AuNP's optical property was based on the surface plasmon resonance (SPR) which corresponds to the overall dimension, shape, and environment of the NPs. The distinctive properties of AuNPs, which include high absorption coefficient, fluorescence, conductivity, generation of higher electromagnetic fields, fluorescence quenching, and a higher rate of catalysis reactions, make them unique and potential candidates to be used for analytical purposes and sensing applications (Fig. 22.1).

The redox potential of metallic AuNPs is useful in developing electrochemical sensors thus enhancing their sensitivity for detecting food-borne microorganisms. The synthesis of these gold NPs mainly requires aqueous or organic solvents and can further be stabilized by chemical bonding or adsorption with the help of a surfactant.

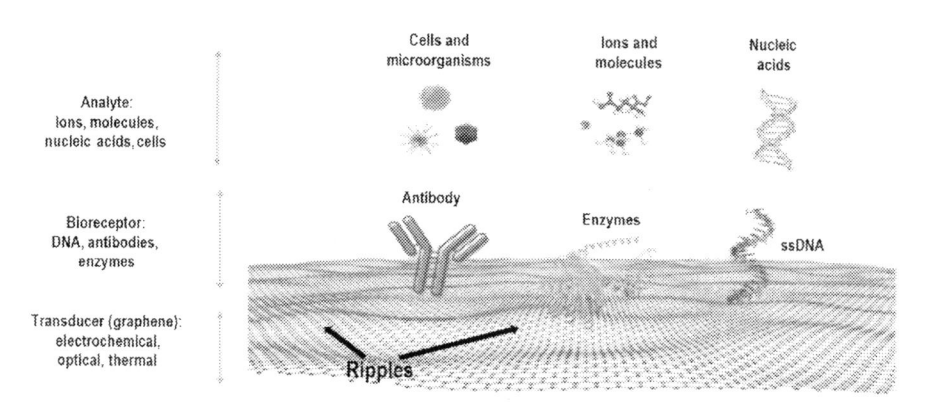

FIGURE 22.2

Graphene-based nanomaterials used as a transducer in different biosensors. Different bioreceptors, e.g., DNA, antibodies, and enzymes, can be immobilized on the surface of graphene-based materials (reproduced with permission from Yildiz et al. (2021)).

Different surfactants can be used to be loaded on gold NPs to avoid the aggregation of these NPs (Sperling et al., 2008). Gold NPs are also combined with redox enzymes, reducing or oxidizing the analyte in the reaction on the basis of substrate present. The analyte detection is based upon the detection of the signal-dependent upon current which will increase with the concentration of the analyte.

22.2.2 Graphene-based nanomaterials

Graphene oxide (GO), graphene, and reduced graphene oxide are the basic components of graphene-based nanomaterials. In a nanobiosensor, graphene-based nanomaterials act as the transducing materials of biosensors (Fig. 22.2).

The main reason for using graphene-based nanomaterials for transducers is their high surface area, electrical conductivity, and capacity to bind to different types of molecules (Rao et al., 2009). Fluorescent transducer-based biosensors were produced using graphene-based nanomaterials as they possess great fluorescent quenching efficiency.

Biosensor's sensitivity and selectivity could be affected by a number of factors: the oxidation state, functional groups, and derivatives used. The sensitivity is mainly affected due to graphene and the bio-receptors and sheet orientation of these graphene oxides further affect the selectivity of the designed biosensors. To detect *Escherichia coli* strain 0157:H7 an electrochemical sensor based on a hybrid chitosan nanocomposite modified with graphene oxide was developed which detects the pathogenic bacterial strain using an oligonucleotide specific probe immobilized on nanocomposite films (Tiwari et al., 2015). The detection limit is around 1×10^{-14} M. The specificity against other bacteria like *E. coli*, *Klebsiella pneumonia*, *Neisseria meningitides*, and *Salmonella typhimurium* is an insignificant signal.

22.2.3 **Carbon nanotubes**

Carbon nanotubes (CNTs) have become one of the most popular nanomaterials in recent years for use in bimolecular and drug delivery systems. These are concentric layers of graphite, multi to single-walled, and cylindrical in shape. Carbon nanotubes possess properties like high thermal conductivity, high surface-to-volume ratio light-weight, electro-catalytic, and mechanical properties (Pandit et al., 2016). The development of biosensors depends mainly on the two most important properties: sensitivity and specificity. A DNA sensor was developed using single-wall surface CNTs. CNTs with a single wall on their surface were used to create the DNA sensor. N-Ethyl-N'-(3-dimethyl aminopropyl) carbodiimide hydrochloride (3-dimethyl aminopropyl) was covalently linked with nanotubes to detect *Salmonella* (Weber et al., 2011). A *salmonella-specific* DNA probe solution was prepared using an electrode and was incubated for 2 hours at room temperature with sensitivity at 1×10^{-9} mol/L DNA. When there is a mismatching nucleotide no fluctuations were recorded on the probe. Indium tin oxide was also employed to create an amino-modified aptasensor employing multiwalled carbon nanotubes (MWCNTs). *S. typhimurium* and *Salmonella enteritidis* were both detected by this sensor and the detection ranges were 5.5 to 101 and 6.7 to 101 cfu/mL, respectively (Hasan et al., 2018).

22.2.4 **Quantum dots**

Smaller than 10 nm NPs are known as quantum dots (QDs). QDs are incredibly small in size and have very high surface area which makes them unique in optronics and electrical fields. They also display fascinating phenomena like a high absorption and emission wavelength and a broad excitation range. Due to their distinctive fluorescence characteristics and biocompatibility, carbon-based QDs are generally employed in biological sectors (Xu et al., 2004; Sun et al., 2006). These are usually formed from graphitic cores with sp^2 hybridization (CDs) and have applications in the fields of optronics, drug delivery, bio-imaging, bio-labelling, and bio-analytics. Further, the properties of these carbon dots can be increased by surface functionalization and doping. Functional groups present on the surface of carbon QDs provide them with improved features in terms of solubility and other optical properties. For the development of a biosensor, surface functionalization is an important attribute. In a study by Wang et al. (2015b) an APTA-sensor was designed where carbon dots are used for the detection of *S. typhimurium* in eggshells and tap water samples with a detection limit of 50 cfu/mL within 2 hours with and without interference from other bacteria. Gold NPs and carbon dots containing specific sensors were developed by Wang et al. (2016) using an aptamer-specific AFB1 detection was done and the limit of detection came to be around 5 pg/mL LOD (pM).

22.2.5 **Magnetic nanomaterials beads**

Magnetic nanomaterials are clusters of magnetic beads of 50–500 nm (Liu et al., 2011; Gonzalez et al., 2015; Huang et al., 2016). These were widely used in biosensors because of their magnetic field. In these biosensors, the magnetic field can be altered. Mainly, these nanomaterials are used in lateral flow assay to detect pathogenic bacteria as they show color that separates both the complex and the target material (Ren et al., 2016; Qiao et al., 2017). Wang et al. (2015a) developed a biosensor for the detection of *Bacillus anthracis* spores. It was based on antibody-covered magnetic beads of around 300 nm. The detection limit in different food samples is 5×10^5 spores/g for baking soda, 2×10^5 spores/g of starch, and 6×10^4 spores/g for powdered milk. In another work, *E. coli* O157:H7 was detected in food samples with the use of magnetic beads and the detection limit was around 12 cfu/mL of broth (Suaifan et al., 2017). No specificity was shown by other pathogenic strains of *Staphylococcus aureus, Pseudomonas aeruginosa*, and *L. monocytogenes*. The sensor showed stability over six months. When magnetic beads were used with gold there was quick recognition of *S. choleraesuis*. The limit of detection in whole milk in 20 hours was around 5×10^5 cfu/mL, whereas the detection limit of colloidal gold without magnetic beads was around 5×10^6 cfu/mL (Xia et al., 2016).

22.3 Analysis of food microorganisms

Nanosensors are devices that detect physical quantities and convert them into measurable signals. Today they have widely been used in the food sector to analyze any microbial contaminants in food samples. For any changes in the quality of food, there will be a biological response that can produce some output signals in the biosensors and help to analyze the contaminations. To prepare nanosensors, proper selection of the nanomaterial is crucial, which depends on the properties of the species to be detected. Nanomaterials provide a large surface area for the immobilization of suitable receptors and analytes and help in sensing with a higher accuracy rate (Gerwen et al., 1998). Various types of nanobiosensors are being used. Immunosensors that can recognize antigen-antibody complex are designed by coating thin films on suitable surfaces for better recognition of the analytes (Pak et al., 2001). One of the highly sensitive devices with nanoscale functions are the nanoelectromechancial systems (NEMS). The NEMS when integrated with a system known as micro-electrochemical systems shows better performance and bio-adhesion properties with improved responses to diverse stimuli. This technology has been proven to be beneficial in the field of optronics where it is related to sensing, display, power generation, energy harvesting, and drug delivery applications. The identification and diagnosis of new diseases become possible by the use of this technology where biomarkers are possibly used to identify the biochemical interactions (Bhushan, 2007). In optical sensors, the sensing nanomaterials should have SPR properties (Zeng et al., 2011). In fluorescence-based sensors, the nanomaterials (mainly QDs) should have suitable emission properties.

Electrochemical nanosensors are frequently used in the food industry for the analysis of food microorganisms. For example, a single single-walled CNT-based biosensor was used to detect *E. coli* K-12 bacterial strain immobilizing

E. coli antibodies on CTNs. The sensor has a detection limit of $\sim 10^2$ cfu/mL (Yamada et al., 2014). Further, the sensor was checked against other pathogenic bacteria but it displays very good selectivity and sensitivity against *E. coli* strain. In another study, AuNPs modified electrode was developed to detect *L. monocytogenes* in spiked blueberries with a limit of detection of around 2 log cfu/g (Davis et al., 2013). Hong et al. (2015) developed an electrochemical biosensor for norovirus detection which was thermally stable during the range of temperatures $4°C-25°C$. Here concanavalin A is used as a recognition element for the detection of norovirus. The nanofabricated biosensor contains a gold electrode which contains the concanavalin A, a lectin originally extracted from jack bean. The norovirus when detected in a realistic environment shows the limit of detection around 60 copies/mL. Several researchers developed and designed different aptasensors for detecting *S. typhimurium* in pork samples (Wang et al., 2015b).

AuNPs with excellent SPR properties are used as optical sensor for the detection of several food pathogens. For example, *E. coli* O157:H7 was detected in milk samples with the help of AuNPs conjugated with *E. coli* antibodies (Lin et al., 2008). The developed sensor was able to detect the concentrations of *E. coli* in the range from 10^2 cfu/mL to 10^7 cfu/mL. Chemoluminiscene assay was also performed to detect the *E. coli* pathogens using porous silicon nanomaterials and the limit of detection was 10^1-10^2 cfu/mL (Mathew and Alocilja, 2005).

22.4 Analysis of food antimicrobials and antioxidants

Food antimicrobials and antioxidants are the compounds used to protect food from spoilage or any kind of toxification. Antimicrobial substances are the agents that help in the restoration of food quality by preventing it from any kind of microbes, i.e., fungi or bacteria thus helping in maintaining its overall nutrient content. Some antimicrobials have been increasingly used for killing the pathogenic microorganisms present in foods. For example, Nisin and lysozyme are used against the food poisoning bacteria, *Clostridium botulinum* in processed cheese, and lactate and diacetate to inactivate *Listeria monocytogenes* in processed meats. Antioxidants, on the other hand, are compounds that help in preventing the food from oxidative damage. The effectiveness and the mode of action of compounds to determine the overall microbial spectrum depends primarily on the physical and chemical attributes of the compounds. One such attribute is the polarity which determines the solubility of the compound in water where microorganisms thrive (Robach, 1980). Another property is the boiling point of a compound which can also directly affect the activity. Heating of the food during processing results in the loss of the volatile compounds present in it, e.g., during the process of cooking certain phenolic compounds are vaporized and lost. So, a proper analysis is always recommended to ensure the efficacy of the compounds in food.

22.4.1 **Optical sensing assays**

Some scientist uses optical sensing assays and utilizes metal NPs like gold, silver, and QDs as devices for the detection of antimicrobial and antioxidant compounds present in food samples. Metal NPs are synthesized by polyphenols which are known as antioxidant compounds. Vasilescu et al. (2012) reviewed different strategies (e.g., electrochemical and optical) to study the role of nanostructures in food antioxidants detection and analysis. This can be achieved by either exploiting the reduction ability of antioxidants toward metal ions and thus resulting in nanoparticle formation through seed-mediated growth or by harnessing the ability of antioxidants for the development of NPs having colorimetric and chromatic changes, which are employed as a measure of antioxidant activity. These features including SPR, chromatic, and colorometric transitions, aid in determining the absorbance of colloidal NPs solutions (Vasilescu et al., 2012). Assays using NPs that rely on changes in the optical characteristics of the particles for optical detection are summarized in Table 22.2. These nanoparticle-based optical approaches are easy and cheap methods for the rapid analysis of food samples.

Gold and silver NPs were generally used for the colorimetric determination of different compounds. One such example is the determination of carvacrol (CA) and thymol (TY) which were extracted from oregano and thyme plants (Memar et al., 2017; Daferera et al., 2002). Carvacrol and thymol essential oils possess excellent antimicrobial, antifungal, and antioxidant properties which can be harnessed by the food industry (Rodriguez et al., 2016; Abd El-Hack et al., 2016; Memar et al., 2017; Nabavi et al., 2015; Friedman et al., 2014). A gold NPs-based colorimetric sensor was developed for the analysis of thymol and carvacrol isomers using pH-dependent formation of gold NPs. Two different approaches based on colorimetric assay (absorption at 550 nm) were developed: one at pH 12 for the determination of total CA and TY, and the other at pH 9 for differential quantification of TY and CA. At pH 12, total determination of CA and TY was performed in accordance with the Folin–Ciocalteu method, whereas, the latter (pH 9) provides a simple way for calculation of TY/CA ratio. Believably, it can be used to assess the quality of essential oils and may become a valuable alternative to more sophisticated, laborious, and highly time-consuming methods. Several techniques were proposed for quantitative analysis exploiting the larger surface area and its catalytic properties of AuNPs. The formation of characteristic LSPR peak of AuNPs indicates the ability of endogenous food polyphenols to reduce Au (III) to AuNPs (0) (Wang et al., 2007). The growth of AuNPs seeds was reported by three model flavonoids (quercetin, daizcol, and puerarin) where CTAB, and $HAuCl_4$ were used as growth solutions (Scampicchio et al., 2006). To detect phenolic compounds antioxidant evaluation of two plant extracts, i.e., of Radix Astragali and soybean were used. A hypoxanthine sensor was developed by Lin et al. (2016) where gold nanorods were used. The generation of different colors like green, blue, brown, purple, pink, yellow, etc. entirely depends on the concentration of hypoxanthine.

Table 22.2 Nanoparticles-based optical sensing assays for the detection of food antioxidants.

Principle	Detection method	Matrix/application	References
Catalytic synthesis of AuNPs	Spectrophotometric detection at 555 nm	Different concentrations of black tea, green tea, and orange juice	Ma et al. (2010)
Reduction of Au^{3+} to Au^0 nanoparticles	Spectrophotometric detection at 555 nm	To check the purity of virgin argan oil based on total phenolic acids	Zougagh et al. (2011)
AuNPs generation by reduction of Au^{3+} ions	Spectrophotometric detection at 539 nm (amla), 543 nm (methanolic plant extract), 545 nm (aqueous plant extract)	Antioxidant activities of the stem bark of the plant, Indian Rosewood (*Dalbergia sissoo*)	McFarland et al. (2003)
Formation of anisotropic silver nanoparticles	Detection at 427 nm	Ascorbic Acid in pharmaceutical products (tablets)	Song et al. (2011)
Reduction of Au^{3+} to Au^0 nanoparticles	Colorimetric at 568 nm	Water samples and pharmaceutical ointments	Dai et al. (2005)
Aggregation of AuNPs	Spectrophotometric at 520 nm	Buffer solution	Chen et al. (2009)
Gold nanoshells growth over SiO_2/AuNPs	Redshift of plasmon resonance peak	Buffer solution	Mock et al. (2003)
Growth of Au nanoshells on SiO_2/AuNPs nanocomposites preadsorbed on ITO substrates	Spectrophotometric (shift of characteristic peak wavelength and increase in absorbance)	Buffer solution	Zhao et al. (2008)
Inhibition of H_2O_2-induced growth of Au nanoshells on SiO_2/AuNPs precursor nanocomposite	Spectrophotometric (shift of characteristic peak wavelength)	Tea and herb extracts	Nezhad et al. (2008)
Ce (IV) reduction	Visual, Optical (LSPR-based)	Rapeseed teas, medicinal mushrooms	Sharpe et al. (2013)
Ce (IV) reduction	Optical (LSPR-based)	Rapeseed and its by-products	Tulodziecka et al. (2016)
AgNPs seed-growth	Optical (LSPR-based)	Fruit juices, olive oil infusion/LLE	Özyürek et al. (2012)

In comparison with the gold NPs, the research on the silver NPs were less reported. AuNPs are more stable in comparison to AgNPs and also their oxidizing potential under the presence of oxygen is lesser. When polyphenols-based AgNPs were synthesized aggregation and un-stability of NPs were observed in suspensions. A colorimetric method was proposed (Vilela et al., 2015) based on the citrate-stabilized silver "seed growth" mediated by polyphenols.

22.4.2 Quantum dots-based fluorescence assays

Quantum dots-based fluorescence assays utilize semiconductor QDs having a particle size less than 10 nm. In a study by Özyürek et al. (2012), L-cysteine-capped CdTe QDs were synthesized for antioxidant activity assay utilizing the inhibitory effect of antioxidant on the UV-induced bleaching of CdTe QDs. Again, CdTe QDs were utilized (Hemmateenejad et al., 2012) for the polyphenols quantification with the help of the enzymatic reaction of laccase. Polyphenols were converted into mono/polyquinones that resulted in fluorescence quenching of QDs with the help of the enzyme laccase. The charge transfer process from quinones to QDs was exploited as optical labels for polyphenols. This method detects polyphenols with good recovery. Graphene QDs (GRQDs) obtained by pyrolysis of citric acid were used to determine the polyphenol fraction of olive oil extracts (Dwiecki et al., 2017). Glutathione, thiamine, ascorbic acid, retinoic acid, and B vitamins in food items have also been detected using different QD-based fluorescence sensors.

22.4.3 Electrochemical assays

Electrochemical biosensors based on various types of electrodes are valued tools to assess antimicrobial and antioxidant activity in food (Benítez-Martínez et al., 2014). A general scheme for the nanomaterials-based electrochemical biosensor is presented in Fig. 22.3. Different metals and their oxides are used as electrode materials and different enzymes act as receptors. For example, glutathione oxidase was used to detect glutathione, whereas, peroxidases and polyphenol oxidases were used to detect hydrogen peroxide and phenolics. The assays are based on a convenient policy to use electrochemical sensors for the oxidative disruption caused by lipids, proteins, and DNA. These sensing mechanism are based on the sensing of different molecules which helps in the detection of reactive oxygen species mainly utilizing cytochrome c, superoxide dismutase, and DNA probes.

Indium tin oxide-coated glasses functionalized with titanium nanotubes (TiNTs) were utilized as an electrochemical sensor for the measurement of antioxidant and free radicals (Petryayeva et al., 2013). Due to its substantial surface area, TiNTs were thought to be an efficient oxygen carrier and catalyst for oxygen conversion. Electrochemical hydrogen peroxide nanosensors (sensitivity $\leq 1\ \mu M$) were developed employing different electrocatalyst, such as multiwalled CNTs/Ag-nanohybrids/Au (Ozkan et al., 2018), Ag-DNA/GCE (Zhao et al., 2009), Ag NPs/CILE

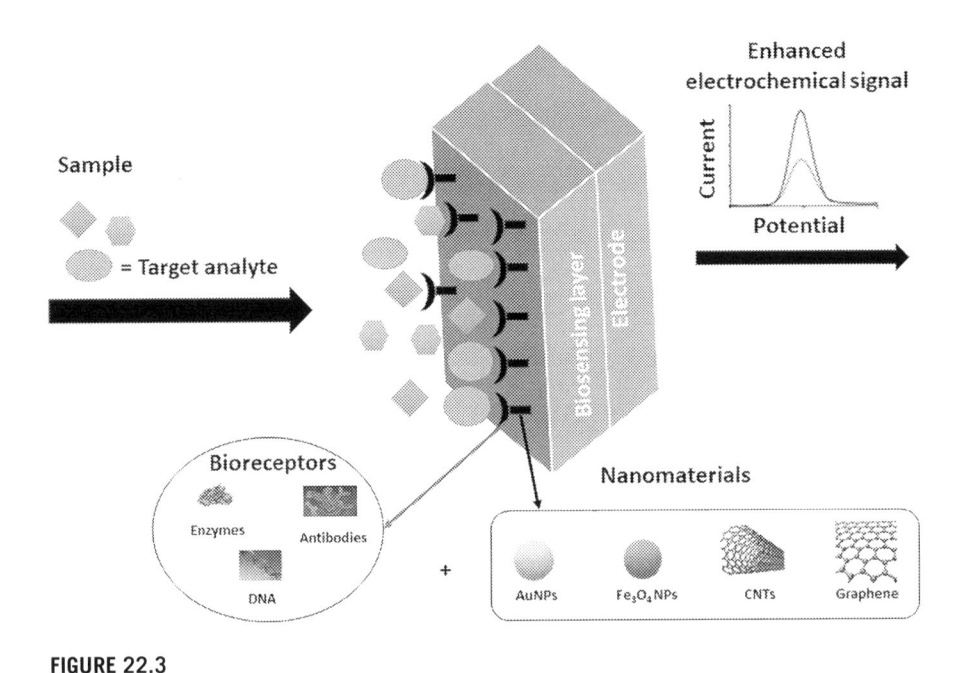

FIGURE 22.3

Main constituents of nanomaterials-based electrochemical biosensor.

(Wu et al., 2006), Ag NPs/GCE (Safavi et al., 2009), CuO NPs/carbon ionic liquid electrode (Welch et al., 2005), and Co_3O_4 NPs/GCE (Ping et al., 2010).

Eugenol, present in different plat products, is employed as a flavorful ingredient in food products. It has also interesting antiviral, antioxidant, and anti-inflammatory characteristics (Ogata et al., 2000; Benencia et al., 2000). The electrochemical behavior of eugenol can also be exploited for its detection in different food samples. For example, the cyclic voltammetry (CV) method was used to explore the electrochemical behavior of eugenol at the Cu@AuNPs/GCE and was successful in analyzing eugenol in food samples (Ziyatdinova et al., 2013). There have also been reports of several ethylene detection sensors. A chemo-resistive sensor was constructed by Esser et al. (2012) using single-walled carbon nanotubes (SWNTs) combined with Cu (I) complex and sandwiched between gold electrodes. A shift in the resistance occurred as a consequence of interaction with ethylene, and a commensurate voltage output was found.

22.4.4 Enzyme electrodes assays

Today one of the leading biosensing devices is the enzyme electrodes that not only possess potential selectivity but are easy, rapid, and high catalytic action-orientated electrodes in the market. Nanomaterials offer unique and creative outlines for the enzyme electrodes where electrochemical reactions are combined with direct wiring of enzymes to the electrode surface (Del Carlo et al., 2016). Enzyme electrodes for polyphenols are based on polyphenol oxidase (PPO) activity. PPO has copper as a prosthetic group and uses molecular oxygen as a co-substrate. The polyphenol electrodes work by two mechanisms; atfirst, cresolase activity occurs where monophenol is converted into an o-diphenol by the addition of a hydroxyl group at the ortho position followed by the catecholase activity where o-diphenol is converted into quinone (Gul et al., 2017). Generated quinones can be reduced at low potential onto the transducer surface producing a current signal proportional to the phenolic compound. Furthermore, the o-diphenols generated can be re-oxidized by the enzymes creating an enzymatic/electrochemical amplification cycle. Hypoxanthine, one of the main compounds in meat products can be quantified using, several enzymatic biosensors with colorimetric (Cunningham et al., 1978; Chen et al., 2017; Berti et al., 1988) or electrochemical detection which in turn quantify the level of hypoxanthine using the enzyme xanthine oxidase. In a study by Agüí et al. (2006), an electrochemical biosensor was developed to detect hypoxanthine by the immobilization of enzymexanthine oxidase onto a carbon-electrode modified with gold NPs. The sensor was then tested on meat samples and found to be sensitive up to a limit detection of 2.2×10^{-7} M of hypoxanthine. The sensor works by first converting the hypoxanthine to xanthine and then oxidized into uric acid. The sensitivity achieved was highest in the xanthine oxidase when it was immobilized on a carbon paste electrode modified with gold NPs. Yan et al. (2017) developed a method for the detection of xanthine based on a colorimetric sensor where it uses a nanocluster formed from copper that carries peroxidase-mimicking property.

22.5 **Stability of food antimicrobials and antioxidants**

Encapsulation methods are proven to be most effective not only for delivering the bioactive components of food antimicrobials but also for food antioxidants thus increasing their bioavailability and most essentially maintaining food stability (Augustin et al., 2006; Champagne and Fustier, 2007; McClements et al., 2009). Due to the increased surface-to-volume area of nanoencapsulated systems, microencapsulation is the best-suited strategy as it provides protection against any kind of adverse conditions. This strategy helps in protecting the food antimicrobials and food antioxidants against degradation, providing them with longer shelf life and also helping in improving the overall pharmokinetic profile of the compound. Moreover, due to the microencapsulation method the toxicity of the antimicrobial compound is significantly reduced (Ansari et al., 2014). A list of different food antimicrobials and antioxidants with nanocarriers are given in Tables 22.3 and 22.4. The common structures of the nanomaterials which are used to encapsulate the antimicrobials and antioxdiants are nanoemulsions, NPs, nanoliposomes, nanofibers, etc., as shown in Fig. 22.4.

Table 22.3 Different food antimicrobials encapsulated in different nanomaterials.

Antimicrobials	Carrier system	Material employed	Target organism	References
Carvacrol	Nanoemulsion	Miglyol 812N sunflower oil	*E. coli C 600, Listeria innocua, Lactobacillus delbrueckii, Saccharomyces cerevisiae*	Terjung et al. (2012) Donsi et al. (2012)
	Nanosphers	Chitosan zein	*Staphylococcus aureus Bacillus cereus Escherichia coli*	Keawchaoon and Yoksan (2011) Wu et al. (2012)
Basil oil	Nanoemulsion	Basil/oil water	*Escherichia coli*	Ghosh et al. (2013)
Carvone and anethole	Nanospheres	PLGA	*Salmonella Typhi*	Esfandyari-Manesh et al. (2013)
Cinnamaldehyde	Nanospheres nanoemulsion	PLGA Sunflower oil	*Salmonella* spp. *Listeria* spp. *Lactobacillus delbrueckii Saccharomyces cerevisiae Escherichia coli*	Gomes et al. (2011) Wu et al. (2012)

22.5.1 Nanoemulsions

These are lipophilic stable colloidal systems comprised of a combination of oil, water, and emulsifier within a size range of <100 nm (Burguera and Burguera, 2012). These nanoemulsions show better shelf life, and stability as compared to the microemulsion system (Fathi et al., 2012). There are different techniques to prepare the nanoemulsions based on the structure and functionality of the material used. The approaches used for preparation are micro-fluidization, homogenization, ultra-sonication, high pressure, and solvent diffusion. Commonly used nanoemulsions are oil-in-water emulsion (O/W). Here the oil droplets are dispersed in water and the emulsifiers stabilize the interface. These nanoemulsions are mainly required for delivering the less soluble food antimicrobial compounds thereby increasing their distribution. The oil-in-water-in-oil (O/W/O) and water-in-oil-in-water (W/O/W) also known as multiple emulsion, where different nanometric-sized polyelectrolytes surround the oil droplets. These nanoemulsion systems help the active compounds to interact more with the cell membrane of bacteria thud helps in releasing intracellular constituents. So, nanoemulsions can be the model system of targeted delivery in the food industry for optimal delivery of food antioxidant and antimicrobial substances. Some representative food antimicrobials and antioxidants encapsulated in nanoemulsions are presented in Tables 22.3 and 22.4.

Microorganisms commonly result in food spoilage. Some essential oils have been proven to be safe and effective to control it. Encapsulation of these essential

Table 22.4 Different food antioxidants encapsulated in different nanomaterials.

Class	Compounds	Systems	References
Flavonols	Fisetin	Polymeric micelles Polymeric nanoparticles	Sechi et al. (2016)
	Myrcetin	Phosphatidylcholine-based solid lipid	Bazylinska et al. (2014)
	Quercetin	Nanospheres PLGA nanocapsules Nanosuspensions	El-Gogary et al. (2014) Karadag et al. (2014) Gonclaves et al. (2015)
	Rutin	Nanoemulsions	Macedo et al. (2014) Sharma et al. (2015)
Flavones	Apigenin	Liposomes	Paini et al. (2015)
	Luteolin	Nanostructured lipid carriers	Liu et al. (2014)
	Baicalin	Phosphatidylcholine-based solid lipid nanocarriers Nanostructured lipid carriers	Bazylinska et al. (2014) Luan et al. (2015)
Flavonones	Nobiletin	Self-nanoemulsion	Ban et al. (2015)
	Naringin	Polymeric nanoparticles	Cordenonsi et al. (2016)
	Naringenin	Edible oil-based lipid nanoparticles Chitosan nanoparticles Polymeric nanoparticles Nanoliposomes	Winarti et al. (2015) Cordenonsi et al. (2016) Wang et al. (2016)
Flavonols	Catechin extract from tea	Nanoliposomes	Zou et al. (2014)
	Catechin extract from green tea leaf waste Epicatechin	Nanoemulsion Bovine serum albumin nanoparticles	Tsai and Chen (2016) Yadav et al. (2014)
	(-)-epigallocatechin gallate	Nanoparticles prepared from chitosan and aspartic acid β-Lactoglobulin nanoparticles Ovalbumin—dextran conjugate nanoparticles	Hong et al. (2014) Lestringant et al. (2014) Li and Gu (2014)
Anthocyanins	Anthocyanin from strawberry fruit	Nanochitosan-based coating with and without copper	Eshghi et al. (2014)
	Anthocyanin rich extract	Multilayered liposomes	Gibis et al. (2014)
	Anthocyanin from blackberry waste extract	Chitosan-coated liposomes	Gultekin-Ozguve et al. (2016)

(continued on next page)

Table 22.4 Different food antioxidants encapsulated in different nanomaterials—cont'd

Class	Compounds	Systems	References
Phenolic acids	Gallic acid	Ionotropic gelation Cross linking-gallic acid charging desolvation Enzyme induced cross linking-cold set gelation	Lamarra et al. (2016) Nourbakhsh et al. (2016) Nourbakhsh et al. (2016)
	Caffeic acid	Reverse phase evaporation	Katuwavila et al. (2016)
	Chlorogenic acid	β-Cyclodextrin Ionic gelation	Ramírez-Ambrosi et al. (2014) Nallamuthu et al. (2015)
Vitamins	Vitamin B2	Ionotropic polyelectrolyte pregelation	Azevedo et al. (2014)
	Vitamin E	Spray drying	Hatege et al. (2015)
	Vitamin C	Ionic gelation	Jiménez et al. (2014)
	Thiamine	Nanoliposomes	Juveriya et al. (2016)
	Folic acid	Nanoemulsion	Assadpour et al. (2016)

oils is important to reduce any kind of mass transfer resistance to the target site, which also increases their stability. Eugenol, an essential oil, was proven to be effective against the bacterial cell membrane of the pathogenic bacteria *S. aureus* (Ghosh et al., 2014). In another study (Terjung et al., 2012), carvacol nanoemulsions were formed by using ultrasonication and high-pressure homogenization approaches. Carvacol mixed with eugenol and triacylglyceride or tween was effective against the *E. coli* C 600 and *L. innocua* strains. When a concentration of 800 ppm was used with a mean droplet size emulsion of 300 nm the complete inhibition of *L. innocua* pathogen was achieved, while a delay in growth was observed at a lower mean droplet size emulsion. Basil oil (*Ocimum basilicum*) nanoemulsions were formed using the ultrasonic emulsification method. The formulation comprising 88% of estragole, tween 80, and water displays high antibacterial activity against *E. coli*. Different dilutions of the emulsion were used which resulted in complete inactivation of the bacteria after 45 minutes. Further fluorescence microscopy was done to see the damage that occurred inside the bacterial cell membrane and the result established that nanoemulsions promoted the damage (Ghosh et al., 2013). A lemongrass oil (LO)-based nanoemulsion was developed by high-pressure homogenization technique where carnauba shellac wax was used for encapsulating the lemongrass oil. The resulting nanoemulsion was effective against *E. coli* and *L. monocytogenes* in 2 hours. Further, these developed nanoemulsions were used after a period of five months where the emulsions were applied to apples and plums which resulted in a significant decrease of aerobic bacteria in the range of 0.8–1.4 log cfu/g (Kim et al., 2013; Jo et al., 2014). A sunflower-based O/W nanoemulsion was developed by Joe et al. (2012),

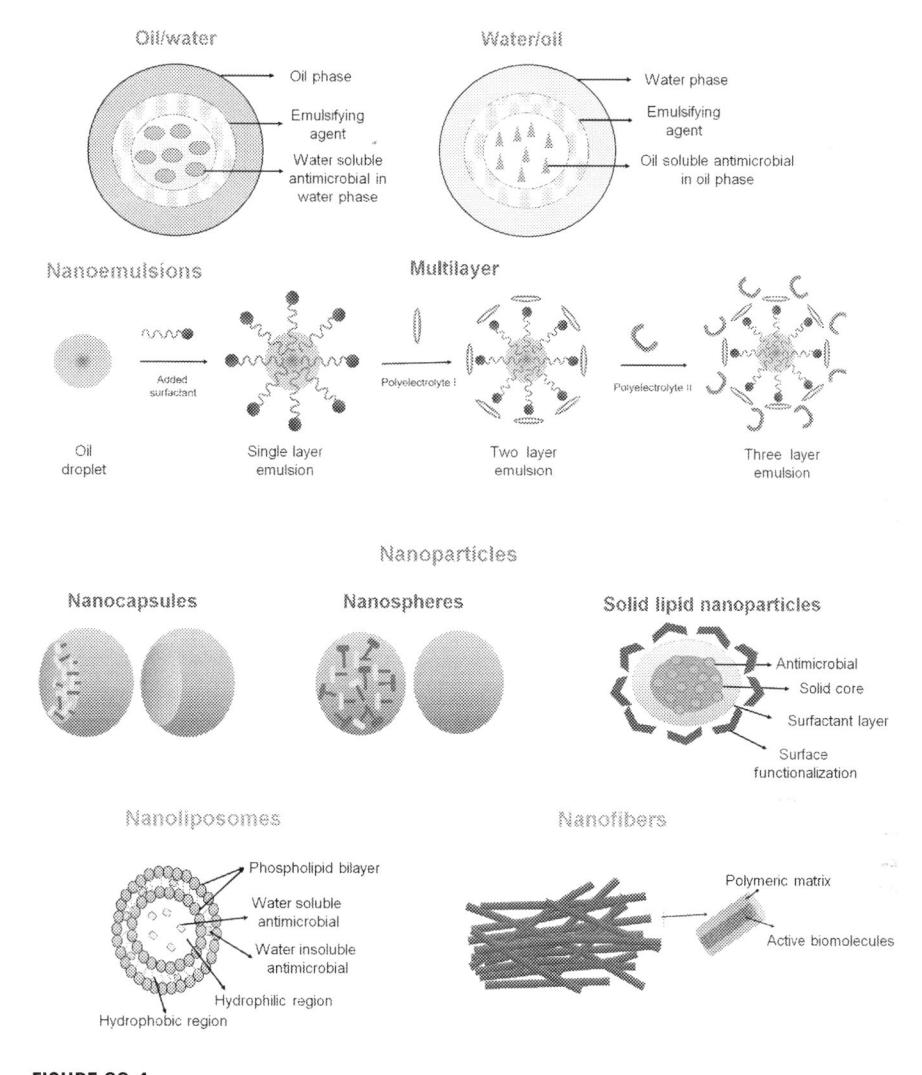

FIGURE 22.4

Structures of nanomaterial used to encapsulate food antimicrobials.

where the surfactant was a cyclic lipopeptide produced by *B. Subtilis*. The antibacterial activity of the formulated nanoemulsion was higher against the *S. aureus* and *L. monocytogenes* as compared with the antibiotic streptomycin. It also determines increased fungicidal activity against the fungus *Penicillium* sp., *Rhizopus nigricans*, and *Aspergillus niger*, in comparison to the positive control sodium benzoate. The nanoemulsion displays exceptional sporicidal activity against *Bacillus cereus* and *Bacillus circulans* strains (3-fold greater than positive control). These nanoemulsions when used against food products like apple juice, milk, and raw chicken result in the reduction of fungal and bacterial colonies. A W/O/W nanoemulsion was formed using bovine lactoferrin, an iron-binding protein with lecithin and poloxamers by homogenization method (Pandit et al., 2016). This nanoemulsion drastically reduces the growth of iron-dependent bacteria. *S. aureus* and *L. innocua* and displayed a MIC of 2000 mg/mL whereas the MIC of 200 mg/mL was found in the case of *Candida albicans* using encapsulated as well as the free lactoferrin.

Food antioxidants are also encapsulated in nanoemulsions thus maintaining their bioavailability and stability. One of the commonly used food antioxidants is flavonoids. Varoious classes of flavonoids include anthocyanins, flavonones, isoflavones, and flavonols. Among the studied compounds, quercetin (a flavonol) is most commonly used. Dian et al. (2014) studied quercetin-loaded polymeric micelles which were stable for longer duration and possess sustainable release property. Encapsulated quercetin is more efficient than free quercetin in terms of its antioxidant properties (Goncalves et al., 2015; Souza et al., 2014). Several other research groups also worked on the permeability of rutin during exvivo gut sac studies. They found that nanoemulsified rutin have better and higher permeability than rutin suspension. The antioxidant property of rutin nanoemulsion was greater than free rutin suspension. Liu et al. (2014) showed the high bioavailability/bioaccessibility of flavones like luteolin and nobiletin after their nanoencapsulation. Fernandez et al. (2016) in an invitro release study conducted on the stomach and intestine through simulation. The study demonstrated proanthocyanidins efficient release through encapsulated grape seed and skin samples. A pharmokinetic study (Tripathi et al., 2016) established that isoflavonone and genistein oral bioavailability in rats was significantly improved, i.e., 2.8-fold by nanoemulsification in comparison to free genistein. In a different study on nanoencapsulation of anthocyanins, displayed that encapsulation makes the compound more stable than the free ones even at high temperatures and pH encapsulation helps them in increasing their storage stability (Ravanfar et al., 2016). During a study by Eshghi et al. (2014), the nanochitosan-based coating was prepared and then its effect was observed on the bioactive components of strawberries. These coatings were produced with or without copper loading and the effect was observed up to 3 weeks during storage. The results obtained were interesting as there was a major reduction seen in the antioxidant potential of strawberries where there was no nanochitosan coating whereas slow reduction was observed in the free nanochitosan without copper followed by loaded copper nanochitosan. To deliver ascorbic acid to marine settings, chitosan-based nanoencapsulation was widely studied. Jiménez-Fernández et al. (2014) conducted a study on the stability of NPs in seawater and found that the NPs that are positively charged (30–35 mV) and have a size of less than <300 are stable. Studies on nanoencapsulation of chlorogenic acid (a phenolic acid) recommended that β-cyclodextrin nanosponges are promising materials for the bioavailability and stability of these acids (Ramírez-Ambrosi et al., 2014).

Vitamins are one of the most important and fundamental elements required for maintaining human health. The vitamins control the development and metabolism and thus aid in disease prevention (Wildman et al., 2016). Therefore, these need to be protected from degradation and there are several studies on the nanoencapsulations of vitamins. For example, nanoemulsion was applied to folic acid to protect them from any kind of detrimental conditions (Assadpour et al., 2016).

22.5.2 Nanoparticles

Nanoparticles including the nanosphere and nanocapsules have at least one-dimension lessor than 100 nm. Nanospheres can be made up of a polymeric matrix of spherical shape, whereas a nanocapsule is a nanovascular system confining of target molecule to the inner cavity (Reis et al., 2006; Chen et al., 2006). The efficiency of these synthesized nanocapsules and nanospheres depends on different attributes like the physical state, charge, morphology, and size (Ahsan et al., 2002; Rodriguez et al., 2004). Different methods have been employed to synthesize NPs for food applications. These are emulsification, polymerization, nanoprecipitation, electrospraying, and solvent evaporation (Weiss et al., 2006). Comparative studies reveal that NP-based stabilization does have several advantages over nanoemulsions. For example, organic solvents are not required for the preparation of NPs and their encapsulation efficiencies are comparatively much higher (Ahsan et al., 2002; Rodriguez et al., 2004) as compared to nanoemulsions. Several compounds have been encapsulated in NPs with antimicrobial and antifungal activities. During a study by Gomes et al. (2011), spherical NPs using the emulsion-evaporation technique were synthesized using poly (lactide-co-glycolide) (PLGA) doped with cinnamaldehyde and eugenol, and polyvinyl alcohol as a surfactant. Due to the nanoencapsulation technique, the solubility of both the doped compounds increased. When the antibacterial activity of these NPs was observed they had effectively inhibited the growth of *Listeria* and *Salmonella* pathogenic bacterial strains. The minimum inhibitory concentration (MIC) was observed around 10–20 mg/mL. Using nanoprecipitation and emulsification methods PLGA nanospheres were synthesized where the nanospheres were loaded with carvone and anethole. The MIC exhibited by loaded NPs were studied against pathogenic bacterial strains like *S. typhi*, *S. aureus*, and *E. coli* and found to be around 227 mg/mL, 182 mg/mL, and 374 mg/mL, respectively. Zhang et al. (2014) developed carvacoral–loaded chitosan NPs which have MIC and MBC of 0.257 mg/mL, 4.11 mg/mL, 2.0 mg/mL, and 8.25 mg/mL, respectively against infectious bacteria (viz. *E. coli, S. aureus*, and *B. cereus*). Various other experiments were performed where the release of carvacrol was seen by adjusting different pH conditions. 52.6% of carvacrol was released in acetate buffer solution (lower pH conditions), whereas, at higher pH conditions 22.5% and 33.1% were released in phosphate buffer solution. Antibacterial activity was seen against *E. coli* bacteria using NPs made of carvacrol and thymol encapsulated in zein nanospheres. Bacterial concentration in comparison with control was decreased from 0.8 log cfu/mL to almost 1.8 log cfu/mL (Wu et al., 2012). Further properties of the NPs were improved using stabilizers like chitosan hydrochloride and sodium caseinate (Zhang et al., 2014).

Antimicrobial peptides are potential substitutes for conventional antibiotics (Marcos and Gandia, 2009). One of the approved antimicrobial peptides nisin known to be thermally-resilient is produced by *Lactococcus lactis*. This is active against several food pathogens like *B. cereus*, *L. monocytogenes*, and *S. aureus* (Joerger, 2007). This peptide works by generating pores on the cell membrane surface resulting in the cytoplasmic component leakage (mainly lipids and proteins) present in the cell (Breukink and Kruijff, 2006). Antimicrobial compounds work by two different mechanisms: either can disrupt the cell membrane resulting in the functionality change or create pores resulting in disruption of the bilayer and leakage of internal components due to the loss in proton motive force (Magalhaes and Nitschke, 2013). Chitosan/carrageenan nanocapsules encapsulated with nisin can be synthesized by ionic complexation method and the resulting nanocapsuleas showed excellent antibacterial effect against *Enterobacter aerogenes*, *Micrococcus luteus*, *P. aeruginosa*, and *S. enterica* and the effect lasted up to 20 days (Lv et al., 2014).

Edible oil-derived lipid NPs have enhanced the oral bioaccessibility of hesperetin and naringenin in the small intestine (Ban et al., 2015). Hydrolysis of these synthesized lipid NPs took place in the small intestine during a stimulated digestion process where the incorporated hesperetin and naringenin components were successfully protected. Another study on polymeric NPs which were synthesized from catechin extract from white tea showed very good antioxidant properties even after nanoencapsulation (Sanna et al., 2015).

Phenolic acid-based NPs have been produced by different nanotechnological approaches. Gallic acid, an important polyphenols, was utilized to assess the stability of the synthesized NPs. Gallic acid was coated on the surface through the emulsion solvent evaporation method and these NPs were stable at a storage temperature of $-20°C$ up to 12 weeks. The in vitro release studies showed enhanced sustained gallic acid release from NPs (Alves et al., 2016). Chitosan nanocarriers loaded with rosmarinic acid were used for a drug delivery profile study and found that there was no chemical interaction between the NPs and rosmarinic acid even after encapsulation. So, it turns out to be an effective matrix as it protects the encapsulated material from different conditions. Besides, a properly controlled release of active materials was found (da Silva et al., 2015).

22.5.3 Nanoliposomes

Nanoliposomes are nanometric bilayer concentric vesicles composed of phospholipids. These are water-soluble, lipid-soluble, and amphiphilic molecules mainly used for encapsulating the material to be delivered at the site. The use of nanoliposomes is applicable in the biomedical field for drug delivery and in the food industry for encapsulation of food constituents. The method of synthesis of these nanoliposomes is through heating or microfluidization (Mozafari et al., 2008). Nanoliposomes interact by different mechanisms with the target cell. It can either interact by adsorption with the target cell, fusion through endocytosis, or through micropinocytosis

(Torchilin et al., 2005). Homegenization technique under high pressure was followed for the synthesis of nisin nanoliposomes as it is present in both lamellar and core phases. So nisin in lecithin-soybean nanoliposomes were developed by a micro-fluidizer. Nisin has the ability to produce pores in the nanoliposomes which affect the release of antimicrobial compound as revealed by transmission electron microscopy. To further slow down the release of nisin these nanoliposomes were attached to a matrix of hydroxypropyl methylcellulose. Those films revealed greater antimicrobial activity for the bacteria *L. monocytogenes* in a time duration of 10 hours compared to the encapsulated nisin in lecithin-soybean (Imran et al., 2012).

Carotenoids were used and encapsulated with nanoliposomes. β-carotene, can-thaxanthin, lutein, and lycopene were the majorly used carotenoids. Different charac-teristics of liposome-encapsulated carotenoids were evaluated based on lipid degrad-ability, stability, and release potential. Canthaxanthin-loaded liposome and lycopene showed micelle formation which helps in a thorough examination of the structural properties of the encapsulated particles. Bioaccessibility of liposome-encapsulated carotenoids was observed to be varied in the order of lutein $>$ β-carotene $>$ lycopene $>$ canthaxanthin. In vitro study demonstrated that the encapsulation potential of fitosterol ester in the oil phase with sodium alginate was much higher when α-lipoic acid and β-carotene was present, whereas the efficiency was decreased in the absence of α-lipoic acid and β-carotene (Gupta and Ghosh, 2012). Again, β-carotene is readily oxidized due to which its encapsulation efficiency is reduced in comparison to α-lipoic acid.

Some other studies demonstrated the formation of nanoliposomes for preventing vitamins from detrimental conditions. For example, upon encapsulation in nanolipo-somes, unstable thiamine has shown 97% encapsulation efficiency and can be used as a carrier (Fathima et al., 2016). Interestingly, those nanoliposomes were stable even up to 300°C and can be stored for up to 3 months at different temperatures.

22.5.4 **Nanofibers**

Nanofibers are materials that possess ultrathin diameters of less than 100 nm. Carbo-hydrates, lipids, and proteins are generally used for the synthesis of these nanofibers. The nanofibers are synthesized via an electrospinning process based on the generation of electric current. Recently, these nanofibers due to their greater surface-to-volume ratios have been immensely employed in the food industry. This surface volume attribute of these nanofibers makes them useful in various applications like the development of efficient delivery systems and using them in edible films. This will be beneficial for food packaging as they do have antimicrobial properties. Electro-spun nanofibers were developed using hydroxyethyl cellulose diallyl dimethyl dimethyl ammonium chloride, polyvinyl alcohol (PVA), copolymer with the polymeric mixture of polyquaternium-4 cellulose (PQ-4). These nanomaterials when used against an-timicrobial pathogenic strains like *S. aureus* and *E. coli* have shown effective activity (Jia et al., 2011). Chitosan, mainly a polysaccharide derivative of chitin produces by N-acetylation of the polymer chitin. These chitosan derivatives are proven to be effective against both bacteria and fungi by interacting with their membranes (Ignatova et al., 2006). PVA, polyvinyl pyrrolidone, and quaternized chitosan were used to develop electrospun nanofibers.

In terms of biodegradability, good carrier ability, and antibacterial properties cellulose-based products are the major food products. For example, a cellulose-based product produced by *Acetobacter xylinum* has wide antibacterial activity due to high fiber content and high water-holding capacity. Insoluble nanofibers were developed by the electrospun method and showed very good antibacterial activities against *S. aureus* at a pH of 4.6. Here, a polyamine protein isolated from corn which is of hydrophobic character is mixed with zein to develop the nanofibers (Giner et al., 2009). In a different study, nanofibers were produced by the cross-linking technique where the cross-linkers are bacterial cellulose, polylysine, a naturally occurring peptide with antimicrobial attributes, and procyanidins. These nanofibers are effective against the pathogenic bacteria *E. coli* and *S. aureus* (Gao et al., 2014).

22.6 Conclusion

Nanobiotechnology has been demonstrated as an innovative approach for rapid and precise analysis of food pathogens, food antimicrobials, and antioxidants. Optical, electrochemical, and fluorescence-based methods are mainly described using various nanostructures. Electrochemical detection recommends several advantages for on-site analysis using user-friendly formats. Over the last decade, various assemblies immobilizing biomolecules and different nanostructures have acted as signal enhancers to produce effective nanobiosensors. The use of magnetic NPs and mixing them into microfluidic systems has been observed as easy, quick, and easy-to-use nanobiosensors. Several nanobiotechnology-based encapsulation techniques are discussed to provide stabilities to the food antimicrobials and antioxidants. These techniques offer novel qualities to food products by improving the overall food taste, appearance, texture, and its nutritional value. In the past few decades, although, a large variety of nanomaterials-based biosensors and stabilization techniques were reported in the literature for food analysis, the majority of them are limited to laboratory research. Commercialization of the processes needs to be more focused on the betterment of food safety.

Acknowledgment

The authors would like to express deep gratitude for the support of the Department of Science and Technology, Government of India for financial assistance through the INSPIRE Faculty Award (IFA16-MS81).

References

Agüí, L., Manso, J., Yáñez-Sedeño, P., Pingarrón, J.M., 2006. Amperometric biosensor for hypoxanthine based on immobilized xanthine oxidase on nanocrystal gold–carbon paste electrodes. Sens. Actuat. B Chem. 113 (1), 272–280.

Ahsan, F., Rivas, I.P., Khan, M.A., Suárez, A.I.T., 2002. Targeting to macrophages: role of physicochemical properties of particulate carriers—liposomes and microspheres—on the phagocytosis by macrophages. J. Control. Release 79 (1-3), 29–40.

Alves, A.D.C.S., Mainardes, R.M., Khalil, N.M., 2016. Nanoencapsulation of gallic acid and evaluation of its cytotoxicity and antioxidant activity. Mater. Sci. Eng. C 60, 126–134.

Ansari, M., Khan, H., Khan, A., Singh, S., Saquib, Q., Musarrat, J., 2014. Gum arabic capped-silver nanoparticles inhibit biofilm formation by multi-drug resistant strains of *Pseudomonas aeruginosa*. J. Basic Microbiol. 54, 1–12.

Assadpour, E., Maghsoudlou, Y., Jafari, S.M., Ghorbani, M., Aalami, M., 2016. Optimization of folic acid nano-emulsification and encapsulation by maltodextrin-whey protein double emulsions. Int. J. Biol. Macromol. 86, 197–207.

Augustin, M.A., Sanguansri, L., Bode, O., 2006. Maillard reaction products as encapsulants for fish oil powders. J. Food Sci. 71 (2), E25–E32.

Azevedo, M.A., Bourbon, A.I., Vicente, A.A., Cerqueira, M.A., 2014. Alginate/chitosan nanoparticles for encapsulation and controlled release of vitamin B2. Int. J. Biol. Macromol. 71, 141–146.

Ban, C., Park, S.J., Lim, S., Choi, S.J., Choi, Y.J., 2015. Improving flavonoid bioaccessibility using an edible oil-based lipid nanoparticle for oral delivery. J. Agric. Food Chem. 63 (21), 5266–5272.

Bazylińska, U., Pucek, A., Sowa, M., Matczak-Jon, E., Wilk, K.A., 2014. Engineering of phosphatidylcholine-based solid lipid nanocarriers for flavonoids delivery. Colloids Surf. A 460, 483–493.

Benencia, F., Courreges, M.C., 2000. In vitro and in vivo activity of eugenol on human herpesvirus. Phytother. Res. 14 (7), 495–500.

Benítez-Martínez, S., Valcárcel, M., 2014. Graphene quantum dots as sensor for phenols in olive oil. Sens. Actuat. B Chem. 197, 350–357.

Berti, G., Fossati, P., Tarenghi, G., Musitelli, C. and Melzi d'Eril, G.V., 1988. Enzymatic colorimetric method for the determination of inorganic phosphorus in serum and urine. J. Clin. Chem. Clin. Biochem. 26 (6), 399–404.

Bhushan, B., 2007. Nanotribology and nanomechanics of MEMS/NEMS and BioMEMS/BioNEMS materials and devices. Microelectron. Eng. 84 (3), 387–412.

Breukink, E., de Kruijff, B., 2006. Lipid II as a target for antibiotics. Nat. Rev. Drug Discov. 5 (4), 321–323.

Burguera, J.L., Burguera, M., 2012. Analytical applications of emulsions and microemulsions. Talanta 96, 11–20.

Champagne, C.P., Fustier, P., 2007. Microencapsulation for the improved delivery of bioactive compounds into foods. Curr. Opin. Biotechnol. 18 (2), 184–190.

Chen, H., Weiss, J., Shahidi, F., 2006. Nanotechnology in nutraceuticals and functional foods. Food Technol. 60 (3), 30–36.

Chen, J., Wang, C., Irudayaraj, J.M., 2009. Ultrasensitive protein detection in blood serum using gold nanoparticle probes by single molecule spectroscopy. J. Biomed. Opt. 14 (4), 040501.

Chen, Z., Lin, Y., Ma, X., Guo, L., Qiu, B., Chen, G., Lin, Z., 2017. Multicolor biosensor for fish freshness assessment with the naked eye. Sens. Actuat. B Chem. 252, 201–208.

Cordenonsi, L.M., Bromberger, N.G., Raffin, R.P., Scherman, E.E., 2016. Simultaneous separation and sensitive detection of naringin and naringenin in nanoparticles by chromatographic method indicating stability and photodegradation kinetics. Biomed. Chromatogr. 30 (2), 155–162.

Cunningham, S.K., Keaveny, T.V., 1978. A two-stage enzymatic method for determination of uric acid and hypoxanthine/xanthine. Clin. Chim. Acta 86 (2), 217–221.

Daferera, D.J., Tarantilis, P.A., Polissiou, M.G., 2002. Characterization of essential oils from Lamiaceae species by Fourier transform Raman spectroscopy. J. Agric. Food Chem. 50 (20), 5503–5507.

Dai, Z., Xu, X., Wu, L., Ju, H., 2005. Detection of trace phenol based on mesoporous silica derived tyrosinase-peroxidase biosensor. Electroanalysis 17 (17), 1571–1577.

Dasgupta, N., Ranjan, S., Mundekkad, D., Ramalingam, C., Shanker, R., Kumar, A., 2015. Nanotechnology in agro-food: from field to plate. Food Res. Int. 69, 381–400.

da Silva, S.B., Amorim, M., Fonte, P., Madureira, R., Ferreira, D., Pintado, M., Sarmento, B., 2015. Natural extracts into chitosan nanocarriers for rosmarinic acid drug delivery. Pharm. Biol. 53 (5), 642–652.

Davis, D., Guo, X., Musavi, L., Lin, C.S., Chen, S.H., Wu, V.C., 2013. Gold nanoparticle-modified carbon electrode biosensor for the detection of *Listeria monocytogenes*. Ind. Biotechnol. 9 (1), 31–36.

Del Carlo, M., Capoferri, D., Gladich, I., Guida, F., Forzato, C., Navarini, L., Compagnone, D., Laio, A., Berti, F., 2016. In silico design of short peptides as sensing elements for phenolic compounds. ACS Sens. 1 (3), 279–286.

Dian, L., Yu, E., Chen, X., Wen, X., Zhang, Z., Qin, L., Wang, Q., Li, G., Wu, C., 2014. Enhancing oral bioavailability of quercetin using novel soluplus polymeric micelles. Nanoscale Res. Lett. 9 (1), 1–11.

Donsì, F., Annunziata, M., Vincensi, M., Ferrari, G., 2012. Design of nanoemulsion-based delivery systems of natural antimicrobials: effect of the emulsifier. J. Biotechnol. 159 (4), 342–350.

Dwiecki, K., Nogala-Kałucka, M., Polewski, K., 2017. Determination of total phenolic compounds in common beverages using CdTe quantum dots. J. Food Process. Preserv. 41 (2), e12863.

El-Gogary, R.I., Rubio, N., Wang, J.T.W., Al-Jamal, W.T., Bourgognon, M., Kafa, H., Naeem, M., Klippstein, R., Abbate, V., Leroux, F., Bals, S., 2014. Polyethylene glycol conjugated polymeric nanocapsules for targeted delivery of quercetin to folate-expressing cancer cells in vitro and in vivo. ACS Nano 8 (2), 1384–1401.

Esfandyari-Manesh, M., Ghaedi, Z., Asemi, M., Khanavi, M., Manayi, A., Jamalifar, H., Atyabi, F., Dinarvand, R., 2013. Study of antimicrobial activity of anethole and carvone loaded PLGA nanoparticles. J. Pharm. Res. 7 (4), 290–295.

Eshghi, S., Hashemi, M., Mohammadi, A., Badii, F., Mohammadhoseini, Z., Ahmadi, K., 2014. Effect of nanochitosan-based coating with and without copper loaded on physicochemical and bioactive components of fresh strawberry fruit (*Fragaria x ananassa* Duchesne) during storage. Food Bioprocess Technol. 7 (8), 2397–2409.

Esser, B., Schnorr, J.M., Swager, T.M., 2012. Selective detection of ethylene gas using carbon nanotube-based devices: utility in determination of fruit ripeness. Angew. Chem. Int. Ed. 51 (23), 5752–5756.

Ezhilarasi, P.N., Karthik, P., Chhanwal, N., Anandharamakrishnan, C., 2013. Nanoencapsulation techniques for food bioactive components: a review. Food Bioprocess Technol. 6 (3), 628–647.

Ezzat Abd El-Hack, M., Alagawany, M., Ragab Farag, M., Tiwari, R., Karthik, K., Dhama, K., Zorriehzahra, J., Adel, M., 2016. Beneficial impacts of thymol essential oil on health and production of animals, fish and poultry: a review. J. Essent. Oil Res. 28 (5), 365–382.

Fakhouri, F.M., Casari, A.C.A., Mariano, M., Yamashita, F., Mei, L.I., Soldi, V., Martelli, S.M., 2014. Effect of a gelatin-based edible coating containing cellulose nanocrystals (CNC) on the quality and nutrient retention of fresh strawberries during storage. IOP Conf. Ser.: Mater. Sci. Eng. 64, 012024.

Fathi, M., Mozafari, M.R., Mohebbi, M., 2012. Nanoencapsulation of food ingredients using lipid based delivery systems. Trends Food Sci. Technol. 23 (1), 13–27.

Fathima, S.J., Fathima, I., Abhishek, V., Khanum, F., 2016. Phosphatidylcholine, an edible carrier for nanoencapsulation of unstable thiamine. Food Chem. 197, 562–570.

Fernández, K., Aburto, J., von Plessing, C., Rockel, M., Aspé, E., 2016. Factorial design optimization and characterization of poly-lactic acid (PLA) nanoparticle formation for the delivery of grape extracts. Food Chem. 207, 75–85.

Friedman, M., 2014. Chemistry and multibeneficial bioactivities of carvacrol (4-isopropyl-2-methylphenol), a component of essential oils produced by aromatic plants and spices. J. Agric. Food Chem. 62 (31), 7652–7670.

Galindo-Rodriguez, S., Allemann, E., Fessi, H., Doelker, E., 2004. Physicochemical parameters associated with nanoparticle formation in the salting-out, emulsification-diffusion, and nanoprecipitation methods. Pharm. Res. 21 (8), 1428–1439.

Gao, C., Yan, T., Du, J., He, F., Luo, H., Wan, Y., 2014. Introduction of broad spectrum antibacterial properties to bacterial cellulose nanofibers via immobilising ε-polylysine nanocoatings. Food Hydrocoll. 36, 204–211.

Ghosh, V., Mukherjee, A., Chandrasekaran, N., 2013. Ultrasonic emulsification of food-grade nanoemulsion formulation and evaluation of its bactericidal activity. Ultrason. Sonochem. 20 (1), 338–344.

Ghosh, V., Mukherjee, A., Chandrasekaran, N., 2014. Eugenol-loaded antimicrobial nanoemulsion preserves fruit juice against, microbial spoilage. Colloids Surf. B 114, 392–397.

Gibis, M., Zeeb, B., Weiss, J., 2014. Formation, characterization, and stability of encapsulated hibiscus extract in multilayered liposomes. Food Hydrocoll. 38, 28–39.

Gomes, C., Moreira, R.G., Castell-Perez, E., 2011. Poly (DL-lactide-co-glycolide)(PLGA) nanoparticles with entrapped trans-cinnamaldehyde and eugenol for antimicrobial delivery applications. J. Food Sci. 76 (2), N16–N24.

Gonçalves, V.S.S., Rodríguez-Rojo, S., De Paz, E., Mato, C., Martín, Á., Cocero, M.J., 2015. Production of water soluble quercetin formulations by pressurized ethyl acetate-in-water emulsion technique using natural origin surfactants. Food Hydrocoll. 51, 295–304.

Gul, I., Ahmad, M.S., Naqvi, S.S., Hussain, A., Wali, R., Farooqi, A.A., Ahmed, I., 2017. Polyphenol oxidase (PPO) based biosensors for detection of phenolic compounds: a review. J. Appl. Biol. Biotechnol. 5 (3), 72–85.

Gültekin-Özgüven, M., Karadağ, A., Duman, Ş., Özkal, B., Özçelik, B., 2016. Fortification of dark chocolate with spray dried black mulberry (Morus nigra) waste extract encapsulated in chitosan-coated liposomes and bioaccessability studies. Food Chem. 201, 205–212.

Guo, S., Wang, E., 2007. Synthesis and electrochemical applications of gold nanoparticles. Anal. Chim. Acta 598 (2), 181–192.

Gupta, S.S., Ghosh, M., 2012. In vitro study of anti-oxidative effects of β-carotene and α-lipoic acid for nanocapsulated lipids. LWT-Food Sci. Technol. 49 (1), 131–138.

Hasan, M.R., Pulingam, T., Appaturi, J.N., Zifruddin, A.N., Teh, S.J., Lim, T.W., Ibrahim, F., Leo, B.F., Thong, K.L., 2018. Carbon nanotube-based aptasensor for sensitive electrochemical detection of whole-cell *Salmonella*. Anal. Biochem. 554, 34–43.

Hategekimana, J., Masamba, K.G., Ma, J., Zhong, F., 2015. Encapsulation of vitamin E: effect of physicochemical properties of wall material on retention and stability. Carbohydr. Polym. 124, 172–179.

Hemmateenejad, B., Shamsipur, M., Khosousi, T., Shanehsaz, M., Firuzi, O., 2012. Antioxidant activity assay based on the inhibition of oxidation and photobleaching of L-cysteine-capped CdTe quantum dots. Analyst 137 (17), 4029–4036.

Hong, S.A., Kwon, J., Kim, D., Yang, S., 2015. A rapid, sensitive and selective electrochemical biosensor with concanavalin A for the preemptive detection of norovirus. Biosens. Bioelectron. 64, 338–344.

Hong, Z., Xu, Y., Yin, J.F., Jin, J., Jiang, Y., Du, Q., 2014. Improving the effectiveness of (−)-epigallocatechin gallate (EGCG) against rabbit atherosclerosis by EGCG-loaded nanoparticles prepared from chitosan and polyaspartic acid. J. Agric. Food Chem. 62 (52), 12603–12609.

Huang, X., Aguilar, Z.P., Xu, H., Lai, W., Xiong, Y., 2016. Membrane-based lateral flow immunochromatographic strip with nanoparticles as reporters for detection: a review. Biosens. Bioelectron. 75, 166–180.

Ignatova, M., Starbova, K., Markova, N., Manolova, N., Rashkov, I., 2006. Electrospun nano-fibre mats with antibacterial properties from quaternised chitosan and poly (vinyl alcohol). Carbohydr. Res. 341 (12), 2098–2107.

Imran, M., Revol-Junelles, A.M., René, N., Jamshidian, M., Akhtar, M.J., Arab-Tehrany, E., Jacquot, M., Desobry, S., 2012. Microstructure and physico-chemical evaluation of nano-emulsion-based antimicrobial peptides embedded in bioactive packaging films. Food Hydrocoll. 29 (2), 407–419.

Jia, B., Zhou, J., Zhang, L., 2011. Electrospun nano-fiber mats containing cationic cellulose derivatives and poly (vinyl alcohol) with antibacterial activity. Carbohydr. Res. 346 (11), 1337–1341.

Jiménez-Fernández, E., Ruyra, A., Roher, N., Zuasti, E., Infante, C., Fernández-Díaz, C., 2014. Nanoparticles as a novel delivery system for vitamin C administration in aquaculture. Aquaculture 432, 426–433.

Jo, W.S., Song, H.Y., Song, N.B., Lee, J.H., Min, S.C., Song, K.B., 2014. Quality and microbial safety of "Fuji" apples coated with carnauba-shellac wax containing lemongrass oil. LWT-Food Sci. Technol. 55 (2), 490–497.

Joe, M.M., Bradeeba, K., Parthasarathi, R., Sivakumaar, P.K., Chauhan, P.S., Tipayno, S., Benson, A., Sa, T., 2012. Development of surfactin based nanoemulsion formulation from selected cooking oils: evaluation for antimicrobial activity against selected food associated microorganisms. J. Taiwan Inst. Chem. Eng. 43 (2), 172–180.

Joerger, R.D., 2007. Antimicrobial films for food applications: a quantitative analysis of their effectiveness. Packag. Technol. Sci. Int. J. 20 (4), 231–273.

Karadag, A., Ozcelik, B., Huang, Q., 2014. Quercetin nanosuspensions produced by high-pressure homogenization. J. Agric. Food Chem. 62 (8), 1852–1859.

Katuwavila, N.P., Perera, A.D.L., Karunaratne, V., Amaratunga, G.A., Karunaratne, D., 2016. Improved delivery of caffeic acid through liposomal encapsulation. J. Nanomater. 2016, 21–28.

Keawchaoon, L., Yoksan, R., 2011. Preparation, characterization and in vitro release study of carvacrol-loaded chitosan nanoparticles. Colloids Surf. B 84 (1), 163–171.

Kim, I.H., Lee, H., Kim, J.E., Song, K.B., Lee, Y.S., Chung, D.S., Min, S.C., 2013. Plum coatings of lemongrass oil-incorporating carnauba wax-based nanoemulsion. J. Food Sci. 78 (10), E1551–E1559.

Lamarra, J., Rivero, S., Pinotti, A., 2016. Design of chitosan-based nanoparticles functionalized with gallic acid. Mater. Sci. Eng. C 67, 717–726.

Lamprecht, A., Saumet, J.L., Roux, J., Benoit, J.P., 2004. Lipid nanocarriers as drug delivery system for ibuprofen in pain treatment. Int. J. Pharm. 278 (2), 407–414.

Lestringant, P., Guri, A., Gülseren, I., Relkin, P., Corredig, M., 2014. Effect of processing on physicochemical characteristics and bioefficacy of β-lactoglobulin–epigallocatechin-3-gallate complexes. J. Agric. Food Chem. 62 (33), 8357–8364.

Li, Z., Gu, L., 2014. Fabrication of self-assembled (−)-epigallocatechin gallate (EGCG) ovalbumin–dextran conjugate nanoparticles and their transport across monolayers of human intestinal epithelial Caco-2 cells. J. Agric. Food Chem. 62 (6), 1301–1309.

Lin, Y.H., Chen, S.H., Chuang, Y.C., Lu, Y.C., Shen, T.Y., Chang, C.A., Lin, C.S., 2008. Disposable amperometric immunosensing strips fabricated by Au nanoparticles-modified screen-printed carbon electrodes for the detection of foodborne pathogen *Escherichia coli* O157: H7. Biosens. Bioelectron. 23 (12), 1832–1837.

Lin, Y., Zhao, M., Guo, Y., Ma, X., Luo, F., Guo, L., Qiu, B., Chen, G., Lin, Z., 2016. Multicolor colormetric biosensor for the determination of glucose based on the etching of gold nanorods. Sci. Rep. 6 (1), 1–7.

Liu, C., Jia, Q., Yang, C., Qiao, R., Jing, L., Wang, L., Xu, C., Gao, M., 2011. Lateral flow immunochromatographic assay for sensitive pesticide detection by using Fe_3O_4 nanoparticle aggregates as color reagents. Anal. Chem. 83 (17), 6778–6784.

Liu, Y., Wang, L., Zhao, Y., He, M., Zhang, X., Niu, M., Feng, N., 2014. Nanostructured lipid carriers versus microemulsions for delivery of the poorly water-soluble drug luteolin. Int. J. Pharm. 476 (1-2), 169–177.

Luan, J., Zheng, F., Yang, X., Yu, A., Zhai, G., 2015. Nanostructured lipid carriers for oral delivery of baicalin: in vitro and in vivo evaluation. Colloids Surf. Physicochem. Eng. 466, 154–159.

Lv, Y., Yang, F., Li, X., Zhang, X., Abbas, S., 2014. Formation of heat-resistant nanocapsules of jasmine essential oil via gelatin/gum arabic based complex coacervation. Food Hydrocoll. 35, 305–314.

Ma, X., Qian, W., 2010. Phenolic acid induced growth of gold nanoshells precursor composites and their application in antioxidant capacity assay. Biosens. Bioelectron. 26 (3), 1049–1055.

Ma, Z., Haddadi, A., Molavi, O., Lavasanifar, A., Lai, R., Samuel, J., 2008. Micelles of poly (ethylene oxide)-b-poly (ε-caprolactone) as vehicles for the solubilization, stabilization, and controlled delivery of curcumin. J. Biomed. Mater. Res. 86 (2), 300–310.

Macedo, A.S., Quelhas, S., Silva, A.M., Souto, E.B., 2014. Nanoemulsions for delivery of flavonoids: formulation and in vitro release of rutin as model drug. Pharm. Dev. Technol. 19 (6), 677–680.

Magalhaes, L., Nitschke, M., 2013. Antimicrobial activity of rhamnolipids against *Listeria monocytogenes* and their synergistic interaction with nisin. Food Control 29, 138–142.

Marcos, J.F., Gandía, M., 2009. Antimicrobial peptides: to membranes and beyond. Expert Opin. Drug Discov. 4 (6), 659–671.

Mathew, F.P., Alocilja, E.C., 2005. Porous silicon-based biosensor for pathogen detection. Biosens. Bioelectron. 20 (8), 1656–1661.

McClements, D.J., Decker, E.A., Park, Y., Weiss, J., 2009. Structural design principles for delivery of bioactive components in nutraceuticals and functional foods. Crit. Rev. Food Sci. Nutr. 49 (6), 577–606.

McFarland, A.D., Van Duyne, R.P., 2003. Single silver nanoparticles as real-time optical sensors with zeptomole sensitivity. Nano Lett. 3 (8), 1057–1062.

Mead, P.S., Slutsker, L., Dietz, V., McCaig, L.F., Bresee, J.S., Shapiro, C., Griffin, P.M., Tauxe, R.V., 1999. Food-related illness and death in the United States. Emerg. Infect. Dis. 5 (5), 607.

Memar, M.Y., Raei, P., Alizadeh, N., Aghdam, M.A., Kafil, H.S., 2017. Carvacrol and thymol: strong antimicrobial agents against resistant isolates. Rev. Med. Microbiol. 28 (2), 63–68.

Mock, J.J., Smith, D.R., Schultz, S., 2003. Local refractive index dependence of plasmon resonance spectra from individual nanoparticles. Nano Lett. 3 (4), 485–491.

Nabavi, S.M., Marchese, A., Izadi, M., Curti, V., Daglia, M., Nabavi, S.F., 2015. Plants belonging to the genus Thymus as antibacterial agents: from farm to pharmacy. Food Chem. 173, 339–347.

Nallamuthu, I., Devi, A., Khanum, F., 2015. Chlorogenic acid loaded chitosan nanoparticles with sustained release property, retained antioxidant activity and enhanced bioavailability. Asian J. Pharm. Sci. 10 (3), 203–211.

Nezhad, M.R.H., Alimohammadi, M., Tashkhourian, J., Razavian, S.M., 2008. Optical detection of phenolic compounds based on the surface plasmon resonance band of Au nanoparticles. Spectrochim. Acta Part A Mol. Biomol. Spectrosc. 71 (1), 199–203.

Nourbakhsh, H., Madadlou, A., Emam-Djomeh, Z., Wang, Y.C., Gunasekaran, S., 2016. One-pot nanoparticulation of potentially bioactive peptides and gallic acid encapsulation. Food Chem. 210, 317–324.

Ogata, M., Hoshi, M., Urano, S., Endo, T., 2000. Antioxidant activity of eugenol and related monomeric and dimeric compounds. Chem. Pharm. Bull. 48 (10), 1467–1469.

Ozkan, G., Kamiloglu, S., Capanoglu, E., Hizal, J., Apak, R., 2018. Use of nanotechnological methods for the analysis and stability of food antioxidants. In: Impact of Nanoscience in the Food Industry. Academic Press, pp. 311–350.

Özyürek, M., Güngör, N., Baki, S., Güçlü, K., Apak, R., 2012. Development of a silver nanoparticle-based method for the antioxidant capacity measurement of polyphenols. Anal. Chem. 84 (18), 8052–8059.

Paini, M., Daly, S.R., Aliakbarian, B., Fathi, A., Tehrany, E.A., Perego, P., Dehghani, F., Valtchev, P., 2015. An efficient liposome based method for antioxidants encapsulation. Colloids Surf. B Biointerfaces 136, 1067–1072.

Pak, S.C., Penrose, W., Hesketh, P.J., 2001. An ultrathin platinum film sensor to measure biomolecular binding. Biosens. Bioelectron. 16 (6), 371–379.

Pandit, J., Aqil, M., Sultana, Y., 2016. Nanoencapsulation technology to control release and enhance bioactivity of essential oils. In: Encapsulations. Academic Press, pp. 597–640.

Pandit, S., Dasgupta, D., Dewan, N., Prince, A., 2016. Nanotechnology based biosensors and its application. Pharma Innovat. 5 (6), 18.

Petryayeva, E., Algar, W.R., Medintz, I.L., 2013. Quantum dots in bioanalysis: a review of applications across various platforms for fluorescence spectroscopy and imaging. Appl. Spectrosc. 67 (3), 215–252.

Ping, J., Ru, S., Fan, K., Wu, J., Ying, Y., 2010. Copper oxide nanoparticles and ionic liquid modified carbon electrode for the non-enzymatic electrochemical sensing of hydrogen peroxide. Microchim. Acta 171 (1), 117–123.

Pinto Reis, C., Neufeld, R.J., Ribeiro, A.J., Veiga, F., 2006. Nanoencapsulation I. Methods for preparation of drug-loaded polymeric nanoparticles. Nanomed. Nanotechnol. Biol. Med. 2 (1), 8–21.

Pitcher, D.G., Fry, N.K., 2000. Molecular techniques for the detection and identification of new bacterial pathogens. J. Infect. 40 (2), 116–120.

Pradhan, N., Singh, S., Ojha, N., Shrivastava, A., Barla, A., Rai, V., Bose, S., 2015. Facets of nanotechnology as seen in food processing, packaging, and preservation industry. Biomed. Res. Int. 2015, 18–22.

Qiao, Z., Lei, C., Fu, Y., Li, Y., 2017. Rapid and sensitive detection of *E. coli* O157: H7 based on antimicrobial peptide functionalized magnetic nanoparticles and urease-catalyzed signal amplification. Anal. Methods 9 (35), 5204–5210.

Qiu, Y., Park, K., 2001. Environment-sensitive hydrogels for drug delivery. Adv. Drug. Deliv. Rev. 53 (3), 321–339.

Quesada-González, D., Merkoçi, A., 2015. Nanoparticle-based lateral flow biosensors. Biosens. Bioelectron. 73, 47–63.

Rai, M., Yadav, A., Gade, A., 2009. Silver nanoparticles as a new generation of antimicrobials. Biotechnol. Adv. 27 (1), 76–83.

Ramírez-Ambrosi, M., Caldera, F., Trotta, F., Berrueta, L.A., Gallo, B., 2014. Encapsulation of apple polyphenols in β-CD nanosponges. J. Inclusion Phenom. Macrocyclic Chem. 80 (1-2), 85–92.

Ravanfar, R., Tamaddon, A.M., Niakousari, M., Moein, M.R., 2016. Preservation of anthocyanins in solid lipid nanoparticles: optimization of a microemulsion dilution method using the placket–Burman and box–Behnken designs. Food Chem. 199, 573–580.

Ren, W., Cho, I.H., Zhou, Z., Irudayaraj, J., 2016. Ultrasensitive detection of microbial cells using magnetic focus enhanced lateral flow sensors. Chem. Commun. 52 (27), 4930–4933.

Reza Mozafari, M., Johnson, C., Hatziantoniou, S., Demetzos, C., 2008. Nanoliposomes and their applications in food nanotechnology. J. Liposome Res. 18 (4), 309–327.

Robach, M.C., 1980. Use of preservatives to control microorganisms in food. Food Technol. 34 (10), 81–84.

Rodriguez-Garcia, I., Silva-Espinoza, B.A., Ortega-Ramirez, L.A., Leyva, J.M., Siddiqui, M.W., Cruz-Valenzuela, M.R., Gonzalez-Aguilar, G.A., Ayala-Zavala, J.F., 2016. Oregano essential oil as an antimicrobial and antioxidant additive in food products. Crit. Rev. Food Sci. Nutr. 56 (10), 1717–1727.

Safavi, A., Maleki, N., Farjami, E., 2009. Electrodeposited silver nanoparticles on carbon ionic liquid electrode for electrocatalytic sensing of hydrogen peroxide. Electroanalysis 21 (13), 1533–1538.

Sanna, V., Lubinu, G., Madau, P., Pala, N., Nurra, S., Mariani, A., Sechi, M., 2015. Polymeric nanoparticles encapsulating white tea extract for nutraceutical application. J. Agric. Food Chem. 63 (7), 2026–2032.

Scampicchio, M., Wang, J., Blasco, A.J., Sanchez Arribas, A., Mannino, S., Escarpa, A., 2006. Nanoparticle-based assays of antioxidant activity. Anal. Chem. 78 (6), 2060–2063.

Sechi, M., Syed, D.N., Pala, N., Mariani, A., Marceddu, S., Brunetti, A., Mukhtar, H., Sanna, V., 2016. Nanoencapsulation of dietary flavonoid fisetin: formulation and in vitro antioxidant and α-glucosidase inhibition activities. Mater. Sci. Eng. C Mater. Biol. Appl. 68, 594–602.

Sharma, S., Sahni, J.K., Ali, J., Baboota, S., 2015. Effect of high-pressure homogenization on formulation of TPGS loaded nanoemulsion of rutin–pharmacodynamic and antioxidant studies. Drug Deliv. 22 (4), 541–551.

Sharpe, E., Frasco, T., Andreescu, D., Andreescu, S., 2013. Portable ceria nanoparticle-based assay for rapid detection of food antioxidants (nanocerac). Analyst 138 (1), 249–262.

Shi, S., Wang, W., Liu, L., Wu, S., Wei, Y., Li, W., 2013. Effect of chitosan/nano-silica coating on the physicochemical characteristics of longan fruit under ambient temperature. J. Food Eng. 118 (1), 125–131.

Siragusa, G.R., Cutter, C.N., Dorsa, W.J., Koohmaraie, M., 1995. Use of a rapid microbial ATP bioluminescence assay to detect contamination on beef and pork carcasses. J. Food Prot. 58 (7), 770–775.

Smith, D.F., Diack, H.L., Pennington, T.H., Pennington, T.H., Russell, E.M., 2005. Food Poisoning, Policy, and Politics: Corned Beef and Typhoid in Britain in the 1960s. Boydell Press.

Song, W., Li, D.W., Li, Y.T., Li, Y., Long, Y.T., 2011. Disposable biosensor based on graphene oxide conjugated with tyrosinase assembled gold nanoparticles. Biosens. Bioelectron. 26 (7), 3181–3186.

Souza, M.P., Vaz, A.F., Correia, M.T., Cerqueira, M.A., Vicente, A.A., Carneiro-da-Cunha, M.G., 2014. Quercetin-loaded lecithin/chitosan nanoparticles for functional food applications. Food Bioprocess Technol. 7 (4), 1149–1159.

Sperling, R.A., Gil, P.R., Zhang, F., Zanella, M., Parak, W.J., 2008. Biological applications of gold nanoparticles. Chem. Soc. Rev. 37 (9), 1896–1908.

Suaifan, G.A., Alhogail, S., Zourob, M., 2017. Based magnetic nanoparticle-peptide probe for rapid and quantitative colorimetric detection of *Escherichia coli* O157: H7. Biosens. Bioelectron. 92, 702–708.

Sun, Y.P., Zhou, B., Lin, Y., Wang, W., Fernando, K.S., Pathak, P., Meziani, M.J., Harruff, B.A., Wang, X., Wang, H., Luo, P.G., 2006. Quantum-sized carbon dots for bright and colorful photoluminescence. J. Am. Chem. Soc. 128 (24), 7756–7757.

Tang, F., Li, L., Chen, D., 2012. Mesoporous silica nanoparticles: synthesis, biocompatibility and drug delivery. Adv. Mater. 24 (12), 1504–1534.

Terjung, N., Löffler, M., Gibis, M., Hinrichs, J., Weiss, J., 2012. Influence of droplet size on the efficacy of oil-in-water emulsions loaded with phenolic antimicrobials. Food Funct. 3 (3), 290–301.

Tiwari, I., Singh, M., Pandey, C.M., Sumana, G., 2015. Electrochemical genosensor based on graphene oxide modified iron oxide–chitosan hybrid nanocomposite for pathogen detection. Sens. Actuat. B 206, 276–283.

Torchilin, V.P., 2005. Recent advances with liposomes as pharmaceutical carriers. Nat. Rev. Drug Discov. 4 (2), 145–160.

Torres-Giner, S., Ocio, M.J., Lagaron, J.M., 2009. Novel antimicrobial ultrathin structures of zein/chitosan blends obtained by electrospinning. Carbohydr. Polym. 77 (2), 261–266.

Tripathi, S., Kushwah, V., Thanki, K., Jain, S., 2016. Triple antioxidant SNEDDS formulation with enhanced oral bioavailability: implication of chemoprevention of breast cancer. Nanomed. Nanotechnol. Biol. Med. 12 (6), 1431–1443.

Tsai, Y.J., Chen, B.H., 2016. Preparation of catechin extracts and nanoemulsions from green tea leaf waste and their inhibition effect on prostate cancer cell PC-3. Int. J. Nanomed. 11, 1907.

Tułodziecka, A., Szydłowska-Czerniak, A., 2016. Determination of total antioxidant capacity of rapeseed and its by-products by a novel cerium oxide nanoparticle-based spectrophotometric method. Food Anal. Methods 9 (11), 3053–3062.

Ubbink, J., Kruger, J., 2006. Physical approaches for the delivery of active ingredients in foods. Trends Food Sci. Technol. 17, 244–254. https://doi.org/10.1016/j.tifs.2006.01.007.

Van Gerwen, P., Laureyn, W., Laureys, W., Huyberechts, G., De Beeck, M.O., Baert, K., Suls, J., Sansen, W., Jacobs, P., Hermans, L., Mertens, R., 1998. Nanoscaled interdigitated electrode arrays for biochemical sensors. Sens. Actuat. B 49 (1-2), 73–80.

Vasilescu, A., Sharpe, E., Andreescu, S., 2012. Nanoparticle-based technologies for the detection of food antioxidants. Curr. Anal. Chem. 8 (4), 495–505.

Vilela, D., Castañeda, R., González, M.C., Mendoza, S., Escarpa, A., 2015. Fast and reliable determination of antioxidant capacity based on the formation of gold nanoparticles. Microchim. Acta 182 (1-2), 105–111.

Wang, B., Chen, Y., Wu, Y., Weng, B., Liu, Y., Lu, Z., Li, C.M., Yu, C., 2016. Aptamer induced assembly of fluorescent nitrogen-doped carbon dots on gold nanoparticles for sensitive detection of AFB1. Biosens. Bioelectron. 78, 23–30.

Wang, R., Xu, Y., Zhang, T., Jiang, Y., 2015b. Rapid and sensitive detection of *Salmonella typhimurium* using aptamer-conjugated carbon dots as fluorescence probe. Anal. Methods 7 (5), 1701–1706.

Wang, J., Zhou, N., Zhu, Z., Huang, J., Li, G., 2007. Detection of flavonoids and assay for their antioxidant activity based on enlargement of gold nanoparticles. Anal. Bioanal. Chem. 388 (5), 1199–1205.

Wang, D.B., Tian, B., Zhang, Z.P., Wang, X.Y., Fleming, J., Bi, L.J., Yang, R.F., Zhang, X.E., 2015a. Detection of *Bacillus anthracis* spores by super-paramagnetic lateral-flow immunoassays based on "road closure". Biosens. Bioelectron. 67, 608–614.

Weber, J.E., Pillai, S., Ram, M.K., Kumar, A., Singh, S.R., 2011. Electrochemical impedance-based DNA sensor using a modified single walled carbon nanotube electrode. Mater. Sci. Eng. C 31 (5), 821–825.

Weiss, J., Takhistov, P., Mcclements, D.J., 2006. Functional materials in food nanotechnology. J. Food Sci. 71 (9), R107–R116.

Welch, C.M., Banks, C.E., Simm, A.O., Compton, R.G., 2005. Silver nanoparticle assemblies supported on glassy-carbon electrodes for the electro-analytical detection of hydrogen peroxide. Anal. Bioanal. Chem. 382 (1), 12–21.

Wildman, R.E., 2016. Handbook of Nutraceuticals and Functional Foods. CRC Press.

Winarti, L., Sari, L.O.R.K., Nugroho, A.E., 2015. Naringenin-loaded chitosan nanoparticles formulation, and its *In Vitro* evaluation against T47D breast cancer cell line, Indonesian J. Pharm. 26 (3), 147–157.

Wu, S., Zhao, H., Ju, H., Shi, C., Zhao, J., 2006. Electrodeposition of silver–DNA hybrid nanoparticles for electrochemical sensing of hydrogen peroxide and glucose. Electrochem. Commun. 8 (8), 1197–1203.

Wu, Y., Luo, Y., Wang, Q., 2012. Antioxidant and antimicrobial properties of essential oils encapsulated in zein nanoparticles prepared by liquid–liquid dispersion method. LWT-Food Sci. Technol. 48 (2), 283–290.

Xia, S., Yu, Z., Liu, D., Xu, C., Lai, W., 2016. Developing a novel immunochromatographic test strip with gold magnetic bifunctional nanobeads (GMBN) for efficient detection of *Salmonella choleraesuis* in milk. Food Control 59, 507–512.

Xu, X., Ray, R., Gu, Y., Ploehn, H.J., Gearheart, L., Raker, K., Crivens, W.A., 2004. Electrophoretic analysis and purification of fluorescent single-walled carbon nanotube fragments. J. Am. Chem. Soc. 126, 12736–12737.

Yadav, R., Kumar, D., Kumari, A., Yadav, S.K., 2014. Encapsulation of catechin and epicatechin on BSA NPs improved their stability and antioxidant potential. EXCLI J. 13, 331.

Yamada, K., Kim, C.T., Kim, J.H., Chung, J.H., Lee, H.G., Jun, S., 2014. Single walled carbon nanotube-based junction biosensor for detection of *Escherichia coli*. PLoS One 9 (9), e105767.

Yan, Z., Niu, Q., Mou, M., Wu, Y., Liu, X., Liao, S., 2017. A novel colorimetric method based on copper nanoclusters with intrinsic peroxidase-like for detecting xanthine in serum samples. J. Nanopart. Res. 19 (7), 1–12.

Yildiz, G., Bolton-Warberg, M., Awaja, F., 2021. Graphene and graphene oxide for bio-sensing: general properties and the effects of graphene ripples. Acta Biomater. 131, 62–79.

Yoshizaki, Y., Yuba, E., Sakaguchi, N., Koiwai, K., Harada, A., Kono, K., 2014. Potentiation of pH-sensitive polymer-modified liposomes with cationic lipid inclusion as antigen delivery carriers for cancer immunotherapy. Biomaterials 35 (28), 8186–8196.

Yu, Y., Zhang, S., Ren, Y., Li, H., Zhang, X., Di, J., 2012. Jujube preservation using chitosan film with nano-silicon dioxide. J. Food Eng. 113 (3), 408–414.

Zeng, S., Yong, K.T., Roy, I., Dinh, X.Q., Yu, X., Luan, F., 2011. A review on functionalized gold nanoparticles for biosensing applications. Plasmonics 6 (3), 491–506.

Zougagh, M., Salghi, R., Dhair, S., Rios, A., 2011. Visual detection of ascorbic acid via alkyne–azide click reaction using gold nanoparticles as a colorimetric probe. Anal. Bioanal. Chem. 399 (7), 2395–2405.

Zhang, Y., Niu, Y., Luo, Y., Ge, M., Yang, T., Yu, L.L., Wang, Q., 2014. Fabrication, characterization and antimicrobial activities of thymol-loaded zein nanoparticles stabilized by sodium caseinate–chitosan hydrochloride double layers. Food Chem. 142, 269–275.

Zhao, W., Brook, M.A., Li, Y., 2008. Design of gold nanoparticle-based colorimetric biosensing assays. ChemBioChem 9 (15), 2363–2371.

Zhao, W., Wang, H., Qin, X., Wang, X., Zhao, Z., Miao, Z., Chen, L., Shan, M., Fang, Y., Chen, Q., 2009. A novel nonenzymatic hydrogen peroxide sensor based on multi-wall carbon nanotube/silver nanoparticle nanohybrids modified gold electrode. Talanta 80 (2), 1029–1033.

Ziyatdinova, G., Ziganshina, E., Budnikov, H., 2013. Voltammetric sensing and quantification of eugenol using nonionic surfactant self-organized media. Anal. Methods 5 (18), 4750–4756.

Zou, L.Q., Liu, W., Liu, W.L., Liang, R.H., Li, T., Liu, C.M., Cao, Y.L., Niu, J., Liu, Z., 2014. Characterization and bioavailability of tea polyphenol nanoliposome prepared by combining an ethanol injection method with dynamic high-pressure microfluidization. J. Agric. Food Chem. 62 (4), 934–941.

Toxicity, environmental risks, and ingestion of nanomaterials leaching from the food packaging

23

Olaniyan Olugbemi[a] and Charles Oluwaseun Adetunji[b]

[a]*Laboratory for Reproductive Biology and Developmental Programming, Department of Physiology, Rhema University Aba, Abia State, Nigeria,* [b]*Applied Microbiology, Biotechnology and Nanotechnology Laboratory, Department of Microbiology, Edo State University Uzairue, Iyamho, Edo State, Nigeria*

23.1 Introduction

Studies have revealed that bionanomaterials are utilized for diverse applications particularly in green technology like the biomedical sector and food industry owing to their physiochemical properties such as interconnected porous networks, nontoxic by-products, biodegradability, immune biocompatibility, and bioavailability (Adetunji et al., 2021a). Food packaging is known to play an important role in maintaining food safety and quality through the utilization of appropriate packaging materials. Through nanotechnology, nanomaterials are generated through the utilization of structures very small in size like 0.1–100 nm, which have been shown to offer tremendous opportunities in the agricultural and food industry. Nanomaterials are currently being utilized for food coloring, nutritional additives, flavoring, antibacterial and antioxidant ingredients for food packaging, biofertilizers, and agrochemicals. Ghada (2020) reported that food packaging is greatly transformed through the utilization of nanoparticles which has improved food safety, packaging, and quality. Nanoparticles are able to prevent pathogenic organisms, reduce light and oxygen entry owing to their properties such as barrier properties, flexibility, stability, antimicrobial activity, and mechanical strength. Through radio frequency identification tags, proper monitoring can be achieved with nanoparticles-embedded nanosensors, nanodevices, and nanotubes to effectively monitor the quality, freshness, contamination, integrity, and packaging conditions. Nanoparticles-based food packages are able to extend the shelf life of food products, improve quality, hygiene, and offer some health benefits. Ranjha et al. (2022) reported that across different food industries, biocompatible nanomaterials are being utilized to reduce the toxicity, enhance immune response, and decrease gastrointestinal tract adverse effects of food toxins. The authors reported that nanomaterials can interact with nutraceutical and functional foods to promote

Nanobiotechnology for Food Processing and Packaging. DOI: https://doi.org/10.1016/B978-0-323-91749-0.00008-3

food quality, safety, packaging, processing, and labeling. Also, nanomaterials are known to detect toxins; pests and pathogens; generation of biocompatible packaging; improvement in color, aroma, and flavor; handle edible film; and authenticate food products. Today silver nanoparticles are generally and widely acceptable due to their significant potential in analysis, preservative, processing, delivery, and safety of food products. The advancement in nanotechnology has facilitated knowledge and applications in different sectors of the economy such as in the manufacturing industry like textiles, cosmetics, toothpaste, kitchen utensils, detergents, soaps, building materials, and toys. Also, nanoparticles appear in the food industry for packaging, which has influenced the environment and human health negatively (Peidaei et al., 2021). The formation of good nutritive value-added food products is the leading innovation in the food industry. For instance, the development of high impermeable nanomaterial packaging is being utilized for the protection of food against ultraviolet radiation, heat, gases, pathogenic organisms, and more strength to the shelf life of the food. This form of packaging today is smart packaging but is not yet widely accepted by consumers due to the risk factors regarding health and the environment. Hence, there is a need to provide toxicological assessment and safety measures, to reduce the risk and improve the safety of the consumers. Even though the toxicological assessment of nanoparticles on human and environmental health is limited, there are reports on the general negative impact on biochemical pathways particularly in the gastrointestinal tract, lungs, brain, and the reproductive system (Ahari et al., 2018). Nanoparticles can cross the cell barriers resulting in oxidative stress and inflammation. The entire process of nanoparticle pathogenesis is dependent on the physiochemical properties of the materials such as the size, shape, as well as physiological concern of the organ. Various investigations into the applications and functions of nanoparticles in the food industry have resulted in human exposure to the nanoparticles. This situation is generating so much concern due to the potential toxicity and health risks of the nanomaterials in the food. Different studies have revealed that due to the toxic nature of some nanoparticles, they may promote allergic lung inflammation due to generation of reactive oxygen species, immune response, allergen-specific Th2-type induction, and activation of immunoglobulins (Xiaojia and Huey-Min, 2016). This chapter provides a detailed overview of nanoparticles' utility for food packaging and processing, along with the toxicity and environmental risk of nanomaterials.

23.2 Nanomaterials utilized for food package

Sujithra and Manikkandan (2019) reported that the numerous advantages of nanopackaging have resulted in increased utilization in recent years due to their protective roles and increase in the shelf life of many food products. The authors revealed that these nanomaterials can function as coating, edible films creating a barrier against pathogenic organisms, thus suppressing the complexity and enhancing recyclability. In an attempt to eradicate nonenvironmentally friendly traditional materials, scientists are beginning to develop nano-polymer films from natural sources or waste materials in order to fabricate nanomaterials for food packaging. The authors gave various approaches and techniques currently utilized to design and fabricate nanomaterials for food packaging such as nano-sensor, chemical release nanofiber, nano tracking packaging, nano-biodegradable packaging, nanoantimicrobial packaging, and nanoantioxidant packaging. Sujithra and Manikkandan (2019) revealed that nanoparticles are utilized in the fabrication of food packages with greater efficiency in food protection from light, fire, thermal, and gas absorption. Furthermore, different nanoparticles for structural films can wade off bacteria and other pathogenic microorganisms that can cause food spoilage and thereby, increasing food safety. Nanocomposites are also able to reduce the release of carbon dioxide from bottles and food storage bins. Also, they can improve mechanical strength, increase heat resistance, reduce weight, improve barrier against oxygen, ultraviolet radiation, carbon dioxide, and volatiles of food package materials and moisture. Many of these products are coated with silver nanoparticles that have greater efficiency in killing bacteria. In plastic packaging, nanosensors are utilized for the detection of spoilage gases, detection of bacteria, and by-products released from contaminated food. Many of the nanoparticles have the potential to enable food to stay fresher for a longer period of time. Silicate nanoparticles are utilized in this regard, thus facilitating freshness in fruit and food products through the reduction in exposure to oxygen and prevention of moisture leakages. Thermoplastic polymers as a form of polymer nanocomposites such as polymeric resin, oxides, and nanoclays are gaining tremendous attention in the beverage, brewery dairy, and food packaging industry. Globally, several countries have adopted the use of nanoparticles in the packaging of food products. Even public awareness and acceptability are still very poor, as a result of many concerns like toxicity, environmental hazards, and others, the potential applications of nanomaterials in the food industry can still not be overemphasized. Many stakeholders have emphasized that the safety of food remains the key concern and not the extension of the shelf life. Thus, there is a need to facilitate research that will ensure more efficient, less toxic, and sustainable food production packaging and process. Vivek et al. (2018) revealed that in food science, nanotechnology has revamped many aspects such as smart packaging through the adoption and utilization of nanoparticles. The authors revealed some of the important nanoparticles such as metal oxides and inorganic metals and their nanocomposites to nano-organic materials with bioactive agents which are widely utilized in the food industry. Nanomaterials are known to prevent the movement of gases, thereby, preventing oxygen and water vapors from causing decarbonization and oxidation. Bazila et al. (2018) revealed that food packaging is a critical area in food science and technology that has witnessed tremendous revolution in the past few years. This is because food packaging can prevent food spoilage, maintain food safety, and quality during production, distribution, and transport. A recent study indicates that the market value for nanotechnology in the beverage and food industry has been projected to reach about \$15.0 billion by 2020 according to the authors. This is so because of the addition of nanoparticles in food packages such as nanocomposites which are biodegradable, nontoxic, and serve as a protective barrier. Different nanoparticles like silver, titanium oxide, zinc oxide, and titanium nitride are incorporated into food packages which serve as antimicrobial agents. These nanoparticles act by causing the degradation of the bacteria membrane through photocatalytic reactions. The authors reported that nanoparticles like montmorillonite, polycaprolactone, polylactic acid, polyhydroxybutyrate, nanosilicates, zein, chitin, and nanolaminates are useful for increasing the shelf life of food products. These are edible coatings on food that form films against carbon dioxide, water vapor, and oxygen. Reports on nanosensors and nanowires have shown that they can detect insects, odors, color, allergenic proteins, and pest invasion in food. Also, nanobarcodes and radiofrequency identification chips like nanotag devices are utilized for anticounterfeit and tracking of delivery of goods and to prevent package contamination. Nanoparticles are utilized to trace authentic products in the supply chain which contains all the necessary product-related information. Mironescu et al. (2021) reported that novel starch-based packaging materials are designed as green nanotechnology packaging for sustainable environmental and human health. These green packaging systems have grown tremendously owing to their unique properties such as eco-friendly nature, biodegradability, cost-effectiveness, and readily available such as natural plants. Natural plants derived materials are developed into smart intelligent packages that extend the shelf life of food, offer protection from external contamination, and maintain quality and safety. Some of the bioactive molecules that are derived from plants which can be developed into nanoparticles for food packages are vitamins, essential oils, polyphenols, amino acids, plant extracts, amylopectin and amylose, and lipids. These materials can interfere with the microbiota/microbiome of humans and immunity, resulting in several health promoting activities like enhanced immunity, suppressed inflammation, blood–brain barrier integrity, and improved functioning of the gut. Polyphenols are immunomodulatory, properly functioning intracellular signals, development of muscular systems, blood clotting, activation of several enzymes, antimicrobial and antioxidant properties in innovative food packaging systems. In the flexible packaging industry, nanoparticle-based meat packaging is growing rapidly, with an estimate of \$38 billion market globally. Currently, advancement in the development of nanoparticle innovative biofilms for food packages has enhanced packaging performance and production. Due to the growing concern of nondegradable packaging waste, advancements in the use of nanomaterials in biodegradable nontoxic food packages have witnessed rapid progress. Food products must remain fresh, easy to handle, healthy, and safe to be accepted by consumers, thus the need for an environmentally friendly nanomaterial-based packaging system. Their unique properties like heat resistance, mechanical strength, improved temperature performance, flexible functionalities, and antimicrobial activity are of great interest in the food industry (Akhileash et al. 2012). High-quality grapheme-based nanomaterials are also utilized in the food industry for biosensors, food packaging, immunosuppressors, antimicrobial coatings, and DNA detection (Ashok and Sundaram, 2014).

23.3 **Toxicity and risk of nanomaterials utilized for food package**

Zainazor et al. (2020) reported that the toxicity of nanoparticles in food packages has become a thing of global concern. The authors describe the negative health effects of nanoparticles from food packages on the body's physiology, thus affecting the intracellular organs, vascular disease, pulmonary system, and the reproductive system. The authors described the toxicological effect and the risk associated with the use of nanoparticles in food packaging. So many reports have been released concerning the negative effects of nanomaterials released into the food from nano-based food packages. Notable is the report from Li et al. (2008), who stated that nanoparticle titanium dioxide is a toxic particle currently utilized in food, drugs, and cosmetics which can cause abnormal sedimentation, hemagglutination, and hemolysis. Also, Handy et al. (2008) reported that the viscera and gills of fish for the production of gelatin-based nanoparticles like nanoclay, nanometals, and organic fillers can cause significant toxicity to the gastrointestinal microflora. Also, de Abreu et al. (2010) reported the harmful impact of trans, trans-1,4-diphenyl-1,3-butadiene trans, trans-1,4-diphenyl-1,3-butadiene (DPBD) and caprolactam, 5-chloro-2-(2,4-dichlorophenoxy) phenol (triclosan) from polyamide and polyamide nanoclays on the body system such as the reproductive, brain, and the cardiovascular system. Vivek et al. (2018) revealed the negative aspects of the adoption of nanoparticles in the food industry such as accumulations of the materials in the body, thus causing harmful effects on different systems and the environment. Ghada (2020) reported the toxicity and environmental effects of nanoparticles in food packaging as a source of public health concern. Despite the enormous benefits, the general acceptance and public health awareness of the toxicity is still of global concern. Many data and reports have been generated to demonstrate the health and environmental implications of the utilization of nanoparticles in the food industry. Bazila et al. (2018) reported that nanoparticles' unique characteristics have caused their wide acceptance and utilization in the food industry resulting in safety concerns which has generated a lot of pressure and rejection by many stakeholders in the industry. Many researchers are investigating the negative effects of nanoparticles on human health through in vivo studies in cells and tissues. The potential applications of nanoparticles in food science and technology are based on cost-effectiveness but public perception and acceptance has always been an issue. Presently, it seems that the public prefers the use of natural constituents as food additives rather than nanostructured molecules that have devastating consequences on human health and the environment. Bazila et al. (2018) highlighted the toxicity of nanoparticles in the food industry. The authors revealed that the toxicity effects of nanomaterials are due to the unexpected behavior of nanoparticles on the human body system. The interaction between the human tissue and engineered nanoparticles has been proposed by many researchers to cause nanoparticles to cross the cell membrane in different organs, thus activating inflammatory and other immune responses. Bazila et al. (2018) showed that five major factors govern the toxicity effects of nanoparticles on any tissue such as dose of exposure, surface area, chemical reactivity, shape and size, and charge distribution. Therefore, physical interaction between tissue and nanomaterials can result in disruption of membranes, disruption of cellular activity, aggregation, dysfunctional transport system, and protein folding. Furthermore, chemical interaction may induce reactive oxygen species production and oxidative damage. The authors also showed different routes where nanoparticles can penetrate and enter the body system such as the skin through the hair follicles, the lungs through the intensive air-blood, and the gastrointestinal system through the circulation system. The study conducted by Nel et al. (2012) reported that investigation on different nanosized particles from carbon soot were investigated for their effect on diverse body systems. The authors showed that nanoparticles caused the generation of reactive oxygen species, membrane damage, inflammation, and nanotoxicity due to the long-term bioavailability of nanoparticles. It is reported that these effects can lead to neurodegeneration such as Parkinson's syndrome, impairment of the DNA and Alzheimer's disease, liver, kidney, cancer, neutrophil apoptosis, decreased cell viability, immunotoxicity, genotoxicity, epigenetic changes, and gastrointestinal damage. Thiruvengadam et al. (2018) reported that nanoparticles produced toxic effects on human blood, lipid peroxidation, and cell death. Nanoparticles cause damage to cellular mitochondria, liposomes, and the immune system. The nanoparticles from food packages protect food from external vibration, microbial infestation, as well as from temperature and shocks. Rapid advancements in nanotechnology in developed nations have provided a great impact in terms of waste management, food security, and environmental health through the utilization of nanoparticles in food packaging. Notably, due to the adverse health and environmental concerns, regulatory bodies in these countries have put several measures in place to reduce the environmental risk, toxicokinetic, and toxicodynamic of these nanomaterials. The authors revealed that nanoparticles are one of the latest technologies in food science and technology which is revolutionizing food security. They revealed that through imaging techniques, separation techniques, and characterization such as size, zeta potential, morphology, shape, dimension, size, charge, surface area, porosity, structure, solubility, and polydispersity indices, the toxicity level can be monitored. They highlighted some of the major techniques that can be used for characterization such as spectroscopy techniques, advanced microscopic, and nuclear magnetic resonance (Akbar et al., 2021). Several studies have indicated that nanoparticles can penetrate the skin or through the gastrointestinal tract across cellular barriers resulting in inflammation and oxidative stress. It is known that nanoparticles can migrate from the packaging materials into the food and also the workers in the food industry can suffer from skin penetration of inhalation of these nanoparticles from the environment leading to metabolic degeneration, infertility, renal failure, and liver damage through oxidative stress (Insoo et al., 2022). Loutfy et al. (2021) wrote extensively on the negative impact of nanoparticles on human health and the environment. Exposure to nanoparticles from food has been reported by the authors to cause atherosclerosis, metastasis and tumor growth, alteration in immune response, liver cancer and pulmonary toxicity through oxidative stress, apoptosis, and activation of the caspase. The authors described the apoptotic pathway of nanoparticle toxicity as destruction of mitochondria membrane, decrease in mitochondrial membrane potential, activation of caspase 9, p38, JNK, and phosphorylation of p53. Many studies have linked exposure to nanoparticles to the formation of colitis, colon cancer, obesity, food allergies, immune dysfunction, diabetes, and microinflammation of the colon. Amra (2020) demonstrated that over the last few decades, nanomaterials application in food packaging and processing has increased tremendously which has resulted in serious toxicity concerns. The author showed that despite the remarkable progress in the application of nanoparticles in the food industry, the toxicity effects are huge and of public concern. Exposure through the circulatory system results in cancer and genotoxicity. Through in vivo and in vitro studies, the toxicity effects of nanoparticles in the food industry have been established, such as damage to human alveolar cells, causing leukemia, renal failure, genotoxicity, DNA strand break and chromosomal damage, and inflammation. Ranjha et al. (2022) have revealed that the migration of nanoparticles into foodstuffs is generating serious concerns about human and environmental safety. The authors suggested that the environmental and health hazards could be resolved by the utilization of biocompatible nanomaterials with nontoxicity and biodegradability polymers such as polysaccharides like carboxymethyl cellulose, chitosan, cellophane, or starch. Consumers are very skeptical when it comes to issues of new technology that have to do with health and environmental concerns. Thus, nanotechnology incorporation into food packaging is generating so much attention. Gokularaman et al. (2017) noted nanoparticles utilized for food packages can migrate into the food and when ingested, increase its circulatory time in the human body due to hydrophilic and positive charge nature. These effects may generate serious negative outcomes on microcirculation, generation of reactive oxygen species, genotoxicity, carcinogenicity, and infertility.

23.4 Nanobiotechnological toxicity evaluation method: methods and techniques used

Leudjo et al. (2021) reported that due to the constant exposure of humans and the environment to nanoparticles, there is a need to analyze the nanotoxicity of many of the nanomaterials being utilized across different fields of science such as medical, pharmaceuticals, industrial, and food. The authors revealed that the physicochemical nature of the nanomaterials can be analyzed with possible nanotoxicological assays such as the conventional methods using in vitro approaches which are known to be generally less time-consuming and cheap to verify the physiological parameters such as oxidative stress, inflammation, cytotoxicity, necrosis, apoptosis, mutation, and DNA damage. Cytokines like interleukin 8 or 6, monocyte chemotactic protein-1, and interleukin-10 can be measured using the ELISA technique. Also, caspases assays, Ames assays, TUNEL assays, COMET assays, and annexin V assays can be utilized to determine the various physiochemical parameters. Again in in vivo methods, various animal species can be utilized to carry out experimental studies such as zebrafish, rabbits, rats, mice, and drosophila. In this method of experiments, the toxicity level in various organs is determined after exposure to nanomaterials via different routes. Various advanced methods are available to analyze the toxicity level of various nanoparticles such as atomic force microscopy, carbon fiber microelectrodes, biomimetic 3-D lung-on-a-chip, fluidic-based cell-on-chip, lateral flow immunoassay, high-throughput nanotoxicity screening, organ-on-chip, and precision-cut tissue slices. Bryce et al. (2009) reported that it is estimated that over 800 consumer products are integrated with nanoparticles containing materials with potential health and environmental safety issues. The authors revealed that various methods and techniques are available to analyze the toxicity level of nanomaterials such as the use of transmission electron microscopy for investigating in vitro uptake and localization characterization. The elemental analysis of nanomaterials will provide mass concentration with techniques like inductively coupled plasma atomic emission spectroscopy. Also, the authors demonstrated that fluorescence spectroscopy is important for quantitative assessment of nanomaterials. Fröhlich (2017) demonstrated the importance of omics technology in the assessment of the nanotoxicity of nanoparticles. The author revealed that various biological tests can be carried out utilizing epigenome, proteome, transcriptome, and metabolome to reveal novel targets. Collins et al. (2017) demonstrated that nanomaterial toxicity can be evaluated using high throughput screening such as flow cytometry, confocal laser scanning spectroscopy, magnetic resonance imaging, positron emission tomography and single-photon emission computer tomography, multiplex analysis of secreted products, impedance-based spectroscopy, high content screening coupled with Epifluorescence (EPi) or confocal laser scanning microscopy (CLSM), and high-throughput omics.

23.5 Toxicity effects on human health

Saniha et al. (2022) reported that nanotechnology is rapidly growing with exponential consequences on the environment and human health due to its characteristic physiochemical properties. Nanoparticles affect various human systems upon exposure such as the respiratory tract, dermal, gastrointestinal system, immune system, the central nervous system causing cancer, physiological changes, inflammation, apoptosis, necrosis, and disorders to death. The authors demonstrated that nanoparticles exist as organic and inorganic metals, aerosol particles, flame retardants, quantum dots, carbon nanotubes, oxides, and coatings. The toxicological profile of nanoparticles on the human system is on the increase due to increased nanomaterials and products in the environment via air, water, medical products, and food chains. Megha et al. (2013) reported that nanoparticles have caused various hazardous effects on the environment and human health with increased exposure resulting in public health concerns. The authors demonstrate that due to their nano-size, they can easily penetrate and uptake through the body system and cause harm. Muhammad et al. (2015) demonstrated that despite the uniqueness of nanomaterials, the health and environmental effects have been serious public health concerns. Jamuna and Ravishankar, (2014) demonstrated that the unique physicochemical properties of nanoparticles are linked to their biological characteristics. The authors highlighted that the application of nanoparticles cuts across various fields of science with increased application and use. Recently, various human health effects such as toxicity have been a major concern limiting its use. Thus, prolonged exposure has adverse consequences on the body's organs like cytotoxicity and genotoxicity. The ability to penetrate blood tissue barriers increases the bioaccumulation and long-term effects resulting in enzyme and protein alteration, gene expression changes, biological dysfunctions, inflammation, necrosis, and apoptosis. The physiochemical properties of nanoparticles like the surface properties, aggregation behavior, solubility, bio-persistence, and biokinetics are associated with nanoparticles' health hazards. Bahadar et al. (2016) revealed that the cellular toxicity, genotoxicity, and immunotoxicity of nanoparticles are caused as a result of cellular inflammation causing the release of biomarkers like IL-8, IL-10, IL-6, and tumor necrosis factor. These biomarkers are cytokines which can be measured using ELISA technique to assess the level of toxicity in the organs.

23.6 Nanofoods safety regulations

Seatonn et al. (2010) reported that the regulation of nanoparticles is seriously linked to their toxicity. There are many regulatory bodies across the globe, particularly in the United States saddled with the responsibility of carrying out risk assessment evaluation on nanoparticles. Of course, the cost of conducting toxicity tests on nanoparticles is not cheap, thus, emerging industries will need robust collaboration with several research institutes for support. From simple to complex or advanced analysis on nanotoxicity in the United States may cost from $249 million to $1.18 billion. Other strategies like the European Union's reach legislation for regulating toxic chemicals may also be a sought of regulatory framework for nanotechnologies. Over the past few years, inventory on nanotechnology use and applications have been reported. For instance, in Australia, the New Zealand government's approach is to regulate food and any danger that nanoparticles can cause to human health and the environment through proper legislation.

23.7 Environmental risks of nanomaterials utilized for food package

Bazila et al. (2018) demonstrated the growing concern about nanoparticle contamination in the environment which is being released in substantial amounts into the environment. The authors revealed that many of these nanomaterials end up in the aquatic and terrestrial ecosystems through natural, intentional, and unintentional routes. The natural ways are through ocean spray, volcanic eruption, dust storms, forest fires, soil erosion, and clouds. The unintentional ways are a result of welding, smoking, metal smelting, mining, industrial waste, vehicle exhaust, and fossil fuel burning. The ecological toxicity of these nanoparticles is based on the physiochemical characteristics of these materials such as chemical composition, surface charge, surface chemistry, particle size, shape, size distribution, crystal structure, porosity, and agglomeration state. Anju and Parayanthala (2021) showed that the advancement in the application of nanoparticles has resulted in serious environmental concerns due to the released toxic metal ions. Due to the increased usage of nanotechnology across different sectors of the economy, there is an unprecedented increased release of metallic ions into the environment causing circulation into food and different ecosystems. The bioaccumulation in aquatic environments is of major concern which can interact with different living systems in aquatic life. The authors revealed that the discharge of nanoparticles into the environment produces increased bioaccumulation in soil, aquatic ecosystem, and air causing an imbalance in the ecosystem through deprivation of essential enzyme activity, increased oxidative stress, biomarker expression, and inflammation in the aquatic ecosystem. These will cause damage to aquatic embryos, growth inhibition, DNA strain break, and infertility. Nanoparticle toxicity affects soil pH, composition, and nutritional status. Soil microbiome is destroyed and protein denaturing, inhibition of the proton pumps. The phytotoxicity of nanoparticles has also been documented and mediated through clathrin-dependant endocytosis resulting in caveolae and lipid-raft mediated uptake, phagocytosis, fluid-phase endocytosis, and disruption in plant metabolic pathways, photosynthesis pathways, and vascular tissue proliferation. Paresh et al. (2009) revealed that many scientists considered nanoparticles as a novel tool in the food industry which has seriously improved the market value but resulted in environmental and public health concerns. The exposure of the populace to nanoparticles is on the rise, due to bioaccumulation, intracellular changes, and disruption of cellular organelles. Matthew et al. (2009) showed that carbon nanoparticles caused aquatic toxicity and impurity resulting in disruption in the aquatic ecosystem. The authors revealed using various molecular techniques like inductively coupled plasma mass spectroscopy and scanning electron microscopy with energy dispersive X-ray spectroscopy and demonstrated that the aquatic hardness, pH, and salinity altered greatly the death of aquatic living organisms, plants, and animals. Studies have shown that very few packaging materials are subjected to health risk assessment, thus posing a risk to the environment and human health. These packages can induce hormonal disruption, mutagenicity, cancer, and infertility. Due to the bisphenol-A, phthalates, perchlorate styrene, nonylphenol, nanoparticles, and fluorochemicals, found in these food packages, can impact negatively disrupting the endocrine system, immunotoxicity, obesity, diabetes, and cancer. Priyanka et al. (2018) reported that the physiochemical properties such as size, surface charge, coating, and morphology of nanoparticles are responsible for the cytotoxicity effects like reactive oxygen species generating oxidative stress, nonoxidant pathways, and inflammation. Abreu et al. (2015) investigated silver nanoparticles with starch composite and silver nanoparticle/ammonium starch/salt composite migration. The authors reported that the migration permitted limit from the components of nanostructured starch films was below 60 mg/kg. The authors showed that several proteins and chemicals utilized in the development of the nanostructured packaging material could cause allergic reactions, endocrine disruption, and cancer. Several products derived from nanomaterials present a huge threat and burden to public health through the deposition of harmful and toxic chemicals to the environment and human body, thus resulting in ecotoxicity and diseases. Ikjot et al. (2018) reported the nanotoxicity of ingested engineered nanomaterials. In their study, they discovered that nanotoxicology in recent years has gained serious attention using various test approaches and systems of evaluation such as in vitro monoculture cell models and in vitro models of gut microbiome. These approaches are cheap, easy, and physiologically relevant. Antul et al. (2021) revealed the toxicity, health risks, and drawbacks of the utilization of nanoparticles in food packaging. The authors noted that ingestion, accidental exposure, and inhalation of nanoparticles from the environment and food packages will cause serious health and environmental effects such as lysosomal damage, lung tumors, cardiovascular diseases, blood clotting, induced cytotoxicity, and infertility. The authors revealed that nanoparticle exposure will cause induction of oxidative stress, disruption of signaling pathways, cytoskeletal damage, chromosome aberrations, mitochondrial disruption, inhibition of cell proliferation, lysosomal damage, release of cytokines, apoptosis, genotoxicity, neurotoxicity, and cancer. Over the years, nanoparticles have been utilized across different sectors of the economy ranging from medical, food, cosmetics, electronics, drinks, nutraceuticals, fabrics, and agriculture. There are too many risk factors associated with exposure to nanoparticles like genotoxicity, carcinogenicity, and mutagenicity. Sudarshan et al. (2014) revealed that in recent years, the rapid evolution in the advancement of nanoparticles has generated a lot of concerns due to their distribution, occurrence, transportation, and fate in the environment. Much of the environmental pollution of nanoparticles has been reported to be from medical sources, consumer products, engineering operations, food, drugs, and agriculture. Many of them are released into the environment as by-products and waste, posing a risk to the ecological system, and human health. Adetunji et al. (2021b) reported that several disciplines have adopted the utilization of nanoparticles for various applications. The authors reported that due to the widespread application of nanoparticles, it has resulted in increased environmental bioaccumulation and toxicity against human health. Adetunji et al. (2021c) also revealed that in the agricultural sector, due to innovative approaches in order to meet food security and sustainable agriculture, the utilization of nanomaterials has become increasingly popular. The utilization of nanoparticles to manage pests, pathogenic organisms, improve soil quality, and generate bio-renewable waste has resulted in environmental pollution and the accumulation of nanoparticles. Even though many of these nanomaterials are reported to be eco-friendly and nontoxic, bioaccumulation in large amounts over time has been reported to be detrimental to health and the environment as shown in Table 23.1.

Table 23.1 Summary of nanomaterials utilized in the food industry.

S. no.	Applications	Nanomaterials	Toxicity	References
1	Food packaging	Silver nanoparticles	Safe	Ghada (2020)
2	Food quality, food quality, packaging, processing labeling	Nanoclays, gold, silver nanoparticles	Toxic	Ranjha et al. (2022)
3	Smart packaging, increase shelf life	Nanocomposites	Safe	Suyithra and Manikkandam (2019)
4	Food packaging	Metal oxides inorganic metals	Less toxic	Vivek et al. (2018)
5	Food packaging	Nanocomposites starch-based nanoparticles	Nontoxic	Bazila et al. (2018), Mironescu et al. (2021)
6	Food, drug, and cosmetics	Nanotitanium dioxide	Toxic	Li et al. (2008)
7	Fish production	Gelatin-based nanoparticles	Toxic	Handy et al. (2008)
8	Food processing	Trans-1,4-diphenyl-1,3-butadiene triclosan	Toxic	de Abreu et al. (2010)
9	Food industry	Carbon soot	Toxic	Nel et al. (2012)
10	Zinc-layered hydroxychloride β-glucan	Zinc-layered hydroxychloride	Safe	Velazquez-Carriles et al. (2018)

(*continued on next page*)

11	Food processing	Carbon dots	Low toxicity and high biocompatibility	Bi et al. (2017)
12	β-glucan-based coating on gold nanoparticles	Gold nanoparticles	Enhance the growth and activity of gut microbiota, boost innate immunity	Li et al. (2019)
13	Biosynthesis of silver nanoparticles utilizing crustacean β-glucan binding protein	Silver nanoparticles	Limit toxicity effects	Anjugam et al. (2018)
14	Chitosan coating on food	Lipid nanoparticles	Nontoxic and biocompatibility	Ramalingam et al. (2016)
15	Packaging material	Polysaccharide-based metallic nanoparticles	Nontoxic and biocompatibility	Tan et al. (2016)
16	Food coatings	Protein-based silver nanoparticles	Low toxicity	Pandey et al. (2020)
17	Food industry	Palladium nanoparticles	Nontoxic	Gnanasekar et al. (2017)
18	Natural antioxidant	Chitosan/alginate nanoparticles	Nontoxic	Aluani et al. (2017)

23.8 Conclusion

Various applications and toxicity of nanoparticles in the food industry have been highlighted in this review. Despite the progress made in nanoparticle research, some of the harmful and hazardous effects have been enumerated on human health and the environment. Different studies have been conducted to address the health and environmental implications of applications of nanomaterials in food packaging, particularly its safety effects. Several research studies are currently ongoing to fully elucidate the molecular pathways involved in the nanoparticle's biochemical and pathophysiological reactions in the body. In the livestock industry, foods have short shelf life, thus, this sector requires suitable nanoparticle packaging materials to assist in extending the shelf life. The future of nanoparticles in food packaging is promising but novel materials and their safety and toxicity must be properly elucidated to prevent harmful side effects on the environment and human health.

References

Abreu, A.S., Oliveira, M., de Sá, A., et al., 2015. Antimicrobial nanostructured starch-based films for packaging. Carbohydr. Polym. 129, 127–134.

Adetunji, C.O., Olaniyan, O.T., Anani, O.A., Inobeme, A., Ukhurebor, K.E., Bodunrinde, R.E., Adetunji, J.B., Singh, K.R., Nayak, V., Palnam, W.D., Singh, R.P., 2021a. Bionanomaterials for Green Bionanotechnology. IOP Publishing.

Adetunji, C.O., Olaniyan, O.T., Anani, O.A., Olisaka, F.N., Inobeme, A., Bodunrinde, R.E., Adetunji, J.B., Singh, K.R.B., Palnam, W.D., Singh, R.P., 2021b. Current scenario of nanomaterials in the environmental, agricultural, and biomedical fields. In: Nanomaterials in Bionanotechnology: Fundamentals and Applications. CRC Press, pp. 129–158.

Adetunji, C.O., Olaniyan, O.T., Singh, K.R.B., Inobeme, A., Nayak, V., Singh, J., Singh, R.P., 2021c. Role of biopesticides derived from bionanomaterials for enhanced food security and sustainable agriculture. Bionanomaterials for Environmental and Agricultural Applications. IOP Publishing, pp. 5–13.

Ahari, H., Karim, G., Anvar, A.A., Pooyamanesh, M., Sajadis, A., Mostaghim, A., Heydari, S., 2018. Synthesis of the silver nanoparticle by chemical reduction method and preparation of nanocomposite based on AgNPS. In: Proceedings of the 4th World Congress on Mechanical, Chemical, and Material Engineering.

Akbar, H., Skalicky, M., Brestic, M., Mahari, S., George Kerry, R., Maitra, S., Sarkar, S., Saha, S., Bhadra, P., Popov, M., Islam, Mst.T., Hejnak, V., Vachova, P., Gaber, A., Islam, T., 2021. Application of nanomaterials to ensure quality and nutritional safety of food. Hindawi J. Nanomater. 2021, 9336082. https://doi.org/10.1155/2021/9336082.

Akhileash, K., Verma, V.P., Singh, Pathak, V., 2012. Application of nanotechnology as a tool in animal products processing and marketing: an overview. Am. J. Food Technol. 7 (8), 445–451.

Aluani, D., Tzankova, V., Kondeva-Burdina, M., Yordanov, Y., Nikolova, E., Odzhakov, F., et al., 2017. Evaluation of biocompatibility and antioxidant efficiency of chitosan-alginate nanoparticles loaded with quercetin. Int. J. Biol. Macromol. 103, 771–782. https://doi.org/10.1016/j.ijbiomac.2017.05.062.

Amra, B., 2020. Nanomaterials in food processing and packaging, its toxicity and food labeling. Acta Sci. Nutr. Health 4 (9), 07–13.

Anju, S., Parayanthala, V.M., 2021. Impact of nanoparticles in balancing the ecosystem. Biointerface Res. Appl. Chem. 11 (3), 10461–10481. https://doi.org/10.33263/BRIAC113.1046110481.

Anjugam, M., Vaseeharan, B., Iswarya, A., Divya, M., Prabhu, N.M., Sankaranarayanan, K., 2018. Biological synthesis of silver nanoparticles using β-1, 3 glucan binding protein and their antibacterial, antibiofilm and cytotoxic potential. Microb. Pathog. 115, 31–40. https://doi.org/10.1016/j.micpath.2017.12.003.

Antul, K., Choudhary, A., Kaur, H., Mehta, S., Husen, A., 2021. Metalbased nanoparticles, sensors, and their multifaceted application in food packaging. J. Nanobiotechnol. 19, 256. https://doi.org/10.1186/s12951-021-00996-0.

Ashok, K.S., Sundaram, G., 2014. Applications of graphene in quality assurance and safety of food. Trends Anal. Chem. 60, 36–53.

Bahadar, H., Maqbool, F., Niaz, K., Abdollahi, M., 2016. Toxicity of nanoparticles and an overview of current experimental models. Iran Biomed. J. 20 (1), 1–11. https://doi.org/10.7508/ibj.2016.01.001.

Bazila, N., Srivastava, G., Qadri, O.S., Faridi, S.A., Islam, R.U., Younis, K., 2018. Importance and health hazards of nanoparticles used in the food industry. Nanotechnol. Rev. 7 (6), 623–641.

Bi, J., Li, Y., Wang, H., Song, Y., Cong, S., Li, D., et al., 2017. Physicochemical properties and cytotoxicity of carbon dots in grilled fish. New J. Chem. 41, 8490–8496. https://doi.org/10.1039/C7NJ02163A.

Bryce, J.M., Love, S.A., Braun, K.L., Haynes, C.L., 2009. Analytical methods to assess nanoparticle toxicity. Anal. Royal Soc. Chem. 134, 425–439. https://doi.org/10.1039/b818082b.

Collins, A.R., Annangi, B., Rubio, L., Marcos, R., Dorn, M., Merker, C., Estrela-Lopis, I., Cimpan, M.R., Ibrahim, M., Cimpan, E., Ostermann, M., Sauter, A., Yamani, N.E., Shaposhnikov, S., Chevillard, S., Paget, V., Grall, R., Delic, J., de-Cerio, F.G., Suarez-Merino, B., Fessard, V., Hogeveen, K.N., Fjellsbø, L.M., Pran, E.R., Brzicova, T., Topinka, J., Silva, M.J., Leite, P.E., Ribeiro, A.R., Granjeiro, J.M., Grafström, R., Prina-Mello, A., Dusinska, M., 2017. High throughput toxicity screening and intracellular detection of nanomaterials. Wiley Interdiscip. Rev. Nanomed. Nanobiotechnol 9 (1), e1413, Epub 2016 Jun 7. PMID: 27273980; PMCID: PMC5215403. https://doi.org/10.1002/wnan.1413.

de Abreu, D.A.P, Cruz, J.M., Angulo, I., Losada, P.P., 2010. Mass transport studies of different additives in polyamide and exfoliated nanocomposite polyamide films for food industry. Packag. Technol. Sci. 59–68. https://doi.org/10.1002/pts.879.

Fröhlich, E., 2017. Role of omics techniques in the toxicity testing of nanoparticles. J. Nanobiotechnol. 15, 84. https://doi.org/10.1186/s12951-017-0320-3.

Ghada, AlS., 2020. Nanotechnology in food packaging and food safety. J. Adv. Res. Food Sci. Nutr. 3 (1), 24–33.

Gnanasekar, S., Murugaraj, J., Dhivyabharathi, B., Krishnamoorthy, V., Jha, P.K., Seetharaman, P., et al., 2017. Antibacterial and cytotoxicity effects of biogenic palladium nanoparticles synthesized using fruit extract of *Couroupita guianensis* Aubl. J. Appl. Biomed. 16, 59–65. https://doi.org/10.1016/j.jab.2017.10.001.

Gokularaman, S., Stalin Cruz, A., Pragalyaashree, M.M., Nishadh, A., 2017. Nanotechnology approach in food packaging: review. J. Pharm. Sci. Res. 9 (10), 1743–1749.

Handy, R.D., von der Kammer, F., Lead, J.R., Hassellöv, M., Owen, R., Crane, M., 2008. The ecotoxicology and chemistry of manufactured nanoparticles. Ecotoxicol. 17, 287–314. https://doi.org/10.1007/s10646-008-0199-8.

Ikjot, S.S., O'Fallon, K.S., Gaines, P., Demokritou, P., Bello, D., 2018. Ingested engineered nanomaterials: state of science in nanotoxicity testing and future research needs. Part. Fibre Toxicol. 15, 29. https://doi.org/10.1186/s12989-018-0265-1.

Insoo, K., Viswanathan, K., Kasi, G., Thanakkasaranee, S., Sadeghi, K., Seo, J., 2022. ZnO nanostructures in active antibacterial food packaging: preparation methods, antimicrobial mechanisms, safety issues, future prospects, and challenges. Food Rev. Int. 38 (4), 537–565. https://doi.org/10.1080/87559129.2020.1737709.

Jamuna, B.A., Ravishankar, R.V., 2014. Environmental risk, human health, and toxic effects of nanoparticles. In: Kharisov, B.I., Kharissova, O.V., Rasika Dias, H.V. (Eds.), Nanomaterials for Environmental Protection, 1st ed., John Wiley & Sons, pp. 523–535. https://doi.org/10.1002/9781118845530.ch31.

Leudjo Taka, A., Tata, C.M., Klink, M.J., Mbianda, X.Y., Mtunzi, F.M., Naidoo, E.B., 2021. A review on conventional and advanced methods for nanotoxicology evaluation of engineered nanomaterials. Molecules 26, 6536. https://doi.org/10.3390/molecules26216536.

Li, X., Chi, P., Cheung, K., 2019. Application of natural β-glucans as biocompatible functional nanomaterials-NC-ND. Food Sci. Hum. Wellness 8, 315–319. https://doi.org/10.1016/j.fshw.2019.11.005.

Li, Q., Mahendra, S., Lyon, D.Y., Brunet, L., Liga, M.V., Li, D., Alvarez, P.J.J., 2008. Antimicrobial nanomaterials for water disinfection and microbial control: potential applications and implications. Water Resour. 42 (18), 4591–4602. https://doi.org/10.1016/j.watres.2008.08.015.

Loutfy, H.M., 2021. A review: metal nanoparticles and their safety processing in functional foods. J. Chem. Sci. Chem. Eng. 2 (1), 19–37.

Matthew, S., Kennedy, A.J., Steevens, J.A., Bednar, A.J., Weiss Jr., C.A., Vikesland, P.J., 2009. Release of metal impurities from carbon nanomaterials influences aquatic toxicity. Environ. Sci. Technol. 43, 4169–4174.

Megha, A., Murugan, M.S., Sharma, A., Rai, R., Sharma, A.K.H., Roy, S.K., 2013. Nanoparticles and its toxic effects: a review. Int. J. Curr. Microbiol. App. Sci. 2 (10), 76–82.

Mironescu, M., Lazea-Stoyanova, A., Barbinta-Patrascu, M.E., Virchea, L.-I., Rexhepi, D., Mathe, E., Georgescu, C., 2021. Green design of novel starch-based packaging materials sustaining human and environmental health. Polymers 13, 1190. https://doi.org/10.3390/polym13081190.

Muhammad, S., Ilyas, M., Basheer, C., Tariq, M., Daud, M., Baig, N., Shehzad, F., 2014. Impact of nanoparticles on human and environment: review of toxicity factors, exposures, control strategies, and future prospects. Environ. Sci. Pollut. Res. 22, 4122–4143. https://doi.org/10.1007/s11356-014-3994-1.

Nel, A., Xia, T., Meng, H., Wang, X., Lin, S., Ji, Z., Zhang, H., 2012. Nanomaterial toxicity testing in the 21st century, use of a predictive toxicological approach and high-throughput screening. Acc. Chem. Res. 46, 607–621.

Pandey, S., De Klerk, C., Kim, J., Kang, M., Fosso-Kankeu, E., 2020. Eco friendly approach for synthesis, characterization and biological activities of milk protein stabilized silver nanoparticles. Polymers 12, 1418. https://doi.org/10.3390/polym12061418.

Paresh, C.R., Hongtao, Y.U., Peter, F.U., 2009. Toxicity and environmental risks of nanomaterials: challenges and future needs. J. Environ. Sci. Health Part C Environ. Carcinog. Ecotoxicol. Rev. 27 (1), 1–35. https://doi.org/10.1080/10590500802708267.

Peidaei, F., Ahari, A., Anvar, A., Ataei, M., Sadeghian, A.A., 2021. Nanotechnology in food packaging and storage: a review. Iran. J. Vet. Med. 15 (2), 122–154.

Priyanka, G., Breen, A., Pillai, S.C., 2018. Toxicity of nanomaterials: exposure, pathways, assessment, and recent advances. ACS Biomater. Sci. Eng. 4, 2237–2275. https://doi.org/10.1021/acsbiomaterials.8b00068.

Ramalingam, P., Yoo, S.W., Ko, Y.T., 2016. Nanodelivery systems based on mucoadhesive polymer coated solid lipid nanoparticles to improve the oral intake of food curcumin. Food Res. Int. 4, 113–119. https://doi.org/10.1016/j.foodres.2016.03.031.

Ranjha, M.M.A.N., Shafique, B., Rehman, A., Mehmood, A., Ali, A., Zahra, S.M., Roobab, U., Singh, A., Ibrahim, S.A., Siddiqui, S.A., 2022. Biocompatible nanomaterials in food science, technology, and nutrient drug delivery: recent developments and applications. Front. Nutr. 8, 778155. https://doi.org/10.3389/fnut.2021.778155.

Saniha, A.A., Mohamed, O., Sabouni, R., Husseini, G., Karami, A., Bai, R.G., 2022. Toxicological impact of nanoparticles on human health: a review. Mater. Express 12, 1–23. https://doi.org/10.1166/mex.2022.2161.

Seaton, A., Tran, L., Aitken, R., Donaldson, K., 2010. Nanoparticles, human health hazard and regulation. J. R. Soc. Interface 7, S119–S129.

Sudarshan, K., Pugh, K., Gupta, A., Ingole, S., 2014. Nanoparticles in the environment: occurrence, distribution, and risks. J. Hazard. Toxic Radioact. Waste 04014039, 1–9. https://doi.org/10.1061/(ASCE)HZ.2153-5515.0000258.

Sujithra, S., Manikkandan, T.R., 2019. Application of nanotechnology in packaging of foods: a review. Int. J. Chem. Tech. Res. 12 (4), 7–14.

Tan, C., Xie, J., Zhang, X., Cai, J., Xia, S., 2016. Polysaccharide-based nanoparticles by chitosan and gum Arabic polyelectrolyte complexation as carriers for curcumin. Food Hydrocoll. 57, 236–245. https://doi.org/10.1016/j.foodhyd.2016.01.021.

Thiruvengadam, M., Rajakumar, G., Chung, I.M., 2018. Nanotechnology: current uses and future applications in the food industry. 3 Biotech 8, 1–13.

Velazquez-Carriles, C., Macias-Rodríguez, M.E., Carbajal-Arizaga, G.G., Silva-Jara, J., Angulo, C., Reyes-Becerril, M., 2018. Immobilizing yeast β-glucan on zinc-layered hydroxide nanoparticle improves innate immune response in fish leukocytes. Fish Shellfish Immunol. 82, 504–513. https://doi.org/10.1016/j.fsi.2018.08.055.

Vivek, K.B., Kamle, M., Shukla, S., Mahato, D.K., Chandra, P., Hwang, S.K., Kumar, P., Huh, Y.S., Han, Y.-K., 2018. Prospects of using nanotechnology for food preservation, safety, and security. J. Food Drug Anal 26, 1201–1214.

Xiaojia, H., Huey-Min, H., 2016. Nanotechnology in food science: functionality, applicability, and safety assessment. J. Food Drug Anal. 1–11.

Zainazor Tuan, T.C., Fisal, A., Goh, E.G., Che Sulaiman, N.F., Sarbon, N.M., 2020. Emerging of bio-nano composite gelatine-based film as bio-degradable food packaging: a review. Food Res. 4 (4), 944–956.

Application of the novel nontoxic nanobiomaterials for the management of food packaging and preservation

24

Olaniyan Olugbemi[a] and Charles Oluwaseun Adetunji[b]

[a]*Laboratory for Reproductive Biology and Developmental Programming, Department of Physiology, Rhema University Aba, Abia State, Nigeria,* [b]*Applied Microbiology, Biotechnology and Nanotechnology Laboratory, Department of Microbiology, Edo State University Uzairue, Iyamho, Edo State, Nigeria*

24.1 Introduction

Studies have demonstrated that nanomaterials are utilized for diverse applications particularly in food science and technology owing to their physiochemical properties such as interconnected porous networks, nontoxic by-products, biodegradability, immune biocompatibility, and bioavailability (Adetunji et al., 2021a). Duncan (2011) reported that nanomaterials such as nanocomposites clay/polymer, silver nanoparticles, and nanomaterials-based assays are currently being utilized in the food industry for the detection of food analytes. Purva et al. (2021) reported that nanotechnology can be utilized to improve food shelf life. Edible coatings made of nontoxic nanoparticles are readily available as carriers of functional ingredients in food packages. The authors revealed that nanocapsules can carry functional ingredients and food additives such as drugs, colorants, vitamins, gelatin, and albumin antimicrobial agents to target places some of which can serve as extracellular antioxidants. In food science, nanoparticles offer a unique opportunity in the food packaging system through nanoadditives and nanoingredients so as to maintain and enhance quality, delivery, safety, and health. Several nanomaterial delivery systems are available such as micelles, nanoemulsions, liposomes, biopolymeric nanoparticles, cubosomes, colloidosomes, nanocochleates, and nanocapsules. Agnishwar et al. (2021) reported that the field of nanotechnology has grown tremendously, particularly in the agricultural sector. The application in food and dairy products processing, transport, and packaging has increased. The authors revealed that currently, nanoparticles are utilized for the detection of dairy product toxins through nanobiosensor. It is known that many of these toxins are generated as a result of microorganisms' activity, predators, pests, and insects causing extreme

Nanobiotechnology for Food Processing and Packaging. DOI: https://doi.org/10.1016/B978-0-323-91749-0.00012-5

climatic conditions and mold infestation which are toxic to humans. The mechanisms involved in the formation of these toxins are binding to the cell-specific receptors or cleaving the molecules thereby forming pores, crossing the intracellular domains, and delivery by pathogenic organisms. Through nanoparticle biosensors, the analytes like antigen, DNA, toxin, RNA, protein, amino acid, enzyme-substrate, microRNA various disease biomarkers, and cell-specific antigens released from these food products are detectable. Prakash et al. (2019) showed that nanoparticles are currently being utilized for food processing and packaging. This is to improve the shelf life, quality, and preservation of food products by the use of bioactive molecules encapsulated into nanoparticles as antimicrobial agents. Pallavi et al. (2020) revealed the importance of nanoparticles in food packaging and several future prospects. In their study, the authors demonstrated that biopolymers are nontoxic and biodegradable materials currently utilized in the development of nano-based coated film packaging with potential antibacterial, heat resistance, cost-effectiveness, and mechanical stability. The authors revealed that the mechanism of the antimicrobial action of nanoparticles is through the formation of reactive oxygen species, disruption of enzyme activity, interruption of DNA synthesis, and damage to internal cell organelles. Fatima and Silvana (2020) demonstrated that the food packaging and sensing nanobiotechnology approach can successfully address challenges in the food industry through the extension of food shelf life, reduction in waste production, and improvement of food quality and safety. The authors revealed that nanoparticles are incorporated into materials to improve mechanical strength, enhance water-repellant activity, increase scavenging activity, enhance antimicrobial properties, and increase the gas barrier properties. Nanoparticle active and functional packaging can save about 1.3 billion tons of food wasted as a result of food spoilage which affects the economy, and human health and increases medical expenses. Traditional packaging is made up of inactive materials but with the advancement in nanotechnology, the authors revealed that nanoparticle-based materials can extend the shelf life, prevent moisture, and oxygen penetration, and are biodegradable. Agriopoulou et al. (2020) reported that nanomaterials are very applicable to food packaging systems and there has been enormous interest in their utilization by food technologists and scientists over the past few years. Many examples of nanomaterials were identified by the authors such as nanocomposites, nanoemulsions, nanoclays, nanostructures, and nanosensors that are currently being utilized in the food industry for food packaging. Lamri et al. (2021) revealed that nanoparticles are gaining more recognition in the food industry due to their cost-effective nature, safety, and sustainable approach to tackling food challenges. In their study, they provided an overview of nanoparticles' preservative nature of meat products and reinforcement of packaging materials. The authors demonstrated that nanoparticles have superior properties like surface energy and optical, mechanical, and electrical properties. Mir et al. (2019) reported that advancement in nanotechnology for food packaging has attracted serious attention over the past few decades. This may be attributed to their physiochemical properties such as improving the mechanical, gas, and thermal barrier characteristics, biodegradable and nontoxic. This provides protection, resistance, label information, safety, nutritional support, and acceptability for consumers. Adetunji et al. (2021b) reported that several disciplines have adopted the utilization of nanoparticles for various applications. The authors reported that due to the widespread application of nanoparticles; it has resulted in increased environmental bioaccumulation and toxicity against human health. Adetunji et al. (2021c) also revealed that in the agricultural sector, due to innovative approaches in order to meet food security and sustainable agriculture, the utilization of nanomaterials has become increasingly popular. The utilization of nanoparticles to manage pests, and pathogenic organisms, improve soil quality, and generate biorenewable waste has resulted in environmental pollution and accumulation of nanoparticles. Even though many of these nanomaterials are reported to be eco-friendly and nontoxic, bioaccumulation in large amounts over time has been reported to be detrimental to health and the environment, thus a need for safety and toxicity assessment. This chapter provides an overview of the application of nanoparticles for food packaging and preservation.

24.2 Nanobiomaterials utilized for food packages and preservatives

Kiss (2020) reported that recently, nanotechnology has gained tremendous attention in the food industry through its utilization for food packaging, enhancement of food functionality, texture modification, and nutritional enhancement. Many nanoencapsulated nanobioactive agents like vitamins, antioxidants, and antibacterial agents are known to enhance the nutritional value of many food products in nanopackages. Nanoparticles are known to protect food from environmental contamination, bacteria invasion, and enhancement of nutraceutical bioavailability. The nanobioactive compounds in nanopackaging have the potential to promote health and wellness, enhance functionality, microbial repelling ability, and safety shelf-life monitoring. The use of nanomaterials in food packaging is considered as smart food packaging with cost-effectiveness and health safety and promotion. Generally, consumers are expected to be provided with food products that are well packaged with packages that are safe and have the ability to maintain food freshness and taste. Also, information regarding the condition and quality should be provided through nanostructural biosensors built into the smart packaging materials. These smart packaging materials contain different components for monitoring food conditions and quality. Oxygen in the food package can cause serious oxidation in the food through reactive oxygen species, thereby affecting the enzyme activity resulting into changes in the taste, coloration, texture, flavor, and rancidity. Thus, the utilization of nanomaterials will help to increase the safety and quality of food products by reducing gas permeability. Through continuous research, discovery, and provision of nanosensitive materials can be used for quality preservation, and monitoring through fast, ultrasensitive, and accurate detection of foodborne toxins and pathogens. Akbar et al. (2021) reported that nanomaterials recently are emerging as novel techniques for enhancing quality preservation of food, enhancing food safety, and extending the shelf life. The unique physiochemical properties of engineered nanoparticles have tremendous applications in the food industry for innovative and smart food packaging systems. Insoo et al. (2020) reported that food packaging using ZnO nanostructures as an active antimicrobial agent is very important in the food supply chain. This nanoparticle is known to extend the shelf life of food products and enhance packaging performance through its antimicrobial mechanism, thus reducing the effects of cross-contamination. The authors revealed that the antimicrobial mechanism is utilized to control and inhibit microbial growth, ensure safety, retain moisture, resist gas or liquid penetration, and maintain shelf life. Several nanoparticles are utilized as antimicrobial agents such as quaternary ammonium salts, phenols, halogenated compounds, chitosan, inorganic materials, chitin, metals, essential oils, and metal oxides (Insoo et al., 2020). Sunita (2018) highlighted the role of nanotechnology in agriculture and the food industry. In the study, the author reported that there has been a drastic growth in the field of nanotechnology which has resulted in the transformation of the food sector. Through nanotechnology, there has been the development of smart food packaging, processing, and sensing. Nanoparticles are known to produce improved food quality, environmental protection and monitoring, safety, and the addition of nutritional supplements. Renata (2014) showed that nanocomposites are currently being applied to food packaging which is seen to enhance the quality and shelf life of food products. The author believed that this innovation would result in the production of lower-weight nanopackages, a reduction in packaging cost, and a reduction in waste generation.

Nanostructured antimicrobials have important physiochemical properties that will enable them to function well as antimicrobial agents. The physiochemical properties include a higher surface area-to-volume ratio, enhanced surface reactivity, heat resistance, and improved mechanical strength. Sergio et al. (2020) revealed that nanomaterials can improve food quality, health, and safety through nanofabrication and nanoencapsulation, thus providing fortification of foods with many bioactive ingredients and preservatives.

24.3 Application of the novel nontoxic nanobiomaterials utilized for food packages and preservatives

Zainazor et al. (2020) reported that municipal waste such as poultry and fish byproducts can be converted into valuable added products like alternative gelatine base film for nanobiotechnology for food packaging. The authors revealed that this alternative gelatine base film is a bio-degradable packaging material with low health risk on humans compared to other nano-sized components with enormous health implications. Gelatin nanoparticles exhibit great antimicrobial activity against pathogenic organisms. Also, gelatine-based nanocomposite films can eliminate bacteria invaders, thus enhancing the shelf life and quality of food. Zainazor et al. (2020) discovered that fish waste like head, trimmings, skin, fins, viscera, roe, scales, bones, and frames can be converted into oils, gelatine, collagen, hydrolysates, bioactive peptides, and protein hydrolysates and subsequently utilized in the production of nanomaterials for food packaging and preservatives. Moreover, poultry waste like chicken skin and bones has garnered interest from biomedical scientists for the production of environmentally friendly Nanobioparticles for food packaging. Zainazor et al. (2020) described the potential of gelatine-based nanocomposite films in the food packaging industry. The authors further explained that gelatin coatings have excellent properties such as film-forming ability, biodegradability, functionality, high optical, and transparency. It is interesting to know that food packaging is growing due to the increase in population across the globe and the need to meet up with food supply. Thus, the demand for food has increased exponentially owing to an explosive increase in population growth. The application of nanotechnology in food packaging is to enhance the quality of food and acceptability by consumers. Packages are meant to provide safe products, extended shelf life, and superior quality. This is achieved through the prevention of light, oxygen, gas, pathogenic organisms, and moisture from entering the food, thus preventing food spoilage. Zainazor et al. (2020) stated that the application of nanoparticles for food packaging can be categorized into three main domains like improvement of food packaging material, active packaging technique, and intelligent/smart packaging. Over the years, researchers have contributed immensely to the development of nanobiotechnology in food industry where over time there has been consistent adoption and utilization of nanoparticles in food packaging and replacement of nonbiodegradable petroleum-based products so as to protect the health of humans. Food industry is multitechnological sectors that utilize materials of high biosafety index. These functional materials are very effective in food additives, supplements, and packaging (Alejandro et al., 2016). Peidaei et al. (2021) reported that packaging needs quality control, long storage time, and safety measures. The authors noted that nanoparticles using polymer materials can meet the entire requirement and solve food challenges such as storage, marketing, protection, communications, and distribution. Generally, nanoparticles in contact with the food will indicate the condition of the food and the package. This technology will detect certain toxins, compounds, metabolites, and pathogenic microbes and provide information about the expiry date. Sampathkumar et al. (2020) reported that nanoparticles are useful in diverse ways particularly in the agricultural sector for developing functional food nanomaterials. The authors reported that several natural polymers are useful in this regard and can be adopted for sustainable food packaging and processing. This approach will provide safety, and quality, thereby promoting environmental sustainability and health. Natural polymers are derived from plants, food waste, animals, and biological sources containing microbial extracellular polymeric materials that can be applied in food preservation, food fortification, delivery systems, and food packaging. Food already has natural polymers such as polysaccharides, fats, and proteins, thus these natural polymers can be used to alter the functional and structural characteristics of food. Animal proteins such as gelatin, milk proteins-casein, egg protein, lactoglobulin as well as plant proteins like zein can be utilized to modify food properties in nanoparticle-based food packaging and processing. It is also worth noting that in the last few decades, nanodelivery system nanomaterials-based biopolymers are attracting significant attention for their application in food processing and packaging owing to their biocompatible and negligible toxic properties.

Sampathkumar et al. (2020) revealed that many biological compounds found in food have the ability to inhibit the growth of pathogenic organisms, thus serving as food preservatives. Notably are essential oils like cinnamon oil, garlic oil, citrus oil, and thymol with the ability to hinder the growth of pathogenic organisms. Many of the essential oils as natural antimicrobial agents are utilized in developing encapsulated nanodelivery systems for active packaging and food processing, thereby actively extending the shelf life of food. Also, the edible films containing nanofiber cellulose combined with citric or ginger oil have antioxidant and antimicrobial activity and can improve the shelf life. Nanoemulsions with essential oils added to edible coatings can improve the quality, safety, and shelf life of food. Other tested nanoparticles with essential oils that have antimicrobial properties are Chitosan nanoparticles, biopolymeric nanoparticles, zein-coated chitosan nanoparticles, Nanoparticles of b-glucan, nanohydrogel system, and dextran nanoparticles. In developing countries, food fortification seems to be a way of providing the needed balanced nutritional support for many deprived children from deficiencies in diet, thus serving as care for the aging population and maintaining proper health. Nanoparticle-encapsulated systems can enhance the fortification of food, bioavailability of bioactive ingredients, and protection for pathogenic organisms. Nanotechnology using nanomaterials can be adopted to maintain the functionality of food products and prevention of pathogenic organisms owing to their antimicrobial activity thus preventing food deterioration and extension of shelf life. Some of the nanoparticles are metal oxides, silver nanoparticles, nanocomposites, nanolaminates, nanofibers, nano-CoQ10 systems, and Polymeric nanoparticles. Gokularaman et al. (2017) noted that nanotechnology has advanced the science of food packaging, thus changing the consumer's preferences and global trends resulting in enhanced quality, safety, and nutritional support. The authors reported that nanotechnology advancement is changing the face of intelligent packaging which will enhance the extension of shelf life through improvement in mechanical strength, antimicrobial properties, and barrier protection against resistance. In smart packaging, nanoparticles are used as reactive agents to notify consumers about the condition of the product. Therefore, these nanoparticles can react with degradation products, microbial contaminants, environmental conditions, metabolites, and food components. The response can be produced which will correlate with the condition of the product. Wesley et al. (2014) revealed that nanotechnology in food packaging and safety is gaining consideration attention. The authors revealed that these nanomaterials possess the ability to control food safety through inhibition of microbial growth, improving tamper visibility, convenience, and delaying oxidation. Thus, the aim of innovative packaging is to extend the shelf life of food products, increase safety measures, and reduce spoilage and contamination. There are different ways to utilize nanoparticles in the food industry. Some of them are:

24.3.1 Nanoencapsulation

In the food industry, the utilization of nano-size materials represents an important technique referred to as nanoencapsulation which is usually implemented to enhance flavor and maintain the taste of the food. This process involves the use of nanocarriers to deliver nanofood additives to the food products altering the morphological nature of the food. Nanoencapsulated compounds are found to release the content in a very slow manner but a longer period of time. Some of the nanoencapsulated compounds are vitamins, omega three fatty acids, essential oils, antioxidants, antibacterial and other nutraceutical ingredients. Nanoliposomes and cochleates delivers lipid in an aqueous medium in fusion like manner to the target cell. Antonio and Francisco (2006) reported that the use of polymeric nanoparticles for nanoencapsulation has increased in the past few years owing to their physiochemical properties such as biocompatibility, shape and size. The authors described the method in the preparation of this nanoparticles as polymerization reaction processes for use as nanoencapsulation. Nanocapsules confines drugs to a particular space with a polymeric membrane for the delivery of therapeutic molecules. Ezhilarasi et al. (2013) revealed that the controlled protection and release of biomolecules using nanoencapsulation is the best and most promising technique in drug delivery system. Drugs or bioactive molecules are trapped and specifically delivered to the target sites in nanoencapsulation. Various techniques in nanoencapsulation are emulsification, inclusion, coacervation, complexation nano-precipitation, supercritical fluid and emulsification–solvent evaporation. Pateiro et al. (2021) reported that nanoencapsulation is a method where bioactive molecules are enclosed for preservation, thereby improving the stability, and physiological properties, and regulating the release at the target sites. Cezarotto et al. (2023) demonstrated that nanoencapsulation is utilized for food ingredients for the food industry. Some bioactive extract of *Vaccinium ashei* was combined with nanoparticles as nanoencapsulation in antioxidant activity to scavenge reactive oxygen species. The authors revealed that bioactive molecules like rutin, resveratrol, hydroxytyrosol, and lycopene possess good encapsulation capacity for antioxidant activity to manage depression. Also, biocides, drugs, perfumes, vitamins, enzymes, and other molecules can be packaged in liquid, gas, or solid nanoencapsulation packages to deliver their core content. Martins et al. (2023) demonstrated that the application of nanoencapsulation has enhanced the utilization of biomolecules in nutraceutical, agricultural, pharmaceutical, and food industries. This technology has offered significant opportunities for the controlled release of vitamins, essential fatty acids, minerals, polyphenols, flavors, colorants, antioxidants, and antimicrobial agents in the gastrointestinal tract. The application of nanoencapsulation has greatly improved the texture, taste, shelf life, and bioavailability of nutrients in the body system.

24.3.2 Nanoemulsions

Nanoemulsions can carry hydrophobic components of food through an aqueous medium due to their polarity. Some examples include nanosurfactant micellar, lipid liposome systems, polymers, ultrafine emulsions, unstable microemulsions, emulsoids, submicrometer emulsions, and miniemulsions. Nanoemulsions have a high surface area ratio, high kinetic, and physical stability, high bioavailability, and high optical clarity. Nanoemulsions are very good in carrying antimicrobial, anti-brown, antioxidant, aromatic, and coloring agents in food products, leading to positive effects on food safety, quality, and improved consumer acceptance. Nanoemulsions are important in several areas such as drug delivery, cosmetics, food, material synthesis, and pharmaceuticals. Several techniques are utilized in the preparation of nanoemulsions such as high-pressure homogenization, phase inversion temperature, ultrasonication, emulsion inversion point, and bubble bursting method. Shah et al. (2010) and Hitendra and Sagar (2016) showed that nanoemulsions are very important drug carriers for the delivery of therapeutic agents in systemic circulations. In nanoemulsions, oil and water which are immiscible solvents form single-phase surfactants, and their use cuts across cosmetics, drug therapies, biotechnologies, and diagnostics. Currently, nanoemulsions are used in the treatment of cancer, the development of vaccines, and other drug delivery systems. Nanoemulsions are generally stable for long periods of time, and they do not undergo creaming, flocculation, and sedimentation. Padmadevi et al. (2016) demonstrated that nanoemulsion has generated significant attention in the food, pharmaceutical, and cosmetics industries for novel delivery of bioactive agents such as lipophilic materials like fatty acids, flavor, colors, and essential oils, drugs, and biological agents like active ingredients and therapeutic molecules. Nanoemulsions are characterized by small droplet-like size, translucent nature, interfacial area, low viscosity, solubility efficiency, and high stability. Himesh and Sarvesh (2021) showed that nanoemulsions are thermodynamically stable against sedimentation for optimal delivery of drugs and other agents and as nanoreactors. Daniela et al. (2011) demonstrated that the importance of nanoemulsions in the cosmetics and skin care industry is gaining significant attention for the prevention and treatment of various skin conditions and diseases. The authors utilized rice bran oil as a nanoemulsion in the treatment of skin diseases and further evaluated the irritation, physical stability, and potential moisturizing properties of the products. Sharma and Nitin (2012) revealed that nanoemulsions are advanced nanoparticles with a thermodynamically stable system for systemic delivery of agents at a controlled rate. Gunjan et al. (2021) gave consolidated information on the application of nanoemulsions as effective bioactive delivery compounds, particularly in the food and cosmetics industry. The authors utilized various spectral analyses like differential scanning calorimetry, transmission electron microscopy, atomic force microscope (AFM), scanning electron microscope (SEM), X-ray diffraction (XRD), scanning tunneling microscope (STM), and Fourier-transform infrared spectroscopy to characterize the efficacy of nanoemulsions.

24.3.3 Nanocomposite

Nanocomposites are hybrid materials with several phases of diverse solid materials and one phase in nanoscale. Nanocomposites are very useful in food industry due to their antimicrobial action against bacterial, yeast and fungi. Nanocomposites are also powerful in reducing gas permeability. Nanotechnology has been described as an important tool for the advancement, improvement and development of science and technology. Nanocomposite materials have various applications across different fields of science. There are different types of nanocomposites in terms of matrix phase like ceramic matrix, metal matrix and polymer matrix nanocomposites. Charles (2013) and

Balaji et al. (2017) demonstrated in their study that that nanocomposites can be utilized for various applications. They further revealed that ceramics matrix nanocomposites possess ceramic fiber materials while metallic matrix nanocomposites are reinforced with metallic carbon nanotube materials or in combination with metal oxides. The polymer nanocomposites may be like nanofiller polymer composites containing nanoparticles like graphene, molybdenum disulfide, carbon nanotubes, and tungsten disulfide. The polymer nanocomposites are important for tissue engineering, cellular therapies and drug delivery using materials like starch, alginate, cellulose, chitosan, gelatin, fibrin, collagen, poly(vinyl alcohol), poly(caprolactone), poly(ethylene glycol), poly(glycerol sebacate) and poly(lactic-co-glycolic acid) for the desired properties. Parameswaranpillai et al. (2015) highlighted the production processes involved in the manufacturing of nanocomposites. In their study, the authors revealed that many nanocomposites are combined with nanoparticles to reinforce their properties. Singh et al. (2021) revealed that nanocomposites are materials for the 21st century industrial revolution with broad-based applications.

24.3.4 Edible nanocoatings and nanocoating materials

Over the years, significant growth and progress have been witnessed in the application of edible materials in the preparation of nanomaterials for food packaging. These nanomaterials' edible coating is generally utilized in the maintenance of fruit and vegetables in the postharvest technology. This technology can help to reduce moisture, gaseous exchange, respiration, and physiological disorders. de Oliveira et al. (2021) showed that perishable food items have been under investigation on how to extend their shelf lives and preservation to maintain their taste, quality, and texture. Owing to various research on edible coatings, increased preservation of fruits and vegetables has been achieved. Further research has also been conducted on how to improve upon the properties of these coatings using nanotechnology. Recently, nanoemulsions have been revealed to possess good water barrier, optical, physiological like antioxidants, antimicrobial and mechanical properties. Eman et al. (2018) demonstrated that edible coatings can be improved upon in terms of their physiochemical properties using nanomaterial chitosan-thyamol/(TPP)tripolyphosphate, chitosan-methyl cellulose/silica (SiO_2), gelatin-coconut fiber/titanium dioxide (TiO_2), gelatin-anthocyanin/kafirin and gelatin–chitosan/(Ag/ZnO), Azhar et al. (2021) demonstrated that nutritional physiology is an important aspect to life owing to the beneficial role of phytonutrients derived from fruits and vegetables. The authors showed that increased consumption of fruits and vegetables have increased over the years but yet the issue of storage is a major challenge in this sector. Thus, they revealed that polysaccharide-based material like chitosan, guar gum, cellulose, sodium alginate and xanthan gum as edible nanocoating is gaining attention for its important role in protection of perishable items from microbes due to the strong antimicrobial properties. Haron et al. (2021) revealed that food quality can be maintained for a long time with the application of nano-edible coatings. This technology will prevent food wastage, change in taste and texture, and improve shelf life. Some of the edible nanoparticle coatings are polysaccharides-based materials like chitosan, cellulose, and pullulan-based materials. Mahela et al. (2020) reported that postharvest infrastructure is a challenge, particularly in developing countries where electricity is a major problem. The use of nanoedible coatings has provided a unique opportunity for enhancing the shelf life, taste, and quality of several postharvest products. The antimicrobial, hydrophilic, and antioxidant characteristics facilitate this quality in the nano-edible coating. The authors revealed that different types of nanocoatings exist such as nanoemulsion, solid lipid nanoparticles, polymeric nanoparticles, nanotubes, lipid nanocarriers, and nanofibers. Some of the disadvantages of the application of nano-edible coatings include the generation of allergic reactions, and not being economically viable, thus further research is still needed in these areas. Hayam et al. (2022) showed that in many parts of low-middle-income countries, a large amount of waste is generated such as peel waste from vegetables and fruits. The authors revealed that this may result in environmental, nutritional, and economic loss if not properly handled. Recently, edible nano-coatings like chitosan-based have shown promising results in postharvest food preservation. The result obtained from the experiment using banana and pomegranate peels has alluded to the low cost, antifungal, and antioxidant nature of the edible nano-coatings.

24.3.5 Nanofibers

Nanofibers have unique physiochemical properties that can be helpful in development of artificial food and targeted delivery system. Nanofibers based biopolymer are very eco-friendly food packages. Xiaomin et al. (2015) reported that electrospinning is a technique utilized in the development of nanostructured fibers for various applications in energy sector. Gouda et al. (2022) demonstrated that chitosan-based nanofibers is a novel delivery system developed through electrospinning technique. Other available techniques described by the authors include template method, phase separation and vapor grown. Ahmed et al. (2019) reported that nanofibers are important solutions to many biomedical challenges. The technology is an emerging area in the field of energy, environment, and agriculture. Thandavamoorthy et al. (2005) revealed that there is enormous potential in the field of nanotechnology for the advancement of technology and science. Nanofiber production using electrospinning method with nanostructured fibrous materials has provided opportunity in energy, biomedical, clothing and engineering sector. Alsaid et al. (2017) noted that nanofibers fabrication can be carried out through template synthesis, phase separation, self-assembly, melt blowing, centrifugal spinning and drawing, electrospinning.

24.3.6 Nanocapsules

Nanocapsules are drug carriers which can give protection against moisture content, pH value, temperature, oxidation– reduction and light. An example is casein micelle-based capsules which is very stable, bioavailable in the gastro intestine. Kovrigina et al. (2023) revealed that nanocapsule are currently utilized for cancer treatment as a cutting-edge technology through a magnetic field-guided targeted drug delivery. Nanocapsule are stable, effective, efficient with higher capacity for drug loading. Ananda et al. (2021) utilized curcumin which possess antiinflammatory and antioxidant properties for nanoencapsulation technique through bioavailability of the active ingredient. Nilewar et al. (2017) reported that nanocapsules can be prepared using two types of polymers; synthetic polymers and natural polymers via solvent evaporation, emulsification/solvent diffusion, dialysis, salting out, super critical fluid and nano precipitation methods. The role of nanocapsules has increased tremendously across therapeutic, biomedical, diagnostic, sensing and imaging fields. Nanocapsules are nanocarriers for transport and protection of cargos at the target sites. Catherine and Hatem (2008) demonstrated that membrane contactor can be utilized in the production of nanocapsules for industrial use. The authors showed that scale-up ability of the membrane contactor is a great advantage over other processes. Poletto et al. (2011) revealed that polymeric nanocapsules are currently gaining much attention in the cosmetics and dermatology sector owing to their ability for controlled release rate, permeability and penetration over the skin. Pavankumar et al. (2012) noted that nanocapsules attracted attention of scientific community due to their size and physiochemical properties. Using various spectral analysis such as X-ray diffraction, transmission electron microscopy, scanning electron microscopy, high-resolution transmission electron microscopy, superconducting quantum interference device, X-ray photoelectron spectroscopy, spectroscopic techniques and multi angle laser light scattering, the authors were able to characterize nanocapsules. Four methods can be utilized in the preparation of nanocapsules such as interfacial polymerization, interfacial deposition, interfacial precipitation, and self-assembly approaches. Malam et al. (2000) reported from their study conducted on nanocapsules in the protection of insulin degradation in the intestinal epithelium via oral administration.

24.4 Toxicological aspects of nanoparticles in food packaging

Helen et al. (2022) reported that nanotechnology application in food and agricultural sectors using nanomaterials for the production or food packaging has received significant attention in the last few years. Nanoparticles in food processing and packaging can improve the shelf life, quality and taste of food through the prevention of contamination from microorganisms. The toxicity level of the nanomaterials due to long term accumulation in the body from food and agricultural products has now generated serious concerns. Appropriate regulation in the use of nanoparticles in agricultural and food industry must be ensured to prevent nanotoxicity and other risk factors. Rahul et al. (2022) revealed that application of nanoparticles in food industry has paved way for improved food packaging to prevent microorganisms, increase shelf life and environmental conditions like oxygen, carbon dioxide and moisture through the use of nano-film coatings. The authors showed that currently human health is at risk due to the migration effects from nanoparticles to human through ingestion, inhalation and cutaneous exposure resulting into generation of reactive oxygen species, inflammation and cancer. Sabarish et al. (2021) demonstrated that food packaging with nanoparticles film is a novel innovation in the food industry, however there are growing concerns of toxicity derived from exposure to the nanoparticles on human and environmental health. The authors noted that nanoclay, nanotitania, nanosilver, nano-zinc, and nanosilica impact negatively on human health such as inflammatory and reactive oxygen species diseases. Stuparu-Cretu et al. (2023) reported that current innovative food and beverage packages involve the use of metal oxides, however some of these metal oxides can travel through the gastrointestinal tract to the cell to cause toxicity. Utkarsh et al. (2022) reported that mechanical barriers, microorganisms and environmental conditions can be improved upon in food packaging by the utilization of nanoparticles. This will protect the food, improve the quality, enhances the nutritional status, flavor, appearance, taste, storage, texture, properties and shelf life. The harmful effects of nanoparticles use in food packaging is numerous. Tschiche et al. (2022) highlighted some of the detrimental effects of biological accumulation of nanoparticles to human health such as cancer, organ failure and dysfunctions. Mahmoud (2015) demonstrated that the revolution of nanoparticles and technology to food industry through the use of nano-biodegradable packaging materials, bioactive encapsulation and biosensors with antimicrobial and heat resistance properties.

24.5 Intelligent food packaging

Yam et al. (2005) and Yezza (2008) reported that smart packaging and intelligent packaging have different interpretation. Food packaging stem from the traditional method where products are communicated to the consumers through protection from environment and as marketing tool. Today, the traditional method is no longer sufficient enough to for food distribution systems for protection and marketing strategy. Thus, intelligent and smart packaging is receiving great attention in the food industry. Intelligent packaging carries sensor, preservatives and phyto ingredient that migrate into the food for safety and improved quality. The authors have referred to intelligent packaging as packaging that can communicate, trace, detect, record, monitor and sense conditions of packaged food to provide information about the quality while smart packaging combine both active packaging with intelligent packaging. Intelligent packaging contains innovative technology like internal and external indicators to carry out the intelligent functions like integrity indicator, time–temperature indicator, rancidity indicator, freeze-damage indicator and food spoilage indicator. Smart packaging possesses the ability to monitor changes across the environment, product and packaging. Intelligent packaging can combine also different technology needed to improve on the performance, convenient, cost effectiveness in the food supply chain. Furthermore, intelligent packaging can provide information on the origin, ingredients, manufacturing date, expiration date, history, composition, storage conditions, temperature, pH, integrity, and microbial growth (Lee and Rahman, 2014).

24.6 **Active food packaging**

Active packaging systems are mainly designed to facilitate increase in shelf life for foods and still maintain the desired quality. The active packaging technologies are physical, biological and chemical ways by which alterations interactions with package or the product. This system is able to eliminate oxygen or oxidation from the products or package via incorporation of active ingredients or components into the packaging material for example polyethylene terephthalate or inside the package. Active packaging can be utilized to control release of antimicrobial agents or polysaccharide particles with encapsulated antimicrobial agents incorporated into the packaging material thereby extending the shelf life of the food products (De Jong et al., 2005).

24.7 **Future prospects and challenges**

In recent years, active, smart and intelligent packaging systems have received massive attention, but a lot of research is still needed in these areas. The discovery of many innovative molecules and advancement in nanotechnology with novel compounds having greater capacity for food safety, improvement, convenient and quality will transform the field of food packaging and become generally acceptable for all.

24.8 **Conclusion**

Several attempts have been made to maintain the quality and freshness of food products particularly from food borne pathogenic organisms. Nanoparticles have been seen to play a major role in this regard, thus serving as antimicrobial packaging systems. Incorporation of nanoparticles into packaging materials provides an alternative approach to solving the problem of ROS generation, cell rupture and cellular metabolism destabilization. Nanoparticles are incorporated into materials to improve the mechanical strength, enhance water repellant activity, increase scavenging activity, enhance the antimicrobial properties and increase the gas barrier properties. Thus, this chapter provides detailed information on the application nanoparticles for food packaging and preservation.

References

Adetunji, C.O., Olaniyan, O.T., Anani, O.A., Inobeme, A., Ukhurebor, K.E., Bodunrinde, R.E., Adetunji, J.B., Singh, K.R., Nayak, V., Palnam, W.D., Singh, R.P., 2021a. Bionanomaterials for green bionanotechnology. In: Bionanomaterials. IOP Publishing.

Adetunji, C.O., Olaniyan, O.T., Anani, O.A., Olisaka, F.N., Inobeme, A., Bodunrinde, R.E., Adetunji, J.B., Singh, K.R.B., Palnam, W.D., Singh, R.P., 2021b. Current scenario of nanomaterials in the environmental, agricultural, and biomedical fields. In: Nanomaterials in Bionanotechnology: Fundamentals and Applications. CRC Press, pp. 129–158.

Adetunji, C.O., Olaniyan, O.T., Singh, K.R.B., Inobeme, A., Nayak, V., Singh, J., Singh, R.P., 2021c. Role of biopesticides derived from bionanomaterials for enhanced food security and sustainable agriculture. In: Bionanomaterials for Environmental and Agricultural Applications. IOP Publishing, pp. 5-1–5-13.

Agnishwar, G., Ghosh, M.M., Pallavi, P., Ramesh, S., Girigoswami, K., 2021. Nanotechnology in detection of food toxins: focus on the dairy products. Biointerface Res. Appl. Chem. 11 (6), 14155–14172. https://doi.org/10.33263/BRIAC116.1415514172.

Agriopoulou, S., Stamatelopoulou, E., Skiada, V., Tsarouhas, P., Varzakas, T., 2020. Emerging nanomaterial applications for food packaging and preservation. Safety issues and risk assessment. Proc. AMIA Annu. Fall Symp. 70, 7. https://doi.org/10.3390/foods_2020-07747.

Ahmed, B., Pal, K., Rahier, H., Uludag, H., Kim, I.S., et al., 2019. Nanofibers as new-generation materials: From spinning and nano-spinning fabrication techniques to emerging applications. Appl. Mater. Today 17, 1–35. 10.1016/j.apmt.2019.06.015.hal-03243158.

Akbar, H., Skalicky, M., Brestic, M., Mahari, S., George Kerry, R., Maitra, S., Sarkar, S., Saha, S., Bhadra, P., Popov, M., Islam, M.T., Hejnak, V., Vachova, P., Gaber, A., Islam, T., 2021. Application of nanomaterials to ensure quality and nutritional safety of food. Hindawi J. Nanomater. 2021, 9336082. https://doi.org/10.1155/2021/9336082.

Alejandro, J.P., Asencio, C.M., Manuel, L.J., Allemandi, D.A., Palma, S.D., 2016. Nanoencapsulation in the food industry: manufacture, applications and characterization. J. Food Bioeng. Nanoproc. 1 (1), 56–79.

Alsaid, A.A., El-Sakhawy, M., Elshakankery, M.H., Kasem, M.H., 2017. Technology of nanofibers: production techniques and properties – critical review. J. Textile Assoc. 5–14.

Ananda, N.T., Rafiq, N., Yousaf, S., Abbas, S., 2021. Assessment of oral bioavailability of nanocapsules loaded-curcumin in-vivo. World J. Adv. Res. Rev. 9 (2), 005–017 2021.

Antonio, J.R., Francisco, V., 2006. Nanoencapsulation I. Methods for preparation of drug-loaded polymeric nanoparticles. Nanomed.: Nanotechnol. Biol. Med. 2 (2006), 8–21.

Azhar, N.I., Asli, N.A., Hajar, N., 2021. Edible nanocoating: the effect of different types of nanoparticles incorporated with polysaccharide-based materials on antibacterial and antifungal activity. In: Inaugural Symposium of Research and Innovation for Food (SoRIF), 2021, pp. 72–74.

Balaji., V., Manasa, B., Aakash, N., Chandrakaanth, B.S., Kiran Kumar, K.C., 2017. Nanocomposites and their applications. Int. J. Eng. Res. Mech. Civil Eng. 2 (11), 81–85.

Catherine, C., Hatem, F., 2008. A new process for drug loaded nanocapsules preparation using a membrane contactor. Drug Dev. Ind. Pharm. 31 (10), 987–992. https://doi.org/10.1080/03639040500306237.

Cezarotto, V.S., Franceschi, E.P., Stein, A.C., Emanuelli, T., Maurer, L.H., Sari, M.H.M., Ferreira, L.M., Cruz, L., 2023. Nanoencapsulation of *Vaccinium ashei* leaf extract in Eudragit® RS100-based nanoparticles increases its in vitro antioxidant and in vivo antidepressant-like actions. Pharmaceuticals 16, 84. https://doi.org/10.3390/ph16010084.

Charles, C.O., 2013. Nanocomposites: an overview. Int. J. Eng. Res. Dev. 8 (11), 17–23.

Daniela, S.B., Pereira, T.A., Maciel, N.R., Bortoloto, J., Viera, G.S., Oliveira, G.C., Rocha-Filho, P.A., 2011. Formation and stability of oil-in-water nanoemulsions containing rice bran oil: *in vitro* and *in vivo* assessments. J. Nanobiotechnol. 9, 44. 2011 https://doi.org/10.1186/1477-3155-9-44.

De Jong, A.R., Boumans, H., Slaghek, T., Van Veen, J., Rijk, R., Van Zandvoort, M., 2005. Active and intelligent packaging for food: is it the future? Food Addit. Contam.: Part A 22, 975979.

de Oliveira Filho, J.G., Miranda, M., Ferreira, M.D., Plotto, A., 2021. Nanoemulsions as edible coatings: a potential strategy for fresh fruits and vegetables preservation. Foods 10, 2438. https://doi.org/10.3390/foods10102438.

Dholariya, P.K., Borkar, S., Borah, A., 2021. Prospect of nanotechnology in food and edible packaging: A review. J. Pharm. Innov. 10 (5), 197–203.

Duncan, T.V., 2011. Applications of nanotechnology in food packaging and food safety: barrier, materials, antimicrobials and sensors. J. Colloid Interface Sci. 363 (1), 1–24. https://doi.org/10.1016/j.jcis.2011.07.017.

Eman, A.A.K., Nahla, S.Z., Hosam El Din, A.-A., 2018. The effect of nano materials on edible coating and films' improvement. Int. J. Pharm. Res. Allied Sci. 7 (3), 20–41.

Ezhilarasi, P.N., Karthik, P., Chhanwal, N., Anandharamakrishnan, C., 2013. Nanoencapsulation techniques for food bioactive components: a review. Food Bioprocess Technol. 1–21. https://doi.org/10.1007/s11947-012-0944-0.

Fatima, M., Silvana, A., 2020. Nanotechnology-based approaches for food sensing and packaging applications. RSC Adv., 10, 19309–19336. https://doi.org/10.1039/D0RA01084G.

Gokularaman, S., Stalin Cruz, A., Pragalyaashree, M.M., Nishadh, A., 2017. Nanotechnology approach in food packaging: review. J. Pharm. Sci. Res. 9 (10), 1743–1749.

Gouda, M., Khalaf, M.M., Shaaban, S., El-Lateef, H.M.A., 2022. Fabrication of chitosan nanofibers containing some steroidal compounds as a drug delivery system. Polymers 14, 2094. https://doi.org/10.3390/polym14102094.

Gunjan, P.M., Ande, S.N., Chavhan, S.A., Bartare, S.A., Malode, L.L., Manwar, J.V., Bakal, R.L., 2021. A critical reveiw on nanoemulsion: advantages, techniques and characterization. World J. Adv. Res. Rev. 11 (3), 462–473.

Haron, N.H., Asli, N.A., Hajar, N., 2021. Edible nanocoating: the properties of different casting methods and different types of nanoparticles incorporated with polysaccharides-based materials. In: Inaugural Symposium of Research and Innovation for Food (SoRIF), pp. 68–71.

Hayam, M.S., EL-Mokadem, M.T., Mohamed, H.G., ELBaz, G.A., Farroh, K.Y., 2022. Characterization of antifungal edible nano-coating materials prepared by some waste peel extracts. Curr. Sci. Int. 11 (2), 244–260.

Helen, O., Passaretti, P., Miri, T., Al-Sharify, Z.T., 2022. The safety of nanomaterials in food production and packaging. Curr. Res. Food Sci. 5, 763–774. https://doi.org/10.1016/j.crfs.2022.04.005.

Himesh, S., Sarvesh, S., 2021. Current update on nanoemulsion: a review. Sch. Int. J. Anat. Physiol. 4 (1), 6–13.

Hitendra, S.M., Sagar, K.L., 2016. Nanoemulsions: a versatile mode of drug delivery system. Ind. J. Novel Drug Deliv. 8 (3), 123–132.

Insoo, K., Viswanathan, K., Kasi, G., Thanakkasaranee, S., Sadeghi, K., Seo, J., 2020. ZnO nanostructures in active antibacterial food packaging: preparation methods, antimicrobial mechanisms, safety issues, future prospects, and challenges. Food Rev. Int. https://doi.org/10.1080/87559129.2020.1737709.

Kiss, E., 2020. Nanotechnology in food systems: a review. Acta Aliment. 49 (4), 460–474. https://doi.org/10.1556/066.2020.49.4.12.

Kovrigina, E., Poletaeva, Y., Zheng, Y., Chubarov, A., Dmitrienko, E., 2023. Nylon-6-coated doxorubicin-loaded magnetic nanoparticles and nanocapsules for cancer treatment. Magnetochemistry 2023 (9), 106. https://doi.org/10.3390/magnetochemistry9040106.

Lamri, M., Bhattacharya, T., Boukid, F., Chentir, I., Dib, A.L., Das, D., Djenane, D., Gagaoua, M., 2021. Nanotechnology as a processing and packaging tool to improve meat quality and safety. Foods 10, 2633. https://doi.org/10.3390/foods10112633.

Lee, S.J., Rahman, A.T.M.M., 2014. Intelligent packaging for food products. In: Han, J.H. (Ed.), Innovations in Food Packaging. Academic Press, San Diego, pp. 171–209.

Mahela, U., Rana, D.K., Joshi, U., Tariyal, Y.S., 2020. Nano edible coatings and their applications in food preservation. J. Postharvest Technol. 8 (4), 52–63.

Mahmoud, M.B., 2015. Nanotechnology in food industry; advances in food processing, packaging and food safety. Int. J. Curr. Microbiol. App. Sci. 4 (5), 345–357.

Malam, A., Couvreur, P., Pinto-Alphandary, H., Gouritin, B., Lacour, B., Farinotti, R., Puisieux, F., Vauthier, C., 2000. Insulin-loaded nanocapsules for oral administration: In vitro and in vivo investigation. Drug Dev. Res. 49 (2), 109–117. https://doi.org/10.1002/(SICI)1098-2299(200002)49:2≤109::AID-DDR4≥3.0.CO;2-%23.

Martins, V.F.R., Pintado, M.E., Morais, R.M.S.C., Morais, A.M.M.B., 2023. Valorisation of micro/nanoencapsulated bioactive compounds from plant sources for food applications towards sustainability. Foods 12, 32. https://doi.org/10.3390/foods12010032.

Mir, A.S.S., Farhadyar, N., Pourdayhimi, P., Azarakhshi, F., 2019. State of nano technology in novel food packaging and new application opportunities. Int. J. Bio-Inorg. Hybr. Nanomater. 8 (4), 143–152.

Nilewar, G., Mute, P.B., Talhan, P.P., Thakre, S., 2017. Nanocapsules: nano novel drug delivery system. PharmaTutor 5 (6), 14–16.

Padmadevi, C., Ariffin, F.D., Eid, A.M., Almahgoubi, A.A., Mohamed, A.T., Issa, Y.S., Elmarzugiejbps, N.A., 2016. Nanoemulsion for cosmetic application. European J. Biomed. Pharm. Sci. 3 (7), 08–11.

Pallavi, C., Fatima, F., Kumar, A., 2020. Relevance of nanomaterials in food packaging and its advanced future prospects. J. Inorg. Organomet. Polym. Mater. 30, 5180–5192. https://doi.org/10.1007/s10904-020-01674-8.

Parameswaranpillai, J., Joseph, G., Shinu, K.P., Jose, S., Salim, N.V., Hameed, N., 2015. Development of hybrid composites for automotive applications: effect of addition of SEBS on the morphology, mechanical, viscoelastic, crystallization and thermal degradation properties of PP/PS–x GnP composites. RSC Adv. 5, 25634–25641.

Pateiro, M., Gómez, B., Munekata, P.E.S., Barba, F.J., Putnik, P., Kovačević, D.B., Lorenzo, J.M., 2021. Nanoencapsulation of promising bioactive compounds to improve their absorption, stability, functionality and the appearance of the final food products. Molecules 26, 1547. https://doi.org/10.3390/molecules26061547.

Pavankumar, K., Kanumur, H., Ravur, N., Maddu, C., Parasuramrajam, R., Thangavel, S., 2012. Nanocapsules: the weapons for novel drug delivery systems. BioImpacts 2 (2), 71–81. https://doi.org/10.5681/bi.2012.011.

Peidaei, F., Ahari, A., Anvar, A., Ataei, M., Sadeghian, A.A., 2021. Nanotechnology in food packaging and storage: a review. Iran. J. Vet. Med. 15 (2), 122–154.

Poletto, F.S., Beck, R.C.R., Guterres, S.S., Pohlmann, A.R., 2011. Polymeric nanocapsules: concepts and applications. In: Beck, R., Guterres, S., Pohlmann, A. (Eds.), Nanocosmetics and Nanomedicines. Springer, Berlin, Heidelberg. https://doi.org/10.1007/978-3-642-19792-5_3.

Prakash, J., Vignesh, K., Anusuya, T., Kalaivani, T., Ramachandran, C., Sudha Rani, R., Rubab, M., Khan, I., Elahi, F., Oh, D.-H., DevanandVenkatasubbu, G., 2019. Application of nanoparticles in food preservation and food processing. J. Food Hyg. Saf. 34 (4), 317–324. https://doi.org/10.13103/JFHS.2019.34.4.317.

Purva, K.D., Shivdutt, B., and Anjan, B., 2021. Prospect of nanotechnology in food and edible packaging: a review. The Pharma Innova. J. 10(5), 197–203.

Rahul, S., Rawat, D., and Kaushik, B., 2022. Toxicological effects of nanomaterials used in food packaging. In: Annu, T. Bhattacharya, S. Ahmed. (Eds.), Nanotechnology in Intelligent Food Packaging. Wiley. https://doi.org/10.1002/9781119819011.ch10.

Renata, D., 2014. Application of nanotechnology in food packaging. J. Microbiol. Biotech Food Sci. Dobrucka. 3 (5), 353–359.

Sabarish, R., Jasila, K., Shivanna, J.M., Jayakumar, A., Varghese, S.A., Krishnankutty, R.E., Parameswaranpillai, J., Siengchin, S., 2021. Environmental and toxicological aspects of nanostructures in food packaging. In: Parameswaranpillai, J., Krishnankutty, R.E., Jayakumar, A., Rangappa, S.M., Siengchin, S. (Eds.), Nanotechnology-Enhanced Food Packaging. Wiley Online Library. https://doi.org/10.1002/9783527827718.ch15

Sampathkumar, K., Tan, K.X., Loo, J.S.C., 2020. Developing nano–delivery systems for agriculture and food applications with nature–derived polymers. iScience 23 (5), 101055. https://doi.org/10.1016/j.isci.2020.101055.

Sergio, T.-G., Prieto, C., Lagaron, J.M., 2020. Nanomaterials to enhance food quality, safety, and health impact. Nanomaterials 10 (941), 1–6. https://doi.org/10.3390/nano10050941.

Shah, P., Bhalodia, D., Shelat, P., 2010. Nanoemulsion: a pharmaceutical review. Sys. Rev. Pharm. 1 (1), 24–32. https://doi.org/10.4103/0975-8453.59509.

Sharma, P.K., Kumar, N., 2012. Nanoemulsions: a review on various pharmaceutical application. Eng. Technol. Glob. J. Pharmacol. 6 (3), 222–225. https://doi.org/10.5829/idosi.gjp.2012.6.3.65135.

Singh, R.P., Singh, P., Singh, K.R.B., 2021. Introduction to Composite Materials Nanocomposites and their Potential Applications. CRC Press, pp. 1–29. https://doi.org/10.1201/9781003080633-1.

Stuparu-Cretu, M., Braniste, G., Necula, G.-A., Stanciu, S., Stoica, D., Stoica, M., 2023. Metal oxide nanoparticles in food packaging and their influence on human health. Foods 12 (9), 1882. https://doi.org/10.3390/foods12091882.

Sunita, B., 2018. Application of nanotechnology on food and agriculture: a review. J. Emerg. Technol. Innovat. Res. 9 (5), 340–344.

Thandavamoorthy, S., Bhat, G.S., Tock, R.W., Parameswaran, S., Ramkumar, S.S., 2005. Electrospinning of nanofibers. J. Appl. Polym. Sci. 96, 557–569. https://doi.org/10.1002/app.21481.

Tschiche, H.R., Bierkandt, F.S., Creutzenberg, O., Fessard, V., Franz, R., Greiner, R., Gruber-Traub, C., Haas, K.-H., Haase, A., Hartwig, A., Hesse, B., Hund-Rinke, K., Iden, P., Kromer, C., Loeschner, K., Mutz, D., Rakow, A., Rasmussen, K., Rauscher, H., … Laux, P., 2022. Analytical and toxicological aspects of nanomaterials in different product groups: challenges and opportunities. NanoImpact 28, 100416. https://doi.org/10.1016/j.impact.2022.100416.

Tuan Zainazor, T.C., Fisal, A., Goh, E.G., Che Sulaiman, N.F., Sarbon, N.M., 2020. Emerging of bio-nano composite gelatine-based film as bio-degradable food packaging: a review. Food Res. 4 (4), 944–956. https://doi.org/10.26656/fr.2017.4(4).365.

Utkarsh, C., Bhardwaj, P., Selvaraj, S.K., Arasu, K., Praveena, S., Pavan, A., Khanna, M., Singh, P., Singh, S., Chakravorty, A., Badoni, B., Banavoth, M., Sonar, P., Paramasivam, V., 2022. Current trends and future perspectives of nanomaterials in food packaging application. J. Nanomater. 2022, 2745416. https://doi.org/10.1155/2022/2745416.

Wesley, S.J., Raja, P., Raj, A.A.S., Tiroutchelvamae, D., 2014. Review on: nanotechnology applications in food packaging and safety. Int. J. Eng. Res. 3 (11), 645–651.

Xiaomin, S., Zhou, W., Ma, D., Ma, Q., Bridges, D., Ma, Y., Hu, A., 2015. Electrospinning of nanofibers and their applications for energy devices. J. Nanomater. 2015, 140716. https://doi.org/10.1155/2015/140716.

Yam, K.L., Takhistov, P.T., Miltz, J., 2005. Intelligent packaging: concepts and applications. J. Food Sci. 70 (1), R1–10.

Yezza, I.A., 2008. Active/intelligent packaging: concept, applications and innovations. In: Technical Symposium New Packaging Technologies to Improve and Maintain Food Safety, September 18-19, 2008. Toronto.

Index

Page numbers followed by "*f*" and "*t*" indicate figures and tables respectively.

Printed in the United States
by Baker & Taylor Publisher Services